**Reciprocal Identities**

$$\sin\theta = \frac{1}{\csc\theta} \qquad \cot\theta = \frac{1}{\tan\theta}$$

$$\cos\theta = \frac{1}{\sec\theta} \qquad \sec\theta = \frac{1}{\cos\theta}$$

$$\tan\theta = \frac{1}{\cot\theta} \qquad \csc\theta = \frac{1}{\sin\theta}$$

**Quotient Identities**

$$\tan\theta = \frac{\sin\theta}{\cos\theta} \qquad \cot\theta = \frac{\cos\theta}{\sin\theta}$$

**Pythagorean Identities**

$$\sin^2\theta + \cos^2\theta = 1$$
$$\tan^2\theta + 1 = \sec^2\theta$$
$$1 + \cot^2\theta = \csc^2\theta$$

**Sum and Difference Identities**

$$\cos(\alpha - \beta) = \cos\alpha\cos\beta + \sin\alpha\sin\beta$$
$$\cos(\alpha + \beta) = \cos\alpha\cos\beta - \sin\alpha\sin\beta$$
$$\sin(\alpha + \beta) = \sin\alpha\cos\beta + \cos\alpha\sin\beta$$
$$\sin(\alpha - \beta) = \sin\alpha\cos\beta - \cos\alpha\sin\beta$$
$$\tan(\alpha + \beta) = \frac{\tan\alpha + \tan\beta}{1 - \tan\alpha\tan\beta}$$
$$\tan(\alpha - \beta) = \frac{\tan\alpha - \tan\beta}{1 + \tan\alpha\tan\beta}$$

**Double-Angle Identities**

$$\sin 2\theta = 2\sin\theta\cos\theta$$
$$\cos 2\theta = \cos^2\theta - \sin^2\theta$$
$$\cos 2\theta = 1 - 2\sin^2\theta$$
$$\cos 2\theta = 2\cos^2\theta - 1$$
$$\tan 2\theta = \frac{2\tan\theta}{1 - \tan^2\theta}$$

**Half-Angle Identities**

$$\sin\frac{\theta}{2} = \pm\sqrt{\frac{1 - \cos\theta}{2}}$$
$$\cos\frac{\theta}{2} = \pm\sqrt{\frac{1 + \cos\theta}{2}}$$
$$\tan\frac{\theta}{2} = \pm\sqrt{\frac{1 - \cos\theta}{1 + \cos\theta}}$$
$$\tan\frac{\theta}{2} = \frac{\sin\theta}{1 + \cos\theta}$$
$$\tan\frac{\theta}{2} = \frac{1 - \cos\theta}{\sin\theta}$$

**Conversion Identities**

$$\sin\alpha\cos\beta = \frac{1}{2}[\sin(\alpha + \beta) + \sin(\alpha - \beta)]$$
$$\sin\alpha\sin\beta = \frac{1}{2}[\cos(\alpha - \beta) - \cos(\alpha + \beta)]$$
$$\cos\alpha\cos\beta = \frac{1}{2}[\cos(\alpha + \beta) + \cos(\alpha - \beta)]$$
$$\sin x + \sin y = 2\sin\frac{x + y}{2}\cos\frac{x - y}{2}$$
$$\sin x - \sin y = 2\cos\frac{x + y}{2}\sin\frac{x - y}{2}$$
$$\cos x + \cos y = 2\cos\frac{x + y}{2}\cos\frac{x - y}{2}$$
$$\cos x - \cos y = -2\sin\frac{x + y}{2}\sin\frac{x - y}{2}$$

**Law of Sines**

In any triangle with angles $A$, $B$, and $C$, and opposite sides $a$, $b$, and $c$, respectively,

$$\frac{\sin A}{a} = \frac{\sin B}{b} = \frac{\sin C}{c}.$$

**Law of Cosines**

In any triangle with angles $A$, $B$, and $C$, and opposite sides $a$, $b$, and $c$, respectively,

$$c^2 = a^2 + b^2 - 2ab\cos C$$
$$a^2 = b^2 + c^2 - 2ab\cos A$$
$$b^2 = a^2 + c^2 - 2ac\cos B$$

# *College*
# TRIGONOMETRY

# *College* TRIGONOMETRY

TIMOTHY J. KELLY
*Hamilton College*

RICHARD H. BALOMENOS
*University of New Hampshire*

JOHN T. ANDERSON
*Hamilton College*

HOUGHTON MIFFLIN COMPANY  BOSTON
*Dallas  Geneva, Ill.  Lawrenceville, N.J.  Palo Alto*

**Cover photograph by Michel Tcherevkoff**
**Chapter openers by Alan Davis/Omnigraphics**



Printed in the United States of America
Library of Congress Catalog Card Number: 86-81565

ISBN Numbers:
Text: 0-395-33295-8
Instructor's Edition: 0-395-33296-6
Solutions Manual: 0-395-33297-4
Alternate Testing Program: 0-395-33298-2

ABCDEFGHIJ-H-89876

# Contents

# Preface

In view of the large number of college algebra and trigonometry texts that are currently available, one might justifiably question the need for writing this particular COLLEGE ALGEBRA AND TRIGONOMETRY series. The answer is simple. The books in this series have been written *to be read!* Students consistently report that precalculus textbooks are unreadable. In developing the manuscript for the Kelly/Balomenos/Anderson COLLEGE ALGEBRA AND TRIGONOMETRY books, the authors have been guided by specific student comments indicating places where more explanation, illustration, or review is needed in order to assure understanding and maintain the flow in reading. It is hoped that both the student and instructor will come to rely on these books for their sound, complete, yet careful explanations.

## Pedagogical Features of the Text

**Readability** As indicated above, this text represents a very serious effort to produce exposition that is accessible to the precalculus student. At each stage of manuscript development, student comments were seriously considered in making explanations more comprehensible. Ultimately, a Field Test Edition was used in actual classroom trials across the country. Final revisions to the manuscript reflect both student and instructor comments from these trials.

**Organization** Each chapter of this book is divided into sections, and within each section, the material is organized around specific goal statements. These goals clearly identify for the student the purpose of the discussion that is to follow. Furthermore, exercise sets, detailed chapter reviews, and comprehensive chapter tests are all organized by this same goal-referencing system.

**Examples** As a further aid to student understanding, worked examples in the text have been enhanced by annotations that explain how the various steps in the solution were obtained. The intent is to simulate, as closely as possible, the verbal explanation that an instructor would supply in presenting problems at the blackboard. Both students and instructors involved in field testing the text found this feature to be of enormous value.

**Exercise Sets** Exercises following each section are classified according to the goals stated in the section. Under each specific goal, problems are quite simple at the outset, then gradually increase in difficulty. Problems in any odd-even pairing (as exercises 5 and 6, for example) are always at the same level of difficulty. At the end of each exercise set, there are "Superset" problems, which are intended to extend and challenge the student's understanding.

## Content Features of the Text

**Content Emphases** Decisions with respect to topical inclusion, exclusion, and sequencing reflect the authors' beliefs regarding what is most crucial and most beneficial to a mathematics student in a precalculus course. Moreover, some highly useful information came as a result of two national surveys, one in 1982, and the other in 1985. Finally, the judgements of the mathematical advisers and the field-testers of the program served as valuable input in forming the present version of the program.

**Functions** The text presents functions as sets of ordered pairs as well as mappings (or rules). Both approaches seem necessary if students are to be prepared for both discrete and continuous follow-ups to this course. In this text, graphs have been fully exploited as a way of exploring a function's properties. It is hoped that students will leave this course with a significantly enlarged set of visualization skills.

**Transformations** Translations, stretchings, and shrinkings have been used extensively in unifying the algebra and geometry of functions. Students are frequently reminded of how simple changes in the algebraic form of a function correspond to changes in its graphical form.

**Problem Solving** One of the authors, Professor Kelly, wishes especially to acknowledge the insight in this regard that was offered to him by his teacher, George Polya. Above all, Professor Polya emphasized the development of a "repertoire" of problem solving strategies that a student could bring to bear on a problem. It is hoped that such an approach is fostered, in some modest way, in the treatment of Problem Solving strategies in the text.

**Calculators** Calculator exercises are referenced in the Exercise Sets throughout the text, by means of the symbol [=]. In many instances these exercises provide students with the opportunity to discover, through computation, some patterns that are of significance in mathematics. In other cases, they present students with some computational complexities that are sometimes unavoidable in applied problems.

## Acknowledgements

An expression of gratitude is due to the following field testers for their contribution to the shaping of this program:

John Graves, Auburn University at Montgomery
Wayne Mackey, Johnson County Community College
Rosario Diprizio, Oakton Community College
Philip Farmer, College of Marin
M. R. Childers, University of Texas at Arlington
Phyllis Cox, Shelby State Community College
Enrico Serpone, Phoenix College
Arthur Dull, Diablo Valley College

In addition, the following individuals have assisted greatly in making valuable mathematical and pedagogical suggestions at various times in the development of the manuscript.

Larry Knop, Hamilton College
Anne Ludington, Loyola College (Baltimore)
Ellen O'Keefe, Wheaton College (Northampton)
Michael Schramm, Hamilton College

## Supplements

A number of supplements have been created as aids for both students and instructors. There are: an *Instructor's Edition,* which contains answers to all problems; a *Study Guide* for the student which contains a guide to each chapter and complete solutions to the odd-numbered problems; a *Solutions Manual,* which contains complete solutions to all problems; and an *Alternate Testing Program,* which contains a complete battery of goal-referenced chapter tests, cumulative tests, and a final exam, available to the instructor.

T. J. K.
R. H. B.
J. T. A.

# 1

# Graphing Functions—A Brief Review

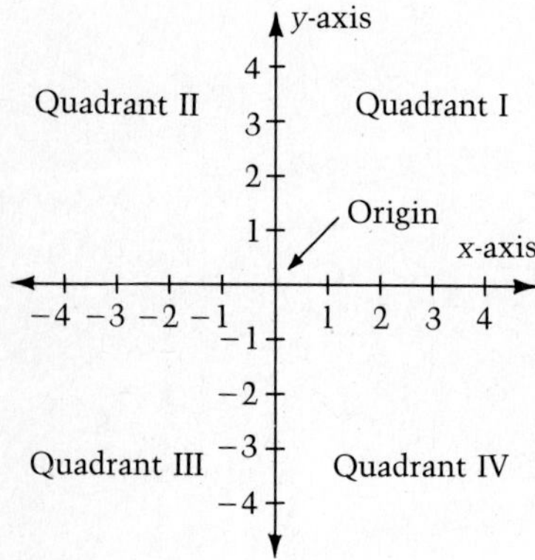


**Figure 1.1** The *xy*-plane.

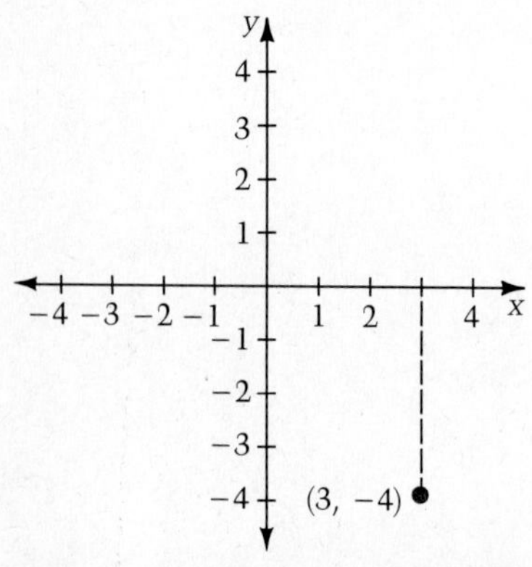


**Figure 1.2**

## 1.1 The Coordinate Plane

**Goal A** *to plot points in the xy-plane*

A graph is a useful tool for displaying information, as it provides a visual representation of how one quantity is related to, or dependent upon some other quantity. As you will recall from algebra, functions are used to express mathematical relationships, and graphs are used to illustrate these relationships. Since functions play an important role in trigonometry, as well as in algebra, the purpose of this chapter is to review techniques for graphing functions.

Let us begin by first considering ordered pairs. In an ordered pair, such as $(3, -4)$, the real numbers 3 and $-4$ are called **coordinates.** Plotting an ordered pair requires two real number lines, one for each coordinate. The two lines, called **axes,** separate the plane into four **quadrants.** We usually call the horizontal axis the **x-axis** and the vertical axis the ***y*-axis.** The point where the axes meet is the **origin,** and the entire system is the ***xy*-plane** (Figure 1.1).

Each ordered pair corresponds to a single point in the plane. Plotting a point requires two moves: one from the origin in the horizontal direction, followed by one in the vertical direction. To plot the point with coordinates $(3, -4)$, start at the origin and move 3 units to the right (right is positive, left negative). Then move 4 units downward (up is positive, down negative). Plot the point by drawing a heavy dot (Figure 1.2).

**Example 1** Plot the following points in the $xy$-plane.

(a) $A(-4, 3)$ (b) $B\left(-\frac{5}{2}, -\frac{1}{2}\right)$ (c) $C(\sqrt{2}, -1)$

***Solution***

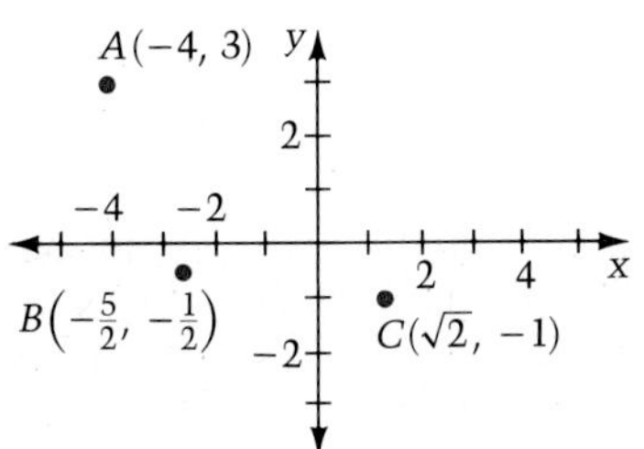


- To plot point $A$, start at the origin and move 4 units to the left and 3 units up.
- To plot point $B$, start at the origin and move $\frac{5}{2}$ units to the left and $\frac{1}{2}$ unit down.
- To plot point $C$, start at the origin and move $\sqrt{2}$ ($\approx 1.4$) units to the right and 1 unit down.

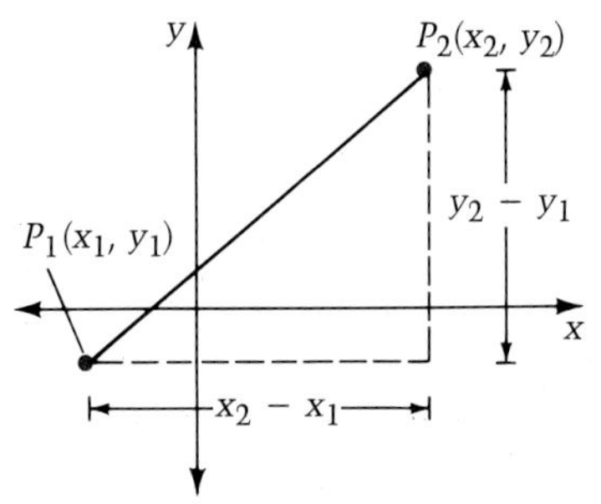


Figure 1.3

**Goal B** *to determine the distance between two points in the plane*

To determine the distance $d(P_1, P_2)$ between two points $P_1$ and $P_2$ in the $xy$-plane, observe that this distance is the length of the hypotenuse of a right triangle (Figure 1.3). The legs of a right triangle are parallel to the axes. The Pythagorean Theorem tells us that the length of the hypotenuse is the square root of the sum of the squares of the legs. We use this fact in stating the following formula.

**The Distance Formula**

The distance between any two points $P_1(x_1, y_1)$ and $P_2(x_2, y_2)$ is denoted $d(P_1, P_2)$ and is given by the formula

$$d(P_1, P_2) = \sqrt{(x_2 - x_1)^2 + (y_2 - y_1)^2}.$$

**Example 2** Determine the distance between $P_1(-2, 5)$ and $P_2(4, -3)$.

***Solution***

$$\begin{aligned} d(P_1, P_2) &= \sqrt{(x_2 - x_1)^2 + (y_2 - y_1)^2} \\ &= \sqrt{(4 - (-2))^2 + (-3 - 5)^2} \\ &= \sqrt{6^2 + (-8)^2} \\ &= \sqrt{36 + 64} = \sqrt{100} = 10 \end{aligned}$$

## Exercise Set 1.1

### Goal A

In exercises 1–20, plot the given points in the $xy$-plane.

**1.** $(3, -4)$

**2.** $(2, 6)$

**3.** $(-3, 4)$

**4.** $(-2, -6)$

**5.** $(0, 5)$

**6.** $(6, 0)$

**7.** $\left(-\frac{2}{3}, 7\right)$

**8.** $(-1, \sqrt{2})$

**9.** $\left(\frac{3}{4}, \frac{1}{3}\right)$

**10.** $\left(-\frac{5}{2}, -\frac{5}{2}\right)$

**11.** $\left(\frac{1}{2}, \frac{\sqrt{3}}{2}\right)$

**12.** $\left(-\frac{1}{2}, -\frac{\sqrt{3}}{2}\right)$

**13.** $\left(\frac{\sqrt{2}}{2}, -\frac{\sqrt{2}}{2}\right)$

**14.** $\left(-\frac{\sqrt{2}}{2}, \frac{\sqrt{2}}{2}\right)$

**15.** $\left(-\frac{\sqrt{3}}{2}, -\frac{1}{2}\right)$

**16.** $\left(-\frac{\sqrt{3}}{2}, \frac{1}{2}\right)$

**17.** $(0, \pi)$

**18.** $(-\pi, -1)$

**19.** $\left(\frac{\pi}{2}, -1\right)$

**20.** $\left(-\frac{3\pi}{2}, -\sqrt{2}\right)$

### Goal B

In exercises 21–36, determine the distance between the given points.

**21.** $(2, 1), (0, 5)$

**22.** $(3, 3), (6, 1)$

**23.** $(3, -4), (3, 4)$

**24.** $(-2, 4), (-2, 7)$

**25.** $(-5, 1), (3, -5)$

**26.** $(4, -1), (-8, -10)$

**27.** $\left(3, \frac{5}{2}\right), \left(-9, -\frac{13}{2}\right)$

**28.** $\left(\frac{1}{2}, -4\right), \left(\frac{5}{4}, 1\right)$

**29.** $(\sqrt{2}, 1), (-1, \sqrt{2})$

**30.** $(2, -\sqrt{3}), (\sqrt{3}, 5)$

**31.** $(0, 0), \left(\frac{1}{2}, \frac{\sqrt{3}}{2}\right)$

**32.** $(0, 0), \left(-\frac{1}{2}, \frac{\sqrt{3}}{2}\right)$

**33.** $\left(\frac{\sqrt{2}}{2}, \frac{\sqrt{2}}{2}\right), \left(-\frac{\sqrt{2}}{2}, -\frac{\sqrt{2}}{2}\right)$

**34.** $\left(\frac{\sqrt{3}}{2}, -\frac{1}{2}\right), \left(-\frac{\sqrt{3}}{2}, \frac{1}{2}\right)$

**35.** $\left(\frac{\sqrt{3}}{2}, \frac{1}{2}\right), (0, 1)$

**36.** $\left(\frac{\sqrt{2}}{2}, \frac{\sqrt{2}}{2}\right), (1, 0)$

### Superset

Recall that a quadrilateral is a **parallelogram** if opposite sides have the same length, and a parallelogram is a **rectangle** if the two diagonals have the same length.

**37.** A quadrilateral $ABCD$ has vertices at $A(-1, -2)$, $B(3, -3)$, $C(5, 1)$ and $D(1, 2)$. Is the quadrilateral a parallelogram? a rectangle?

**38.** A quadrilateral $ABCD$ has vertices at $A(0, 0)$, $B(-4, 4)$, $C(-8, 0)$, and $D(-4, -4)$. Is the quadrilateral a parallelogram? a rectangle?

A triangle is **isosceles** if at least two sides have the same length, and **equilateral** if all three sides have the same length.

**39.** A triangle $ABC$ has vertices at $A(4, 3)$, $B(-3, 1)$, and $C(2, -7)$. Is the triangle isosceles? equilateral?

**40.** A triangle $ABC$ has vertices at $A(2, -3)$, $B(-5, 1)$, and $C(3, 5)$. Is the triangle isosceles? equilateral?

**41.** Show that the point

$$\left(\frac{x_1 + x_2}{2}, \frac{y_1 + y_2}{2}\right)$$

is the midpoint of the segment with endpoints $A(x_1, y_1)$ and $B(x_2, y_2)$.

**42.** Find the midpoints of both diagonals of the quadrilaterals in exercises 37 and 38.

A point $C$ is on the segment with endpoints $A$ and $B$ if $d(A, C) + d(C, B) = d(A, B)$. (Note that $d(X, Y)$ represents the distance between $X$ and $Y$.) In exercises 43 and 44, determine whether the given points lie on the same line.

**43.** $\left(0, \frac{5}{2}\right), (5, 5), \left(-6, -\frac{1}{2}\right)$

**44.** $(0, -4), (2, 0), (5, 6), (-1, -5)$

## 1.2 Functions

**Goal A** *to evaluate functions*

As stated in section 1, we sometimes use a function to describe a relationship between quantities. The concept of function is important in trigonometry as an aid in the study of periodic relationships. In general, a set of ordered pairs is called a **relation.** A function is a special type of relation.

**Definition**

Whenever a process pairs each member of a first set with exactly one member of a second set, the resulting set of ordered pairs is called a **function.** Thus, a function contains no two different ordered pairs having the same first coordinate. The set of all first coordinates is called the **domain** of the function and the set of all second coordinates is called the **range.**

**Example 1** Use a set of ordered pairs to describe the function represented by the caloric graph shown below. Determine the domain and range of the function.

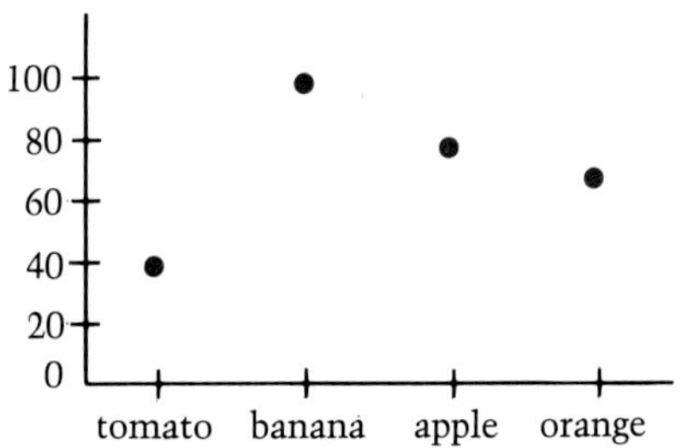


■ In forming the set of ordered pairs, the first coordinates are values from the horizontal axis, and the second coordinates are the corresponding values from the vertical axis.

***Solution*** The function is the set of ordered pairs: {(tomato, 40), (banana, 100), (apple, 80), (orange, 70)}. The domain is {tomato, banana, apple, orange}, and the range is {40, 70, 80, 100}.

At some time you have probably already checked an "ideal weight" chart to determine your "perfect" weight. There are formulas which describe ideal weight as a function of height. For a male over age 25 with a medium frame, the formula might be $w = 4h - 121$, where $w$ represents weight in pounds and $h$ represents height in inches. The formula determines a function since to each value of height there is assigned exactly one ideal weight.

| $h$ (in) | $w$ (lb) | $(h, w)$ |
|---|---|---|
| 62 | 127 | (62, 127) |
| 68 | 151 | (68, 151) |
| 76 | 183 | (76, 183) |

Given a value of height $h$, the ideal weight is found by *evaluating the function,* that is, by replacing $h$ with the given height and then finding the corresponding value of $w$. The chart at the left lists some ordered pairs for this function: (62, 127), (68, 151), and (76, 183). The second coordinates are called **function values,** because they are found by evaluating the function.

**Example 2** The formula for converting the temperature in degrees Fahrenheit (F) to degrees Celsius (C) is given by the function $C = \frac{5}{9}(F - 32)$. Evaluate the function for $F = 5, 14, 32,$ and 50, and display the results in a chart.

***Solution*** For F = 5, we get $C = \frac{5}{9}\cdot(5 - 32) = -15$

$F = 14 \qquad C = \frac{5}{9}\cdot(14 - 32) = -10$

$F = 32 \qquad C = \frac{5}{9}\cdot(32 - 32) = 0$

$F = 50 \qquad C = \frac{5}{9}\cdot(50 - 32) = 10$

| F | C | (F, C) |
|---|---|---|
| 5 | −15 | (5, −15) |
| 14 | −10 | (14, −10) |
| 32 | 0 | (32, 0) |
| 50 | 10 | (50, 10) |

**Example 3** A certain function is described by the formula $y = 7x^2 + 3x - 1$. Evaluate the function for $x = -5, -1, 0,$ and 5, and display the results as ordered pairs.

***Solution***

$$y = 7\cdot x^2 + 3\cdot x - 1$$
$$x = -5 \quad y = 7\cdot(-5)^2 + 3\cdot(-5) - 1 = 159$$
$$x = -1 \quad y = 7\cdot(-1)^2 + 3\cdot(-1) - 1 = 3$$
$$x = 0 \quad y = 7\cdot(0)^2 + 3\cdot(0) - 1 = -1$$
$$x = 5 \quad y = 7\cdot(5)^2 + 3\cdot(5) - 1 = 189$$

The resulting ordered pairs are (−5, 159), (−1, 3), (0, −1), and (5, 189).

In the two previous examples, we have seen that some functions can be described by equations involving two variables. In example 2, C is the **dependent variable,** and F is the **independent variable.** C is dependent upon F because we first select a value for F and then use it to find C. We call F the independent variable because we are free to choose any value for it.

We can use letters to name functions. For example,

$$f = \{(-1, 2), (5, 3), (11, 12), (15, 25)\}.$$

In this case, we use $f(x)$ to denote the second coordinate in the ordered pair whose first coordinate is $x$. Thus, we write

$f(-1)$ is read "$f$ of $-1$," and 2 is the "function value at $-1$."

$$f(-1) = 2, \quad f(5) = 3, \quad f(11) = 12, \quad f(15) = 25.$$

The equation

$$f(x) = 2x^2 + 7x - 4$$

determines the same set of ordered pairs (and therefore the same function) as the equation

$$y = 2x^2 + 7x - 4.$$

You will see $y$ and $f(x)$ used interchangeably, and you should become comfortable with both these usages.

**Example 4** For the function $g(x) = 2x^2 - 4x + 11$ find

(a) $g(3)$ (b) $g(b^3)$ (c) $g(3 + h)$

***Solution***

(a) $g(3) = 2(3)^2 - 4(3) + 11 = 17$ ▪ Replace $x$ with 3.

(b) $g(b^3) = 2(b^3)^2 - 4b^3 + 11$ ▪ Replace $x$ with $b^3$.
$= 2b^6 - 4b^3 + 11$

(c) $g(3 + h) = 2(3 + h)^2 - 4(3 + h) + 11$ ▪ Replace $x$ with $3 + h$.
$= 2(9 + 6h + h^2) - 12 - 4h + 11$
$= 18 + 12h + 2h^2 - 12 - 4h + 11$
$= 2h^2 + 8h + 17$

It is often useful to think of a function $f$ as a rule that assigns a unique range value $f(x)$ to each $x$ in the domain. For example, the function $f(x) = x^2 - 2x - 8$ can be thought of as a rule that assigns range value $-8$ to domain value 0, and range value 7 to domain value $-3$.

Some functions assign the same range value to different domain values $x_1$ and $x_2$. That is, for some functions $f$, $x_1 \neq x_2$, but $f(x_1) = f(x_2)$. For example, for the function $f(x) = x^2 - 2x - 8$, $f(-3) = 7$ and $f(5) = 7$ as well. Such a function is not *one-to-one.* We say that a function is **one-to-one** if $x_1 \neq x_2$ implies $f(x_1) \neq f(x_2)$: different domain values are associated with different range values.

**Example 3** Determine whether each of the following functions is one-to-one.

(a) $h(x) = 2x + 1$ (b) $f(x) = 2x^2 - 5$

***Solution*** (a) $h(x) = 2x + 1$

Let $x_1$ and $x_2$ be two different domain values. Then $2x_1$ and $2x_2$ are different, and, thus, $2x_1 + 1$ and $2x_2 + 1$ are different. Since different domain values produce different range values, the function is one-to-one.

(b)
$$f(x) = 2x^2 - 5$$
$$f(1) = 2(1)^2 - 5 = -3$$
$$f(-1) = 2(-1)^2 - 5 = -3$$

■ To show that $f(x)$ is not one-to-one, we need to find two different numbers in the domain that produce the same range value.

Since $f(1) = f(-1)$, the function is not one-to-one.

## Goal B *to evaluate composite functions*

A function may be viewed as a number processor that takes a domain value as input and produces a range value as output. For example, we can treat the functions $f(x) = x^2 + 4$ and $g(x) = 3x - 1$ as processors.

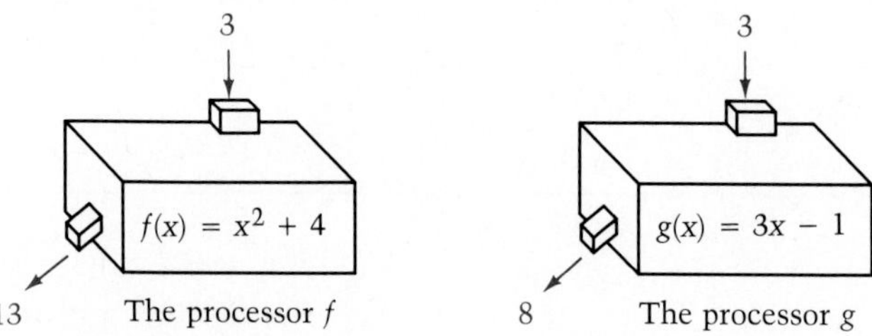


**Figure 1.4**

The function $f$ takes an $x$-value from the domain as input, squares it and adds four to produce the function value $y = f(x)$. The function $g$ takes an $x$-value, multiplies it by three and subtracts one to produce the function value $y = g(x)$.

The functions $g$ and $f$ can be combined to generate the composite function $f \circ g$, read "$f$ circle $g$." This is illustrated below.

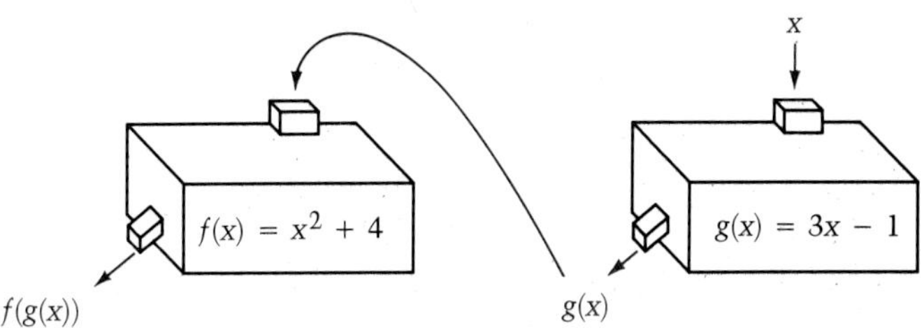


**Figure 1.5**

The function $f \circ g$ is the composition of $g$ followed by $f$. First, $x$ is processed by $g$ and then the output $g(x)$ becomes the input for $f$.

**Definition**
Let $f$ and $g$ be functions with the range of $g$ contained in the domain of $f$. For each $x$ in the domain of $g$, the **composite function** $f \circ g$ is defined as

$$(f \circ g)(x) = f(g(x)).$$

In the definition of $f \circ g$, it is necessary that the range of $g$ be contained in the domain of $f$ so that every output of $g$ is acceptable input for $f$. In general, $f \circ g$ and $g \circ f$ are not the same function. The order of the functions in a composition is important.

**Example 4** Suppose $f(x) = 2x^2 - 3$, $g(x) = |x| - 5$, and $h(x) = \sqrt{x}$. Evaluate the following.

(a) $(f \circ g \circ h)(4)$ (b) $(h \circ f)(1)$

***Solution***

(a) $(f \circ g \circ h)(4) = f(g(h(4)))$ ■ First, evaluate $h(4)$.
$= f(g(2))$ ■ $h(4) = \sqrt{4} = 2$.
$= f(-3)$ ■ $g(2) = |2| - 5 = -3$.
$= 15$ ■ $f(-3) = 2(-3)^2 - 3 = 15$.

(b) $(h \circ f)(1) = h(f(1))$ ■ First evaluate $f(1)$.
$= h(-1)$ ■ $f(1) = 2(1)^2 - 3 = -1$.
$= \sqrt{-1}$

$(h \circ f)(1)$ is not defined because $f(1) = -1$ is not in the domain of $h$.

**Goal C** *to find the inverse function of a one-to-one function*

Sometimes the composite of two functions has the effect of producing the same value $x$ that you started with. That is, $(f \circ g)(x) = x$ and $(g \circ f)(x) = x$ for all $x$ in the domains of $f$ and $g$. Functions that behave this way are called *inverse functions.*

**Definition**

Given any two functions $f$ and $g$, $g$ is the **inverse** of $f$ if

for each $x$ in the domain of $g$, $(f \circ g)(x) = x$, and
for each $x$ in the domain of $f$, $(g \circ f)(x) = x$.

In general, for every one-to-one function $f$, there exists an inverse function, denoted $f^{-1}$, associated with $f$. (Note: the symbol $f^{-1}$ is read "$f$ inverse" and does not mean $\frac{1}{f}$.)

In the next examples we verify that two functions are inverse functions and apply an easy method for finding the inverse of a one-to-one function.

**Example 5** Show that the one-to-one functions $f(x) = -\frac{1}{2}x - 3$ and $g(x) = -2x - 6$ are inverse functions.

***Solution***

$$(f \circ g)(x) = f(g(x)) = f(-2x - 6) = -\frac{1}{2}(-2x - 6) - 3 = (x + 3) - 3 = x$$

$$(g \circ f)(x) = g(f(x)) = g\left(-\frac{1}{2}x - 3\right) = -2\left(-\frac{1}{2}x - 3\right) - 6 = (x + 6) - 6 = x$$

Since $(f \circ g)(x) = x$ for all $x$ in the domain of $g$, and $(g \circ f)(x) = x$ for all $x$ in the domain of $f$, $f$ and $g$ are inverse functions.

**Example 6** The function $f(x) = 3x - 2$ is one-to-one. Find its inverse function.

***Solution***

$y = 3x - 2$ ■ Write the given function in terms of $x$ and $y$.

$x = 3y - 2$ ■ Interchange the roles of $x$ and $y$.

$y = f^{-1}(x) = \dfrac{x + 2}{3}$

## Exercise Set 1.2 ▤ Calculator Exercises in the Appendix

### Goal A

In exercises 1–4, describe the function with a set of ordered pairs. State the domain and range.

1.

2.

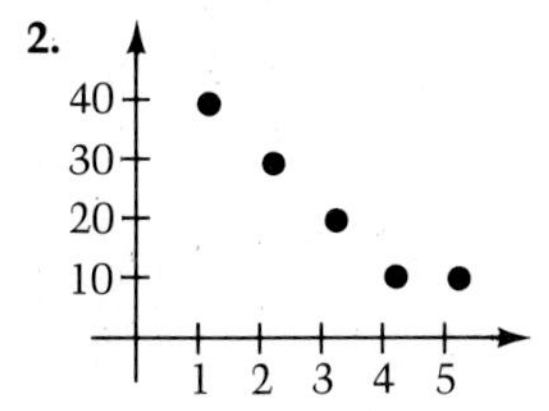

3.

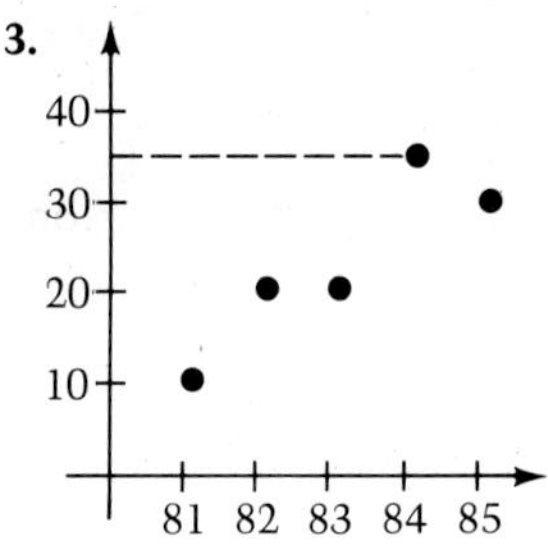

4.

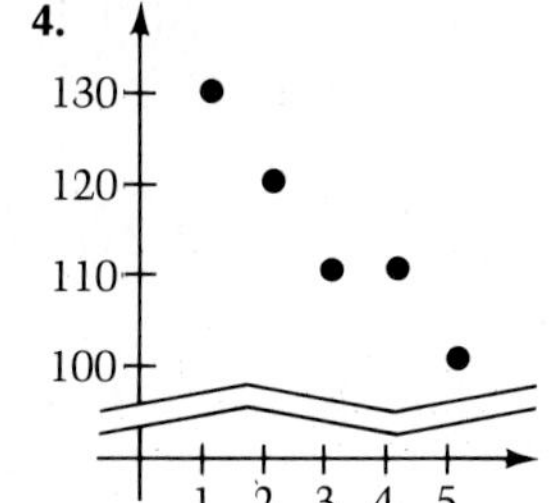

In exercises 5–10, find $f(-2)$, $f(0)$, $f(5)$, $f(c^2)$ and $f(2 + h)$ for each function.

**5.** $f(x) = 8 - 3x$

**6.** $f(x) = \frac{1}{2}x + 4$

**7.** $f(x) = x^2 - 5x + 6$

**8.** $f(x) = x^2 + 5x - 6$

**9.** $f(x) = 10$

**10.** $f(x) = -7$

For exercises 11–14, determine whether the given function is one-to-one.

**11.** $f(x) = 6 - 2x$

**12.** $f(x) = |3x|$

**13.** $f(x) = x^3 + 5$

**14.** $f(x) = x^3 - 2x^2$

### Goal B

In exercises 15–22, evaluate the given expression for the functions $f(x) = 2x - 1$, $g(x) = |x|$, $h(x) = 3$, and $j(x) = 10x - 7$.

**15.** $(f \circ g)(-2)$

**16.** $(g \circ f)(-2)$

**17.** $(g \circ f)(0)$

**18.** $(f \circ g)(0)$

**19.** $(j \circ h)(5)$

**20.** $(j \circ h)(-5)$

**21.** $(h \circ j)(5)$

**22.** $(h \circ j)(-5)$

In exercises 23–26, evaluate the given expression for $f(x) = \frac{1}{x - 4}$, $g(x) = 4x^2 - 8$, and $h(x) = \sqrt{x}$.

**23.** $(g \circ f)(-6)$

**24.** $(f \circ g)(0)$

**25.** $(f \circ f)(2)$

**26.** $(h \circ g)(1)$

### Goal C

In exercises 27–30, show that the given two one-to-one functions are inverse functions.

**27.** $f(x) = 3x - 2$, $g(x) = \frac{x + 2}{3}$

**28.** $f(x) = \frac{1}{2}x - \frac{3}{2}$, $g(x) = 2x + 3$

**29.** $f(x) = 4 - 2x$, $g(x) = \frac{4 - x}{2}$

**30.** $f(x) = x^3 - 1$, $g(x) = \sqrt[3]{x + 1}$

In exercises 31–34, the given function is one-to-one. Find its inverse function.

**31.** $f(x) = 5 - 3x$

**32.** $f(x) = 2x - 7$

**33.** $f(x) = \frac{x + 4}{2}$

**34.** $f(x) = \frac{1}{2}x + 4$

### Superset

**35.** If $F(x) = x^2 + 1$, find (a) $F(3 + F(1))$, (b) $F(3) + F(1)$, and (c) $F(t^2 - F(t - 1))$.

**36.** Suppose $f(x) = \sqrt{x}$. Show that

$$\frac{f(3 + h) - f(3)}{h} = \frac{1}{f(3 + h) + f(3)}.$$

A function $f$ is **even** if for all $x$ in the domain of $f$, $f(-x) = f(x)$, and **odd** if $f(-x) = -f(x)$. In exercises 37–42, classify each function as even, odd, or neither.

**37.** $f(x) = x$

**38.** $g(x) = x^2$

**39.** $h(x) = \sqrt{x}$

**40.** $F(x) = |x|$

**41.** $j(x) = (x - 1)(x + 1)$

**42.** $j(x) = x^5 - x^3 - 6x$

## 1.3 Techniques of Graphing

**Goal A** *to graph functions in the xy-plane*

Since functions are sets of ordered pairs, we can use point-plotting methods to sketch their graphs. The first graph we shall consider is that of a straight line.

**Straight Line**

Any function that can be described by an equation of the form $y = mx + b$, where $m$ and $b$ are real numbers, has a **straight line** as its graph. The graph is completely determined by two points.

**Example 1** Graph the function described by the equation $y = 4x - 5$.

***Solution***

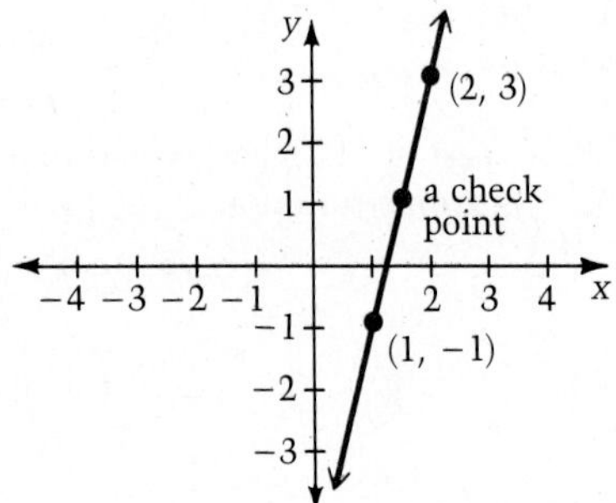


- To begin, find two ordered pairs of the function.

  $x = 1 \quad y = 4(1) - 5 = -1$

  $x = 2 \quad y = 4(2) - 5 = 3$
- Plot and label the points, and draw the line through them.
- Check: Find another ordered pair of the function. It should lie on the line.

Next, we examine parabolas and V-shaped graphs.

**Parabola**

Any function that can be described by an equation of the form $y = ax^2 + bx + c$, where $a$, $b$, and $c$ are real numbers and $a \neq 0$, has a **parabola** as its graph.

**V-Shaped Graph**

Any function that can be described by an equation of the form $y = a|x| + k$, where $a$ and $k$ are real numbers and $a \neq 0$, has a **V-shaped graph.** Such a graph consists of two rays originating from a point and forming a corner.

**Example 2** Graph the functions described by the following equations.

(a) $f(x) = x^2$ (b) $f(x) = -x^2 + 4x + 1$ (c) $y = |x|$ (d) $y = -2|x| + 1$

***Solution*** (a)

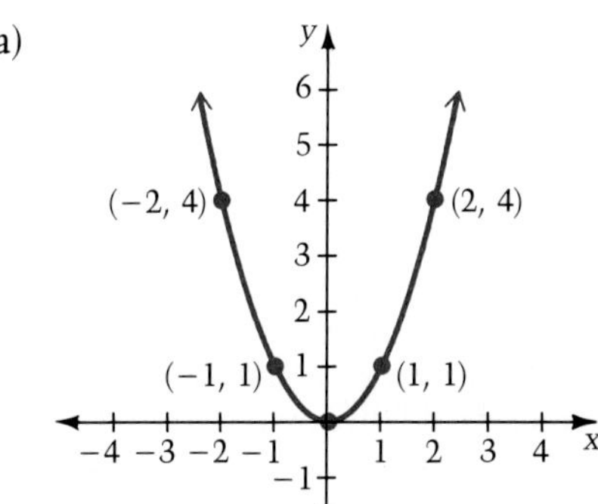


(b)

(c)

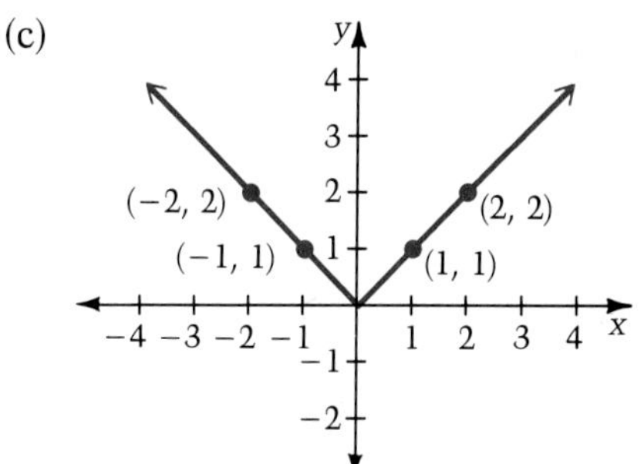


(d)

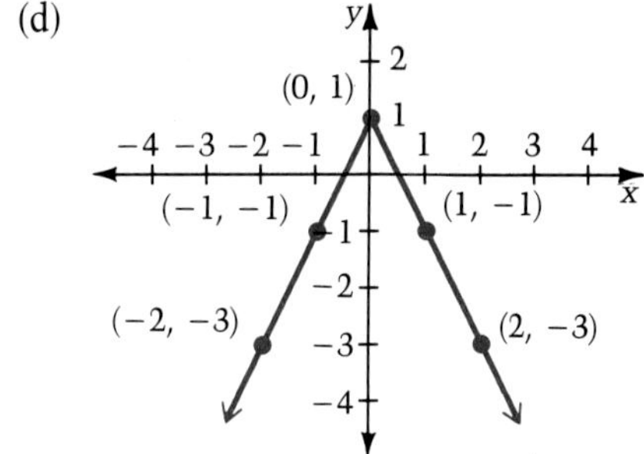


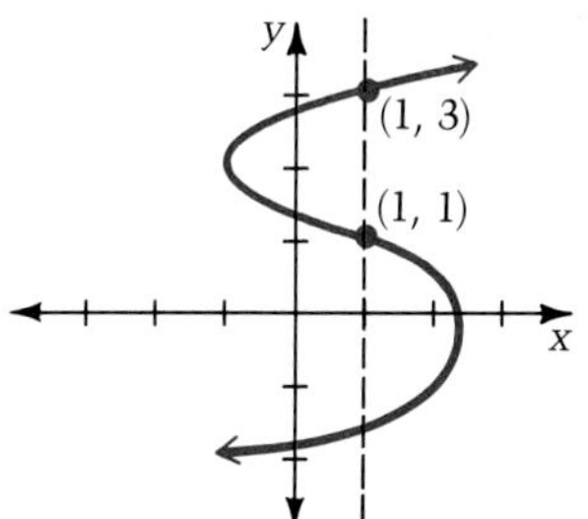


**Figure 1.6**

Recall that a set of ordered pairs is a function if no two different ordered pairs have the same first coordinate. The ordered pairs (1, 1) and (1, 3) have the same first coordinate; thus the curve in Figure 1.6 is not the graph of a function. The fact that such points lie on the same vertical line suggests a quick way of determining when a graph represents a function.

**The Vertical Line Test**

For any graph in the *xy*-plane, if some vertical line can be drawn so that it intersects the graph more than once, then the graph does not represent a function of *x*. If no such line exists, the graph does represent a function of *x*.

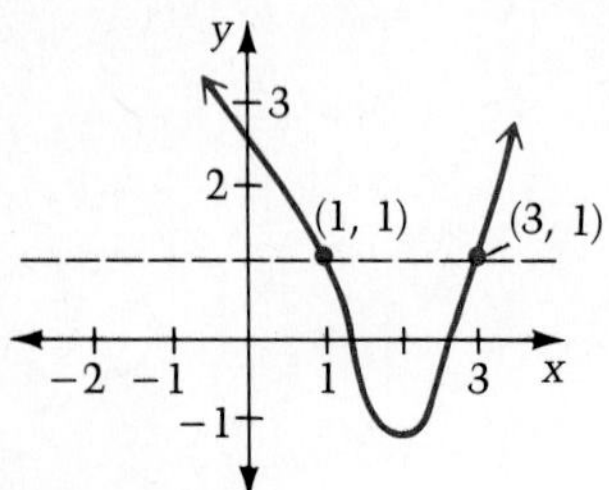

Figure 1.7 Not a one-to-one function.

As we have seen, a function is one-to-one if no two ordered pairs with different first coordinates have the same second coordinate. The ordered pairs (1, 1) and (3, 1) have different first coordinates but the same second coordinate; thus the curve in Figure 1.7 is not the graph of a one-to-one function. The fact that such points lie on the same horizontal line suggests a quick way of determining when a graph represents a one-to-one function.

### The Horizontal Line Test

For any function's graph in the $xy$-plane, if some horizontal line can be drawn so that it intersects the graph more than once, then the graph does not represent a one-to-one function of $x$. If no such line exists, the graph does represent a one-to-one function of $x$.

Thus far we have graphed functions whose domains are the entire set of real numbers: $x$ can be replaced by any real number. However, it is sometimes necessary to restrict the domain of a function in order to exclude values that produce undefined expressions.

$y = \sqrt{x}$ ■ Since the square root of a negative number is not a real number, the domain of this function is the set of all nonnegative real numbers.

$y = \dfrac{1}{x}$ ■ Since division by zero is undefined, the domain of this function is the set of all real numbers except zero.

**Example 3** Graph the function described by the equation $y = \dfrac{1}{x}$.

***Solution***

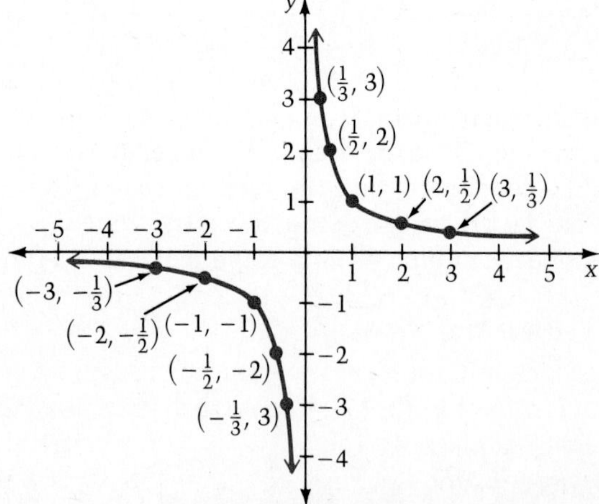


■ The domain is the set of all real numbers except 0. Notice that the graph of this function has two parts. Each part comes close to the $x$- and $y$-axes but never touches them. When a graph approaches a line in this manner, we say that the line is an **asymptote** of the graph. Here, the $x$-axis is a **horizontal asymptote** and the $y$-axis is a **vertical asymptote.**

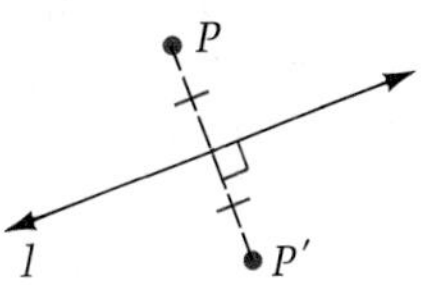


Figure 1.8

**Goal B** *to use symmetry to graph functions*

Graphing functions by plotting points can be very time-consuming. We can save time if we first determine whether the graph exhibits any symmetry.

In general, two points $P$ and $P'$ are symmetric with respect to a line $l$ if the line $l$ is the perpendicular bisector of the segment from $P$ to $P'$. Thus, $P'$ is the reflection of $P$ across the line $l$.

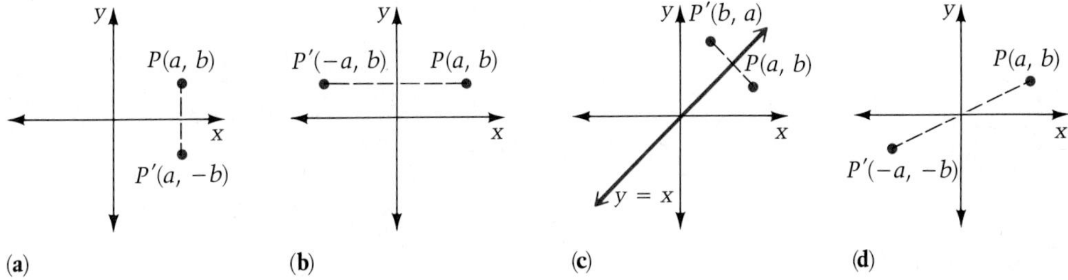


Figure 1.9

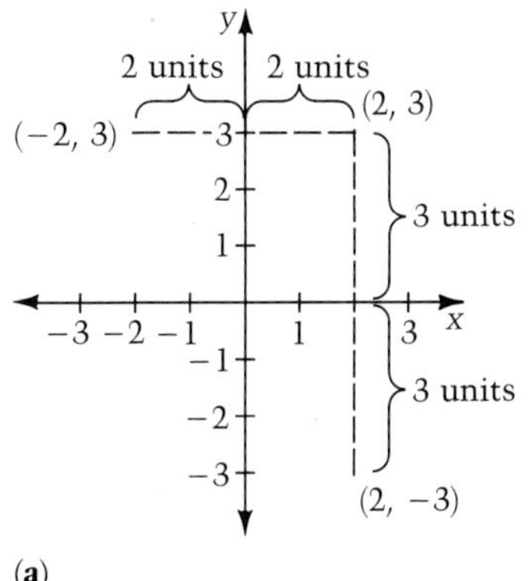


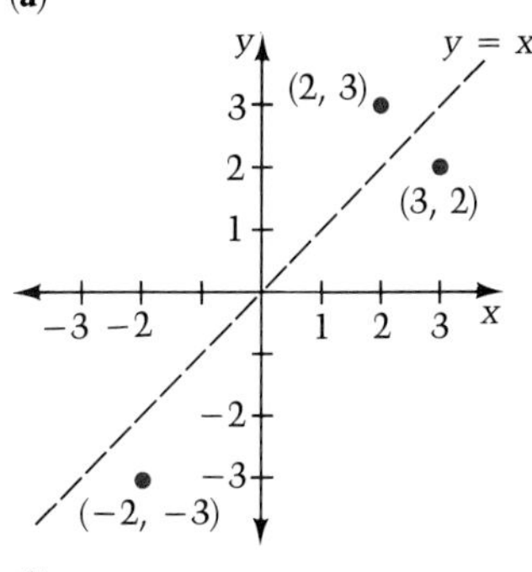


Figure 1.10

Suppose we have two points $P$ and $P'$, and the coordinates of $P$ are $(a, b)$. If $P$ and $P'$ are symmetric with respect to the $x$-axis (Figure 1.9(a)), then the coordinates of $P'$ are $(a, -b)$. If these points are symmetric with respect to the $y$-axis (Figure 1.9(b)), then the coordinates of $P'$ are $(-a, b)$. If $P$ and $P'$ are symmetric with respect to the line $y = x$ (Figure 1.9(c)), the $x$- and $y$-coordinates are interchanged so that the coordinates of $P'$ are $(b, a)$.

Two points $P$ and $P'$ are symmetric with respect to the origin (Figure 1.9(d)) if they are endpoints of a segment having the origin as the midpoint. The coordinates of $P'$, $(-a, -b)$, have the same absolute values as those of point $P$, but opposite signs.

Let us now apply the information above to a specific point, say (2, 3). To reflect (2, 3) across the $x$-axis, we change the sign of the $y$-coordinate; to reflect this point across the $y$-axis, we change the sign of the $x$-coordinate. Thus, as illustrated in Figure 1.10(a), (2, 3) and (2, −3) are symmetric with respect to the $x$-axis, and (2, 3) and (−2, 3) are symmetric with respect to the $y$-axis.

To reflect the point (2, 3) across the line $y = x$, we interchange the $x$- and $y$-coordinates; to reflect it through the origin, we change the signs of the coordinates. Thus, as illustrated in Figure 1.10(b),

(2, 3) and (3, 2) are symmetric with respect to the line $y = x$,

and

(2, 3) and (−2, −3) are symmetric with respect to the origin.

The graph of a function may exhibit symmetry with respect to the $y$-axis or origin. Suppose the graph of the function $f$ is symmetric with respect to the $y$-axis. This means that for every point $(a, b)$ on this graph, the point $(-a, b)$ is also on the graph, that is, $f(-a) = b = f(a)$. We say that a function $f$ is **even** if $f(-x) = f(x)$ for all $x$ in the domain of $f$. Hence, we can conclude that *the graphs of even functions are symmetric with respect to the y-axis.*

Similarly, if a function $f$ is symmetric with respect to the origin, then for every point $(a, b)$ on the graph, the point $(-a, -b)$ is also on the graph, that is, $f(-a) = -b = -f(a)$. A function $f$ is **odd** if $f(-x) = -f(x)$ for all $x$ in the domain. Thus, *the graphs of odd functions are symmetric with respect to the origin.*

As illustrated in the following example, it is easier to graph a function if you test for symmetry first.

**Example 4** Sketch the graph of the function $f(x) = x^3 - 4x$.

***Solution*** First we test for symmetry.

$$\begin{aligned} f(-x) &= (-x)^3 - 4(-x) \\ &= -x^3 + 4x \\ &= -(x^3 - 4x) \\ &= -f(x) \end{aligned}$$

■ Since $f$ is odd, the graph is symmetric with respect to the origin.

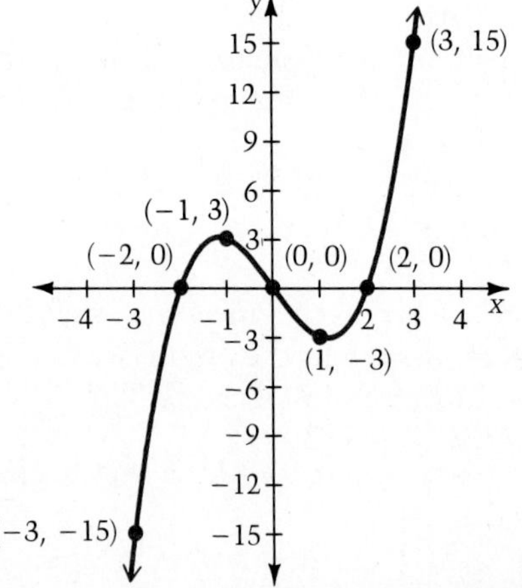


| x | y | (x, y) |
|---|---|---|
| 0 | 0 | (0, 0) |
| 1 | −3 | (1, −3) |
| 2 | 0 | (2, 0) |
| 3 | 15 | (3, 15) |

■ Plot these points, draw a curve through them, and reflect the curve through the origin.

The graphs of a one-to-one function $f$ and its inverse $f^{-1}$ are reflections of each other across the line $y = x$. This fact is useful in determining the graph of $f^{-1}$ from the graph of $f$.

**Example 5** The function $f(x) = x^3$, with restricted domain $-1 \leq x \leq 2$, is one-to-one. Sketch the graph of its inverse function.

***Solution***

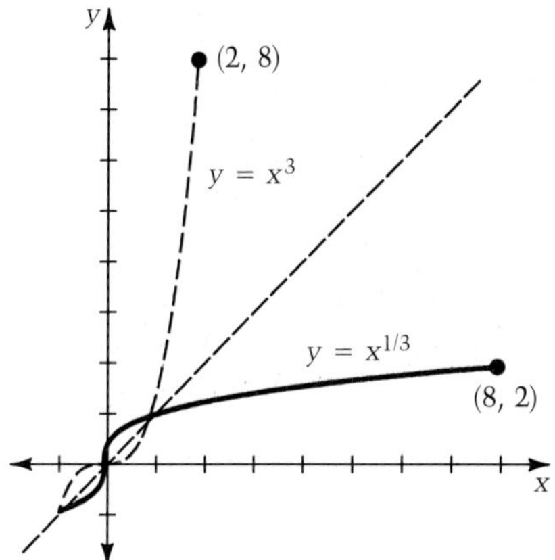


- Sketch the graph of the function $f(x) = x^3$. Then reflect it across the line $y = x$ to obtain the graph of the inverse function $f^{-1}(x) = x^{1/3}$.
- Domain of $f$: $[-1, 2]$
  Range of $f$: $[-1, 8]$
- Domain of $f^{-1}$: $[-1, 8]$
  Range of $f^{-1}$: $[-1, 2]$

In Example 5, we illustrated that

$$\text{domain of } f = \text{range of } f^{-1},$$
$$\text{domain of } f^{-1} = \text{range of } f.$$

It is important to remember that these relationships hold for all functions that have inverses.

**Goal C** *to graph functions by applying transformations to a known graph*

Certain changes in the equation describing a function produce simple changes in the function's graph. These changes in the graph are called **transformations.** There are three general types of transformations: **translations, reflections,** and **changes in shape.**

A translation occurs when a graph is moved vertically or horizontally. The shape and size of the graph remain the same; only its position changes. Reflections have already been discussed under Goal B. A change in shape is a stretching of a graph away from an axis or a shrinking of a graph toward an axis.

Suppose we are given the graph of the function $y = f(x)$, and we know that $a$, $b$, and $k$ are positive real numbers. In the following table, we summarize the relationship between changes in the function's equation and changes in its graph.

| New Equation | Changes to the Graph of $y = f(x)$ |
|---|---|
| $y = f(x) + k$ | graph moves $k$ units up |
| $y = f(x) - k$ | graph moves $k$ units down |
| $y = f(x + k)$ | graph moves $k$ units to the left |
| $y = f(x - k)$ | graph moves $k$ units to the right |
| $y = f(-x)$ | graph is reflected across the $y$-axis |
| $y = -f(x)$ | graph is reflected across the $x$-axis |
| $y = b \cdot f(x)$ | if $b > 1$, graph is stretched away from $x$-axis<br>if $0 < b < 1$, graph is shrunk toward $x$-axis |
| $y = f(ax)$ | if $a > 1$, graph is shrunk toward $y$-axis<br>if $0 < a < 1$, graph is stretched away from $y$-axis |

**Example 6** Sketch the graph of the function $y = -3(x + 1)^2$.

***Solution***

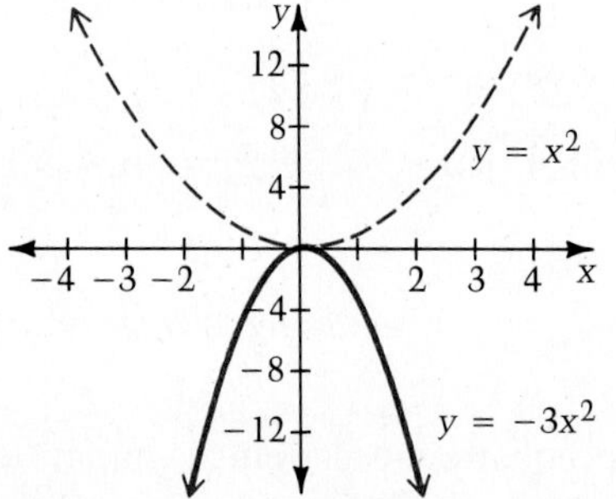


Step 1: This is a transformation of $y = x^2$. We reflect $y = x^2$ across the $x$-axis to obtain $y = -x^2$. The 3 suggests stretching $y = -x^2$ away from the $x$-axis to obtain $y = -3x^2$.

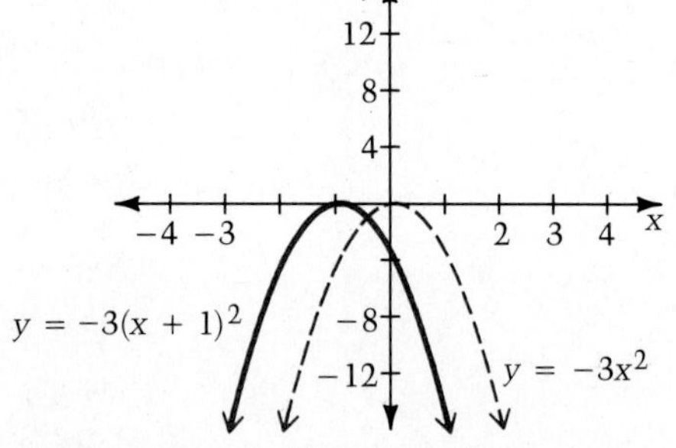


Step 2: The $(x + 1)$ suggests a translation of $y = -3x^2$ one unit to the left to obtain $y = -3(x + 1)^2$.

## Exercise Set 1.3

### Goal A

In exercises 1–12, graph the function described by the given equation.

**1.** $y = 2x$

**2.** $y = \frac{1}{3}x$

**3.** $y = 3 - \frac{x}{2}$

**4.** $y = -5x + 1$

**5.** $y = x^2 - 1$

**6.** $y = 1 - x^2$

**7.** $2y = x^2 - 2x - 1$

**8.** $\frac{1}{4}y = x^2 - 8x + 12$

**9.** $y = |x| + 2$

**10.** $y = \frac{1}{3}|x|$

**11.** $y = 3 - |x|$

**12.** $y = 5 - 3|x|$

In exercises 13–18, graph the function described by the equation and state the domain.

**13.** $f(x) = \sqrt{x + 3}$

**14.** $g(x) = \sqrt{x - 1}$

**15.** $y = \sqrt{4 - x}$

**16.** $y = \sqrt{2 + x}$

**17.** $y = \frac{1}{x + 4}$

**18.** $y = \frac{1}{x - 8}$

### Goal B

In exercises 19–24, plot and label the reflection of each point (a) across the $x$-axis, (b) across the $y$-axis, and (c) through the origin.

**19.** $(3, -2)$

**20.** $(-2, 1)$

**21.** $(4, 3)$

**22.** $(-4, -8)$

**23.** $(0, 5)$

**24.** $(-3, 0)$

For exercises 25–30, state the coordinates of the point that is the reflection across the line $y = x$ of the given point in exercises 19–24.

In exercises 31–34, use symmetry to graph the given equation.

**31.** $y = x^2 - 2$

**32.** $y = x^2 + 3$

**33.** $3y = x^3$

**34.** $3y = x^3 - x$

In exercises 35–40, the function $y = f(x)$ is one-to-one. Find the equation of the inverse function, identify its domain and range, and sketch the inverse.

**35.** $f(x) = 8 - 2x,\ 1 \le x \le 4$

**36.** $f(x) = 4x - 3,\ 0 \le x \le 2$

**37.** $f(x) = (x + 1)^2,\ 0 \le x < 2$

**38.** $f(x) = (x - 1)^3,\ -2 < x < 3$

**39.** $f(x) = \sqrt{5 - x},\ x \le 5$

**40.** $f(x) = \sqrt{x + 4},\ x \ge 4$

### Goal C

In exercises 41–50, use transformations to graph the given function.

**41.** $y = (x - 2)^2$

**42.** $y = (x - 4)^2$

**43.** $y = |x + 4|$

**44.** $y - 3 = |x|$

**45.** $y = \frac{1}{x - 4}$

**46.** $y - 3 = \frac{1}{x}$

**47.** $y = 3(x - 2)^2 + 1$

**48.** $y = -2(x + 1)^2 - 3$

**49.** $y = 4|x + 1| - 4$

**50.** $y = 4|x - 2| + 3$

### Superset

In exercises 51–54, determine the coordinates of the reflection of the point across the line.

**51.** $(3, 2)$, line $x = 4$

**52.** $(2, 7)$, line $x = 6$

**53.** $(4, -3)$, line $y = -1$

**54.** $(-2, -5)$, line $y = -3$

In exercises 55–60, graph each function by transforming the graph of $y = f(x)$ below.

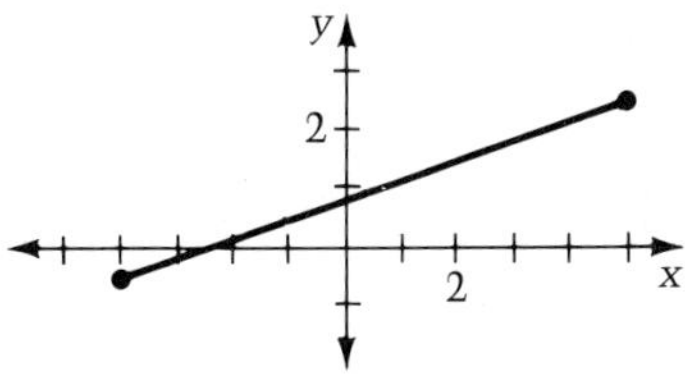


**55.** $y = -f(x)$

**56.** $y = f(x) - 2$

**57.** $y = f(x - 2)$

**58.** $y = f^{-1}(x)$

**59.** $y = f^{-1}(x) - 2$

**60.** $y = f^{-1}(x - 2)$

## 1.4 The Linear Function

**Goal A** *to determine the slope of a line and to sketch the line*

A function described by an equation of the form $y = mx + b$ is called a **linear function.** In section 3, we saw that these functions have straight lines as their graphs. Given any two points on a line, we can determine the line's *slope,* a number that measures the slant of the line.

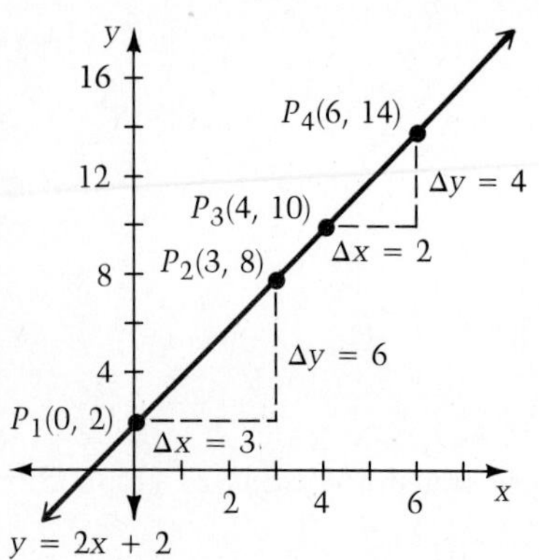

Figure 1.11

**Definition**

If $P_1(x_1, y_1)$ and $P_2(x_2, y_2)$ are two different points on a nonvertical line, then the **slope** $m$ of the line is given by the formula

$$m = \frac{y_2 - y_1}{x_2 - x_1} = \frac{\Delta y}{\Delta x}.$$

The slope of a line can be found geometrically: it is the ratio of the change in vertical direction (difference in $y$-values) to the change in horizontal direction (difference in $x$-values). In Figure 1.11, whether we use the points $P_1$ and $P_2$, or $P_3$ and $P_4$, we obtain the same slope, 2.

**Example 1** Find the slope of the line determined by the following ordered pairs; then sketch the line. (a) (3, 5) and (1, −1) (b) (−2, 3) and (1, −2)

***Solution*** (a) $m = \dfrac{-1 - 5}{1 - 3} = \dfrac{-6}{-2} = 3$

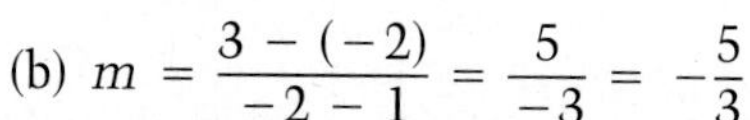

(b) $m = \dfrac{3 - (-2)}{-2 - 1} = \dfrac{5}{-3} = -\dfrac{5}{3}$

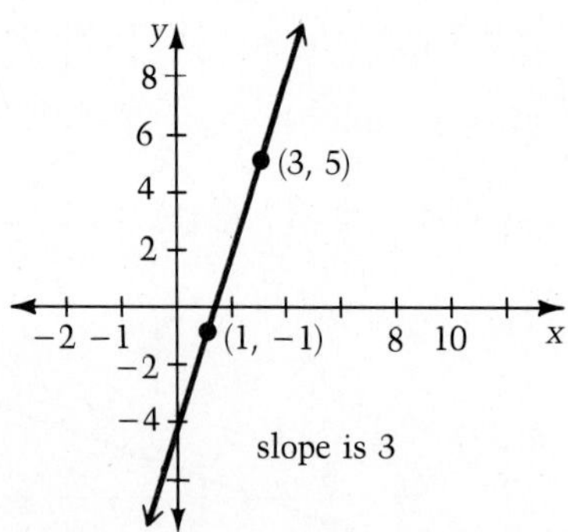


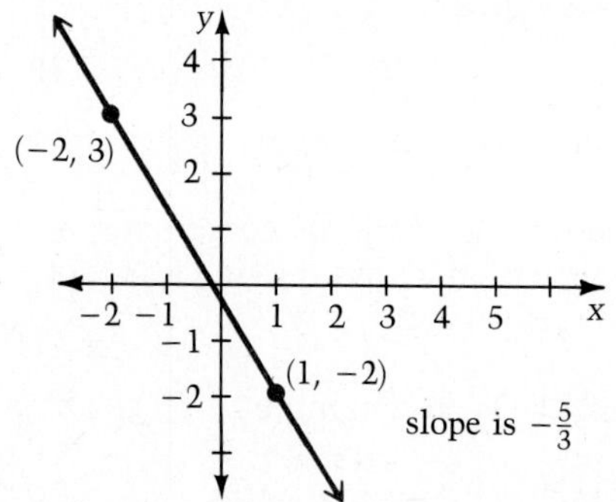


A positive slope means the line slants up from left to right. A negative slope means the line slants down from left to right.

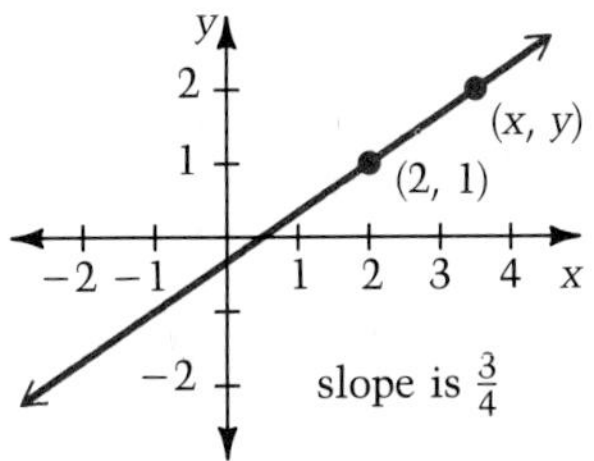

Figure 1.12

## Goal B *to determine the equation of a line*

Suppose we know that the point (2, 1) lies on a line whose slope is $\frac{3}{4}$. How can we find the equation of this line? Thinking of (2, 1) as $(x_1, y_1)$ and $(x, y)$ as $(x_2, y_2)$, we can use the definition of slope to find an equation of the line.

$$\frac{3}{4} = \frac{y - 1}{x - 2}$$

■ Remember $m = \frac{y_2 - y_1}{x_2 - x_1}$.

$$y - 1 = \frac{3}{4}(x - 2)$$

■ This is known as the point-slope form.

**Straight line: point-slope form**

If a line contains the point $(x_1, y_1)$ and has slope $m$, then the **point-slope form** of the equation of a line is

$$y - y_1 = m(x - x_1).$$

**Example 2** (a) Write the point-slope form of the equation of the line containing the point $(3, -1)$ and having slope 5. (b) Find an equation of the line containing $(-2, 4)$ and having slope $-3$. Write it in the form $y = mx + b$.

***Solution*** (a)

$$y - y_1 = m(x - x_1)$$
$$y - (-1) = 5(x - 3)$$

(b)

$$y - 4 = -3(x - (-2))$$
$$y = -3(x + 2) + 4$$
$$y = -3x - 6 + 4$$
$$y = -3x - 2$$

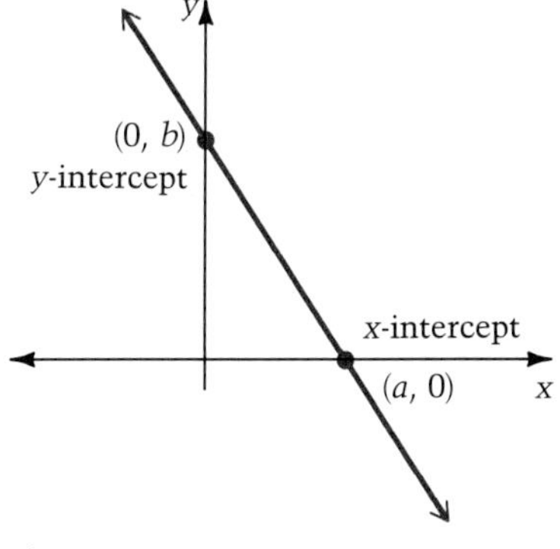

Figure 1.13

An important point on a line is the point where the line intersects the $y$-axis. The line graphed in Figure 1.13 crosses the $y$-axis at the point $(0, b)$. The number $b$ is called the **y-intercept.** If we know the line has slope $m$, we can use the point-slope form to write the equation as $y - b = m(x - 0)$, or simply, $y = mx + b$.

**Straight line: slope-intercept form**

If a line has a $y$-intercept $b$ and slope $m$, then the **slope-intercept form** of the equation of the line is $y = mx + b$.

In Figure 1.13, the line crosses the $x$-axis at $(a, 0)$. The number $a$ is called the **x-intercept.**

**Example 3** Write an equation of the line having $y$-intercept $-1$ and slope $\frac{3}{5}$. Sketch the graph.

***Solution***

$$y = mx + b$$

$$y = \frac{3}{5}x + (-1)$$

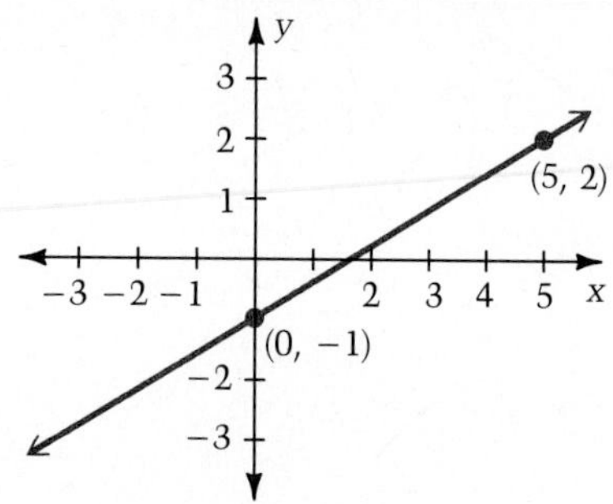


In order to sketch the line, we need two points. We already know $(0, -1)$ is a point on the line. To find a second point, choose a value for $x$ and find $y$. Let $x = 5$; then $y = 2$.

Given two points on a line, you can find an equation of the line by first calculating the slope and then using the *point-slope form.*

**Example 4** Determine an equation of the line containing the points $(-1, 3)$ and $(5, 2)$.

***Solution***

$$m = \frac{3 - 2}{-1 - 5} = -\frac{1}{6}$$ ■ First find the slope.

$$y - y_1 = m(x - x_1)$$

$$y - 2 = -\frac{1}{6}(x - 5)$$ ■ We used $m = -\frac{1}{6}$, and took $(x_1, y_1)$ as $(5, 2)$.

$$y = -\frac{1}{6}x + \frac{5}{6} + 2$$

$$y = -\frac{1}{6}x + \frac{17}{6}$$

The equation found in Example 4 can be rewritten in the form $\frac{1}{6}x + y - \frac{17}{6} = 0$, known as the general form of the equation of this line.

**Straight line: general form**

An equation of the form $Ax + By + C = 0$, where $A$, $B$ and $C$ are real numbers, with $A$ and $B$ not both zero, is called the **general form** of the equation of the line.

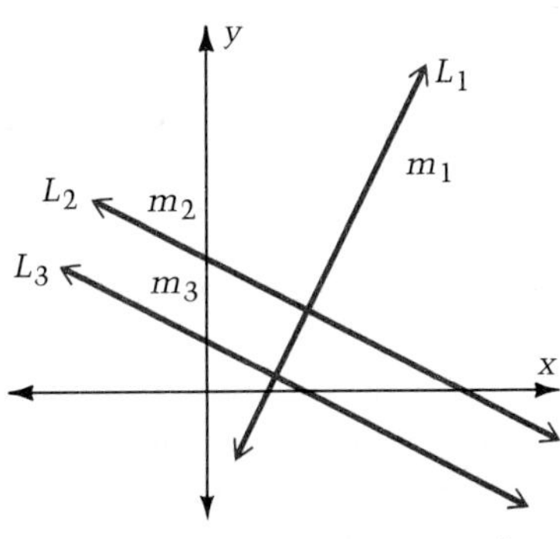


**Figure 1.14** $m_2 = m_3$, $m_2 = -\dfrac{1}{m_1}$, and $m_3 = -\dfrac{1}{m_1}$

Of the three forms of the equation of a line (the point-slope form, the slope-intercept form, and the general form) the slope-intercept form is best suited for comparing lines, because the slope can be read from the equation. In addition, knowing the slope is essential to finding the equation of a line parallel or perpendicular to a given line.

**Fact**

If two nonvertical lines are parallel, then their slopes are equal.

If two nonvertical lines are perpendicular, then their slopes are negative reciprocals. That is, if $m_1$ is the slope of a line, then $-\dfrac{1}{m_1}$ is the slope of any line perpendicular to it.

**Example 5** Line $\mathcal{M}$ is described by the equation $2x + 3y - 5 = 0$. Find an equation of the line $\mathcal{P}$ perpendicular to $\mathcal{M}$ that contains the point $(0, -3)$.

***Solution***

$$2x + 3y - 5 = 0$$

■ First, rewrite the equation for $\mathcal{M}$ in a slope-intercept form.

$$y = -\frac{2}{3}x + \frac{5}{3}$$

■ We can now conclude $\mathcal{M}$ has slope $-\frac{2}{3}$.

Since $\mathcal{P}$ is perpendicular to $\mathcal{M}$, the slope of $\mathcal{P}$ is the negative reciprocal of $-\frac{2}{3}$, that is, $\frac{3}{2}$.

$$y = \frac{3}{2}x + (-3)$$

■ Use the slope-intercept equation $y = mx + b$ because we know the $y$-intercept is $-3$.

$$\frac{3}{2}x - y - 3 = 0$$

■ Rewrite as a general linear equation.

Notice that the equation of a horizontal line has the form

$$y = b, \quad \text{where } b \text{ is a real number,}$$

and the slope of the line is 0. The equation of a vertical line has the form

$$x = c, \quad \text{where } c \text{ is a real number.}$$

A vertical line is not the graph of a function. Furthermore, the slope of a vertical line is undefined, since the difference in $x$-coordinates is zero for any two points on the line.

## Exercise Set 1.4 [=] Calculator Exercises in the Appendix

### Goal A

In exercises 1–12, find the slope of the line determined by the following ordered pairs; then sketch the line.

1. (3, 8), (4, 13)
2. (5, 2), (7, 10)
3. (−5, −6), (3, −8)
4. (8, −5), (−10, 10)
5. (2, 5), (−7, 5)
6. (5, 2), (5, −7)
7. (3, 0), (1, 0)
8. (0, 8), (0, 11)
9. $\left(\frac{1}{2}, \frac{1}{8}\right), \left(\frac{3}{2}, \frac{1}{4}\right)$
10. $\left(\frac{1}{3}, \frac{1}{4}\right), \left(\frac{2}{3}, \frac{3}{8}\right)$
11. (0.1, 0.01), (1, 1)
12. (0.2, 1.3), (1.8, 1.5)

### Goal B

For exercises 13–24, determine an equation of the line passing through the pairs of points given in exercises 1–12. Write your answer in (a) point-slope form, (b) slope-intercept form, and (c) general form.

In exercises 25–36, determine an equation of the line satisfying the stated conditions. Write your answer in (a) point-slope form, (b) slope-intercept form, and (c) general form.

25. slope is −2; passes through (3, 5)
26. slope is $\frac{1}{2}$; passes through (−1, 1)
27. slope is 0; passes through (2, −2)
28. slope is not defined; passes through (2, −2)
29. $y$-intercept is 3; slope is $\frac{1}{2}$
30. $y$-intercept is −2; slope is $\frac{3}{4}$
31. $y$-intercept is 0; slope is −2
32. $y$-intercept is −2; slope is 0
33. line is parallel to the graph of $2x + y - 10 = 0$ and passes through the origin
34. line is parallel to the $x$-axis and passes through (−2, −8)
35. line is perpendicular to the graph of $y = 6x - 10$ and passes through (5, −2)
36. line is perpendicular to the graph of $y = -7$ and passes through the origin

### Superset

37. Show that the three points (1, 3), (5, 1), and (6, 3) are vertices of a right triangle. (Use slopes to show that two of the sides are perpendicular.)
38. Show that the points (1, 4), (2, 7), and (−3, −8) lie on the same line. (Use slopes.)
39. An ant starts at the point (1, 2) when $t = 0$ and follows a path on the $xy$-plane such that its coordinates at any time $t$ are given by the equations $x = 3t + 1$ and $y = 2 - t$. By solving each equation for $t$, and equating these expressions in $x$ and $y$, determine a function of $x$ that describes the ant's path. What is the domain of the function?
40. Follow the procedure outlined in exercise 39 to determine an equation in slope-intercept form that describes the graph given by each of the following pairs of equations.

    (a) $x = 2t - 3$, $y = t + 4$
    (b) $x = t$, $y = 5 - 6t$
    (c) $x = 3t + 2$, $y = 2 - 3t$
    (d) $x = t^2 - 3$, $y = t^2 + 4$
41. A teacher has given a test in which the highest grade was 75, and the lowest grade was 40. Determine a linear function, $y = f(x)$, that can be used to distribute the test scores so that the grade of 40 becomes 60 and 75 becomes 100.
42. We say that $y$ *varies directly* with $x$ provided there exists some nonzero constant $k$ such that $y = kx$. Suppose $V$ varies directly with $t$, and when $t = \frac{1}{2}$, $V = \pi$. For what value of $t$ is $V$ equal to 1?

## 1.5 The Quadratic Function

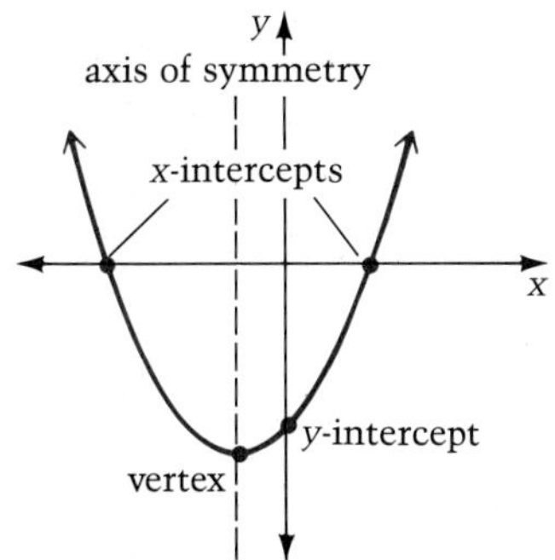


**Figure 1.15**

**Goal A** *to determine the intercepts of the graph of a quadratic function*

An equation of the form $f(x) = ax^2 + bx + c$, with $a$, $b$, and $c$ real numbers and $a \neq 0$, is called the **general form of a quadratic function** and its graph is a parabola. Such parabolas have either a high or low point, called the **vertex.** Passing through the vertex is a line of symmetry called the **axis of symmetry.**

When graphing a parabola it is useful to determine the **x-intercepts,** that is, the values of $x$ for which $f(x) = 0$. Since these are the values of $x$ where $y = 0$, the $x$-intercepts tell us where the graph intersects the $x$-axis. The value of $f(0)$ is called the **y-intercept.** Since $f(0)$ is the function value at $x = 0$, the $y$-intercept tells us where the graph intersects the $y$-axis.

$A \cdot B = 0$ implies $A = 0$ or $B = 0$

To find the $x$-intercepts of the graph of a quadratic function, we solve the equation $ax^2 + bx + c = 0$. The **Principle of Zero Products** may be used to solve some quadratic equations. It states that if the product of two or more factors is zero, then at least one of the factors must be zero.

Recall that the quadratic formula can be used to solve any quadratic equation. That is, the real number solutions of any equation $ax^2 + bx + c = 0$ with $a \neq 0$ can be determined by the formula

$$x = \frac{-b \pm \sqrt{b^2 - 4ac}}{2a}.$$

**Example 1** Find the $x$-intercepts of the graph of each quadratic function.

(a) $f(x) = 2x^2 - x - 6$ (b) $g(x) = x^2 - x - 10$

***Solution*** (a) Factor and apply the Principle of Zero Products.

$$2x^2 - x - 6 = 0$$
$$(2x + 3)(x - 2) = 0$$

| $2x + 3 = 0$ | $x - 2 = 0$ |
|---|---|
| $x = -\frac{3}{2}$ | $x = 2$ |

The $x$-intercepts are $x = -\frac{3}{2}$ and 2.

(b) Use the quadratic formula.

$$x^2 - x - 10 = 0$$
$$x = \frac{-(-1) \pm \sqrt{(-1)^2 - 4(1)(-10)}}{2(1)}$$
$$x = \frac{1 \pm \sqrt{41}}{2}$$

The $x$-intercepts are

$$x = \frac{1 + \sqrt{41}}{2} \text{ and } \frac{1 - \sqrt{41}}{2}.$$

Note that in the previous example, $f(0) = 2(0)^2 - 0 - 6 = -6$ and $g(0) = 0^2 - 0 - 10 = -10$. Thus, the $y$-intercepts of the functions $f$ and $g$ are $-6$ and $-10$, respectively.

**Goal B** *to graph functions of the form* $f(x) = ax^2 + bx + c$

Another form of the equation of a parabola is $f(x) = a(x - h)^2 + k$, where $a$, $h$, and $k$ are real numbers. We call this equation the **vertex form of a parabola,** and from it we can deduce the following features of a parabola:

$$\text{The graph opens} \begin{cases} \text{upward if } a > 0, \\ \text{downward if } a < 0. \end{cases}$$

$$\text{The graph is} \begin{cases} \text{wider than } f(x) = x^2 \text{ if } |a| < 1, \\ \text{narrower than } f(x) = x^2 \text{ if } |a| > 1. \end{cases}$$

$$\text{The vertex } (h, k) \text{ is} \begin{cases} \text{a high point if the parabola opens downward,} \\ \text{a low point if the parabola opens upward.} \end{cases}$$

The axis of symmetry is the line $x = h$.

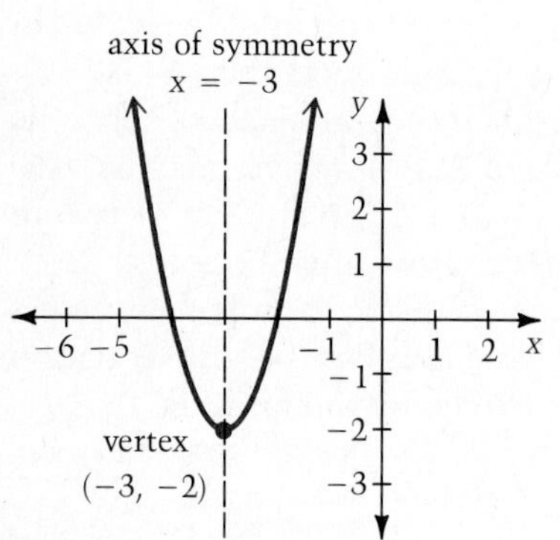


**Figure 1.16** $f(x) = 2(x + 3)^2 - 2$

For example, consider the function

$$f(x) = 2(x + 3)^2 - 2.$$

Its graph opens upward because the coefficient of $(x + 3)^2$ is positive. This parabola is narrower than the parabola $f(x) = x^2$, has vertex $(-3, -2)$, and is symmetric with respect to the line $x = -3$.

It is easy to graph a parabola if its equation is in vertex form. If, however, we are asked to sketch the graph of a quadratic equation which has been given in general form, we can use a method called **completing the square** to convert this equation from general form to vertex form. For example, suppose we need to graph the function $f(x) = 2x^2 - 4x - 6$. We convert this equation to vertex form as follows:

$$f(x) = 2x^2 - 4x - 6$$

$$= 2(x^2 - 2x + \square) - 6$$

■ First, factor the coefficient of $x^2$ out of the $x^2$- and $x$-terms. A space is left for the number that makes the expression in parentheses a perfect square.

$$= 2(x^2 - 2x + \underset{\substack{\blacktriangle \\ \text{the square} \\ \text{of } \left(\frac{1}{2}\cdot(-2)\right)}}{1}) - 6 - 2 \cdot 1$$

■ In the space write the square of half the coefficient of the $x$-term. Since we have added $2 \cdot 1$, we must subtract $2 \cdot 1$.

$$= 2(x - 1)^2 - 8$$

■ $a = 2$, $h = 1$, and $k = -8$.

Thus, the graph of $f(x) = 2x^2 - 4x - 6$ has its vertex (low point) at $(1, -8)$.

**Example 2** Find the vertex form of the quadratic equation $f(x) = 2x^2 + 12x + 16$.

***Solution***

$$f(x) = 2x^2 + 12x + 16$$
$$= 2(x^2 + 6x + \square) + 16$$
$$= 2(x^2 + 6x + 9) + 16 - 18$$

■ Since we added $2 \times 9$ we must subtract 18.

$$= 2(x + 3)^2 - 2$$
$$= 2(x - (-3))^2 + (-2)$$

■ The vertex form.

**Example 3** Graph the function $y = 4x^2 - 16x + 7$. Show the vertex, the axis of symmetry, $x$-intercepts and two other points.

***Solution***

(1) First, find the $x$-intercepts.

$$x = \frac{-(-16) \pm \sqrt{(-16)^2 - 4(4)(7)}}{2(4)}$$

■ Use the quadratic formula.

$$x = \frac{7}{2} \quad \text{and} \quad x = \frac{1}{2}$$

■ The $x$-intercepts are $\frac{7}{2}$ and $\frac{1}{2}$.

(2) Find the vertex form.

$$y = 4(x^2 - 4x + \square) + 7$$

■ Complete the square.

$$y = 4(x^2 - 4x + (-2)^2) + 7 - 4(-2)^2$$
$$y = 4(x - 2)^2 - 9$$

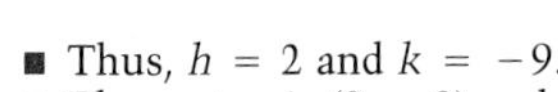

■ Thus, $h = 2$ and $k = -9$.
■ The vertex is $(2, -9)$ and the axis of symmetry is $x = 2$.
■ Since $a = 4$, that is, $a > 0$, the parabola opens upward.
■ Finally, we find two more points: Select points where $x = 0$ and 4.
$x = 0 \quad y = 4(0)^2 - 16(0) + 7 = 7$
$x = 4 \quad y = 4(4)^2 - 16(4) + 7 = 7$
■ The two other points are $(0, 7)$ and $(4, 7)$.

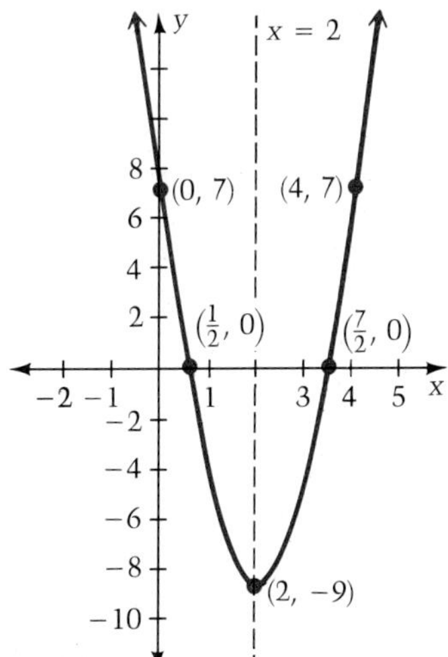

## Exercise Set 1.5 ⊟ Calculator Exercises in the Appendix

### Goal A

In exercises 1–8, find the $x$-intercepts of the graph of the quadratic function.

1. $y = 5x - x^2$
2. $y = 3x^2 - 15x$
3. $y = x^2 - 100$
4. $y = x^2 - 64$
5. $y = x^2 + 9x + 8$
6. $y = x^2 + 5x + 6$
7. $y = x^2 - 3x - 10$
8. $y = x^2 + 4x - 21$

In exercises 9–16, use the quadratic formula to find the $x$-intercepts of the graph of the quadratic function.

9. $y = x^2 + 5x + 4$
10. $y = x^2 + 11x + 18$
11. $y = x^2 - 8x + 16$
12. $y = x^2 + 4x - 60$
13. $y = x^2 + x + 10$
14. $y = x^2 + x - 10$
15. $y = x^2 + x - 1$
16. $y = x^2 - 2x + 2$

### Goal B

In exercises 17–26, write the given equation in vertex form. Find the vertex, the axis of symmetry, and the direction in which the parabola opens.

17. $y - 8 = (x - 3)^2 + 2$
18. $y + 7 = 3(x + 4)^2 + 5$
19. $3y = -(x + 1)^2 - 5$
20. $-2y = -(x - 4)^2 - 7$
21. $3(y + 1) = (x - 9)^2$
22. $4(y - 6) = (x + 2)^2$
23. $7y = -2(x + 6)^2$
24. $10y = 3(x + 1)^2$
25. $y = x^2 + 4$
26. $2y = 3x^2 - 8$

In exercises 27–38, graph the given function. Label the vertex, the $x$-intercepts, the axis of symmetry, and two other points.

27. $f(x) = 5x^2 - 3$
28. $g(x) = 4 - 2x^2$
29. $y = x^2 - 4x - 5$
30. $y = x^2 + 6x + 6$
31. $y = 4x - x^2 - 4$
32. $y = 2 - x^2 - x$
33. $y = x^2 - 6x + 9$
34. $y = x^2 + 4x + 4$
35. $f(x) = 4x^2 - 20x + 25$
36. $f(x) = 9x^2 + 48x + 64$
37. $y = -2x^2 + 20x - 54$
38. $y = 3x^2 - 12x + 20$

### Superset

In exercises 39–42, determine the maximum or minimum function value. State whether this value is a maximum or minimum.

39. $y = -3(x + 1)^2$
40. $y = 5(x - 2)^2$
41. $y = 5 + 4x - x^2$
42. $y = 4x - x^2$
43. Determine $b$ such that the quadratic function $f(x) = x^2 - bx + 6$ has exactly one $x$-intercept.
44. The quadratic function $h(x) = 2x^2 - 3x + C$ has 5 as one of its $x$-intercepts. Determine the value of $C$ and the other $x$-intercept.
45. We say that $y$ varies directly with $x^2$ provided there exists some nonzero constant $k$ such that $y = kx^2$. Suppose $d$ varies directly with $t^2$ and that $d$ is 5 when $t = 3$. Find $d$ when $t = 5$.
46. The height $h(t)$ of a projectile shot upwards with an initial velocity of 352 ft/sec can be described as a function of time $t$:
$$h(t) = 352t - 16t^2.$$
(a) By finding the $y$-coordinate of the vertex, determine the maximum height of the projectile.
(b) When does the projectile achieve its maximum height?
(c) When does the projectile return to the ground?
47. The sum of two numbers is 12.
(a) Describe the two numbers using the variable $x$.
(b) Describe the product of the two numbers as a function of $x$.
(c) Determine the two values that make the product as large as possible.
48. The perimeter of a rectangle is 60 ft.
(a) Describe the area of the rectangle as a function of one variable.
(b) Determine the dimensions of the rectangle that produce the largest area, and state the value of this maximum area.

# Chapter Review & Test

## Chapter Review

### 1.1 The Coordinate Plane (pp. 2–4)

We graph ordered pairs of real numbers on a plane formed by two perpendicular number lines, called *axes.* The axes separate the plane into four *quadrants.* The horizontal axis is called the *x-axis,* and the vertical axis is called the *y-axis.* The axes meet at the *origin,* and the entire system is the *xy-plane.* (p. 2)

The distance between any two points $P_1(x_1, y_1)$ and $P_2(x_2, y_2)$ in the $xy$-plane is given by the formula $d(P_1, P_2) = \sqrt{(x_2 - x_1)^2 + (y_2 - y_1)^2}$. (p. 3)

### 1.2 Functions (pp. 5–11)

A *function* is a set of ordered pairs, with no two of them having the same first coordinate and different second coordinate. The set of all first coordinates is the *domain,* and the set of all second coordinates is the *range.* In general, any set of ordered pairs is a *relation.* (p. 5)

A function $f$ is *one-to-one* if each $y$-value in the range of $f$ corresponds to exactly one $x$-value in the domain. (p. 8)

Let $f$ and $g$ be functions, with the range of $g$ contained in the domain of $f$. For each $x$ in the domain of $g$, the *composite function* $f \circ g$ is defined as $(f \circ g)(x) = f(g(x))$. (p. 9)

Given two functions $f$ and $g$, $g$ is the *inverse* of $f$ if for each $x$ in the domain of $g$, $(f \circ g)(x) = x$, and for each $x$ in the domain of $f$, $(g \circ f)(x) = x$. (p. 10)

### 1.3 Techniques of Graphing (pp. 12–19)

Equation of a *straight line:* $y = mx + b$, where $m$ and $b$ are real numbers.

Equation of a *parabola:* $y = ax^2 + bx + c$, where $a$, $b$, and $c$ are real numbers and $a \neq 0$.

Equation of a *V-shaped graph:* $y = a|x| + k$, where $a$ and $k$ are real numbers and $a \neq 0$. (p. 12)

The graphs of even functions are symmetric with respect to the $y$-axis, and the graphs of odd functions are symmetric with respect to the origin. (p. 16) The graph of any one-to-one function is reflected across the line $y = x$ to obtain the graph of its inverse function. (p. 17)

There are three general types of transformations: *translations, reflections,* and *changes in shape.* (pp. 17–18)

### 1.4 The Linear Function (pp. 20–24)

An equation of the form $y = mx + b$ describes a *linear function.* Its graph is a straight line with slope $m$ and $y$-intercept $b$. (p. 20)

There are three different types of straight line equations (pp. 20–22):

| *Point-slope form* | *Slope-intercept form* | *General form* |
|---|---|---|
| $y - y_1 = m(x - x_1)$ | $y = mx + b$ | $Ax + By + C = 0$ |

### 1.5 The Quadratic Function (pp. 25–28)

An equation of the form $f(x) = ax^2 + bx + c$, with $a$, $b$, and $c$ real numbers and $a \neq 0$, is called the *general form of a quadratic function,* and its graph is a parabola. In general, parabolas have either a high or low point, called the *vertex.* The *axis of symmetry* is a line of symmetry that passes through the vertex. (p. 25)

An equation of the form $f(x) = a(x - h)^2 + k$, where $a$, $h$, and $k$ are real numbers, is called the *vertex form of a parabola.* The method called *completing the square* can be used to convert a quadratic equation from general form to vertex form. (p. 26)

## Chapter Test

**1.1A** Plot the given points in the $xy$-plane.

**1.** $(4, 1)$

**2.** $(-2, 5)$

**3.** $\left(-\frac{\sqrt{3}}{2}, -\frac{1}{2}\right)$

**4.** $\left(3, -\frac{\pi}{2}\right)$

**1.1B** Determine the distance between the given points.

**5.** $(2, 6), (-1, -3)$

**6.** $(0, \sqrt{2}), (\sqrt{2}, -2)$

**7.** $\left(\frac{1}{2}, 1\right), (0, 1)$

**8.** $\left(\frac{\sqrt{2}}{2}, \frac{\sqrt{2}}{2}\right), (0, 0)$

**1.2A** Describe each of the following functions with a set of ordered pairs. State the domain and range of each function.

**9.** **10.**

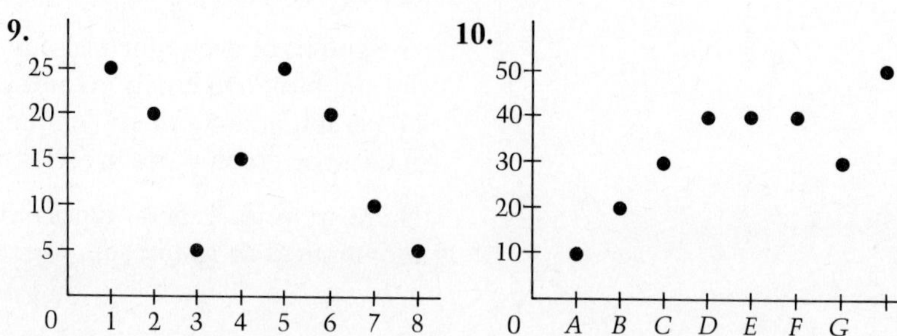

Determine $f(-2)$, $f(0)$, $f(h^2)$, and $f(b - 1)$.

**11.** $f(x) = -x^2 - 5x + 9$ **12.** $f(x) = 4 - x^3$ **13.** $f(x) = \frac{1}{3}x - 5$

Determine whether the given functions are one-to-one.

**14.** $f(x) = 3x - 5$ **15.** $f(x) = x^3 + 4x^2$

**1.2B** Evaluate each of the following expressions given that $f(x) = 3x + 3$, $g(x) = |x| - 2$ and $h(x) = \frac{1}{x - 1}$.

**16.** $(f \circ g)(-2)$ **17.** $(g \circ f)(0)$ **18.** $(f \circ h)(10)$

**19.** $(h \circ f)(10)$ **20.** $(h \circ g)(-3)$

**1.2C** Show that the two given one-to-one functions are inverse functions.

**21.** $f(x) = 7 - 5x$, $g(x) = \frac{7 - x}{5}$

**22.** $f(x) = x^3 + 2$, $g(x) = \sqrt[3]{x - 2}$

**1.3A** Graph the function.

**23.** $y = 10x + 3$ **24.** $f(x) = x^2 - 3x - 8$

**25.** $g(x) = 11 - 4|x|$

**1.3B** Plot and label the reflection of each point (a) across the $x$-axis, (b) across the $y$-axis, (c) across the line $y = x$, and (d) through the origin.

**26.** $(1, -4)$ **27.** $(0, -3)$ **28.** $(2, 0)$

The given function $y = f(x)$ is one-to-one. Find an equation describing the inverse function, identify its domain and range, and sketch its graph.

**29.** $f(x) = 2x + 5, -1 \le x \le 2$ **30.** $f(x) = \sqrt{x - 3}, x \ge 7$

**1.3C** Use transformations to graph each of the following functions.

**31.** $y = -(x + 2)^2 + 1$ **32.** $2y = |x - 1| - 2$

**1.4A** Find the slope of the line determined by each of the following ordered pairs. Then sketch the line.

**33.** $\left(\frac{1}{6}, \frac{1}{2}\right)$ and $\left(\frac{2}{5}, -\frac{2}{3}\right)$ **34.** $(3.4, 0.3)$ and $(-1.2, -2)$

**1.4B** Determine an equation of the line satisfying the stated conditions. Write your answer in (a) point-slope form, (b) slope-intercept form, and (c) general form.

**35.** The line is parallel to $y = 3x$; it passes through $(-2, 4)$.

**36.** The line is perpendicular to $y = 2x$; it passes through $(6, -1)$.

**1.5A** Find the $x$-intercepts of the graph of each quadratic function.

**37.** $y = x^2 + 7x + 12$

**38.** $y = x^2 + 3x - 1$

**1.5B** Write each of the following equations in vertex form. Find the vertex, the axis of symmetry, and the direction in which the parabola opens.

**39.** $y = 6(x + 4)^2 + 9$

**40.** $-2(y + 3) = (x - 5)^2$

Sketch the graph of each of the following functions. Label the vertex, the $x$-intercepts, the axis of symmetry, and two additional points.

**41.** $f(x) = x^2 - 4x - 21$

**42.** $f(x) = 9 - 4x^2$

## Superset

**43.** Determine whether the given points lie on the same line.

(a) $(0, 5), (2, 1), (-3, 10)$

(b) $(0, -2), (3, 7), (-1, -5), (6, 16)$

**44.** For the function $f(x) = 3x^2 + 2x$, find $\dfrac{f(2 + h) - f(2)}{h}$.

**45.** The quadratic function $f(x) = 3x^2 - 5x + C$ has $-2$ as one of its $x$-intercepts. Determine the value of $C$ and the other $x$-intercept.

**46.** Find two numbers whose difference is 20 and whose product is as small as possible.

# 2 Trigonometry: An Introduction

## 2.1 Triangles

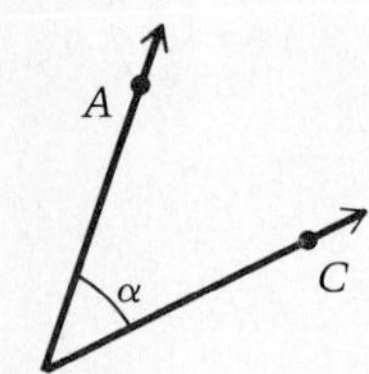

Figure 2.1

**Goal A** *to determine sides and angles of special types of triangles*

Trigonometry is a branch of mathematics that developed as a tool for solving problems involving triangles. In fact, the word "trigonometry" is derived from the Greek words for "measurement of triangles." It is appropriate, therefore, to begin our study of trigonometry by reviewing some basic information about triangles.

An **angle** is a figure consisting of two rays having the same endpoint. This common endpoint is called the **vertex.** For the angle in Figure 2.1, $B$ is the vertex. This angle can be described in three ways: $\angle ABC$, $\angle B$, and $\alpha$. The symbol $\alpha$ is the Greek letter *alpha.* Other Greek letters often used in describing angles are $\theta$ (*theta*), $\varphi$ (*phi*), and $\beta$ (*beta*).

A common unit for measuring angles is the degree (°). Angles of various measures are shown in Figure 2.2. The symbols $m(\angle B)$ and $m(\alpha)$ are used to indicate the measures of angles $B$ and $\alpha$, respectively. However, the "$m$" is often omitted. Thus, you may see $m(\theta) = 45°$ or simply $\theta = 45°$.

An angle with measure 90° is called a **right angle.** The special symbol □, signifies a right angle when placed at a vertex as in Figure 2.2. An angle with measure between 0° and 90° is called **acute,** and an angle with measure between 90° and 180° is called **obtuse.**

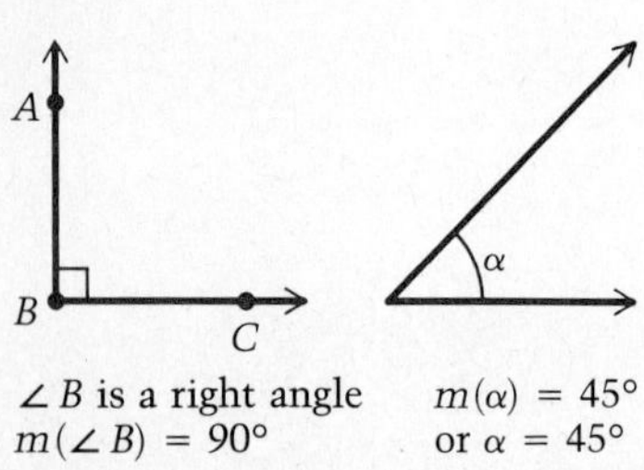


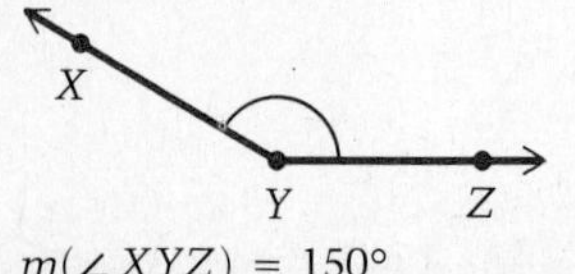

Figure 2.2

Recall that a triangle has 3 angles and 3 sides. The triangle below is referred to as $\triangle ABC$. Points $A$, $B$, and $C$ are called **vertices** of the triangle.

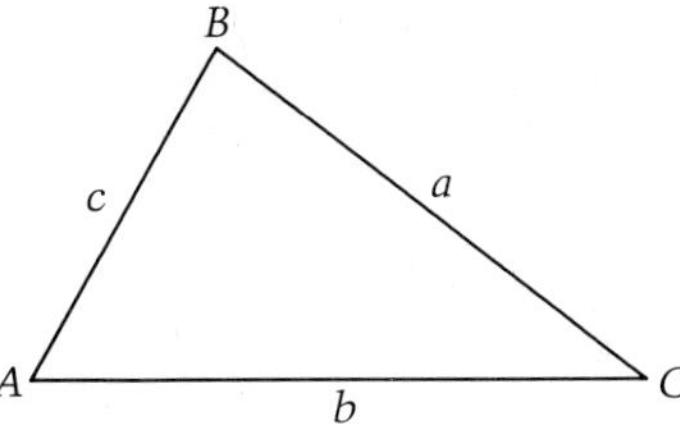


Figure 2.3

■ When we label a triangle, it is common to let $a$ stand for the side opposite vertex $A$, $b$ for the side opposite vertex $B$, and $c$ for the side opposite vertex $C$.

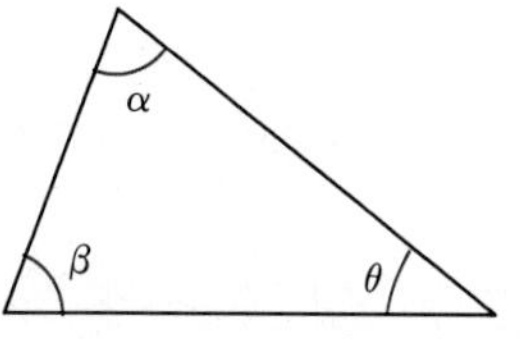


**Figure 2.4** $\alpha + \beta + \theta = 180°$ or $m(\alpha) + m(\beta) + m(\theta) = 180°$.

**Fact**

In any triangle, the sum of the measures of the angles is 180°.

Certain types of triangles are of special importance to the development of trigonometry. An **isosceles** triangle is a triangle with two sides of equal length. In an isosceles triangle, the angles opposite the sides of equal length have the same measure.

**Example 1** In $\triangle ABC$, $a = 5$, $b = 5$, and $m(\angle A) = 65°$. Find $m(\angle C)$.

***Solution***

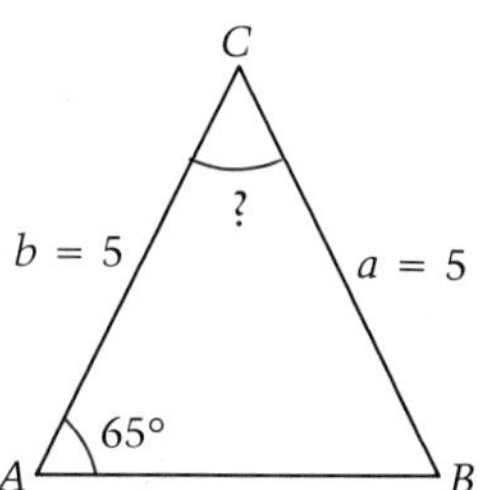


■ Begin by drawing a triangle and labeling its known parts.

■ Since $a = b$, the triangle is isosceles. Thus $m(\angle B) = m(\angle A) = 65°$.

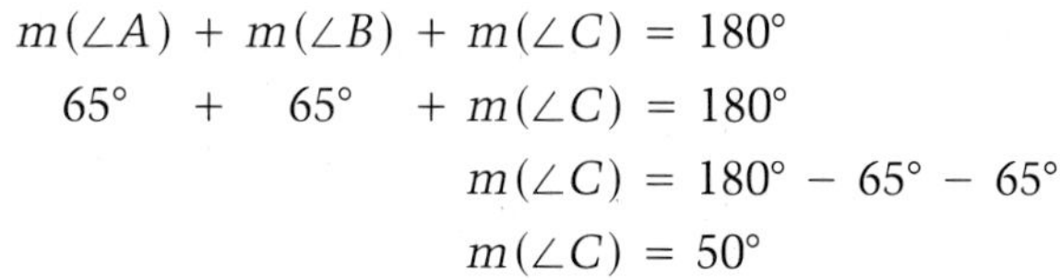

$$
\begin{aligned}
m(\angle A) + m(\angle B) + m(\angle C) &= 180° \\
65° \quad + \quad 65° \quad + m(\angle C) &= 180° \\
m(\angle C) &= 180° - 65° - 65° \\
m(\angle C) &= 50°
\end{aligned}
$$

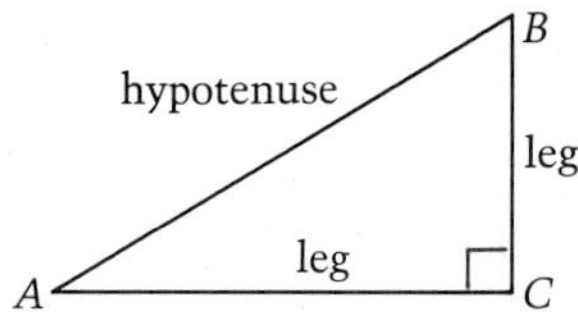


**Figure 2.5** A right triangle.

A **right triangle** is a triangle in which one of the angles is a right angle. In a right triangle, it is common to label the right angle $C$. The longest side of a right triangle is opposite the right angle and is called the **hypotenuse.** The other two sides are called **legs.**

**Example 2** $\triangle ABC$ is an isosceles right triangle with $m(\angle C) = 90°$. Find $m(\angle A)$.

***Solution***

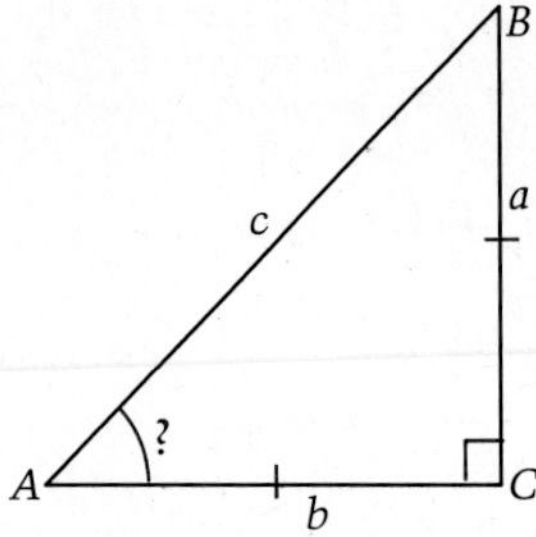


- Begin by drawing a triangle and labeling the known parts.
- Sides (or angles) having equal measure are denoted by the same hash marks. The hash marks in this figure indicate that $a = b$.

$$m(\angle A) + m(\angle B) + m(\angle C) = 180°$$ ■ True for any triangle.

$$m(\angle A) + m(\angle B) + 90° = 180°$$ ■ $C$ is a right angle.

$$m(\angle A) + m(\angle A) = 90°$$ ■ $m(\angle B) = m(\angle A)$ since $\triangle ABC$ is isosceles.

$$m(\angle A) = 45°$$

One useful fact about a right triangle is that the square of the hypotenuse is equal to the sum of the squares of the legs. This fact is known as the Pythagorean Theorem.

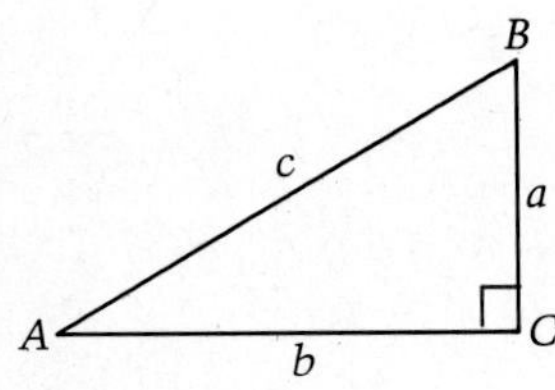


**Figure 2.6** $c^2 = a^2 + b^2$

**Pythagorean Theorem**

If $ABC$ is a right triangle, with $a$ and $b$ the lengths of the legs and $c$ the length of the hypotenuse, then $c^2 = a^2 + b^2$.

A triangle whose three sides have the same length is called an **equilateral triangle.** In an equilateral triangle, each of the angles measures 60°. You should convince yourself that the following statement is true: every equilateral triangle is also isosceles, but not every isosceles triangle is equilateral.

**Example 3** Given an equilateral triangle with side of length 1, find the length of the altitude.

***Solution***

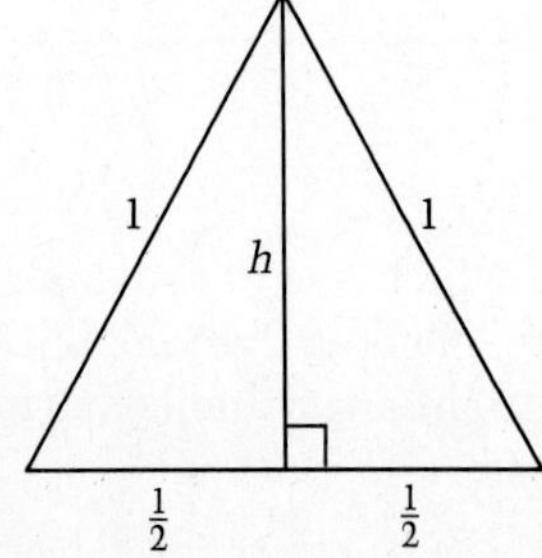


- Begin by drawing an equilateral triangle with altitude $h$. The **altitude** is a perpendicular from a vertex to the opposite side. In an equilateral triangle, the altitude cuts the opposite side into two equal lengths.

$$1^2 = h^2 + \left(\frac{1}{2}\right)^2$$

■ Apply the Pythagorean Theorem to either of the two smaller triangles. They are congruent.

$$1 = h^2 + \frac{1}{4}$$

$$h = \sqrt{\frac{3}{4}} = \frac{\sqrt{3}}{2}$$

■ Length is nonnegative, thus the value $-\frac{\sqrt{3}}{2}$ has been discarded.

The altitude is $\frac{\sqrt{3}}{2}$.

**Goal B** *to solve problems involving similar triangles*

We call triangles **similar** if they have the same shape, even though they may have different sizes. This means that one of the triangles is a "magnification" or a "reduction" of the other.

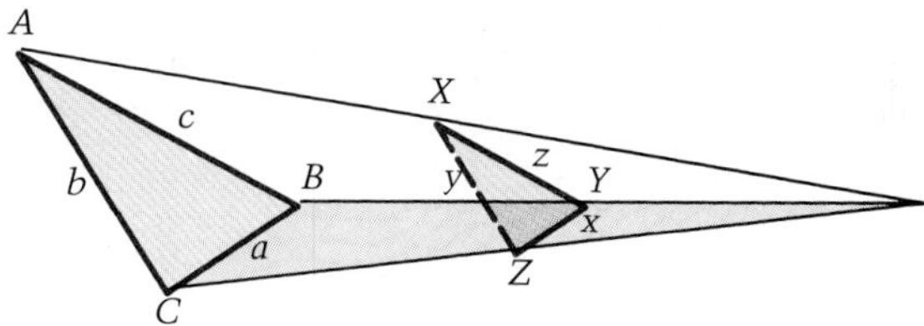


Figure 2.7

In Figure 2.7, triangle $ABC$ and triangle $XYZ$ are similar. This is written $\triangle ABC \sim \triangle XYZ$. When we say that $\triangle ABC$ is similar to $\triangle XYZ$, we imply a special correspondence between the two triangles such that the measures of the corresponding angles are equal and the corresponding sides of the triangles are proportional. Thus,

$$m(\angle A) = m(\angle X), \quad m(\angle B) = m(\angle Y), \quad \text{and} \quad m(\angle C) = m(\angle Z),$$

and

$$\frac{a}{x} = \frac{b}{y} = \frac{c}{z}.$$

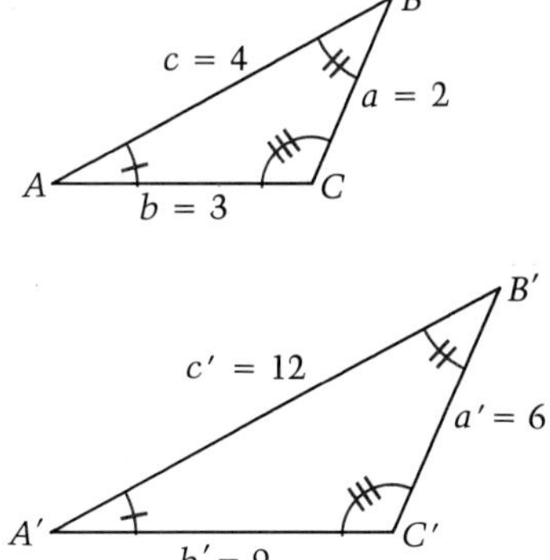


Figure 2.8

In Figure 2.8, $\triangle A'B'C' \sim \triangle ABC$. Notice that

$$\frac{a}{a'} = \frac{2}{6} = \frac{1}{3}, \quad \frac{b}{b'} = \frac{3}{9} = \frac{1}{3}, \quad \frac{c}{c'} = \frac{4}{12} = \frac{1}{3}.$$

Thus, the ratio of the corresponding sides is $\frac{1}{3}$.

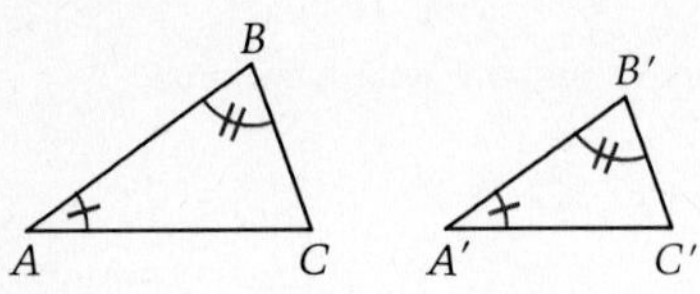

Figure 2.9 $\triangle ABC \sim \triangle A'B'C'$

You may recall from geometry that to prove two triangles are similar, you must find two pairs of corresponding angles having the same measure.

**Angle-Angle (A-A) Similarity Theorem**
Two triangles are similar if two angles of one triangle have the same measure as two angles of the other triangle.

**Example 4** In the triangles below, $m(\angle A) = m(\angle X)$, $m(\angle B) = m(\angle Y)$, and the lengths of some sides are given. Find $c$.

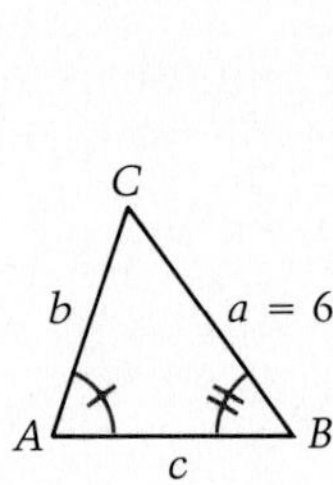


Z
y
x = 12
X
Y
z = 8

***Solution***

$\triangle ABC \sim \triangle XYZ$ ■ The triangles are similar by the A-A Similarity Theorem.

$\frac{6}{12} = \frac{c}{8}$ ■ By similarity, $\frac{a}{x} = \frac{c}{z}$.

$12c = 6 \cdot 8$ ■ The result of cross-multiplication.

$c = 4$

**Example 5** In the right triangles below $m(\angle A) = m(\angle A')$, $a = 3$, $b = 4$, and $a' = 9$. Find $c'$.

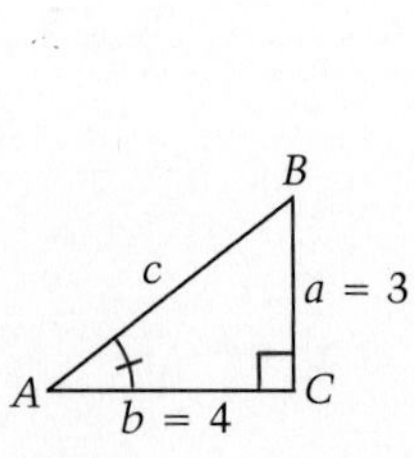


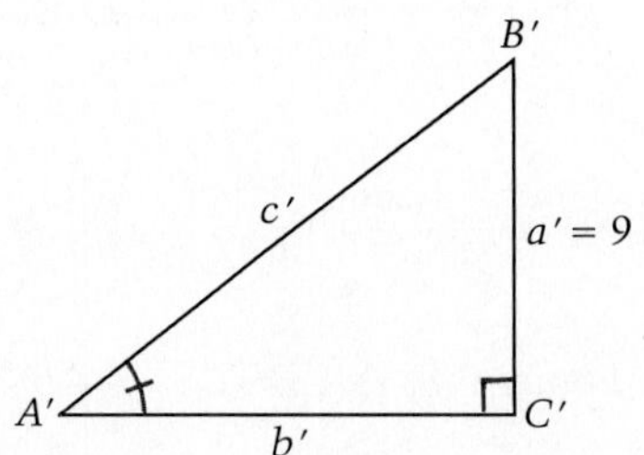


***Solution***

$\triangle ABC \sim \triangle A'B'C'$ ■ The A-A Similarity Theorem is used.

$\frac{a}{a'} = \frac{b}{b'} = \frac{c}{c'}$ ■ This proportion follows from similarity. To find $c'$, we must first find $c$.

$$c^2 = a^2 + b^2$$
$$c = \sqrt{a^2 + b^2} = \sqrt{3^2 + 4^2}$$
$$c = 5$$

■ Now that $c$ has been determined, we use the proportion to find $c'$.

$$\frac{a}{a'} = \frac{c}{c'}$$
$$\frac{3}{9} = \frac{5}{c'}$$
$$c' = 15$$

**Example 6** A building casts a shadow 84 ft long. At the same instant a nearby fence casts a shadow 14 ft long. If the fence is 10 ft high, what is the height of the building?

***Solution***

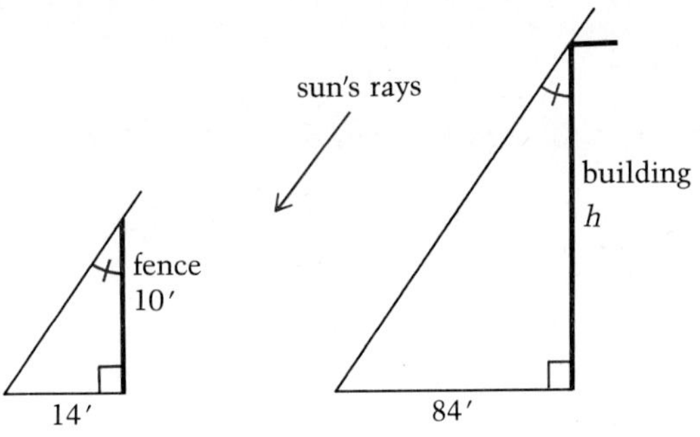


■ Draw a figure. If we assume that the building and fence are perpendicular to the ground, and the sun's rays are parallel, then the two triangles are similar.

$$\frac{h}{10} = \frac{84}{14}$$
$$14h = 10 \cdot 84$$
$$h = 60$$

■ Set up a proportion using corresponding sides of the two triangles.

The height of the building is 60 ft.

## Exercise Set 2.1 ▤ Calculator Exercises in the Appendix

### Goal A

In exercises 1–6, use the given information about $\triangle ABC$ to solve for the missing part.

1. $m(\angle A) = 40°$, $m(\angle B) = 73°$, $m(\angle C) = ?$
2. $m(\angle A) = 58°$, $m(\angle C) = 16°$, $m(\angle B) = ?$
3. $a = b$, $m(\angle A) = 30°$, $m(\angle B) = ?$
4. $a = b$, $m(\angle C) = 82°$, $m(\angle B) = ?$
5. $m(\angle A) = m(\angle B)$, $a = 18$, $b = ?$
6. $m(\angle C) = 90°$, $a = b$, $m(\angle A) = ?$

In exercises 7–18, $\triangle ABC$ is a right triangle, $c$ is the length of the hypotenuse, and $a$ and $b$ are the lengths of the legs. Find the length of the third side given the lengths of two sides.

**7.** $a = 6, b = 8$
**8.** $a = 12, c = 15$
**9.** $b = 15, c = 39$
**10.** $a = 15, c = 17$
**11.** $b = 5, c = 13$
**12.** $a = 12, c = 20$
**13.** $a = 2, b = 3$
**14.** $a = 5, b = 5$
**15.** $b = 1, c = 1\frac{1}{4}$
**16.** $a = \frac{6}{5}, b = \frac{8}{5}$
**17.** $a = 1.4, b = 4.8$
**18.** $a = 7.5, b = 4$

## Goal B

In exercises 19–22, $\triangle ABC$ and $\triangle A'B'C'$ are two triangles such that $m(\angle A) = m(\angle A')$ and $m(\angle B) = m(\angle B')$. Given the lengths of some sides, find the length of the indicated side.

**19.** $a = 10, b = 8, a' = 8, b' = ?$
**20.** $a = 12, b = 9, a' = ?, b' = 6$
**21.** $a = 18, b = ?, a' = 9, b' = 5$
**22.** $a = ?, b = 27, a' = 6, b' = 12$

In exercises 23–28, $\triangle ABC$ is a right triangle and $PQ$ is perpendicular to $AC$ (see the figure below). Find the length of the indicated side.

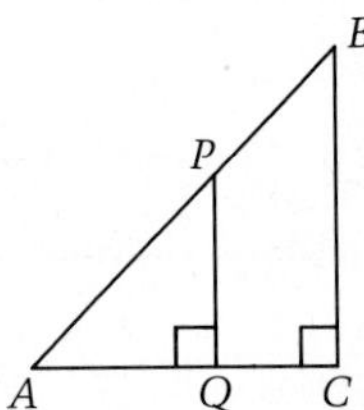


**23.** $AP = 8, AQ = 6, AC = 18, AB = ?$
**24.** $AQ = 10, PQ = 8, BC = 12, AC = ?$
**25.** $AP = 6, PQ = 4, AC = 18, BC = ?$
**26.** $AQ = 10, PQ = 6, BC = 36, AB = ?$
**27.** $\frac{PQ}{AQ} = \frac{1}{2}, AB = 18, BC = ?$
**28.** $\frac{PQ}{AP} = \frac{\sqrt{3}}{3}, BC = 10\sqrt{3}, AC = ?$

**29.** A 6 ft pole perpendicular to the ground casts a shadow 8 ft long at the same time that a telephone pole casts a shadow 56 ft long. What is the height of the telephone pole?

**30.** The top of a church spire casts a shadow 200 ft long. At the same time a nearby 8 ft wall casts a shadow 25 ft long. How high is the top of the spire?

## Superset

**31.** A boat travels 8 mi due east and then 15 mi due north. How far is the boat from its starting point?

**32.** A 40 ft ladder is placed against a wall, with its foot 24 ft from the base of the wall. At what height does the ladder touch the wall?

**33.** A television set has a square picture screen with a 19 in diagonal. What are the dimensions of the screen? What is its area?

In exercises 34–41, $\triangle ABC$ is an equilateral triangle, $BD$ is the altitude to side $AC$ and the length of $AB$ is 5. Find the following.

**34.** $m(\angle ADB)$
**35.** $m(\angle ABD)$
**36.** $m(\angle DBC)$
**37.** $BD$
**38.** $AD$
**39.** $DC$
**40.** area of $\triangle ABC$
**41.** area of $\triangle ABD$

**42.** Find the length of a side of a square having the same area as a rectangle with base of length 12 and diagonal of length 15.

**43.** Find the area of a right triangle with one leg of length 16 and hypotenuse of length 34.

**44.** Find the length of a diagonal of a rectangle with sides of lengths 32 and 24.

**45.** Find the length of a diagonal of a square with side of length 12.

**46.** Find the length of a side of an equilateral triangle with altitude of length $h$.

## 2.2 Trigonometric Ratios of Acute Angles

**Goal A** *to determine the trigonometric ratios of an acute angle in a right triangle*

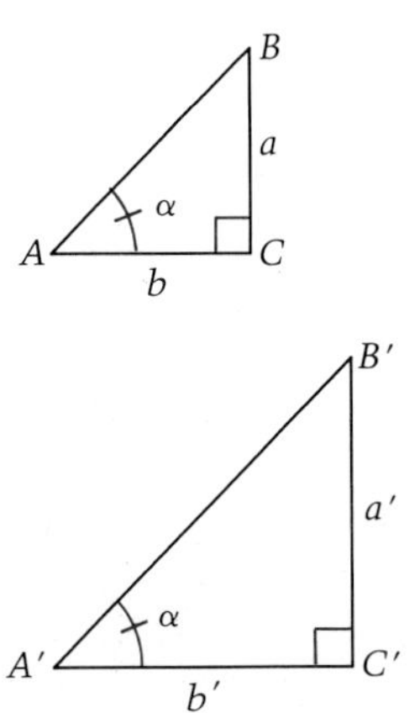

Figure 2.10

Figure 2.10 shows two right triangles $ABC$ and $A'B'C'$ with

$$m(\angle A) = m(\angle A').$$

Since each triangle has a right angle and each triangle has an acute angle with the same measure, the triangles are similar by the A-A Similarity Theorem. Because of the similarity we can write

$$\frac{a}{a'} = \frac{b}{b'}.$$

Multiplying each side of the above equation by $\frac{a'}{b}$ produces

$$\frac{a}{b} = \frac{a'}{b'}.$$

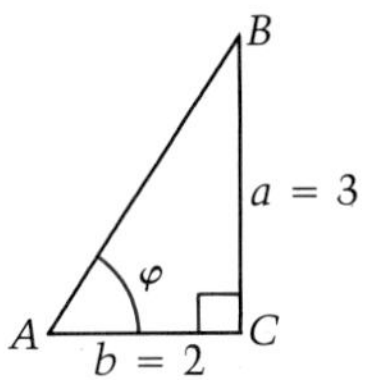

Figure 2.11

This last equation tells us that a ratio of sides of $\triangle ABC$ is equal to the ratio of corresponding sides in $\triangle A'B'C'$. These ratios are equal because the equality of $m(\angle A)$ and $m(\angle A')$ implies that the two right triangles are similar. Moreover, it is the value of $\alpha$ that determines the value of this ratio.

For example, suppose that in the right triangle at the left, $\angle A$ has measure $\varphi$. Notice that the ratio $\frac{a}{b}$ is $\frac{3}{2}$ in this triangle. Now consider the right triangles in Figure 2.12. Because the corresponding acute angle has measure $\varphi$, the corresponding ratio in each of these triangles is also $\frac{3}{2}$.

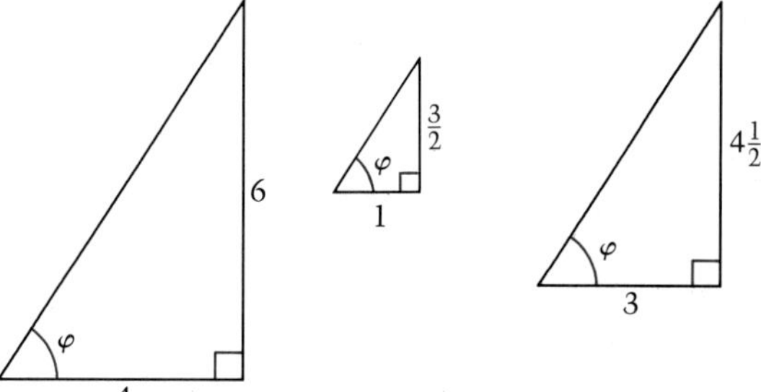

Figure 2.12

As we see in Figure 2.12, a particular acute angle (here $\varphi$) always produces the same ratio of opposite to adjacent sides (in this case the ratio is $\frac{3}{2}$). Therefore, we can think of the angle as a domain value for a function having the ratio as the function value.

Let us adopt a way of describing the sides of a right triangle in terms of an acute angle $\theta$.

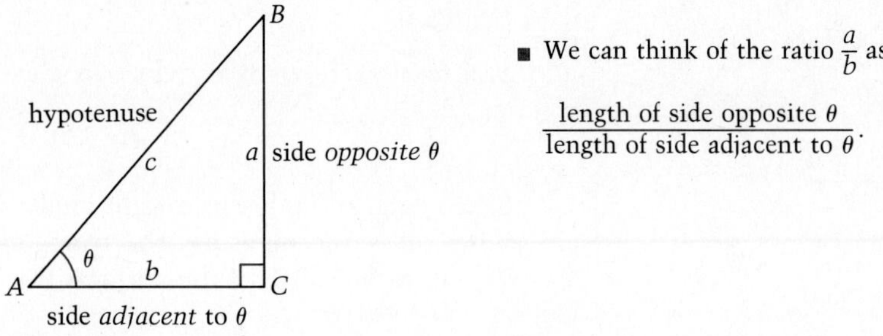


Figure 2.13

■ We can think of the ratio $\frac{a}{b}$ as

$$\frac{\text{length of side opposite } \theta}{\text{length of side adjacent to } \theta}.$$

The opposite/adjacent ratio is called the **tangent** of the angle. For an acute angle $\theta$ we write

$$\text{tangent}(\theta) = \frac{\text{length of side opposite } \theta}{\text{length of side adjacent to } \theta}.$$

For any right triangle $ABC$, there are six possible ratios of sides that can be formed. Each ratio can be defined as a function of an acute angle $\theta$.

**Definition**

Let $\theta$ be an acute angle of a right triangle. The trigonometric functions of $\theta$ are defined by the following ratios:

$$\text{sine } \theta = \frac{\text{length of side } \textbf{opposite } \theta}{\text{length of } \textbf{hypotenuse}}, \qquad \text{denoted } \sin \theta$$

$$\text{cosine } \theta = \frac{\text{length of side } \textbf{adjacent} \text{ to } \theta}{\text{length of } \textbf{hypotenuse}}, \qquad \text{denoted } \cos \theta$$

$$\text{tangent } \theta = \frac{\text{length of side } \textbf{opposite } \theta}{\text{length of side } \textbf{adjacent} \text{ to } \theta}, \qquad \text{denoted } \tan \theta$$

$$\text{cotangent } \theta = \frac{\text{length of side } \textbf{adjacent} \text{ to } \theta}{\text{length of side } \textbf{opposite } \theta}, \qquad \text{denoted } \cot \theta$$

$$\text{secant } \theta = \frac{\text{length of } \textbf{hypotenuse}}{\text{length of side } \textbf{adjacent} \text{ to } \theta}, \qquad \text{denoted } \sec \theta$$

$$\text{cosecant } \theta = \frac{\text{length of } \textbf{hypotenuse}}{\text{length of side } \textbf{opposite } \theta}, \qquad \text{denoted } \csc \theta$$

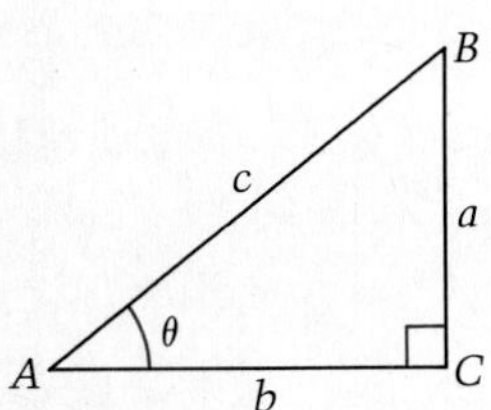


Figure 2.14

$\sin \theta = \frac{a}{c}$ $\quad \csc \theta = \frac{c}{a}$

$\cos \theta = \frac{b}{c}$ $\quad \sec \theta = \frac{c}{b}$

$\tan \theta = \frac{a}{b}$ $\quad \cot \theta = \frac{b}{a}$

**Example 1** Find the six trigonometric ratios of $\alpha$ in the right triangle below.

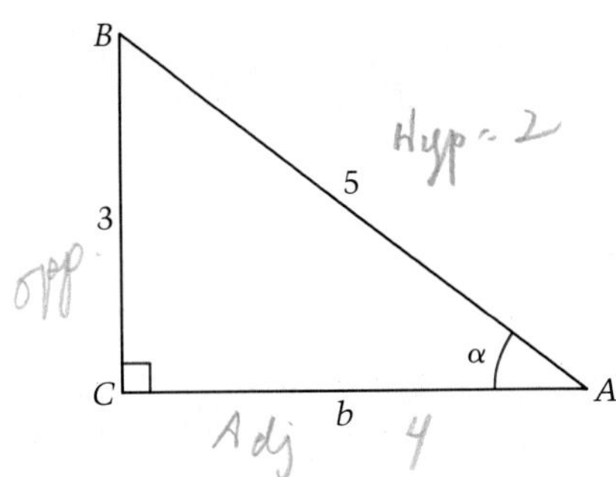


***Solution***

$5^2 = 3^2 + b^2$

$b^2 = 25 - 9 = 16$

$b = 4$

■ To determine the six trigonometric ratios of $\alpha$, we need to know the lengths of the three sides of the triangle. The Pythagorean Theorem is used to find $b$.

$$\sin \alpha = \frac{\text{opposite}}{\text{hypotenuse}} = \frac{3}{5} \qquad \cos \alpha = \frac{\text{adjacent}}{\text{hypotenuse}} = \frac{4}{5} \qquad \tan \alpha = \frac{\text{opposite}}{\text{adjacent}} = \frac{3}{4}$$

$$\csc \alpha = \frac{\text{hypotenuse}}{\text{opposite}} = \frac{5}{3} \qquad \sec \alpha = \frac{\text{hypotenuse}}{\text{adjacent}} = \frac{5}{4} \qquad \cot \alpha = \frac{\text{adjacent}}{\text{opposite}} = \frac{4}{3}$$

Note that in Example 1, the function values of $\sin \alpha$ and $\csc \alpha$ are reciprocals of one another. This is also the case for $\cos \alpha$ and $\sec \alpha$, and for $\tan \alpha$ and $\cot \alpha$. Note also that we do not know the measure of $\alpha$, even though we know its six trigonometric function values.

**Example 2** Angles $\alpha$ and $\beta$ are the two acute angles in the right triangle shown at the right. Show that

$$\sin \alpha = \cos \beta$$

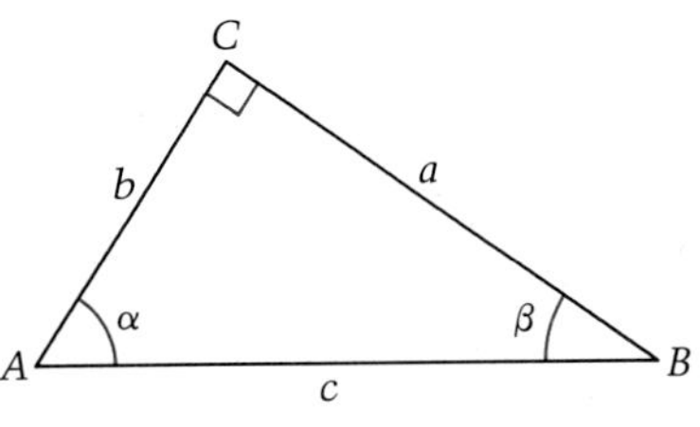


***Solution***

$$\sin \alpha = \frac{\text{side opposite } \alpha}{\text{hypotenuse}} = \frac{a}{c} \qquad \cos \beta = \frac{\text{side adjacent to } \beta}{\text{hypotenuse}} = \frac{a}{c}$$

Since $\sin \alpha$ and $\cos \beta$ each equal $\frac{a}{c}$, $\sin \alpha = \cos \beta$.

**Goal B** *to determine the trigonometric ratios of special angles (angles with measure 30°, 45°, or 60°)*

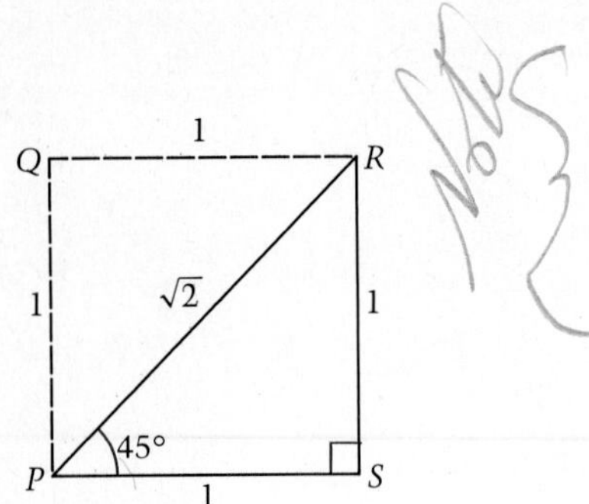

Figure 2.15

To determine the trigonometric ratios of an angle whose measure is 45°, consider the square *PQRS* with sides of length 1. The diagonal of the square bisects the right angle *P* and forms the hypotenuse of an isosceles right triangle. Its length is $\sqrt{2}$ (by the Pythagorean Theorem). We can use $\triangle PRS$ to determine the six trigonometric ratios of 45°.

$$\sin 45° = \frac{1}{\sqrt{2}} = \frac{\sqrt{2}}{2} \qquad \cos 45° = \frac{1}{\sqrt{2}} = \frac{\sqrt{2}}{2} \qquad \tan 45° = \frac{1}{1} = 1$$

$$\csc 45° = \frac{\sqrt{2}}{1} = \sqrt{2} \qquad \sec 45° = \frac{\sqrt{2}}{1} = \sqrt{2} \qquad \cot 45° = \frac{1}{1} = 1$$

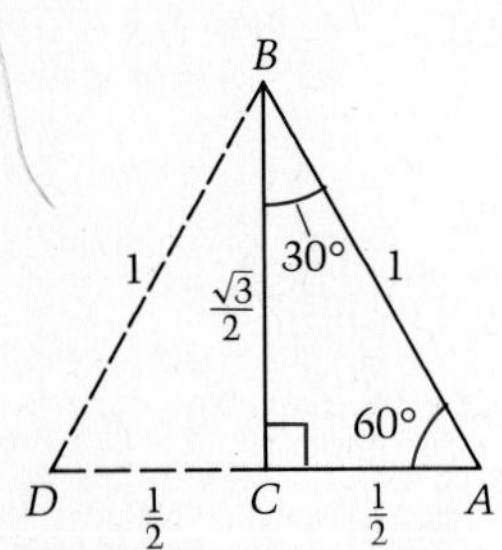

Figure 2.16

The trigonometric ratios of 30° and 60° can be found by using an equilateral triangle with sides of length 1. Recall that the altitude from a vertex to the opposite side bisects both the angle and the side. Thus, we can use $\triangle ABC$ to determine the six trigonometric ratios of 30° and 60°.

$$\sin 30° = \frac{1}{2} \qquad \cos 30° = \frac{\sqrt{3}}{2} \qquad \tan 30° = \frac{1}{\sqrt{3}} = \frac{\sqrt{3}}{3}$$

$$\csc 30° = 2 \qquad \sec 30° = \frac{2}{\sqrt{3}} = \frac{2\sqrt{3}}{3} \qquad \cot 30° = \sqrt{3}$$

$$\sin 60° = \frac{\sqrt{3}}{2} \qquad \cos 60° = \frac{1}{2} \qquad \tan 60° = \sqrt{3}$$

$$\csc 60° = \frac{2}{\sqrt{3}} = \frac{2\sqrt{3}}{3} \qquad \sec 60° = 2 \qquad \cot 60° = \frac{1}{\sqrt{3}} = \frac{\sqrt{3}}{3}$$

| $f(\theta)$ \ $\theta$ | **30°** | **45°** | **60°** |
|---|---|---|---|
| sin $\theta$ | $\frac{1}{2}$ | $\frac{\sqrt{2}}{2}$ | $\frac{\sqrt{3}}{2}$ |
| cos $\theta$ | $\frac{\sqrt{3}}{2}$ | $\frac{\sqrt{2}}{2}$ | $\frac{1}{2}$ |
| tan $\theta$ | $\frac{\sqrt{3}}{3}$ | 1 | $\sqrt{3}$ |
| cot $\theta$ | $\sqrt{3}$ | 1 | $\frac{\sqrt{3}}{3}$ |
| sec $\theta$ | $\frac{2\sqrt{3}}{3}$ | $\sqrt{2}$ | 2 |
| csc $\theta$ | 2 | $\sqrt{2}$ | $\frac{2\sqrt{3}}{3}$ |

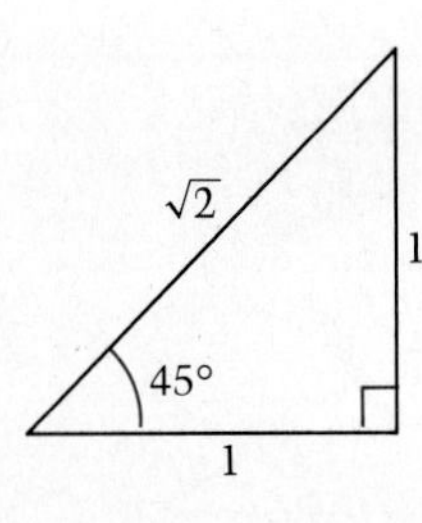


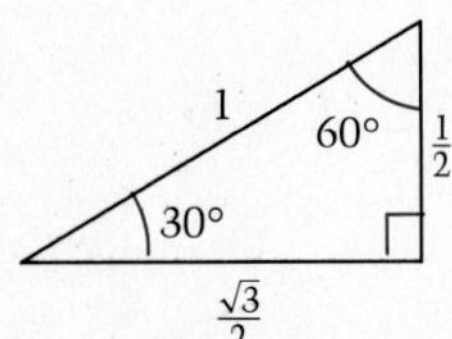

Figure 2.17

**Example 3** Given a right triangle *FED* with $m(\angle F) = 30°$ and hypotenuse of length 40, find the length of side $f$.

***Solution***

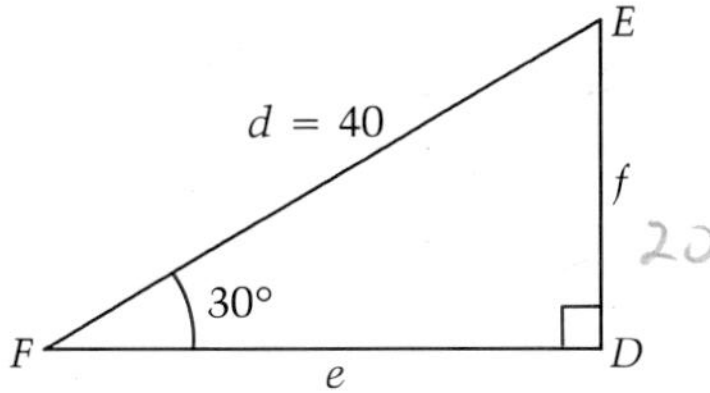


■ Begin by drawing a diagram and labeling the parts.

■ To find $f$, we choose a trigonometric ratio involving $f$ and some known side.

$$\sin 30° = \frac{f}{d} = \frac{1}{2}$$

■ $\sin \theta = \frac{\text{side opposite } \theta}{\text{hypotenuse}}$ and $\sin 30° = \frac{1}{2}$.

$$\frac{f}{40} = \frac{1}{2}$$

■ The known value of $d$ is substituted.

$$f = 20$$

**Goal C** *to determine trigonometric ratios of an angle given one ratio*

The trigonometric ratios depend upon one another. For an acute angle, if one trigonometric ratio is known, we can find the other five ratios by using the properties of right triangles.

**Example 4** Given that $\sin \beta = \frac{5}{7}$ and $\beta$ is an acute angle, find $\cos \beta$.

***Solution***

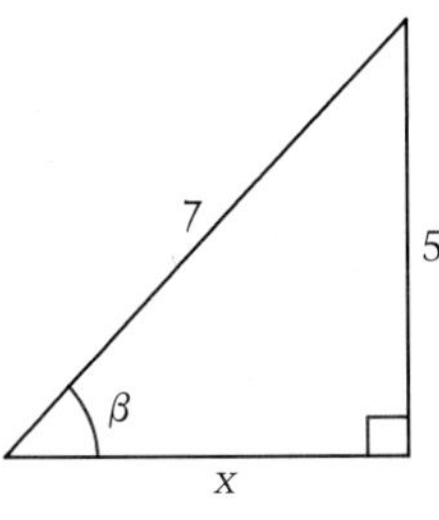


■ Since $\sin \beta$ is the ratio of the side opposite $\beta$ to the hypotenuse, we begin by drawing a right triangle in which this ratio is $\frac{5}{7}$. The side opposite $\beta$ is labeled 5 and the hypotenuse is labeled 7. The side adjacent to $\beta$ is labeled $x$ and will be needed in determining $\cos \beta$.

$$7^2 = x^2 + 5^2$$

■ The Pythagorean Theorem is used.

$$x^2 = 7^2 - 5^2 = 24$$

$$x = \sqrt{24} = 2\sqrt{6}$$

■ Remember to choose the positive value only.

$$\cos \beta = \frac{2\sqrt{6}}{7}$$

■ $\cos \beta = \frac{\text{side adjacent to } \beta}{\text{hypotenuse}}$.

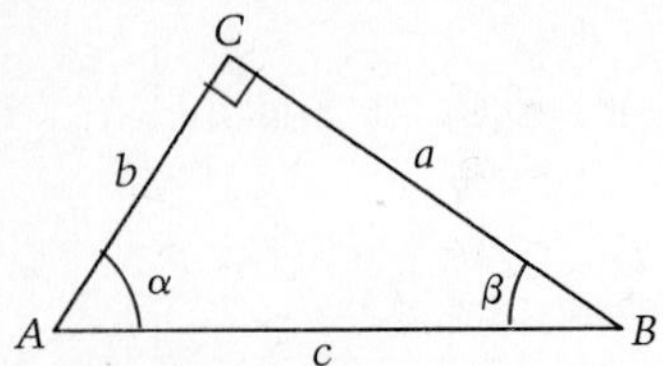

Figure 2.18 $\alpha$ and $\beta$ are complementary.

Two angles are **complementary** if the sum of their measures is 90°. In any right triangle, the acute angles are complementary. This is illustrated in Figure 2.18 which shows that $\alpha + \beta = 90°$. In Example 2, we discovered that $\sin \alpha = \cos \beta$, that is, sine of $\alpha$ = cosine of the complement of $\alpha$:

$$\sin \alpha = \cos(90° - \alpha).$$

**Example 5** If $\alpha$ and $\beta$ are complementary, and $\cos \alpha = \frac{\sqrt{11}}{6}$, find $\cos \beta$ and $\tan \beta$.

***Solution***

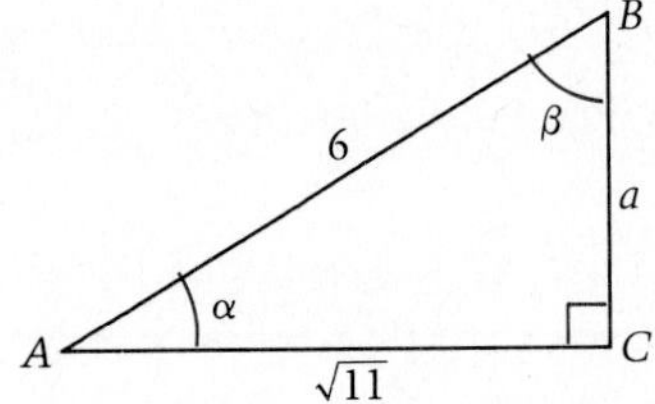


- Since $\alpha$ and $\beta$ are complementary, we can draw a right triangle having $\alpha$ and $\beta$ as acute angles.
- $\cos \alpha = \frac{\text{side adjacent to } \alpha}{\text{hypotenuse}} = \frac{\sqrt{11}}{6}$, so side $AC$ is labeled $\sqrt{11}$ and $AB$ is labeled 6.
- Pythagorean Theorem: $6^2 = (\sqrt{11})^2 + a^2$ implies $a^2 = 36 - 11 = 25$. Thus, $a = 5$.

$$\cos \beta = \frac{\text{side adjacent to } \beta}{\text{hypotenuse}} = \frac{5}{6} \qquad \tan \beta = \frac{\text{side opposite } \beta}{\text{side adjacent to } \beta} = \frac{\sqrt{11}}{5}$$

## Exercise Set 2.2 ⊟ Calculator Exercises in the Appendix

### Goal A

In exercises 1–16, use the given information about the right triangle $ABC$ to find the six trigonometric ratios of $\alpha$,

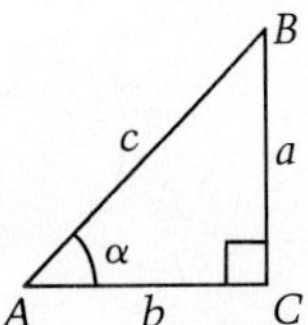


**1.** $a = 6, b = 8$ **2.** $a = 7, b = 24$

**3.** $a = 8, c = 17$ **4.** $b = 10, c = 26$

**5.** $c = 37, a = 12$ **6.** $b = 48, c = 50$

**7.** $a = 21, c = 35$ **8.** $c = 41, b = 9$

**9.** $a = 1, c = \sqrt{2}$ **10.** $a = 2, b = 2$

**11.** $a = 1, c = 2$ **12.** $b = \sqrt{3}, a = 1$

**13.** $b = \frac{1}{2}, a = \frac{\sqrt{3}}{2}$ **14.** $c = 2, b = 1$

**15.** $a = 1, b = \sqrt{3}$ **16.** $c = 2, a = \sqrt{3}$

In exercises 17–20, $\alpha$ and $\beta$ are the acute angles in right triangle $ABC$. Show the following.

**17.** $\tan \alpha = \cot \beta$ **18.** $\csc \alpha = \sec \beta$

**19.** $\sin \beta = \cos \alpha$ **20.** $\tan \beta = \cot \alpha$

### Goal B

In exercises 21–26, given an equilateral triangle $\triangle ABC$ with side of length $s$ and altitude of length $h$, find the indicated part.

**21.** If $s = 6$, find $h$ **22.** If $s = 12$, find $h$

23. If $h = 6$, find $s$

24. If $h = 15$, find s

25. If $h = 5\sqrt{3}$, find the perimeter of $\triangle ABC$

26. If $h = 9$, find the area of $\triangle ABC$

In exercises 27–36, $\triangle ABC$ has right angle at $C$ and sides of length $a$, $b$, and $c$. Given the following information, find the indicated part.

27. $m(\angle A) = 30°$, $b = 21$, find $c$

28. $m(\angle B) = 30°$, $c = 58$, find $a$

29. $m(\angle B) = 45°$, $b = 24$, find $c$

30. $m(\angle A) = 45°$, $c = 50$, find $b$

31. $m(\angle B) = 60°$, $b = 15\sqrt{3}$, find $a$

32. $m(\angle B) = 60°$, $a = 24$, find $b$

33. $a = 12$, $b = 12$, find $m(\angle A)$

34. $b = 30$, $c = 60$, find $m(\angle A)$

35. $b = 24$, $a = 24\sqrt{3}$, find $m(\angle A)$

36. $a = 10\sqrt{3}$, $c = 20$, find $m(\angle A)$

## Goal C

In exercises 37–46, $\alpha$ and $\beta$ are complementary. Given the following information, find the indicated trigonometric ratios.

37. $\cos \alpha = \frac{3}{5}$, find $\sin \alpha$ and $\tan \beta$

38. $\sin \alpha = \frac{5}{13}$, find $\cos \alpha$ and $\cos \beta$

39. $\csc \alpha = \frac{17}{15}$, find $\tan \alpha$ and $\cos \beta$

40. $\sec \alpha = \frac{29}{21}$, find $\tan \beta$ and $\tan \alpha$

41. $\tan \alpha = \frac{3}{4}$, find $\sin \beta$ and $\cos \alpha$

42. $\tan \alpha = \frac{4}{3}$, find $\sin \alpha$ and $\sec \beta$

43. $\sin \alpha = \frac{7}{25}$, find $\cos \alpha$ and $\tan \alpha$

44. $\cos \alpha = \frac{35}{37}$, find $\tan \alpha$ and $\sin \beta$

45. $\cos \alpha = 0.9$, find $\cos \beta$ and $\sin \beta$

46. $\sin \alpha = 0.3$, find $\cos \beta$ and $\tan \alpha$

## Superset

In exercises 47–50, use the trigonometric ratios of special angles to solve each problem.

47. Show that in an equilateral triangle with side of length $s$ that the altitude is $\frac{\sqrt{3}}{2}s$.

48. If the diagonal of a square has length $6\sqrt{2}$, what is the area of the square?

49. In an isosceles right triangle with hypotenuse 12, what is the length of the altitude to the hypotenuse?

50. Find the lengths of the sides of $\triangle ABC$ if $m(\angle A) = 30°$, $m(\angle B) = 60°$, and the altitude from $C$ has length 8.

In exercises 51–52, find the area of the triangle.

51.

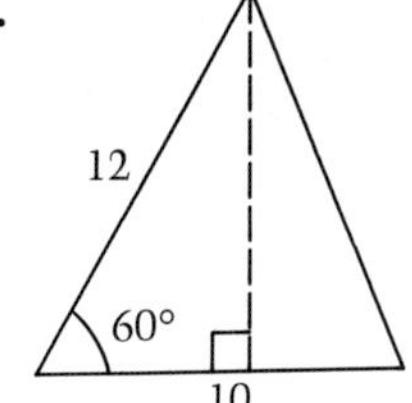


52.

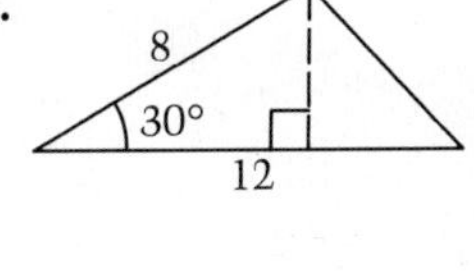


53. The local fire department's longest ladder measures 72 ft. If the angle between the ground and the ladder must be 60°, how high can the ladder reach? How far from a building should the foot of the ladder be?

54. One of the world's tallest flagpoles is 256 ft high. Guy wires are used to support the pole as shown in the diagram below. If the guy wires are anchored 60 ft from the foot of the flagpole, how long is each wire? How high up the pole are the wires fastened?

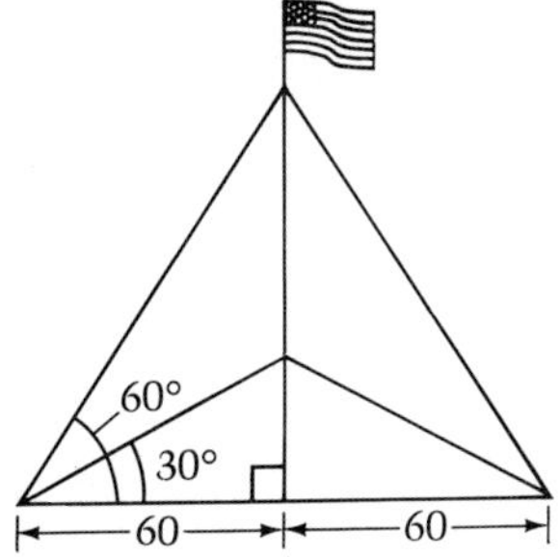

## 2.3 Angles of Rotation

**Goal A** *to sketch an arbitrary angle as a rotation*

Thus far we have defined the trigonometric functions for angles between 0° and 90°. In this section we shall extend our definitions to include *all* angles. In order to do this, it is useful to think of an angle as being formed by rotating a ray.

Figure 2.19

■ Between twelve midnight and 4 A.M. the hour hand of a clock rotates to produce the angle shown in the figure. The ray pointing to 12 is called the **initial side** of the angle, and the ray pointing to 4 is called the **terminal side.**

An angle is in **standard position** in the $xy$-plane, if its vertex is at the origin and its initial side lies along the positive $x$-axis. An angle in standard position is usually named by the quadrant in which the terminal side lies. Each of the angles below is in standard position. Notice that an angle may be the result of one or more complete rotations about the vertex, as shown in Figure 2.20(b).

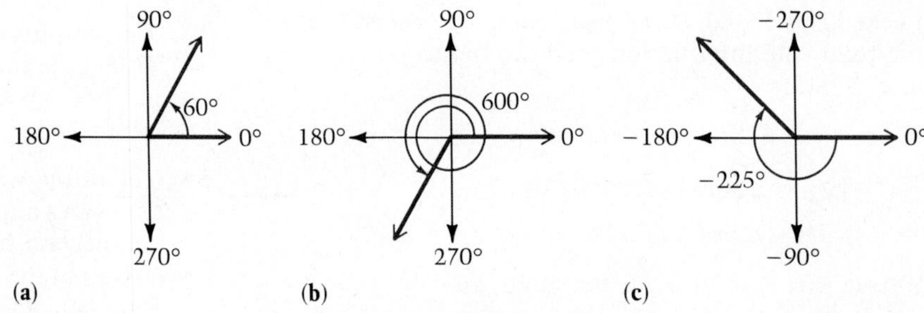


**Figure 2.20** (a) A first quadrant angle. (b) A third quadrant angle. (c) A second quadrant angle.

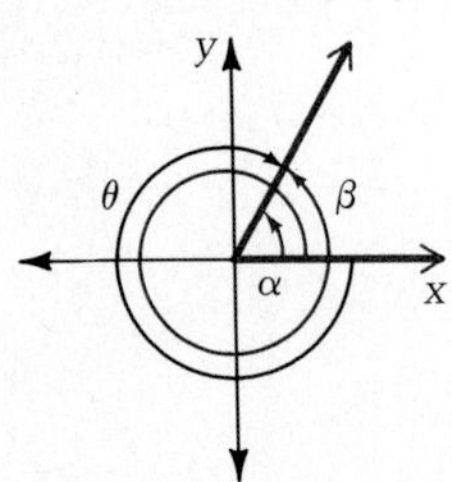


**Figure 2.21**

We agree that a counterclockwise rotation produces a positive angle as shown Figure 2.20(a) and Figure 2.20(b). A clockwise rotation produces a negative angle as shown in Figure 2.20(c).

An angle whose terminal side lies on the $x$- or $y$-axis is called a **quadrantal angle.** When angles in standard position have the same terminal side, they are called **coterminal.** Angles $\alpha$, $\beta$, and $\theta$ in Figure 2.21 are coterminal.

**Example 1** Sketch each of the following angles in standard position.

(a) 120° (b) 210° (c) −60° (d) 315°

***Solution***

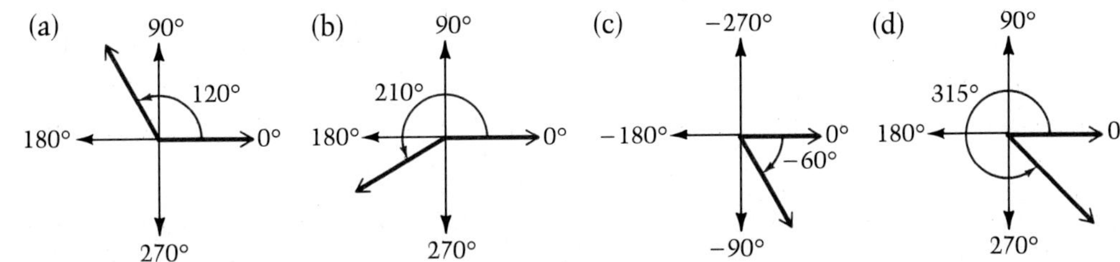


**Example 2** Sketch each of the following angles in standard position. Determine the measure of the angle between 0° and 360° that is coterminal with each angle.

(a) −30° (b) 398° (c) 810°

***Solution***

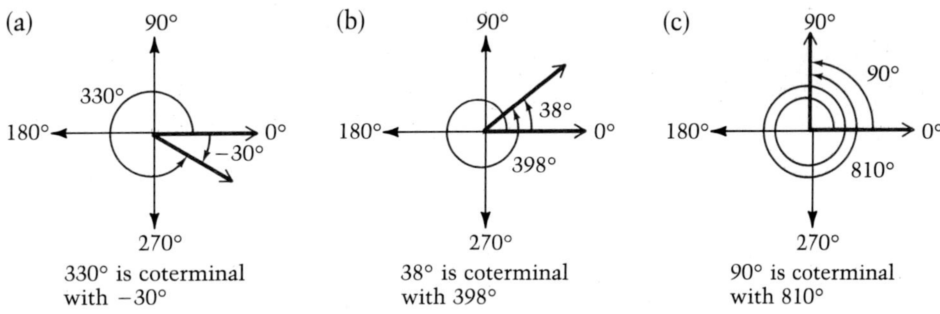


330° is coterminal with −30°

38° is coterminal with 398°

90° is coterminal with 810°

## Goal B *to determine trigonometric function values of arbitrary angles*

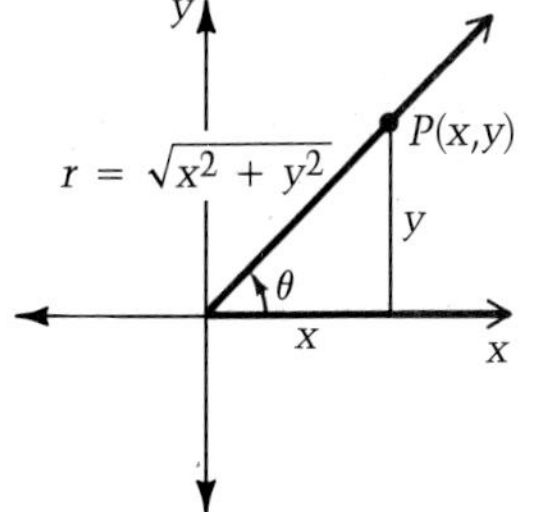


**Figure 2.22**

Suppose we have an acute angle $\theta$ in standard position, and we choose a point $P(x, y)$ on the terminal side of $\theta$ as shown in Figure 2.22. If a perpendicular is drawn from $P$ to the $x$-axis, we can then define the trigonometric functions of $\theta$ in terms of the sides of the resulting right triangle.

$$\sin \theta = \frac{y}{r} \qquad \cos \theta = \frac{x}{r} \qquad \tan \theta = \frac{y}{x}$$

$$\csc \theta = \frac{r}{y} \qquad \sec \theta = \frac{r}{x} \qquad \cot \theta = \frac{x}{y}$$

Note that the trigonometric functions of $\theta$ have been defined in terms of the coordinates of a point on the terminal side.

This method suggests a way of defining the values of the trigonometric functions of an angle in any quadrant. We simply need to know the $x$- and $y$-coordinates of a point on the terminal side of the angle.

**Definition**

Let $\theta$ be an angle in standard position with $P(x, y)$ a point on its terminal side. Then

$$\sin\theta = \frac{y}{r}, \qquad \cos\theta = \frac{x}{r}, \qquad \tan\theta = \frac{y}{x} \quad (x \neq 0),$$

$$\csc\theta = \frac{r}{y} \quad (y \neq 0), \qquad \sec\theta = \frac{r}{x} \quad (x \neq 0), \qquad \cot\theta = \frac{x}{y} \quad (y \neq 0),$$

where $r = \sqrt{x^2 + y^2}$ is the distance from $P$ to the origin.

**Example 3** If $(-3, 4)$ is a point on the terminal side of an angle $\alpha$ in standard position, determine the values of the six trigonometric functions of $\alpha$.

***Solution***

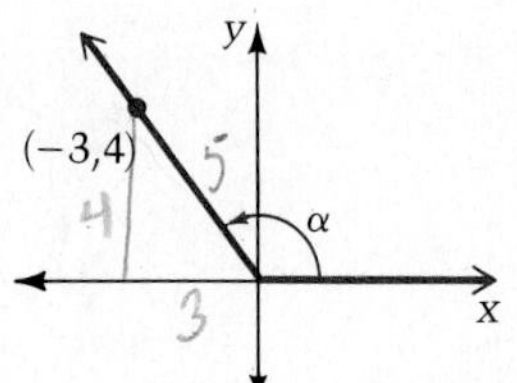


■ Begin by plotting the point $(-3, 4)$ and sketching the angle $\alpha$.

To determine the trigonometric function values, we must first find the values of $x$, $y$, and $r$.

$$x = -3,\ y = 4,\text{ and } r = \sqrt{x^2 + y^2} = \sqrt{(-3)^2 + (4)^2} = \sqrt{25} = 5.$$

$$\sin\alpha = \frac{y}{r} = \frac{4}{5} \qquad \cos\alpha = \frac{x}{r} = -\frac{3}{5} \qquad \tan\alpha = \frac{y}{x} = -\frac{4}{3}$$

$$\csc\alpha = \frac{r}{y} = \frac{5}{4} \qquad \sec\alpha = \frac{r}{x} = -\frac{5}{3} \qquad \cot\alpha = \frac{x}{y} = -\frac{3}{4}$$

Thus far we have not computed the values of the trigonometric functions of 0°, 90°, or any of the other quadrantal angles. To determine these values, first select a point $(x, y)$ on the terminal side of the angle, and then apply the definition of the trigonometric functions.

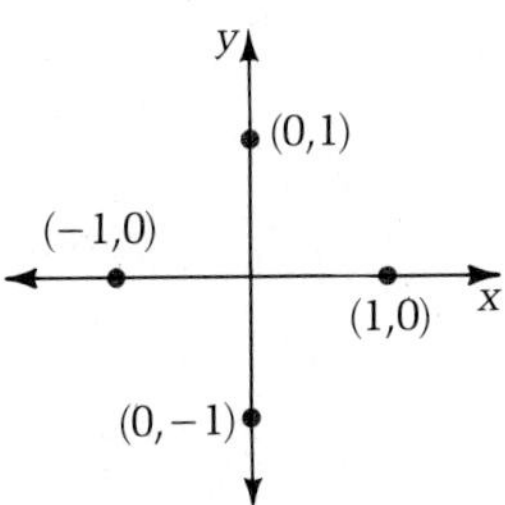


Figure 2.23

■ There are only four possible terminal sides for a quadrantal angle: the positive $x$-axis, the positive $y$-axis, the negative $x$-axis, and the negative $y$-axis. Use the points plotted in Figure 2.23 to determine the values of the trigonometric functions of quadrantal angles.

For example, an angle of 90° has the positive $y$-axis as its terminal side. The point (0, 1) can be used to determine trigonometric function values of 90°: $x = 0$, $y = 1$, $r = \sqrt{0^2 + 1^2} = 1$.

$\sin 90° = \frac{1}{1} = 1$ $\quad$ $\cos 90° = \frac{0}{1} = 0$ $\quad$ $\tan 90° = \frac{1}{0}$ ■ undefined

$\csc 90° = \frac{1}{1} = 1$ $\quad$ $\sec 90° = \frac{1}{0}$ ■ undefined $\quad$ $\cot 90° = \frac{0}{1} = 0$

On the other hand, an angle of 0° has the positive $x$-axis as its terminal (and initial) side, so the point (1, 0) is used to determine the function values: $x = 1$, $y = 0$, $r = \sqrt{1^2 + 0^2} = 1$.

$\sin 0° = \frac{0}{1} = 0$ $\quad$ $\cos 0° = \frac{1}{1} = 1$ $\quad$ $\tan 0° = \frac{0}{1} = 0$

$\csc 0° = \frac{1}{0}$ ■ undefined $\quad$ $\sec 0° = \frac{1}{1} = 1$ $\quad$ $\cot 0° = \frac{1}{0}$ ■ undefined

**Example 4** Compute the following: (a) $\cos 270°$ (b) $\tan(-270°)$ (c) $\sec 540°$

***Solution*** (a) 270° has its terminal side on the negative $y$-axis, so use $(0, -1)$.

$x = 0$, $y = -1$, $r = \sqrt{0^2 + (-1)^2} = 1$; $\cos 270° = \frac{x}{r} = \frac{0}{1} = 0$

(b) $-270°$ has its terminal side on the positive $y$-axis, so use (0, 1).

$x = 0$, $y = 1$, $r = \sqrt{0^2 + 1^2} = 1$; $\tan(-270°) = \frac{y}{x} = \frac{1}{0}$ ■ undefined

(c) 540° has its terminal side on the negative $x$-axis, so use $(-1, 0)$.

$x = -1$, $y = 0$, $r = \sqrt{(-1)^2 + 0^2} = 1$; $\sec 540° = \frac{r}{x} = \frac{1}{-1} = -1$

We add the angles 0° and 90° to our list of special angles, and we expand the table given in the previous section to include these two angles.

Table of Special Angles

| $f(\theta)$ \ $\theta$ | 0° | 30° | 45° | 60° | 90° |
|---|---|---|---|---|---|
| $\sin\theta$ | 0 | $\frac{1}{2}$ | $\frac{\sqrt{2}}{2}$ | $\frac{\sqrt{3}}{2}$ | 1 |
| $\cos\theta$ | 1 | $\frac{\sqrt{3}}{2}$ | $\frac{\sqrt{2}}{2}$ | $\frac{1}{2}$ | 0 |
| $\tan\theta$ | 0 | $\frac{\sqrt{3}}{3}$ | 1 | $\sqrt{3}$ | * |
| $\cot\theta$ | * | $\sqrt{3}$ | 1 | $\frac{\sqrt{3}}{3}$ | 0 |
| $\sec\theta$ | 1 | $\frac{2\sqrt{3}}{3}$ | $\sqrt{2}$ | 2 | * |
| $\csc\theta$ | * | 2 | $\sqrt{2}$ | $\frac{2\sqrt{3}}{3}$ | 1 |

*undefined

## Exercise Set 2.3 ⊟ Calculator Exercises in the Appendix

### Goal A

In exercises 1–12, sketch each of the following angles in standard position.

**1.** 60° **2.** 135° **3.** 330°

**4.** 255° **5.** 390° **6.** 580°

**7.** −30° **8.** −240° **9.** −390°

**10.** −415° **11.** −725° **12.** −1142°

In exercises 13–20, determine the measure of the angle between 0° and 360° that is coterminal with each angle.

**13.** 405° **14.** 485° **15.** 723°

**16.** 990° **17.** −38° **18.** −180°

**19.** −660° **20.** −1689°

In exercises 21–34, determine a positive angle between 0° and 360° inclusive and a negative angle between 0° and −360° inclusive that are coterminal with the given angle.

**21.** 20° **22.** 102° **23.** 225°

**24.** 270° **25.** −410° **26.** −450°

**27.** −1351° **28.** 1300° **29.** 0°

**30.** 1080° **31.** 1575° **32.** −940°

**33.** $\frac{1}{2}^{\circ}$ **34.** $-\frac{1}{2}^{\circ}$

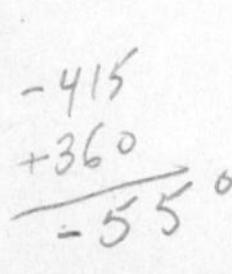


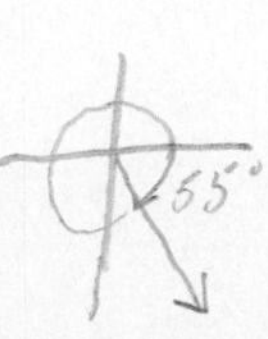

### Goal B

In exercises 35–38, determine the values of the six trigonometric functions of $\alpha$.

35.
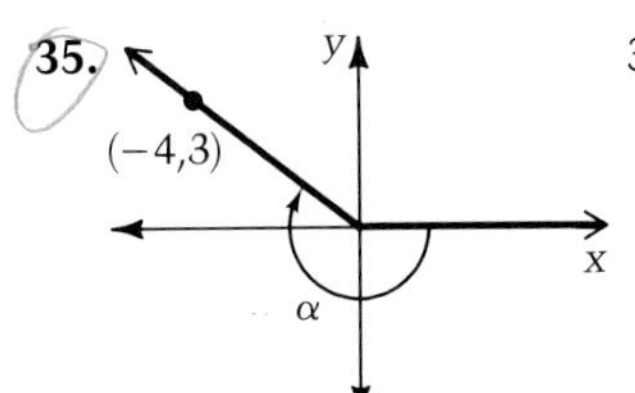


36.
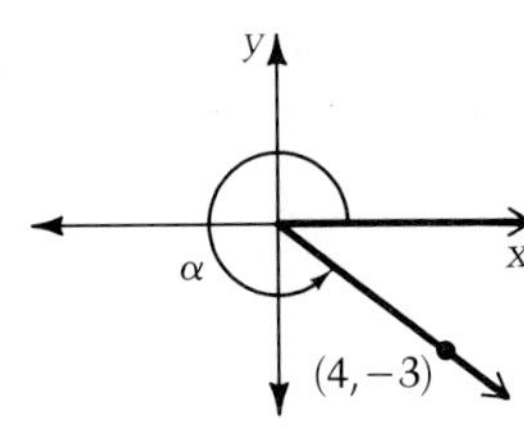


37.
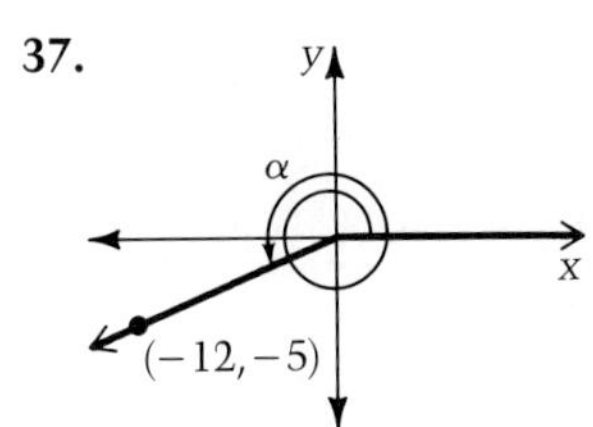


38.
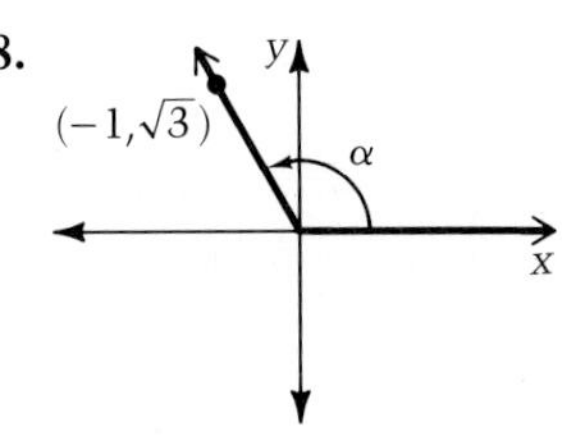


In exercises 39–48, the given point is on the terminal side of an angle $\alpha$. Determine the values of the six trigonometric functions of $\alpha$.

**39.** $(-8, 6)$ **40.** $(-5, 12)$ **41.** $(1, 1)$

**42.** $(2, 6)$ **43.** $(3, -3)$ **44.** $(-3, -9)$

**45.** $(1, -2)$ **46.** $(-\sqrt{3}, -1)$ **47.** $(\sqrt{5}, -2)$

**48.** $(2\sqrt{3}, 4)$

In exercises 49–60, evaluate each of the following.

**49.** $\sin 90°$ **50.** $\cos 180°$

**51.** $\cot 180°$ **52.** $\tan 270°$

**53.** $\sec 450°$ **54.** $\sin 810°$

**55.** $\cos(-90°)$ **56.** $\csc(-630°)$

**57.** $\tan(-540°)$ **58.** $\sec(-1350°)$

**59.** $\csc(-720°)$ **60.** $\cot(-900°)$

### Superset

In exercises 61–64, determine the measure of the given angle.

**61.** $\varphi$ is half of a complete revolution in a counterclockwise direction.

**62.** $\varphi$ is half of a complete revolution in a clockwise direction.

**63.** $\varphi$ is three tenths of a complete revolution in a counterclockwise direction.

**64.** $\varphi$ is one and one quarter complete revolutions in a clockwise direction.

In exercises 65–72, the value of one trigonometric function of $\varphi$ is given. Assuming that $\varphi$ is a second quadrant angle, find the values of the other five trigonometric functions of $\varphi$.

**65.** $\tan \varphi = -5$ **66.** $\sin \varphi = \frac{4}{5}$

**67.** $\csc \varphi = \frac{5}{3}$ **68.** $\cos \varphi = -\frac{3}{4}$

**69.** $\cot \varphi = -\frac{5}{3}$ **70.** $\tan \varphi = -1$

**71.** $\cos \varphi = -\frac{3}{5}$ **72.** $\sec \varphi = -\sqrt{3}$

In exercises 73–80, the value of one trigonometric function of $\varphi$ is given. If the terminal side of $\varphi$ lies in the given quadrant, find the values of the other five trigonometric functions of $\varphi$.

**73.** third quadrant; $\sin \varphi = -\frac{2}{5}$

**74.** second quadrant; $\sin \varphi = \frac{3}{5}$

**75.** third quadrant; $\cos \varphi = -\frac{3}{5}$

**76.** fourth quadrant; $\tan \varphi = -1$

**77.** fourth quadrant; $\csc \varphi = -\frac{5}{3}$

**78.** third quadrant; $\tan \varphi = 1$

**79.** fourth quadrant; $\cos \varphi = 0.7$

**80.** second quadrant; $\cot \varphi = -1.5$

## 2.4 Identities and Tables

**Goal A** *to use basic trigonometric identities*

An **identity** is an equation that is true for all values of the variables for which both sides of the equation are defined. Several trigonometric identities are easily derived from the definitions of the trigonometric functions.

| **Reciprocal Identities** | Derivation |
|---|---|
| (1) $\sin\theta = \dfrac{1}{\csc\theta}$ | $\sin\theta = \dfrac{y}{r} = \dfrac{1}{\frac{r}{y}} = \dfrac{1}{\csc\theta}$ |
| (2) $\cos\theta = \dfrac{1}{\sec\theta}$ | $\cos\theta = \dfrac{x}{r} = \dfrac{1}{\frac{r}{x}} = \dfrac{1}{\sec\theta}$ |
| (3) $\tan\theta = \dfrac{1}{\cot\theta}$ | $\tan\theta = \dfrac{y}{x} = \dfrac{1}{\frac{x}{y}} = \dfrac{1}{\cot\theta}$ |
| (4) $\cot\theta = \dfrac{1}{\tan\theta}$ | $\cot\theta = \dfrac{x}{y} = \dfrac{1}{\frac{y}{x}} = \dfrac{1}{\tan\theta}$ |
| (5) $\sec\theta = \dfrac{1}{\cos\theta}$ | $\sec\theta = \dfrac{r}{x} = \dfrac{1}{\frac{x}{r}} = \dfrac{1}{\cos\theta}$ |
| (6) $\csc\theta = \dfrac{1}{\sin\theta}$ | $\csc\theta = \dfrac{r}{y} = \dfrac{1}{\frac{y}{r}} = \dfrac{1}{\sin\theta}$ |

| **Quotient Identities** | Derivation |
|---|---|
| (7) $\dfrac{\sin\theta}{\cos\theta} = \tan\theta$ | $\dfrac{\sin\theta}{\cos\theta} = \dfrac{\frac{y}{r}}{\frac{x}{r}} = \dfrac{y}{r}\cdot\dfrac{r}{x} = \dfrac{y}{x} = \tan\theta$ |
| (8) $\dfrac{\cos\theta}{\sin\theta} = \cot\theta$ | $\dfrac{\cos\theta}{\sin\theta} = \dfrac{\frac{x}{r}}{\frac{y}{r}} = \dfrac{x}{r}\cdot\dfrac{r}{y} = \dfrac{x}{y} = \cot\theta$ |

**Example 1** Given $\sin\beta = 0.324$ and $\cos\beta = 0.946$, find the remaining trigonometric functions of $\beta$. Round the answer to three decimal places.

***Solution***

$$\tan\beta = \frac{\sin\beta}{\cos\beta}$$

$$= \frac{0.324}{0.946} \approx 0.342$$

■ 0.342 is an approximation for tan $\beta$, and we write $\tan\beta \approx 0.342$.

$$\cot\beta = \frac{1}{\tan\beta}$$

$$= \frac{1}{0.342} \approx 2.924$$

■ If the identity $\frac{\cos\beta}{\sin\beta} = \cot\beta$ is used, the answer is 2.920. The difference is due to rounding error.

$$\sec\beta = \frac{1}{\cos\beta}$$

$$= \frac{1}{0.946} \approx 1.057$$

$$\csc\beta = \frac{1}{\sin\beta}$$

$$= \frac{1}{0.324} \approx 3.086$$

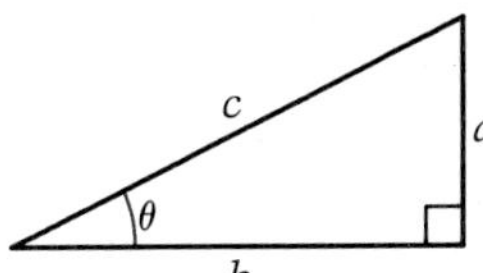


**Figure 2.24**

The next group of identities that we will consider is referred to as the Pythagorean Identities because their derivation depends on the Pythagorean Theorem. Using the right triangle at the left, we can make the following statements.

$$a^2 + b^2 = c^2$$

■ This is the Pythagorean Theorem.

$$\frac{a^2}{c^2} + \frac{b^2}{c^2} = \frac{c^2}{c^2}$$

■ Both sides divided by $c^2$.

$$\left(\frac{a}{c}\right)^2 + \left(\frac{b}{c}\right)^2 = 1$$

$$(\sin\theta)^2 + (\cos\theta)^2 = 1$$

■ $\sin\theta = \frac{a}{c}$; $\cos\theta = \frac{b}{c}$.

It is common to write $(\sin\theta)^2$ as $\sin^2\theta$ and $(\cos\theta)^2$ as $\cos^2\theta$.

| **Pythagorean Identities** | | Derivation |
|---|---|---|
| (9) | $\sin^2\theta + \cos^2\theta = 1$ | Shown above. |
| (10) | $\tan^2\theta + 1 = \sec^2\theta$ | Divide both sides of (9) by $\cos^2\theta$. Then use the Quotient and Reciprocal Identities. |
| (11) | $1 + \cot^2\theta = \csc^2\theta$ | Divide both sides of (9) by $\sin^2\theta$. Use the Quotient and Reciprocal Identities. |

**Example 2** In each of the following, $\theta$ is an acute angle.

(a) If $\sin\theta = \frac{1}{3}$, find $\cos\theta$. (b) If $\tan\theta = 1.6$, find $\sec\theta$.

***Solution***

(a)
$$\sin^2\theta + \cos^2\theta = 1$$
■ Identity (9).

$$\left(\frac{1}{3}\right)^2 + \cos^2\theta = 1$$

$$\cos^2\theta = 1 - \frac{1}{9}$$

$$\cos\theta = +\sqrt{\frac{8}{9}} = \frac{2\sqrt{2}}{3}$$
■ For an acute angle all trigonometric function values are positive.

(b)
$$\tan^2\theta + 1 = \sec^2\theta$$
■ Identity (10).

$$(1.6)^2 + 1 = \sec^2\theta$$

$$\sec^2\theta = 3.56$$

$$\sec\theta = +\sqrt{3.56} \approx 1.89$$
■ Again, select the positive root.

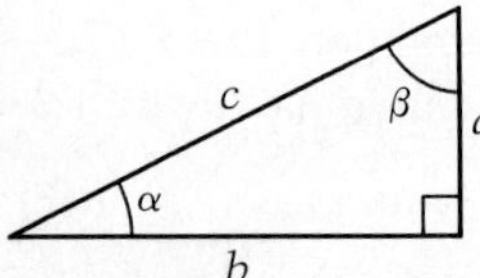


**Figure 2.25** Since $\sin\alpha = \frac{a}{c}$ and $\cos\beta = \frac{a}{c}$, $\sin\alpha = \cos\beta$.

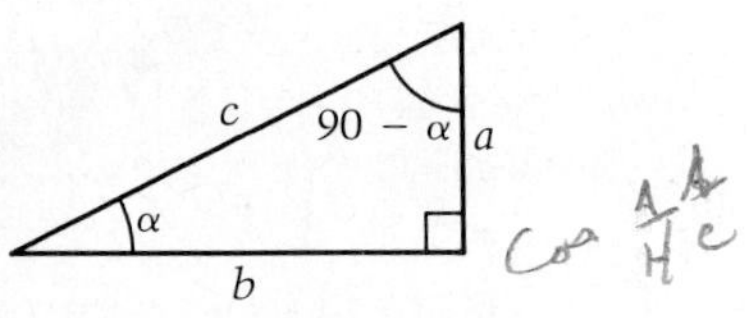


**Figure 2.26** Since $\cos\alpha = \frac{b}{c}$ and $\sin(90° - \alpha) = \frac{b}{c}$, $\cos\alpha = \sin(90° - \alpha)$.

Recall that if $\alpha$ and $\beta$ are complementary angles, then $\alpha + \beta = 90°$. Figure 2.25 shows that in this case $\sin\alpha = \cos\beta$. Since $\beta = 90° - \alpha$, we can write $\sin\alpha = \cos(90° - \alpha)$. This is called a **Cofunction Identity** since the sine and cosine are cofunctions of each other. The tangent and cotangent are also cofunctions of each other, as are the secant and cosecant. We now list the cofunction identities.

| **Cofunction Identities** | Derivation |
|---|---|
| $\sin\alpha = \cos(90° - \alpha)$ | See Figure 2.25. |
| $\cos\alpha = \sin(90° - \alpha)$ | See Figure 2.26. |
| $\tan\alpha = \cot(90° - \alpha)$ | $\tan\alpha = \frac{\sin\alpha}{\cos\alpha} = \frac{\cos(90° - \alpha)}{\sin(90° - \alpha)} = \cot(90° - \alpha)$ |
| $\cot\alpha = \tan(90° - \alpha)$ | $\cot\alpha = \frac{1}{\tan\alpha} = \frac{1}{\cot(90° - \alpha)} = \tan(90° - \alpha)$ |
| $\sec\alpha = \csc(90° - \alpha)$ | $\sec\alpha = \frac{1}{\cos\alpha} = \frac{1}{\sin(90° - \alpha)} = \csc(90° - \alpha)$ |
| $\csc\alpha = \sec(90° - \alpha)$ | $\csc\alpha = \frac{1}{\sin\alpha} = \frac{1}{\cos(90° - \alpha)} = \sec(90° - \alpha)$ |

In words, the cofunction identities say that the trigonometric function of an angle is equal to the cofunction of the complement of the angle.

**Example 3** Complete the following statements with an acute angle.

(a) $\sin 37° = \cos$ ____ (b) $\cot 52° = \tan$ ____ (c) $\csc 45° = \sec$ ____

***Solution***

(a) $\sin 37° = \cos(90° - 37°) = \cos 53°$
(b) $\cot 52° = \tan(90° - 52°) = \tan 38°$
(c) $\csc 45° = \sec(90° - 45°) = \sec 45°$

**Example 4** Given $\sin 70° \approx 0.940$ and $\cos 70° \approx 0.342$, find (a) $\cot 20°$ (b) $\csc 20°$. Round the answer to three decimal places.

***Solution***

(a) $\cot 20° = \tan(90° - 20°) = \tan 70° = \dfrac{\sin 70°}{\cos 70°} \approx \dfrac{0.940}{0.342} \approx 2.749$

(b) $\csc 20° = \dfrac{1}{\sin 20°} = \dfrac{1}{\cos 70°} \approx \dfrac{1}{0.342} \approx 2.924$

**Goal B** *to use the tables of trigonometric function values*

We know the trigonometric function values of the special angles, but what about $\cos 37°$ or $\sin 72°$? Because such values are not easy to compute, Table 4 in the Appendix lists values of the six trigonometric functions of angles between 0° and 90°. Although the values in this table are approximations, we agree to write $\sin 45° = 0.7071$ (from Table 4) even though the exact value of $\sin 45°$ is $\frac{\sqrt{2}}{2}$, and 0.7071 is only an approximation for $\frac{\sqrt{2}}{2}$.

Fractional parts of a degree are typically measured in minutes, with 60 minutes in a degree ($1° = 60'$). For example, $24\frac{1}{2}° = 24°30'$ and $42\frac{1}{3}° = 42°20'$. Sometimes it is necessary to record degree measure in decimal form. For example, $58°15' = 58.25°$, since $15' = \frac{1}{4}° = 0.25°$.

**Example 5** Use Table 4 to determine (a) $\tan 39°20'$ (b) $\cos 62°40'$

***Solutions***

(a) $\tan 39°20' = 0.8195$

- For angles between 0° and 45°, start at the *top left* of Table 4 and read down the extreme left columns until you find 39°20′. Then move across to the column having $\tan \alpha$ at the top.

(b) $\cos 62°40' = 0.4592$

- For angles between 45° and 90°, start at the *bottom right* of Table 4 and read up the extreme right columns until you reach 62°20′. Then move across to the column having $\cos \alpha$ at the bottom.

**Example 6** Use Table 4 to determine acute angle $\alpha$ if $\csc \alpha = 1.211$.

***Solution*** $\alpha = 55°40'$

■ Read up the column in Table 4 having $\csc \alpha$ at the bottom. Move across to the right-hand column to find $55°40'$.

Table 4 only contains angles that are multiples of ten minutes. To compute $\sin 36°27'$, we find $\sin 36°20'$ and $\sin 36°30'$ from the table. The value of $\sin 36°27'$ is between these values. To obtain an approximation, we use a method called **linear interpolation.** This method involves setting up a proportion using the 4 differences $d_1$, $d_2$, $d_3$, and $d_4$ shown below.

| | $\alpha$ | $\sin \alpha$ | |
|---|---|---|---|
| $d_1$ (36°20′ to 36°27′); $d_2$ (36°20′ to 36°30′) | 36°20′ | 0.5925 | $d_3$ (0.5925 to ?); $d_4$ (0.5925 to 0.5948) |
| | 36°27′ | ? | |
| | 36°30′ | 0.5948 | |

■ 36°20′ and 36°30′ are called **bracketing values** for 36°27′.

$$\frac{d_1}{d_2} = \frac{d_3}{d_4}$$

■ The proportion is solved for $d_3$, the only unknown value.

$$\sin 36°27' = 0.5925 + d_3$$

■ $d_3$ is added to the upper trigonometric function value.

**Example 7** Find (a) $\sin 36°27'$ (b) $\cos 73°48'$

***Solution*** (a)

| | $\alpha$ | $\sin \alpha$ | |
|---|---|---|---|
| 7 (36°20′ to 36°27′); 10 (36°20′ to 36°30′) | 36°20′ | 0.5925 | $x$ (0.5925 to ?); 0.0023 (0.5925 to 0.5948) |
| | 36°27′ | ? | |
| | 36°30′ | 0.5948 | |

■ Set up a chart involving the bracketing values. Determine the three differences, and let $x$ represent the unknown difference.

$$\frac{7}{10} = \frac{x}{0.0023}$$

$$x \approx 0.0016$$

■ Set up the proportion and solve. Since the values are given to four decimal places, we approximate the value of $x$ to 4 decimal places.

$$\sin 36°27' = 0.5925 + 0.0016 = 0.5941$$

■ $x$ is added to the upper function value.

(b)

| | $\alpha$ | $\cos\alpha$ | |
|---|---|---|---|
| 8 (to 73°48′), 10 (to 73°50′) | 73°40′ | 0.2812 | $x$ (to ?), $-0.0028$ (to 0.2784) |
| | 73°48′ | ? | |
| | 73°50′ | 0.2784 | |

■ Set up the proportion: $\frac{8}{10} = \frac{x}{-0.0028}$. Solve to get $x = -0.0022$.

$$\cos 73°48' = 0.2812 + (-0.0022) = 0.2790$$

**Example 8** If $\tan\varphi = 0.8915$ and $\varphi$ is an acute angle, determine $\varphi$.

***Solution***

| | $\varphi$ | $\tan\varphi$ | |
|---|---|---|---|
| $x$ (to ?), 10 (to 41°50′) | 41°40′ | 0.8899 | 0.0016 (to 0.8915), 0.0053 (to 0.8952) |
| | ? | 0.8915 | |
| | 41°50′ | 0.8952 | |

■ The value 0.8915 is between 0.8899 and 0.8952.

$$\frac{x}{10'} = \frac{0.0016}{0.0053}$$

■ Set up the proportion and solve.

$$x = 3.019'$$

■ Round to the nearest minute.

$$x \approx 3'$$

$$\varphi = 41°40' + 3' = 41°43'$$

■ Add the value of $x$ to the upper angle value.

## Exercise Set 2.4 ⊟ Calculator Exercises in the Appendix

### Goal A

In exercises 1–4, two approximate trigonometric function values of an angle are given. Find the other four trigonometric function values rounded to two decimal places.

**1.** $\sin 32° = 0.53$, $\cos 32° = 0.85$

**2.** $\tan 55° = 1.43$, $\sin 55° = 0.82$

**3.** $\tan 66° = 2.25$, $\sec 66° = 2.46$

**4.** $\tan 23° = 0.42$, $\cos 23° = 0.92$

In exercises 5–10, complete the following statements with an acute angle.

**5.** $\sin 54° = \cos$ ____

**6.** $\cos 12° = \sin$ ____

**7.** $\tan 38° = \cot$ ____

**8.** $\cot 81° = \tan$ ____

**9.** $\sec 45° = \csc$ ____

**10.** $\csc 71° = \sec$ ____

In exercises 11–18, rewrite each expression as a trigonometric function of an angle between 45° and 90°.

**11.** $\cos 27°$ **12.** $\sec 12°$ **13.** $\cot 38°$

**14.** $\tan 41°$ **15.** $\csc 3°$ **16.** $\cos 43°$

**17.** $\sin 19°$ **18.** $\sin 23°$

In exercises 19–24, approximate the function values given that $\sin 32° = 0.53$, $\cos 32° = 0.85$, $\sin 23° = 0.39$, and $\tan 23° = 0.42$. Round your answer to two decimal places.

**19.** $\sin 58°$ **20.** $\sec 58°$ **21.** $\sec 67°$

**22.** $\tan 58°$ **23.** $\tan 67°$ **24.** $\cot 67°$

In exercises 25–32, use the Reciprocal and Pythagorean Identities to determine the values of the other five trigonometric functions of the acute angle $\alpha$.

**25.** $\sin \alpha = \frac{1}{2}$

**26.** $\cos \alpha = \frac{3}{10}$

**27.** $\tan \alpha = \frac{3}{2}$

**28.** $\csc \alpha = 3$

**29.** $\cot \alpha = \frac{1}{2}$

**30.** $\sec \alpha = 10$

**31.** $\cos \alpha = \frac{4}{5}$

**32.** $\sin \alpha = \frac{7}{10}$

## Goal B

In exercises 33–44, use Table 4 in the Appendix to determine the following.

**33.** sin 31°10′

**34.** cos 43°50′

**35.** tan 21°40′

**36.** tan 49°20′

**37.** sec 56°50′

**38.** sec 10°30′

**39.** cot 48°10′

**40.** csc 15°40′

**41.** cos 52°20′

**42.** sin 80°10′

**43.** csc 45°30′

**44.** cot 78°50′

In exercises 45–52, use Table 4 to determine the acute angle $\alpha$.

**45.** $\sin \alpha = 0.4695$

**46.** $\cos \alpha = 0.8480$

**47.** $\cot \alpha = 1.072$

**48.** $\tan \alpha = 3.078$

**49.** $\cos \alpha = 0.7698$

**50.** $\sin \alpha = 0.4669$

**51.** $\sec \alpha = 1.086$

**52.** $\csc \alpha = 1.061$

In exercises 53–62, find the value of each of the following.

**53.** sin 39°33′

**54.** tan 12°15′

**55.** tan 31°23′

**56.** cos 19°3′

**57.** cos 71°27′

**58.** sin 63°11′

**59.** cot 82°18′

**60.** cot 75°52′

**61.** sec 87°47′

**62.** csc 45°36′

In exercises 63–68, determine the acute angle $\alpha$.

**63.** $\sin \alpha = 0.0600$

**64.** $\cos \alpha = 0.9727$

**65.** $\cos \alpha = 0.3490$

**66.** $\tan \alpha = 1.7030$

**67.** $\tan \alpha = 0.5000$

**68.** $\sin \alpha = 0.8500$

## Superset

In exercises 69–76, determine which statements are identities.

**69.** $\cos \alpha \tan \alpha = \sin \alpha$

**70.** $\cos \alpha \sec \alpha = \cot \alpha$

**71.** $\sec \alpha = \tan \alpha \csc \alpha$

**72.** $\sin \alpha \cot \alpha = \cos \alpha$

**73.** $\dfrac{\tan \alpha}{\sin \alpha} = \sec \alpha$

**74.** $\dfrac{\csc \alpha}{\sec \alpha} = \cot \alpha$

**75.** $\dfrac{1}{\tan \alpha} = \dfrac{\cos \alpha}{\sin \alpha}$

**76.** $\cot \alpha = \dfrac{\sec \alpha}{\csc \alpha}$

In exercises 77–80, evaluate each pair of expressions by using the values of the special angles.

**77.** $\sin(90° - 30°)$, $\sin 90° - \sin 30°$

**78.** $\sin(60° + 30°)$, $\sin 60° + \sin 30°$

**79.** $\cot 90° - \cot 30°$, $\cot(90° - 30°)$

**80.** $\cos(45° + 45°)$, $\cos 45° + \cos 45°$

In exercises 81–84, verify that each of the following is true.

**81.** $\sin(60° + 30°) = \sin 60° \cos 30° + \sin 30° \cos 60°$

**82.** $\sin 60° \cos 30° = \frac{1}{2}(\sin 90° + \sin 30°)$

**83.** $\cos 90° = (\cos 45°)(\cos 45°) - (\sin 45°)(\sin 45°)$

**84.** $\tan 60° = \dfrac{2 \tan 30°}{1 - \tan^2 30°}$

## 2.5 Reference Angles

**Goal A** *to determine the sign of a trigonometric function value*

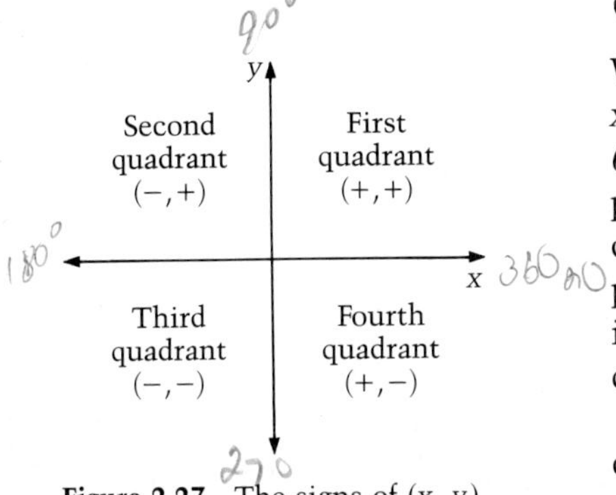


**Figure 2.27** The signs of $(x, y)$

We have defined the six trigonometric functions of an angle $\theta$ in terms of $x$, $y$, and $r$, where $x$ and $y$ are coordinates of a point on the terminal side of $\theta$, and $r$ is the distance between that point and the origin. Since $r$ must be positive, the sign of a trigonometric function value, such as cos 210°, is determined by the signs of $x$ and $y$. For example, since $\cos\theta = \frac{x}{r}$ and $r$ is positive, the sign of $\cos\theta$ is determined by the sign of $x$, which is positive in the first and fourth quadrants and negative in the second and third quadrants.

The chart below summarizes information needed to determine the sign of a trigonometric function value.

90°

**Sin $\theta$** (and csc $\theta$) | **All** function values

180° — 0°/360°

**Tan $\theta$** (and cot $\theta$) | **Cos $\theta$** (and sec $\theta$)

270°

■ The memory device ASTC (Always Study Trigonometry Carefully) is often used as a reminder of which trigonometric functions are positive in which quadrants.

| positive functions | A | S | T | C |
|---|---|---|---|---|
| quadrant | 1st | 2nd | 3rd | 4th |

**Figure 2.28** The quadrants in which the trigonometric functions are positive.

**Example 1** Determine the sign of the following:

(a) sin 190° (b) tan 135° (c) cos(−50°)

***Solution***

(a) 190° is a third quadrant angle. sin 190° is negative.

*Think:*

| A | S | T | C |
|---|---|---|---|
| 1 | 2 | 3 | 4 |

■ In the third quadrant, only tan and cot are positive.

(b) 135° is a second quadrant angle. tan 135° is negative.

*Think:*

| A | S | T | C |
|---|---|---|---|
| 1 | 2 | 3 | 4 |

■ In the second quadrant, only sin and csc are positive.

(c) −50° is a fourth quadrant angle. cos(−50°) is positive.

*Think:*

| A | S | T | C |
|---|---|---|---|
| 1 | 2 | 3 | 4 |

■ In the fourth quadrant, only cos and sec are positive.

**Goal B** *to determine the reference angle for an angle of any measure*

Suppose you wish to evaluate cos 210°. You already know one part of the answer: the *sign* of cos 210° is negative. Thus,

$$\cos 210^\circ = -\_\_\_\_.$$

To complete the statement cos 210° = –____, we use a *reference angle.* A **reference angle** for a given angle $\theta$ is a first quadrant angle whose trigonometric function values are numerically the same as the trigonometric function values of $\theta$—only the signs may differ.

Figure 2.29 presents second, third, and fourth quadrant angles. In each case, the reference angle is found by performing the appropriate reflection of the terminal side necessary to make it a first quadrant angle. We denote the reference angle of any angle $\theta$ by attaching an asterisk ($\theta^*$).

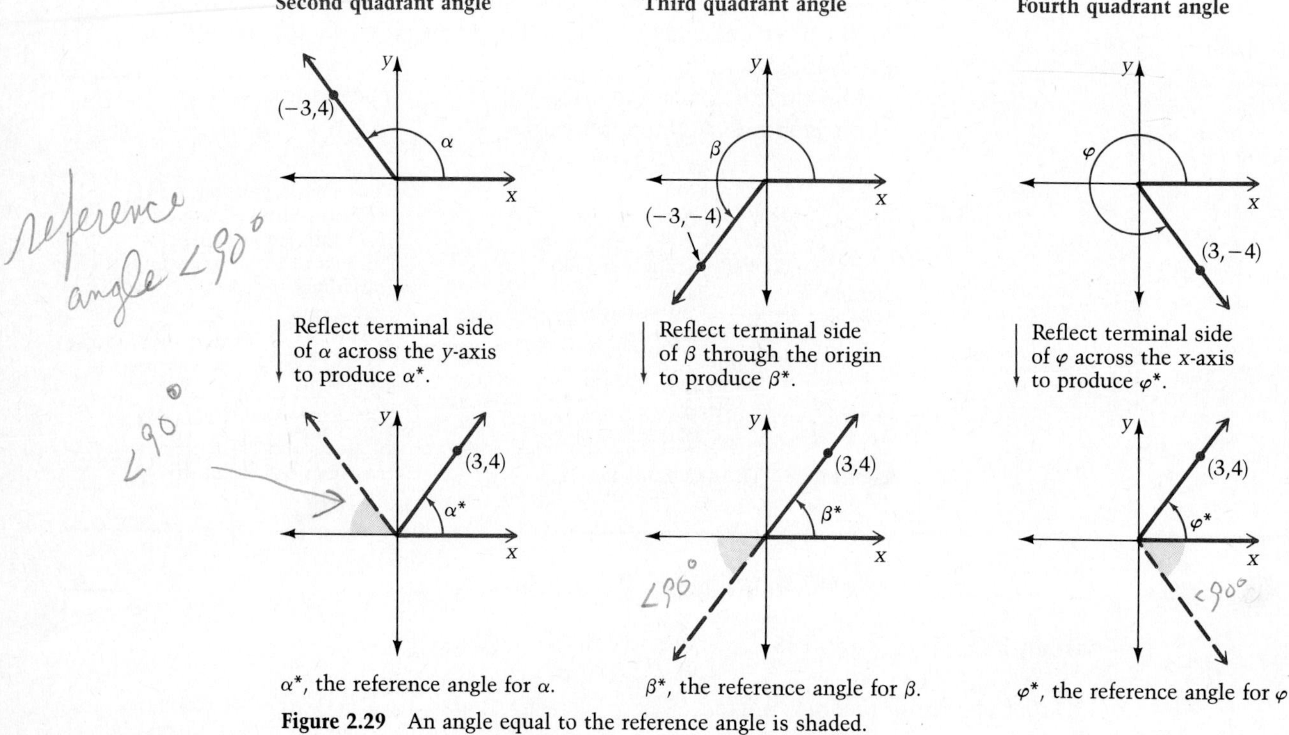


**Figure 2.29** An angle equal to the reference angle is shaded.

Figure 2.29 illustrates the following useful rule.

**Rule for Finding a Reference Angle**

Suppose $\theta$ is an angle between 0° and 360° inclusive and $\theta^*$ is the reference angle of $\theta$.

If $\theta$ is a first quadrant angle, $\theta^* = \theta$.
If $\theta$ is a second quadrant angle, $\theta^* = 180° - \theta$.
If $\theta$ is a third quadrant angle, $\theta^* = \theta - 180°$.
If $\theta$ is a fourth quadrant angle, $\theta^* = 360° - \theta$.

If you are given an angle greater than 360° or less than 0°, add or subtract an appropriate multiple of 360° so that the measure of the resulting angle is between 0° and 360°. Then use the above rule.

**Example 2** Determine the reference angle for each of the following angles.

(a) $\theta = 240°$ (b) $\theta = 1000°$ (c) $\theta = -200°$

***Solution*** (a) 240° is a third quadrant angle. Therefore $\theta^\star = 240° - 180° = 60°$. The reference angle is 60°.

(b) 1000° is greater than 360°. Subtract $2 \cdot 360°$ from 1000° to get 280°. Now, 280° is a fourth quadrant angle. Thus, $\theta^\star = 360° - 280° = 80°$. The reference angle is 80°.

(c) $-200°$ is less than 0°. Add $1 \cdot 360°$ to get 160°. Now, 160° is a second quadrant angle. Thus, $\theta^\star = 180° - 160° = 20°$. The reference angle is 20°.

**Goal C** *to use reference angles to determine trigonometric function values*

Because of the way we defined a reference angle, the trigonometric function values of a given angle are the same as the trigonometric function values of the reference angle, except maybe for the sign. We use the ASTC chart in Goal A to determine the correct sign.

To evaluate an expression like tan 240°, we follow a three step procedure:

**Step 1.** Determine the sign of $\tan \theta$: tan 240° is positive
**Step 2.** Find the reference angle $\theta^\star$: $240° - 180° = 60°$
**Step 3.** Evaluate $\tan \theta^\star$: $\tan 60° = \sqrt{3}$

Thus, $\tan 240° = +\tan 60° = \sqrt{3}$.

**Example 3** Evaluate (a) cos 120° (b) sin 280°20′.

***Solution*** (a) cos 120° = sign cos (reference angle of 120°)

= −cos (reference angle of 120°) ■ 120° is a second quadrant angle; cos is negative there.

= −cos 60° ■ $\theta^\star = 180° - 120° = 60°$.

$= -\frac{1}{2}$

(b) sin 280°20′ = sign sin (reference angle of 280°20′)

= −sin (reference angle of 280°20′) ■ 280°20′ is a fourth quadrant angle; sin is negative there.

= −sin 79°40′ = −0.9838 ■ $\theta^\star = 360° - 280°20' = 79°40'$

## Exercise Set 2.5

### Goal A

In exercises 1–14, determine the sign of the following.

**1.** cot 200° **2.** sec 175°

**3.** cos 315° **4.** tan 285°

**5.** sin 457° **6.** cot 703°

**7.** csc(1210°) **8.** sin 1568°

**9.** tan(−279°) **10.** csc(−112°)

**11.** sec(−763°) **12.** cos(−581°)

**13.** sin 196°36′ **14.** tan(−185°42′)

### Goal B

In exercises 15–28, determine the reference angle for each of the following angles.

**15.** 200° **16.** 185° **17.** 320°

**18.** 197° **19.** 485° **20.** 696°

**21.** −250° **22.** −185° **23.** −444°

**24.** −715° **25.** −1445° **26.** −1081°

**27.** 265°35′ **28.** −95°58′

### Goal C

In exercises 29–60, evaluate the following.

**29.** cos 240° **30.** csc 240°

**31.** tan 300° **32.** sin 120°

**33.** cot 225° **34.** sec 135°

**35.** sec 330° **36.** tan 180°

**37.** csc 300° **38.** sec 300°

**39.** tan 750° **40.** cos 1290°

**41.** sin(−210°) **42.** sin(−1290°)

**43.** tan 158° **44.** sin 345°

**45.** cos 408° **46.** cot 497°

**47.** cot 320° **48.** tan 212°

**49.** cos 1258° **50.** cos 485°

**51.** sec(−280°) **52.** csc(−272°)

**53.** sin(−1134°) **54.** tan(−666°)

**55.** csc(−138°) **56.** sec(−295°)

**57.** sin 95°20′ **58.** cot 245°40′

**59.** cot(−100°30′) **60.** cos 250°10′

### Superset

In exercises 61–70, assume that the angle $\theta$ terminates in the given quadrant. Find the sign of the given trigonometric function value.

**61.** second quadrant; cos $\theta$

**62.** third quadrant; sin $\theta$

**63.** fourth quadrant; tan $\theta$

**64.** fourth quadrant; sec $\theta$

**65.** third quadrant; cos $\theta$

**66.** fourth quadrant; cot $\theta$

**67.** second quadrant; sec $\theta$

**68.** second quadrant; tan $\theta$

**69.** fourth quadrant; sin $\theta$

**70.** second quadrant; csc $\theta$

In exercises 71–76, assume that the angle $\theta$ has measure between 180° and 270°. Find the sign of the following trigonometric function values.

**71.** $\sin \frac{\theta}{2}$ **72.** $\sin 2\theta$

**73.** $\sin(90° + \theta)$ **74.** $\sin(90° - \theta)$

**75.** $\sin(180° - \theta)$ **76.** $\sin(180° + \theta)$

# Chapter Review & Test

## Chapter Review

### 2.1 Triangles (pp. 34–40)

In any triangle, the sum of the measures of the angles is 180°. (p. 35)

An *isosceles triangle* is a triangle with two sides of equal length. In any isosceles triangle the angles opposite the two sides of equal length have the same measure. (p. 35) A *right triangle* is a triangle in which one of the angles is a right angle. (p. 35) An *equilateral triangle* is a triangle with three sides of equal length. Each angle of an equilateral triangle has measure 60°. (p. 36)

In *similar triangles,* corresponding angles have the same measure and corresponding sides are proportional. (p. 37) *A-A Similarity Theorem:* Two triangles are similar if two angles of one triangle have the same measure as two angles of the other triangle. (p. 38)

### 2.2 Trigonometric Ratios of Acute Angles (pp. 41–47)

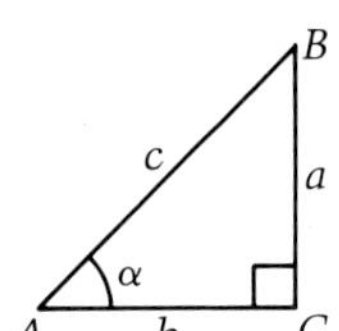


Let $\alpha$ be an acute angle in a right triangle $ABC$. Then the trigonometric functions of $\alpha$ are defined by the following ratios (p. 42):

$$\sin \alpha = \frac{a}{c} \qquad \cos \alpha = \frac{b}{c} \qquad \tan \alpha = \frac{a}{b}$$

$$\csc \alpha = \frac{c}{a} \qquad \sec \alpha = \frac{c}{b} \qquad \cot \alpha = \frac{b}{a}$$

### 2.3 Angles of Rotation (pp. 48–53)

An angle is in *standard position* in the $xy$-plane if its vertex is at the origin, and its initial side lies on the positive $x$-axis. The angle is *positive* if the rotation is counterclockwise and *negative* if the rotation is clockwise. (p. 48)

Let $\theta$ be an angle in standard position with $P(x, y)$ a point on its terminal side. Then

$$\sin \theta = \frac{y}{r}, \qquad \cos \theta = \frac{x}{r}, \qquad \tan \theta = \frac{y}{x} \quad (x \neq 0),$$

$$\csc \theta = \frac{r}{y} \quad (y \neq 0), \qquad \sec \theta = \frac{r}{x} \quad (x \neq 0), \qquad \cot \theta = \frac{x}{y} \quad (y \neq 0),$$

where $r = \sqrt{x^2 + y^2}$ is the distance from $P$ to the origin. (p. 50)

### 2.4 Identities and Tables (pp. 54–60)

**Reciprocal Identities**

$$\sin\theta = \frac{1}{\csc\theta} \quad \cos\theta = \frac{1}{\sec\theta} \quad \tan\theta = \frac{1}{\cot\theta}$$

$$\csc\theta = \frac{1}{\sin\theta} \quad \sec\theta = \frac{1}{\cos\theta} \quad \cot\theta = \frac{1}{\tan\theta}$$

**Quotient Identities**

$$\frac{\sin\theta}{\cos\theta} = \tan\theta$$

$$\frac{\cos\theta}{\sin\theta} = \cot\theta$$

**Cofunction Identities**

$$\sin\theta = \cos(90^\circ - \theta) \quad \cos\theta = \sin(90^\circ - \theta)$$

$$\tan\theta = \cot(90^\circ - \theta) \quad \cot\theta = \tan(90^\circ - \theta)$$

$$\sec\theta = \csc(90^\circ - \theta) \quad \csc\theta = \sec(90^\circ - \theta)$$

**Pythagorean Identities**

$$\sin^2\theta + \cos^2\theta = 1$$

$$\tan^2\theta + 1 = \sec^2\theta$$

$$1 + \cot^2\theta = \csc^2\theta$$

### 2.5 Reference Angles (pp. 61–64)

The sign of a trigonometric function value depends on the quadrant in which the angle terminates. It is found by using the ASTC chart:

| positive functions | A | S | T | C |
|---|---|---|---|---|
| quadrant | 1 | 2 | 3 | 4 |

A reference angle $\theta^*$ for a given angle $\theta$ is a first quadrant angle. The trigonometric function values of $\theta$ and $\theta^*$ are numerically the same. Only the signs may differ. (p. 62)

## Chapter Test

**2.1A** Use the given information about $\triangle ABC$ to solve for the missing part.

**1.** $m(\angle A) = 22^\circ$, $m(\angle B) = 95^\circ$, $m(\angle C) = ?$

**2.** $m(\angle B) = 79^\circ$, $m(\angle C) = 79^\circ$, $m(\angle A) = ?$

Assume that $\triangle ABC$ is a right triangle, $c$ is the length of the hypotenuse and $a$ and $b$ are the lengths of the legs. Find the length of the third side given the lengths of two sides.

**3.** $a = 12$, $b = 16$ **4.** $a = 10$, $c = 26$

**2.1B** Assume that $\triangle ABC$ and $\triangle A'B'C'$ are two triangles with $m(\angle A) = m(\angle A')$ and $m(\angle B) = m(\angle B')$. Given the lengths of some sides, find the length of the indicated side.

**5.** $a = 2$, $b = 5$, $a' = 12$, $b' = ?$

**6.** $a = 7$, $b = 8$, $a' = ?$, $b' = 20$

**2.2A** Use the given information about the right triangle $ABC$ to find the values of the six trigonometric ratios of $\alpha$.

**7.**

**8.**

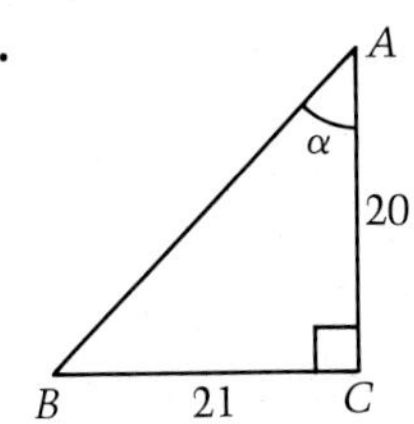


**2.2B** $\triangle ABC$ has right angle at $C$ and sides of length $a$, $b$, and $c$. Given the following information, find the indicated part.

**9.** $m(\angle A) = 30°$, $a = 24$, find $c$ **10.** $m(\angle A) = 60°$, $c = 18$, find $a$

**2.2C** The angles $\alpha$ and $\beta$ are complementary. Given the following information, find $\tan \alpha$ and $\tan \beta$.

**11.** $\sin \alpha = \frac{3}{7}$ **12.** $\cos \alpha = \frac{5}{7}$

**2.3A** Sketch each of the following angles in standard position.

**13.** 150° **14.** −135° **15.** −215° **16.** 325°

Determine the measure of the angle between 0° and 360° which is coterminal with the given angle.

**17.** 435° **18.** 1255° **19.** −435° **20.** −1255°

**2.3B** The given point is located on the terminal side of angle $\theta$. Find $\sin \theta$ and $\tan \theta$.

**21.** $(-10, 24)$ **22.** $(-12, -9)$

Evaluate each of the following.

**23.** $\cos(-270°)$ **24.** $\sin 180°$

**2.4A** Complete the following with an acute angle.

**25.** $\sin 77° = \cos$ ____ **26.** $\cot 37° = \tan$ ____

If $\sin 72° = 0.95$ and $\cos 72° = 0.31$, find the following function values. Round your answer to two decimal places.

**27.** $\sec 18°$ **28.** $\cot 18°$

Use the Reciprocal and Pythagorean Identities to determine the values of the other five trigonometric functions of acute angle $\alpha$.

**29.** $\tan \alpha = \frac{3}{2}$

**30.** $\sin \alpha = \frac{5}{6}$

**2.4B** Use Table 4 in the Appendix to determine the following.

**31.** $\cos 42°23'$

**32.** $\sec 63°48'$

**2.5A** Determine the sign of the following.

**33.** $\cos 200°$

**34.** $\cot(-200°)$

**35.** $\tan(-300°)$

**36.** $\csc 250°$

**2.5B** Determine the reference angle for each of the following angles.

**37.** 215°

**38.** 175°

**2.5C** Evaluate each of the following.

**39.** $\tan 135°$

**40.** $\cot(-240°)$

**41.** $\cos 348°20'$

**42.** $\sin 178°50'$

## Superset

**43.** Find the side of the square that has the same area as a rectangle with one side of length 16 and diagonal 20.

**44.** Between midnight and noon, how many times do the hands of a clock make an angle of 90°? an angle of 180°?

In exercises 45–46, the value of one trigonometric function of $\varphi$ is given. If the terminal side of $\varphi$ lies in the given quadrant, find the values of the other five trigonometric functions.

**45.** third quadrant; $\cos \varphi = -\frac{12}{13}$

**46.** fourth quadrant; $\cot \varphi = -\frac{9}{40}$

In exercises 47–48, verify that each of the following is true.

**47.** $\sin(2 \cdot 120°) = 2\sin 120° \cos 120°$

**48.** $\cos(135° - 45°) = \cos 135° \cos 45° + \sin 135° \sin 45°$

In exercises 49–50, assume that the angle $\theta$ terminates in the given quadrant. Find the sign of the trigonometric function value.

**49.** second quadrant; $\tan \theta$

**50.** fourth quadrant; $\csc \theta$

# 3 Trigonometric Functions

## 3.1 Radian Measure and the Unit Circle

**Goal A** *to determine points and standard arcs on the unit circle*

Up to this point, our study of trigonometry has focused on angles and triangles. In this chapter we shall develop a way of defining the trigonometric functions so that their domains consist of real numbers, not just angle measurements. This allows us to apply trigonometric functions to a much wider variety of problems. Many natural phenomena such as the motion of planets, the ocean's tides, and the beat of a human heart, are *periodic,* that is, certain behavior is repeated at regular intervals. When defined as functions of real numbers, trigonometric functions provide an excellent means of describing periodic behavior in nature.

The circle of radius 1 with center at the origin is called the **unit circle.** In Figure 3.1 we have labeled the horizontal axis the $u$-axis and the vertical axis the $v$-axis. An equation for the unit circle is easily derived by means of the distance formula. Since any point $T(u, v)$ on the unit circle is at a distance 1 from the origin (0, 0), we have, by the distance formula,

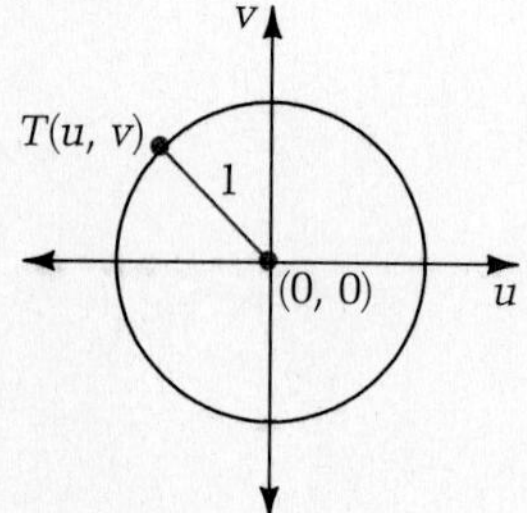


**Figure 3.1** The unit circle in the $uv$-plane.

$$\sqrt{(u - 0)^2 + (v - 0)^2} = 1.$$

Upon squaring both sides of this equation, we get

$$u^2 + v^2 = 1,$$

which is the common way of describing the unit circle in the $uv$-plane.

**Example 1** Determine whether each of the following points is on the unit circle.

(a) $\left(\frac{1}{2}, \frac{\sqrt{3}}{2}\right)$ (b) $\left(-\frac{\sqrt{2}}{2}, \frac{\sqrt{2}}{2}\right)$ (c) $\left(\frac{1}{3}, \frac{2}{3}\right)$

***Solution*** (a) $\left(\frac{1}{2}\right)^2 + \left(\frac{\sqrt{3}}{2}\right)^2 = \frac{1}{4} + \frac{3}{4} = 1$

■ To determine whether a point $(u, v)$ is on the unit circle, check to see if its coordinates satisfy the equation $u^2 + v^2 = 1$.

Thus, $\left(\frac{1}{2}, \frac{\sqrt{3}}{2}\right)$ is on the unit circle.

(b) $\left(\frac{-\sqrt{2}}{2}\right)^2 + \left(\frac{\sqrt{2}}{2}\right)^2 = \frac{2}{4} + \frac{2}{4} = 1$

Thus, $\left(-\frac{\sqrt{2}}{2}, \frac{\sqrt{2}}{2}\right)$ is on the unit circle.

(c) $\left(\frac{1}{3}\right)^2 + \left(\frac{2}{3}\right)^2 = \frac{1}{9} + \frac{4}{9} = \frac{5}{9} \neq 1$

Thus, $\left(\frac{1}{3}, \frac{2}{3}\right)$ is not on the unit circle.

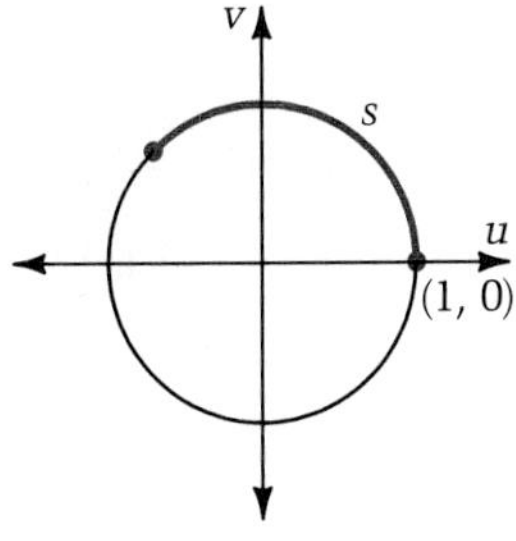


**Figure 3.2** The standard arc $s$.

Let us define a **standard arc s** as an arc on the unit circle which starts at the point (1, 0) and travels $s$ units counterclockwise if $s$ is positive, and $s$ units clockwise if $s$ is negative. Since the circumference $C$ of the unit circle is

$$C = 2\pi r = 2\pi(1) = 2\pi,$$

arcs whose lengths are rational multiples of $\pi$ $\left(\text{e.g., } \frac{\pi}{2}, -\frac{3\pi}{4}, 5\pi\right)$ are easy to visualize.

**Example 2** Represent each of the following real numbers as a standard arc.

(a) $\pi$ (b) $\frac{\pi}{2}$ (c) $\frac{\pi}{3}$ (d) $-\frac{3\pi}{2}$ (e) $-\frac{2\pi}{3}$ (f) $\frac{13\pi}{6}$

***Solution*** (a)

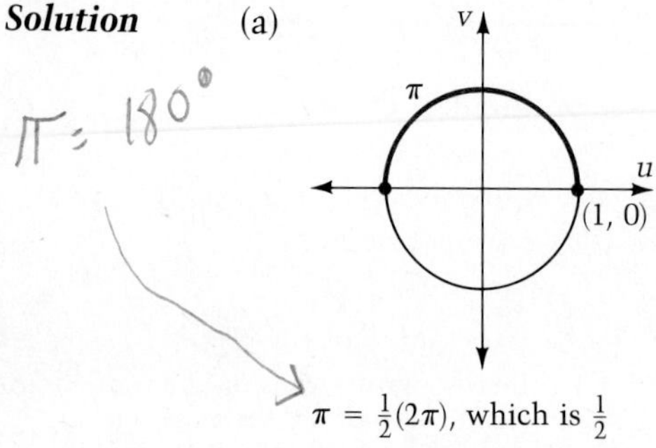


$\pi = \frac{1}{2}(2\pi)$, which is $\frac{1}{2}$ of the circumference.

(b)

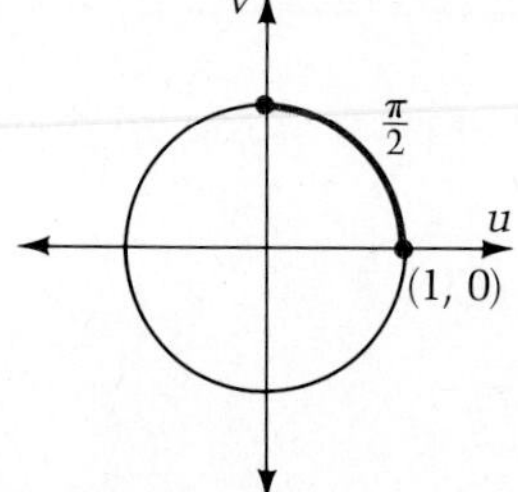


$\frac{\pi}{2} = \frac{1}{4}(2\pi)$, which is $\frac{1}{4}$ of the circumference.

(c)

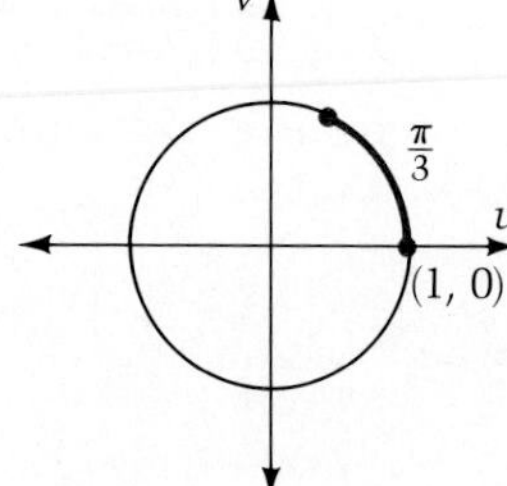


$\frac{\pi}{3} = \frac{1}{6}(2\pi)$, which is $\frac{1}{6}$ of the circumference.

(d)

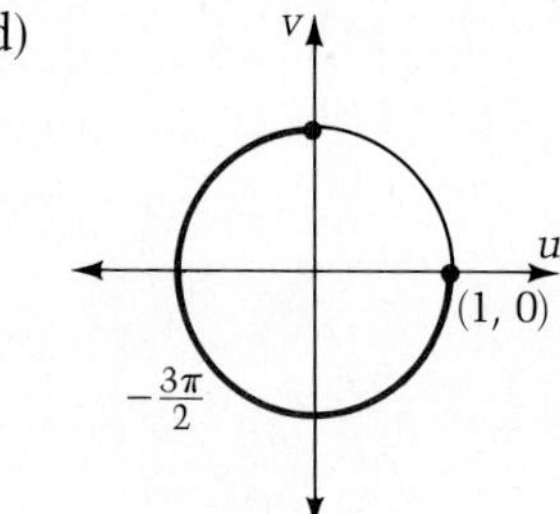


$\frac{3\pi}{2}$ is $\frac{3}{4}$ of the circumference. Since $-\frac{3\pi}{2} < 0$, the arc travels clockwise.

(e)

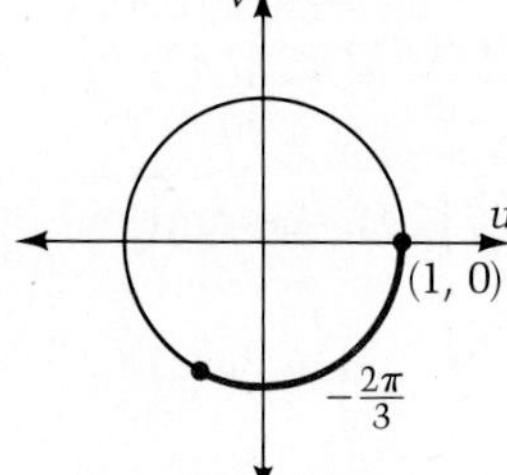


$\frac{2\pi}{3}$ is $\frac{1}{3}$ of the circumference. Since $-\frac{2\pi}{3} < 0$, the arc travels clockwise.

(f)

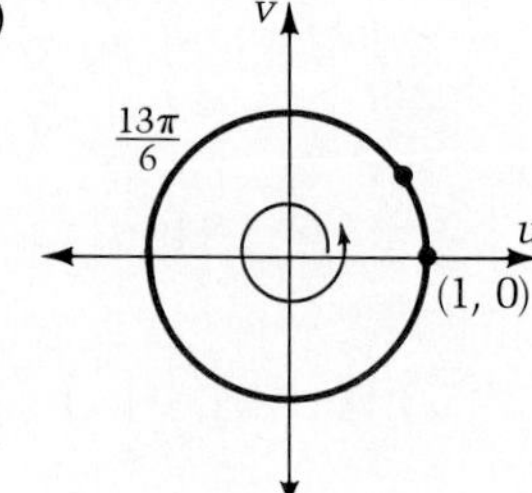


$\frac{13\pi}{6}$ is $1\frac{1}{12}$ of the circumference.

**Goal B** *to determine the radian measure of an angle*

We have discussed the trigonometric functions of angles measured in degrees and used these functions to solve some interesting problems. However, to use trigonometric functions to solve other types of problems, we need to define them in such a way that their domains consist of real numbers (with no units attached).

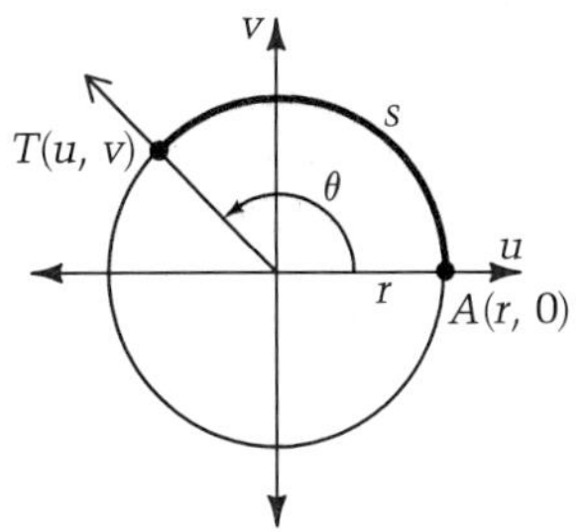


Figure 3.3

Figure 3.3 shows a positive angle $\theta$ in standard position, and a circle of radius $r$ with center at the origin. The angle $\theta$ is called a **central angle** since its vertex is the center of the circle. We say that angle $\theta$ "subtends" or "cuts" an arc of the circle. Let $s$ be the length of the arc.

Notice that $A(r, 0)$ is the point where the initial side of angle $\theta$ intersects the circle. Furthermore, $T(u, v)$ is the point where the terminal side of angle $\theta$ intersects the circle, and $s$ is the *length* of the arc cut by angle $\theta$. We use $s$, $r$, and $\theta$, as just described, to make the following definition.

**Definition**

The **radian measure** of a positive central angle $\theta$ is defined as the ratio of the arclength $s$ to the radius $r$:

$$\theta = \frac{s}{r}.$$

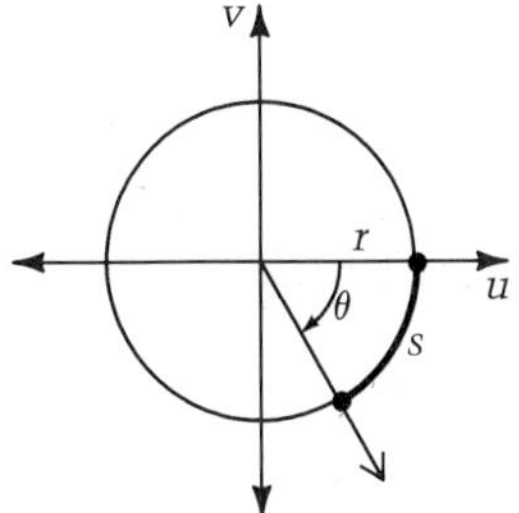


Figure 3.4 $\theta = -\frac{s}{r}$

Since both $s$ and $r$ have the same units of length, their ratio $\theta$ will have no units. However, we will say that the angle has a measure of $\theta$ "radians" to remind us that it was measured by using the radius. If $\theta$ is a negative angle, its radian measure is defined to be negative as Figure 3.4 suggests.

In a unit circle the radian measure of a positive central angle is equal to the *length* of the arc that it cuts. This is shown in Figure 3.5.

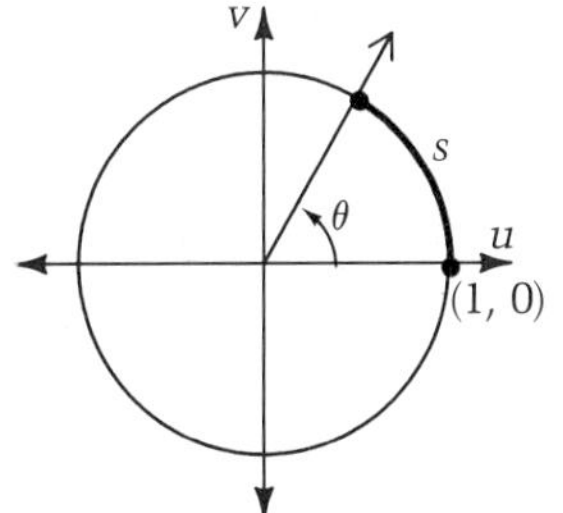


Figure 3.5

■ Since $r = 1$, $\theta = \frac{s}{r} = \frac{s}{1}$. Thus, $\theta = s$.

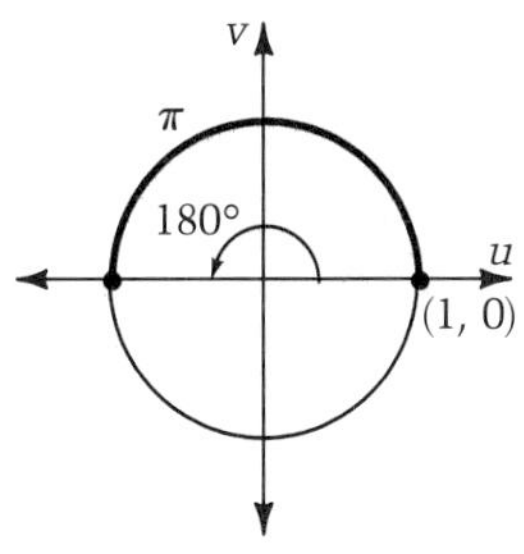


Figure 3.6

Now consider a central angle of 180° in a unit circle as shown in Figure 3.6. Since an angle of 180° cuts an arc of length $\pi$ (half the circumference), the above statement suggests that an angle of 180° has a radian measure of $\pi$. That is,

$$180° = \pi \text{ radians.}$$

We use this relationship in setting up a proportion for converting the degree measure of an angle to radian measure and vice versa.

| degrees | radians |
|---|---|
| 0° | 0 |
| 30° | $\frac{\pi}{6}$ |
| 45° | $\frac{\pi}{4}$ |
| 90° | $\frac{\pi}{2}$ |

**Fact**

If $d$ is the degree measure of an angle and $\alpha$ is its radian measure, then

$$\frac{d}{180^\circ} = \frac{\alpha}{\pi}.$$

**Example 3**

(a) Convert 165° to radians.

(b) Convert $-\frac{3\pi}{4}$ radians to degrees.

(c) Convert 3 radians to degrees.

(d) Convert 40°36′ to radians.

***Solution***

(a) $$\frac{165^\circ}{180^\circ} = \frac{\alpha}{\pi}$$ ■ Set up the proportion.

$$\alpha = \frac{165}{180}\pi$$

$$= \frac{11}{12}\pi \text{ radians}$$

(b) $$\frac{d}{180^\circ} = \frac{-\frac{3\pi}{4}}{\pi}$$

$$d = -\frac{3\pi}{4} \cdot \frac{1}{\pi} \cdot 180^\circ = -135^\circ$$

(c) $$\frac{d}{180^\circ} = \frac{3}{\pi}$$

$$d = \frac{3 \times 180^\circ}{\pi} \approx 172^\circ$$ ■ $\pi \approx 3.14$.

(d) Since $1^\circ = 60'$, we have $36' = \left(\frac{36'}{60'}\right)^\circ = (0.6)^\circ$. Thus, $40^\circ 36' = (40 + 0.6)^\circ = 40.6^\circ$.

$$\frac{d}{180^\circ} = \frac{\alpha}{\pi}$$

$$\frac{40.6^\circ}{180^\circ} = \frac{\alpha}{\pi}$$

$$\alpha = \frac{40.6}{180} \cdot \pi \approx 0.71 \text{ radians}$$

**Goal C** *to solve problems using radian measure*

Since the definition of radian measure depends on the radius and arc of a circle, a variety of problems involving circular arcs can be solved using radian measure.

**Example 4** Determine the length of an arc cut by a central angle of 60° in a circle of radius 2 yd.

***Solution***

$$\theta = \frac{s}{r}$$

■ To use this formula, 60° must first be converted to radians: $\frac{60°}{180°} = \frac{\theta}{\pi}$, thus $\theta = \frac{60}{180}\pi = \frac{\pi}{3}$.

$$\frac{\pi}{3} = \frac{s}{2}$$

■ $\frac{\pi}{3}$ is substituted for $\theta$.

$$s = \frac{2\pi}{3}$$

The arc has length $\frac{2\pi}{3}$ yd ≈ 2.09 yd (rounded to two decimal places).

**Example 5** A wheel of radius 80 cm rolls along the ground without slipping and rotates through an angle of 45°. How far does the wheel move?

***Solution***

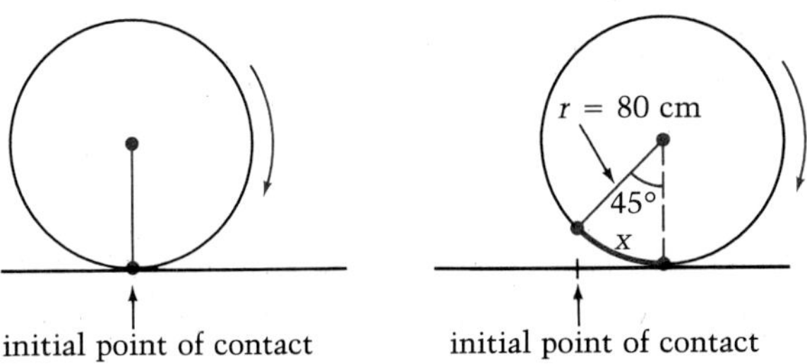


The distance that the wheel advances is equal to the length of the arc shown in the right-hand figure above. We use this fact to find $s$.

$$\theta = \frac{s}{r}$$

■ To use this formula, $\theta$ must be measured in radians. $\frac{45°}{180°} = \frac{\theta}{\pi}$, thus $\theta = \frac{45}{180}\pi = \frac{\pi}{4}$.

$$\frac{\pi}{4} = \frac{s}{80}$$

$$s = 80 \cdot \frac{\pi}{4} = 20\pi$$

The wheel has moved $20\pi$ cm (≈63 cm) to the right.

## Exercise Set 3.1 ⊟ Calculator Exercises in the Appendix

### Goal A

In exercises 1–4, determine whether the point is on the unit circle.

1. $\left(-\frac{\sqrt{3}}{2}, -\frac{1}{2}\right)$
2. $\left(-\frac{\sqrt{3}}{3}, -\frac{1}{3}\right)$
3. $\left(\frac{\sqrt{2}}{2}, -\frac{1}{2}\right)$
4. $\left(-\frac{\sqrt{3}}{2}, \frac{\sqrt{2}}{2}\right)$

In exercises 5–12, represent the real number as a standard arc.

5. $\frac{\pi}{4}$
6. $\frac{3\pi}{2}$
7. $\frac{2\pi}{3}$
8. $\frac{5\pi}{4}$
9. $-\frac{\pi}{4}$
10. $-\frac{7\pi}{4}$
11. $\frac{25\pi}{6}$
12. $-\frac{11\pi}{3}$

### Goal B

In exercises 13–34, convert the measure of the angle from degrees to radians or vice versa.

13. 60°
14. 45°
15. 315°
16. 360°
17. 210°
18. 330°
19. −30°
20. −15°
21. 75°
22. 300°
23. −540°
24. −200°
25. $\frac{3\pi}{2}$
26. $\frac{\pi}{3}$
27. $\frac{5\pi}{6}$
28. $\frac{5\pi}{2}$
29. $-3\pi$
30. $-\frac{3\pi}{4}$
31. $-\frac{25\pi}{6}$
32. $-\frac{19\pi}{9}$
33. 1.5
34. 5

### Goal C

35. Determine the length of an arc cut by a central angle of 90° in a circle of radius 4 in.
36. Determine the length of an arc cut by a central angle of 30° in a circle of radius 2 cm.
37. A central angle of 1.5 radians cuts an arc of length 12 m. Find the radius of the circle.
38. A central angle of 2.5 radians cuts an arc of 25 ft. Find the radius of the circle.
39. What is the measure in radians of the angle through which the minute hand of a clock turns in 42 min?
40. What is the measure of the angle in degrees through which the minute hand of a clock turns between 1:30 P.M. and 2:20 P.M. of the same day?

### Superset

41. Find the number of radians through which each of the hands of a clock move in (a) 12 hours, (b) in one hour, (c) in 30 min, (d) in 5 min.
42. After midnight, at what time are the hands of a clock first perpendicular to each other?
43. The minute hand of a clock is of length 5 cm and the hour hand is of length 3.6 cm. How far does the tip of the minute hand move in 12 hours? How far does the tip of the hour hand move in one hour?
44. The tires of an automobile are 24 in in diameter. If the automobile backs up 20 ft, through what angle does each wheel turn?
45. (Refer to exercise 44) Through what angle does each wheel turn if the automobile travels one mile?
46. Through what angle does a water wheel of radius 32 ft turn if it rotates at the rate of 2 mi per hour for 1 min?
47. Suppose $\theta$ is an angle in standard position. Determine the coordinates of the point of intersection of the terminal side of $\theta$ with the unit circle if (a) $\theta = \frac{\pi}{6}$, (b) $\theta = \frac{\pi}{4}$, (c) $\theta = \frac{\pi}{3}$.

## 3.2 Trigonometric Functions of Real Numbers

**Goal A** *to evaluate trigonometric functions whose domains are real numbers*

One of the central ideas in mathematics is the concept of function. Of special interest to us are those functions having the set of real numbers as domain. Radian measure provides a means for defining the trigonometric functions so that their domains are sets of real numbers.

The transition from angle measurements to real numbers as domain values is made possible by noticing the following:

> every real number $x$ can be associated uniquely with the central angle $\theta$ that cuts standard arc $x$ on the unit circle.

This result is illustrated by the following procedure. To determine the angle $\theta$ associated with the real number $x$, begin by representing $x$ as a standard arc. This arc terminates at a unique point $T(u, v)$, and is cut by a unique central angle $\theta$, having $T(u, v)$ on its terminal side. We have thus associated angle $\theta$ with the real number $x$.

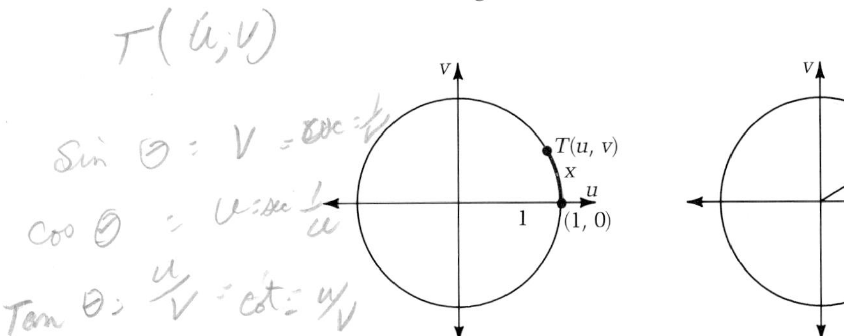


(a) Represent $x$ as a standard arc.

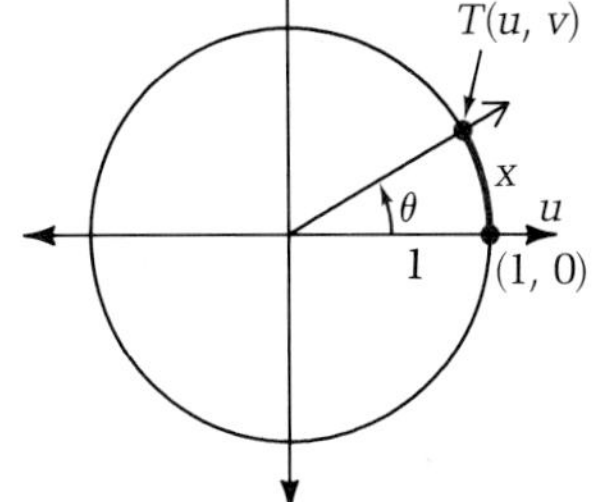


(b) $\theta$ is the unique angle that cuts the standard arc $x$.

**Figure 3.7**

Recall that we previously defined the trigonometric functions of $\theta$ in terms of the coordinates of the point $T(u, v)$. Since $r = 1$, we have

$$\theta = \frac{x}{1} = x.$$

We define the trigonometric functions of the real number $x$ to be the same as the trigonometric functions of angle $\theta$.

**Definition**

Let $x$ be any real number, let $\theta$ be the central angle in standard position in the unit circle having radian measure $x$, and let $T(u, v)$ be the terminal point on standard arc $x$. Then

| | |
|---|---|
| $\sin x = \sin \theta = v$ | domain: all real numbers |
| $\cos x = \cos \theta = u$ | domain: all real numbers |
| $\tan x = \tan \theta = \dfrac{v}{u}$ | domain: all real numbers except odd multiples of $\frac{\pi}{2}$ (which make denominator $u = 0$). |
| $\cot x = \cot \theta = \dfrac{u}{v}$ | domain: all real numbers except integral multiples of $\pi$ (which make denominator $v = 0$). |
| $\sec x = \sec \theta = \dfrac{1}{u}$ | domain: all real numbers except odd multiples of $\frac{\pi}{2}$. |
| $\csc x = \csc \theta = \dfrac{1}{v}$ | domain: all real numbers except integral multiples of $\pi$. |

Figure 3.8

**Example 1** Evaluate each of the following expressions.

(a) $\sin \dfrac{\pi}{2}$ (b) $\cos(-\pi)$ (c) $\tan \dfrac{3\pi}{2}$ (d) $\sin 0$

***Solution*** (a)

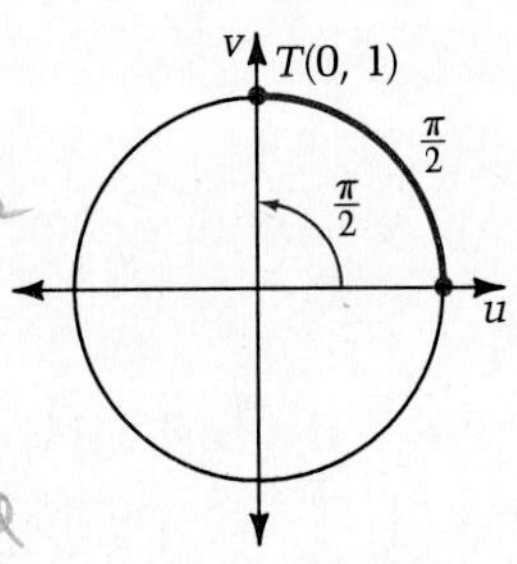


- Draw a unit circle and standard arc $\frac{\pi}{2}$. (Note: $\frac{\pi}{2}$ is $\frac{1}{4}$ of the circumference.)
- $\sin \frac{\pi}{2}$ = the second coordinate of $T(0, 1) = 1$.

(b)

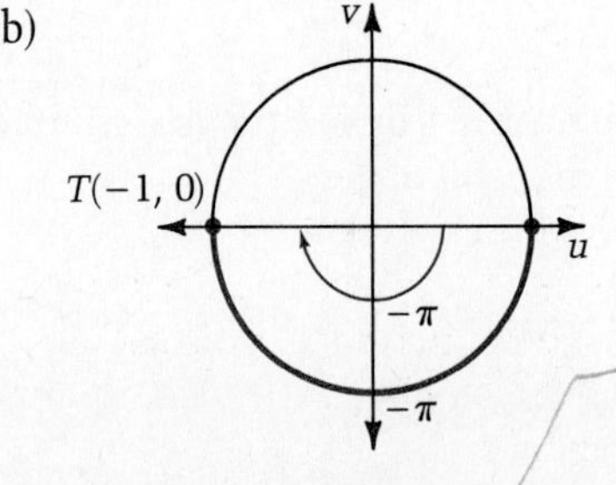


- Draw a unit circle and standard arc $-\pi$.
- $\cos(-\pi)$ = the first coordinate of $T(-1, 0)$, $= -1$.

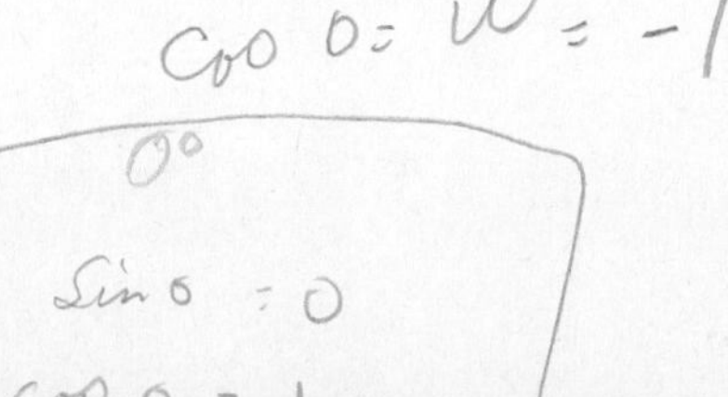

(c)

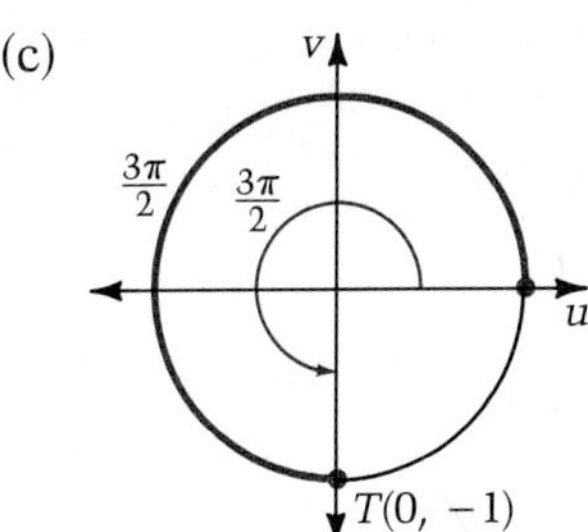


■ Draw a unit circle and standard arc $\frac{3\pi}{2}$.

■ $\tan \frac{3\pi}{2} = \frac{\text{the second coordinate of } T(0, -1)}{\text{the first coordinate of } T(0, -1)} = \frac{-1}{0}$ which is not defined.

(d) The point (1, 0) is the terminal point of standard arc 0 on the unit circle.

$$\sin 0 = \text{the second coordinate of } T(1, 0) = 0$$

In Example 1 we saw that trigonometric function values of integral multiples of $\frac{\pi}{2}$ are easily determined, since the corresponding arcs terminate at one of the four points (1, 0), (0, 1), (−1, 0), or (0, −1). Three other special arcs and their terminal points are shown on the unit circles below.

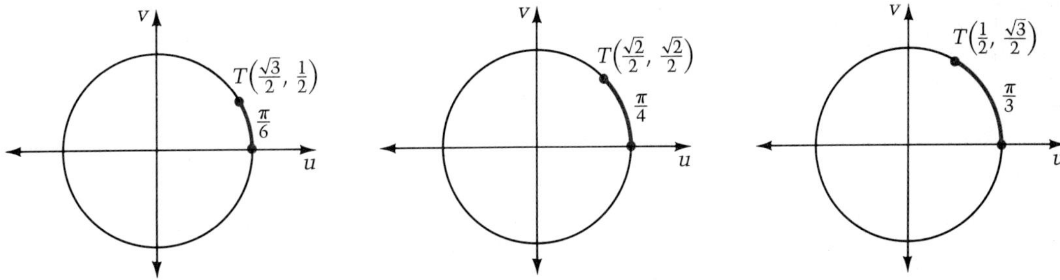


**Figure 3.9**

Knowing the coordinates of these terminal points allows us to determine trigonometric function values of $\frac{\pi}{6}$, $\frac{\pi}{4}$, and $\frac{\pi}{3}$. For example,

$$\sin \frac{\pi}{4} = \text{the second coordinate of } \left(\frac{\sqrt{2}}{2}, \frac{\sqrt{2}}{2}\right) = \frac{\sqrt{2}}{2},$$

$$\sec \frac{\pi}{3} = \frac{1}{\text{the first coordinate of } \left(\frac{1}{2}, \frac{\sqrt{3}}{2}\right)} = \frac{1}{\frac{1}{2}} = 2.$$

In addition, we can determine trigonometric function values of some other real numbers by reflecting these three points through the origin or across either axis. This is demonstrated in the next example.

**Example 2** Evaluate each of the following expressions.

(a) $\cos \frac{5\pi}{6}$ (b) $\sin\left(-\frac{\pi}{4}\right)$ (c) $\tan \frac{4\pi}{3}$

***Solution*** (a)

■ Draw the standard arc $\frac{5\pi}{6}$. The terminal point is found by reflecting the terminal point of $\frac{\pi}{6}$ across the vertical axis. Thus,

$$\cos \frac{5\pi}{6} = \text{the first coordinate of } \left(-\frac{\sqrt{3}}{2}, \frac{1}{2}\right) = -\frac{\sqrt{3}}{2}.$$

Thus, $\cos \frac{5\pi}{6} = -\frac{\sqrt{3}}{2}$.

(b)

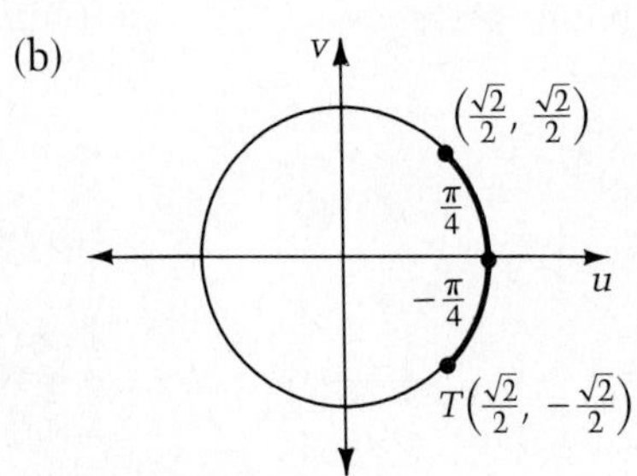


■ Draw the standard arc $-\frac{\pi}{4}$. The terminal point is found by reflecting the terminal point of $\frac{\pi}{4}$ across the horizontal axis. Thus, $\sin\left(-\frac{\pi}{4}\right)$ = the second coordinate of $\left(\frac{\sqrt{2}}{2}, -\frac{\sqrt{2}}{2}\right) = -\frac{\sqrt{2}}{2}$.

Thus, $\sin\left(-\frac{\pi}{4}\right) = -\frac{\sqrt{2}}{2}$.

(c)

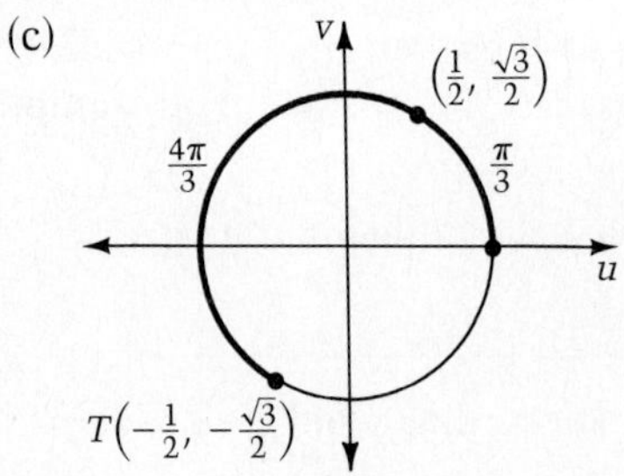


■ Draw the standard arc $\frac{4\pi}{3}$. The terminal point is found by reflecting the terminal point of $\frac{\pi}{3}$ through the origin. Thus,

$$\tan \frac{4\pi}{3} = \frac{\text{the second coordinate of } \left(-\frac{1}{2}, -\frac{\sqrt{3}}{2}\right)}{\text{the first coordinate of } \left(-\frac{1}{2}, -\frac{\sqrt{3}}{2}\right)} = \frac{\frac{-\sqrt{3}}{2}}{-\frac{1}{2}} = \sqrt{3}$$

Thus, $\tan \frac{4\pi}{3} = \sqrt{3}$.

By virtue of the method used in the previous two examples, we can update the Table of Special Angles from the previous chapter.

Table of Special Values

| $f(x)$ \ $x$ | 0 (or 0°) | $\frac{\pi}{6}$ (or 30°) | $\frac{\pi}{4}$ (or 45°) | $\frac{\pi}{3}$ (or 60°) | $\frac{\pi}{2}$ (or 90°) |
|---|---|---|---|---|---|
| sin $x$ | 0 | $\frac{1}{2}$ | $\frac{\sqrt{2}}{2}$ | $\frac{\sqrt{3}}{2}$ | 1 |
| cos $x$ | 1 | $\frac{\sqrt{3}}{2}$ | $\frac{\sqrt{2}}{2}$ | $\frac{1}{2}$ | 0 |
| tan $x$ | 0 | $\frac{\sqrt{3}}{3}$ | 1 | $\sqrt{3}$ | * |
| cot $x$ | * | $\sqrt{3}$ | 1 | $\frac{\sqrt{3}}{3}$ | 0 |
| sec $x$ | 1 | $\frac{2\sqrt{3}}{3}$ | $\sqrt{2}$ | 2 | * |
| csc $x$ | * | 2 | $\sqrt{2}$ | $\frac{2\sqrt{3}}{3}$ | 1 |

* undefined

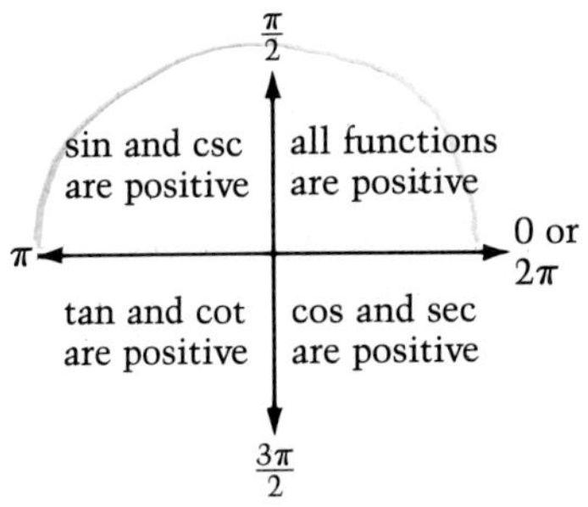

**Figure 3.10**

With a few simple changes, you can use the "reference angle" technique developed in the previous chapter to determine trigonometric function values of numbers outside the interval $[0, \frac{\pi}{2}]$. To find the trigonometric function value, attach the appropriate sign to the trigonometric function value of the reference value of $x$. The sign of your answer will depend on the quadrant in which the standard arc $x$ terminates.

For any $x$ between 0 and $2\pi$, the reference value for $x$ is determined as follows. (Recall that $180° = \pi$ radians.)

| $x$ | reference value for $x$ |
|---|---|
| between 0 and $\frac{\pi}{2}$ | $x$ |
| between $\frac{\pi}{2}$ and $\pi$ | $\pi - x$ |
| between $\pi$ and $\frac{3\pi}{2}$ | $x - \pi$ |
| between $\frac{3\pi}{2}$ and $2\pi$ | $2\pi - x$ |

To determine the trigonometric function value of a number outside the interval $[0, 2\pi)$, begin by adding or subtracting an appropriate multiple of $2\pi$ to produce a value between 0 and $2\pi$.

**Example 3** Determine each of the following function values.

(a) $\cos \frac{7\pi}{6}$ (b) $\tan\left(-\frac{\pi}{4}\right)$ (c) $\csc \frac{20\pi}{3}$

***Solution***

(a) $\cos \frac{7\pi}{6} = \boxed{\text{sign}} \cos \boxed{\text{reference value for } \frac{7\pi}{6}}$

$= -\cos \boxed{\text{reference value for } \frac{7\pi}{6}}$

■ Standard arc $\frac{7\pi}{6}$ terminates in the third quadrant, where the cosine is negative.

$= -\cos \frac{\pi}{6} = -\frac{\sqrt{3}}{2}$

■ Reference value: $\frac{7\pi}{6} - \pi = \frac{\pi}{6}$.

(b) $\tan\left(-\frac{\pi}{4}\right) = \tan \frac{7\pi}{4}$

■ Add $2\pi$ to $-\frac{\pi}{4}$ to produce a value between 0 and $2\pi$.

$= \boxed{\text{sign}} \tan \boxed{\text{reference value for } \frac{7\pi}{4}}$

■ Standard arc $\frac{7\pi}{4}$ terminates in the fourth quadrant, where the tangent is negative.

$= -\tan \frac{\pi}{4} = -1$

■ Reference value: $2\pi - \frac{7\pi}{4} = \frac{\pi}{4}$.

(c) $\csc \frac{20\pi}{3} = \csc \frac{2\pi}{3}$

■ Subtract $3 \cdot 2\pi$ from $\frac{20\pi}{3}$ to produce a value between 0 and $2\pi$.

$= \boxed{\text{sign}} \csc \boxed{\text{reference value for } \frac{2\pi}{3}}$

$= +\csc \boxed{\text{reference value for } \frac{2\pi}{3}}$

■ Standard arc $\frac{2\pi}{3}$ terminates in the second quadrant, where the cosecant is positive.

$= +\csc \frac{\pi}{3} = \frac{2\sqrt{3}}{3}$

■ Reference value: $\pi - \frac{2\pi}{3} = \frac{\pi}{3}$.

To evaluate the trigonometric function values of a real number $x$ that is not associated with a special arc, we can use Table 4. The column headed by the word "radians" contains the values of $x$.

**Example 4** Determine each of the following function values. Round your answer to two decimal places.

(a) $\cos 3.3$ (b) $\sec \dfrac{21\pi}{5}$

***Solution***

(a) $\cos 3.3 = \text{sign}\ \cos \left(\text{reference value for } 3.3\right)$

■ $\cos 3.3$ means "the cosine of 3.3 radians."

$= -\cos \left(\text{reference value for } 3.3\right)$

■ Since 3.3 is between $\pi$ ($\approx$3.14) and $\frac{3\pi}{2}$ ($\approx$4.71), standard arc 3.3 terminates in the third quadrant, where cosine is negative.

$\approx -\cos 0.16$

■ Reference value: $3.3 - \pi \approx 0.16$.

$\approx -0.99$

■ Table 4

(b) $\sec \dfrac{21\pi}{5} = \sec \dfrac{\pi}{5}$

■ Subtract $2 \cdot 2\pi$ from $\frac{21\pi}{5}$ to produce a value between 0 and $2\pi$.

$= \text{sign}\ \sec \left(\text{reference value for } \frac{\pi}{5}\right)$

$= +\sec \left(\text{reference value for } \frac{\pi}{5}\right)$

■ Standard arc $\frac{\pi}{5}$ terminates in the first quadrant where all trigonometric functions are positive.

$= +\sec \dfrac{\pi}{5}$

■ Reference value: $\frac{\pi}{5}$

$\approx \sec 0.63$

■ $\frac{\pi}{5} \approx 0.63$

$\approx 1.24$

■ Table 4

## Exercise Set 3.2 ⊟ Calculator Exercises in the Appendix

### Goal A

In exercises 1–10, determine the function value.

1. $\cot 0$
2. $\csc \frac{\pi}{2}$
3. $\tan 2\pi$
4. $\csc 2\pi$
5. $\sin \frac{\pi}{3}$
6. $\tan \frac{\pi}{4}$
7. $\csc \frac{\pi}{6}$
8. $\sec \frac{\pi}{6}$
9. $\cot \frac{\pi}{4}$
10. $\cos \frac{\pi}{3}$

In exercises 11–30, determine the function value.

11. $\cos \frac{2\pi}{3}$
12. $\cot \frac{5\pi}{3}$
13. $\sec\left(-\frac{\pi}{6}\right)$
14. $\tan\left(-\frac{5\pi}{6}\right)$
15. $\tan \frac{3\pi}{4}$
16. $\csc \frac{5\pi}{4}$
17. $\sin\left(-\frac{3\pi}{4}\right)$
18. $\sec\left(-\frac{5\pi}{4}\right)$
19. $\cot\left(-\frac{7\pi}{6}\right)$
20. $\sin\left(-\frac{4\pi}{3}\right)$
21. $\csc\left(-\frac{7\pi}{4}\right)$
22. $\sec\left(-\frac{11\pi}{6}\right)$
23. $\cos \frac{5\pi}{2}$
24. $\sin 5\pi$
25. $\cot \frac{10\pi}{3}$
26. $\csc \frac{9\pi}{4}$
27. $\tan\left(-\frac{7\pi}{3}\right)$
28. $\csc\left(-\frac{11\pi}{4}\right)$
29. $\sec \frac{25\pi}{6}$
30. $\cos\left(-\frac{21\pi}{4}\right)$

In exercises 31–38, determine the function value. Round to two decimal places.

31. $\sin 3.7$
32. $\tan 5$
33. $\sec 8.5$
34. $\cos 10.5$
35. $\csc \frac{12\pi}{5}$
36. $\sin \frac{5\pi}{7}$
37. $\tan -\frac{\pi}{9}$
38. $\cot -\frac{4\pi}{7}$

### Superset

In exercises 39–44, verify that the given statement is true.

39. $\sin \frac{13\pi}{6} \sec \frac{5\pi}{3} = \tan \frac{9\pi}{4}$

40. $\tan \frac{4\pi}{3} \tan \frac{7\pi}{6} = \tan \frac{13\pi}{4}$

41. $\tan \frac{\pi}{3} = \dfrac{1 - \cos \frac{4\pi}{3}}{\sin \frac{\pi}{3}}$

42. $\tan \frac{\pi}{6} = \dfrac{1 - \cos \frac{\pi}{3}}{\sin \frac{\pi}{3}}$

43. $\cos \frac{5\pi}{6} = \cos \frac{\pi}{2} \cos \frac{\pi}{3} - \sin \frac{\pi}{2} \sin \frac{\pi}{3}$

44. $\sin \frac{\pi}{3} = 2 \sin \frac{\pi}{6} \cos \frac{\pi}{6}$

In exercises 45–50, solve for the smallest positive value of $x$.

45. $\cot x = -1$
46. $\csc x = -\sqrt{2}$
47. $\sec x = -\frac{2\sqrt{3}}{3}$
48. $\tan x = -\sqrt{3}$
49. $\sec(-x) = \sqrt{2}$
50. $\csc(-x) = -2$

## 3.3 The Trigonometric Functions: Basic Graphs and Properties

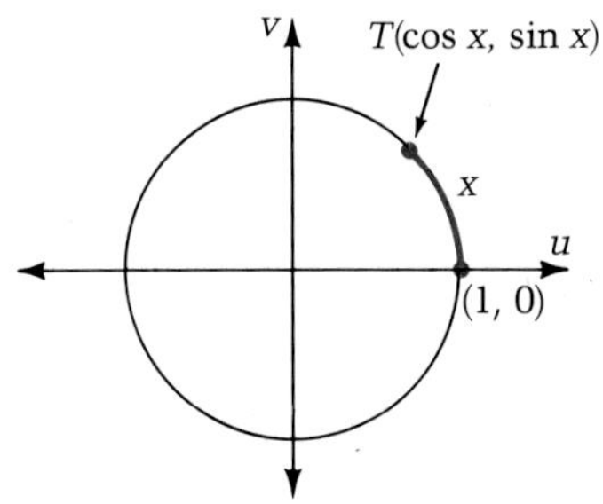


Figure 3.11

**Goal A** *to graph the sine and cosine functions*

We have seen that every real number $x$ uniquely determines a standard arc on the unit circle. The coordinates of the terminal point $T$ of this arc are used to define the six trigonometric functions of the real number $x$. The first coordinate of $T$ is $\cos x$ and the second coordinate is $\sin x$. There are many different standard arcs (and thus many different values of $x$) that are associated with the same terminal point $T$. Fortunately, these values of $x$ occur in a predictable, or *periodic* way.

Figure 3.12 demonstrates that $\frac{\pi}{3}$, $\frac{7\pi}{3}$, and $\frac{13\pi}{3}$ all determine the same terminal point $(\frac{1}{2}, \frac{\sqrt{3}}{2})$. Adding any positive or negative multiple of $2\pi$ to $\frac{\pi}{3}$ adds one or more complete revolutions to the arc. The resulting standard arc still has $(\frac{1}{2}, \frac{\sqrt{3}}{2})$ as its terminal point. This means that $\sin \frac{\pi}{3}$, $\sin \frac{7\pi}{3}$, and $\sin \frac{13\pi}{3}$ are the same: $\frac{\sqrt{3}}{2}$. Thus, the sine function repeats the same value each time the domain value changes by $2\pi$. Functions exhibiting such repetitive behavior are called *periodic.*

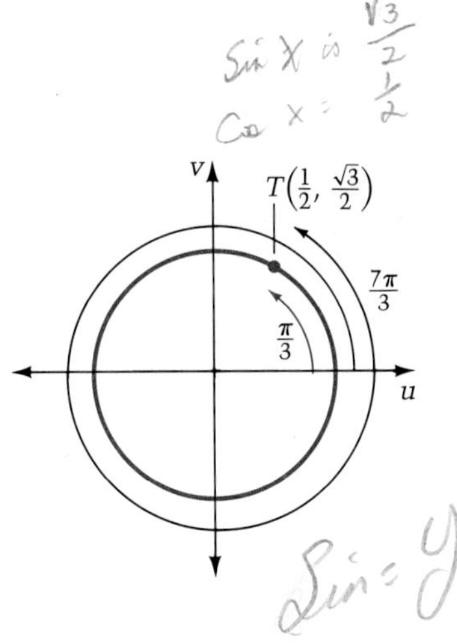


**Definition**

A function $f$ is called a **periodic function** if there is a positive real number $p$ such that

$$f(x) = f(x + p)$$

for all $x$ in the domain of $f$; the smallest such positive number $p$ is called the **period** of the function.

Both the sine and cosine functions are periodic with period $2\pi$. For that reason we write

$$\sin x = \sin(x + n \cdot 2\pi), \quad \text{for any integer } n, \text{ and}$$
$$\cos x = \cos(x + n \cdot 2\pi), \quad \text{for any integer } n.$$

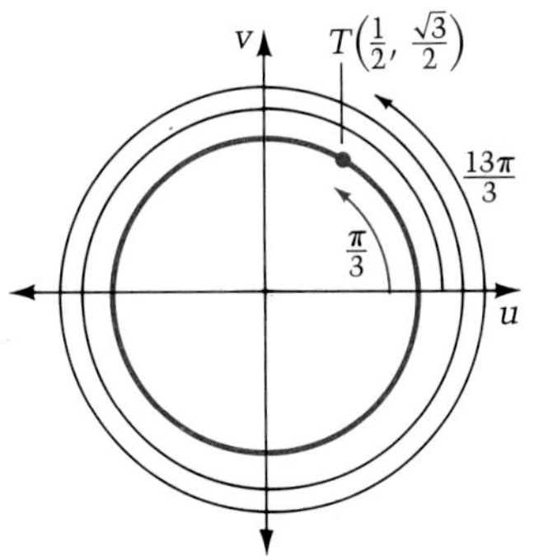


Figure 3.12

To graph the sine function, we shall first study its behavior for $x$-values from 0 to $2\pi$, and thus produce one complete cycle of the graph. Then, since the function has period $2\pi$, the pattern observed in the graph for values between 0 and $2\pi$ will be repeated over and over.

Recall that the sine function is defined by means of the second coordinate of a standard arc's terminal point $T$. Thus, values on the horizontal axis of the graph of the sine function correspond to standard arcs, and values on the vertical axis correspond to the second coordinates of the arcs' terminal points.

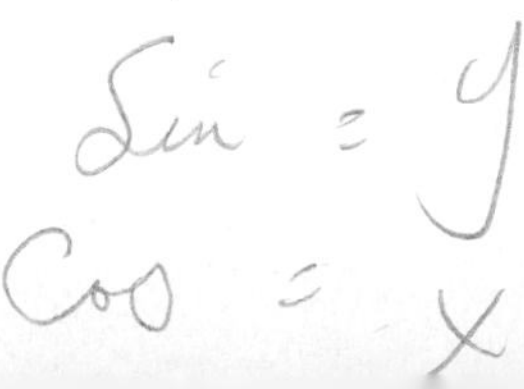

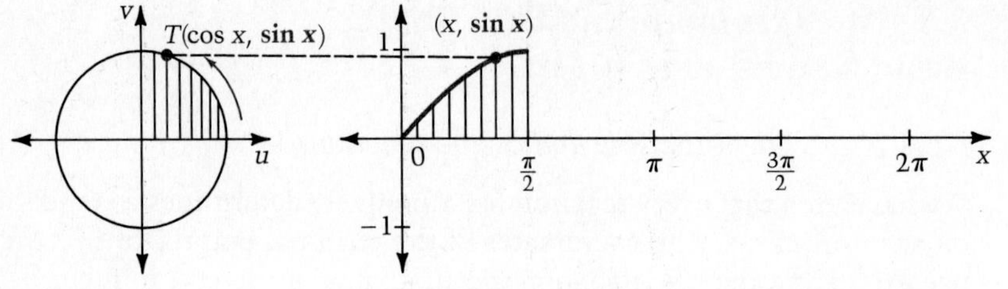


- As $x$ increases from 0 to $\frac{\pi}{2}$, second coordinates ("heights") of terminal points increase from 0 to 1.

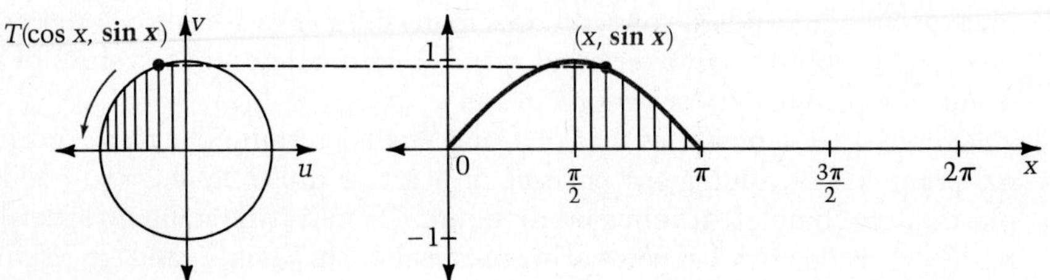


- As $x$ increases from $\frac{\pi}{2}$ to $\pi$, second coordinates decrease from 1 to 0.

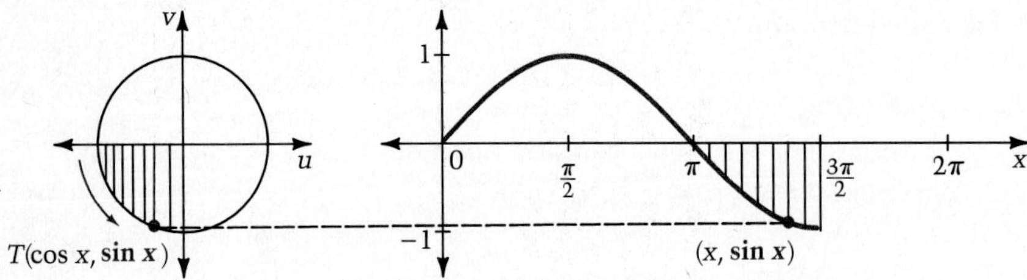


- As $x$ increases from $\pi$ to $\frac{3\pi}{2}$, second coordinates decrease from 0 to $-1$.

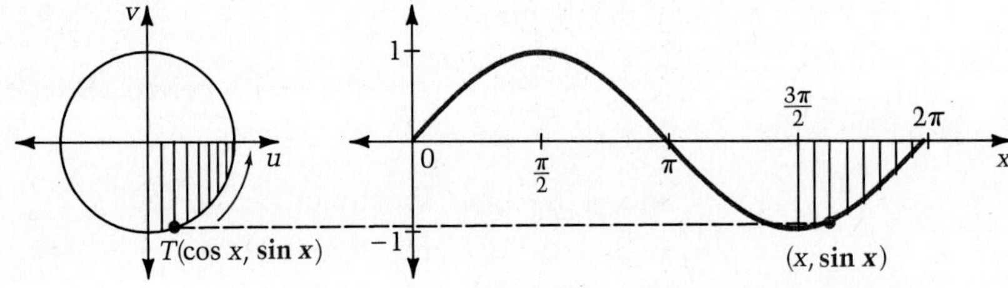


- As $x$ increases from $\frac{3\pi}{2}$ to $2\pi$, second coordinates increase from $-1$ to 0.

**Figure 3.13** Constructing one cycle of the graph of $y = \sin x$.

To get a more precise graph of the sine function, we construct a table of special values of $x$, plot the corresponding ordered pairs, and draw a smooth curve through the points. Since the sine function is periodic with period $2\pi$, the portion of the graph between 0 and $2\pi$ will repeat every $2\pi$ units. The complete graph is shown below. You should be able to sketch this graph quickly from memory.

| $x$ | $0$ | $\frac{\pi}{6}$ | $\frac{\pi}{4}$ | $\frac{\pi}{3}$ | $\frac{\pi}{2}$ | $\frac{3\pi}{4}$ | $\pi$ | $\frac{5\pi}{4}$ | $\frac{3\pi}{2}$ | $\frac{7\pi}{4}$ | $2\pi$ |
|---|---|---|---|---|---|---|---|---|---|---|---|
| $\sin x$ | $0$ | $\frac{1}{2}$ | $\frac{\sqrt{2}}{2}$ | $\frac{\sqrt{3}}{2}$ | $1$ | $\frac{\sqrt{2}}{2}$ | $0$ | $-\frac{\sqrt{2}}{2}$ | $-1$ | $-\frac{\sqrt{2}}{2}$ | $0$ |

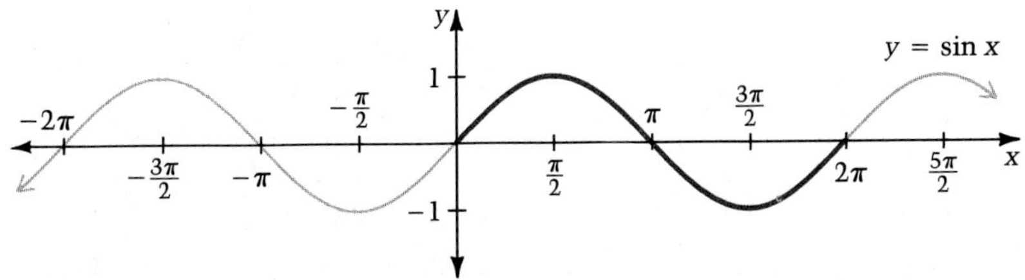


**Figure 3.14**

**Properties of the Sine Function**

**Domain** the set of all real numbers

**Range** the set of real numbers between $-1$ and $1$ inclusive

**Period** $2\pi$

**Symmetry** with respect to the origin; thus sine is an "odd" function, that is $\sin(-x) = -\sin x$.

We now turn our attention to the cosine function. Recall that the cosine is defined by means of the first coordinate of an arc's terminal point. Thus, values on the horizontal axis of the graph of the cosine function correspond to standard arcs, and values on the vertical axis correspond to the first coordinates of the arcs' terminal points. For example, to graph the cosine function for values of $x$ between 0 and $\frac{\pi}{2}$, we notice that as standard arc $x$ increases from 0 to $\frac{\pi}{2}$, first coordinates of the terminal points *decrease* from 1 to 0.

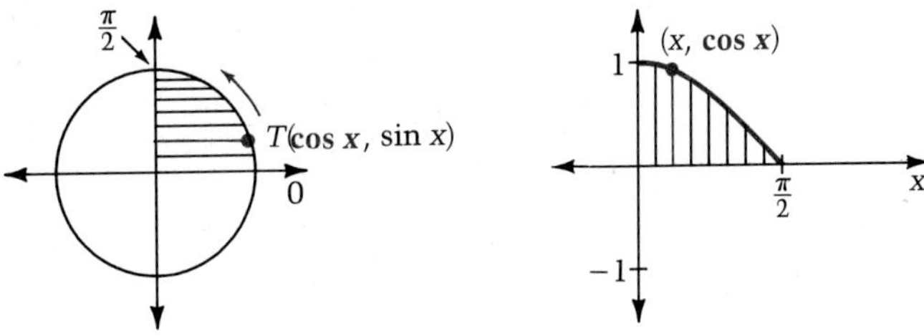


**Figure 3.15** Constructing the graph of $y = \cos x$ for $0 \le x \le \frac{\pi}{2}$.

To get a more precise graph of the cosine function, we construct a table of special values of $x$, plot the corresponding ordered pairs, and draw a smooth curve through the points. We use the fact that the cosine function is periodic with period $2\pi$ to complete the graph. Note that the graph of the cosine function is symmetric with respect to the $y$-axis.

| $x$ | $0$ | $\frac{\pi}{6}$ | $\frac{\pi}{4}$ | $\frac{\pi}{3}$ | $\frac{\pi}{2}$ | $\frac{3\pi}{4}$ | $\pi$ | $\frac{5\pi}{4}$ | $\frac{3\pi}{2}$ | $\frac{7\pi}{4}$ | $2\pi$ |
|---|---|---|---|---|---|---|---|---|---|---|---|
| $\cos x$ | $1$ | $\frac{\sqrt{3}}{2}$ | $\frac{\sqrt{2}}{2}$ | $\frac{1}{2}$ | $0$ | $-\frac{\sqrt{2}}{2}$ | $-1$ | $-\frac{\sqrt{2}}{2}$ | $0$ | $\frac{\sqrt{2}}{2}$ | $1$ |

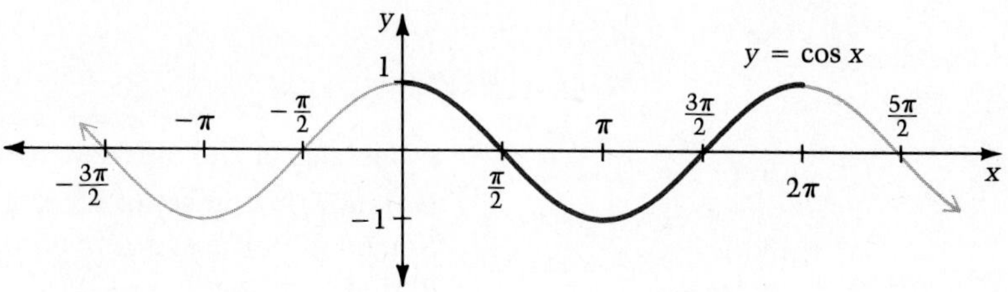


**Figure 3.16**

**Properties of the Cosine Function**

**Domain** the set of all real numbers

**Range** the set of real numbers between $-1$ and $1$ inclusive

**Period** $2\pi$

**Symmetry** with respect to the $y$-axis; thus the cosine is an "even" function, that is, $\cos(-x) = \cos x$.

**Example 1** (a) Graph $y = \sin x$ and $y = \sin(-x)$ on the same set of axes.

(b) Graph $y = \cos x$ and $y = \cos\left(x + \frac{\pi}{2}\right)$ on the same set of axes.

***Solution*** (a) To obtain the graph of $y = \sin(-x)$, reflect the graph of $y = \sin x$ across the $y$-axis.

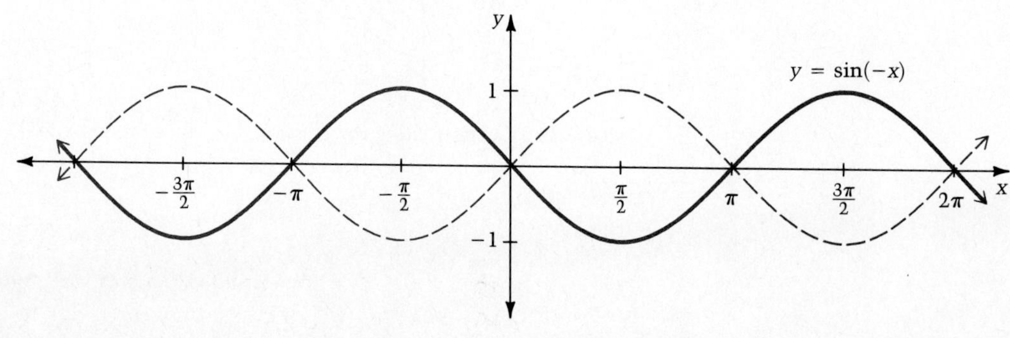

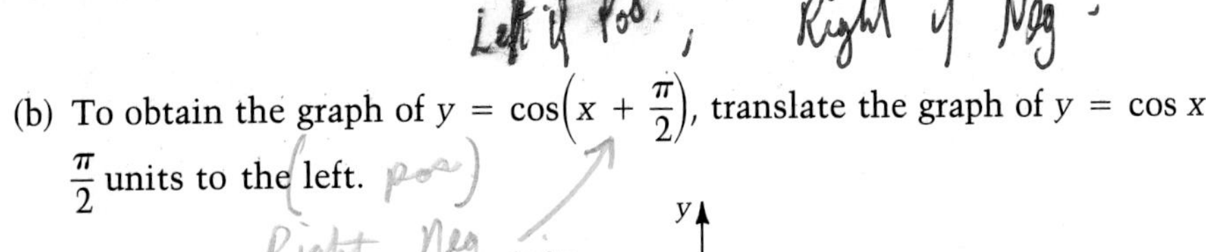


(b) To obtain the graph of $y = \cos\left(x + \frac{\pi}{2}\right)$, translate the graph of $y = \cos x$ $\frac{\pi}{2}$ units to the left.

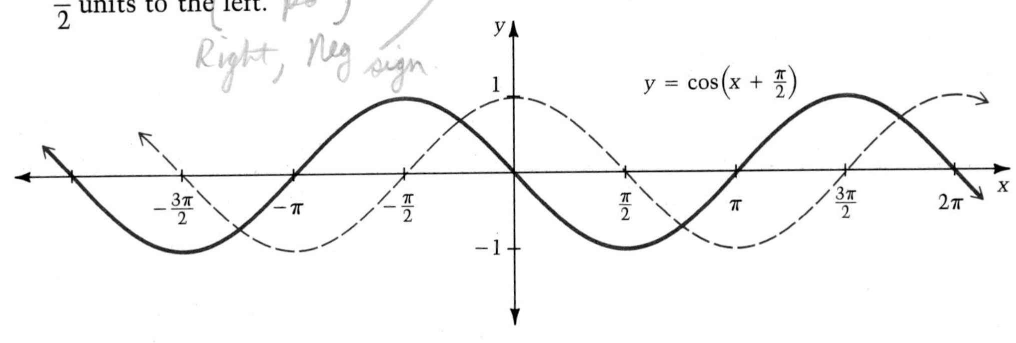


**Goal B** *to graph the other trigonometric functions*

The graphs of the tangent and cotangent functions look very different from those of the sine and cosine functions. To get an idea of what the graph of $y = \tan x$ looks like, recall that

$$\tan x = \frac{\sin x}{\cos x},$$

and thus, the tangent function will not be defined when the denominator, $\cos x$, is 0. Since $\cos x$ is 0 for all $x$ in the set

$$\left\{\cdots -\frac{5\pi}{2}, -\frac{3\pi}{2}, -\frac{\pi}{2}, \frac{\pi}{2}, \frac{3\pi}{2}, \frac{5\pi}{2}, \cdots\right\},$$

the tangent function is not defined for any of these values. We begin the graph of the tangent function by drawing the vertical asymptotes at each of these values.

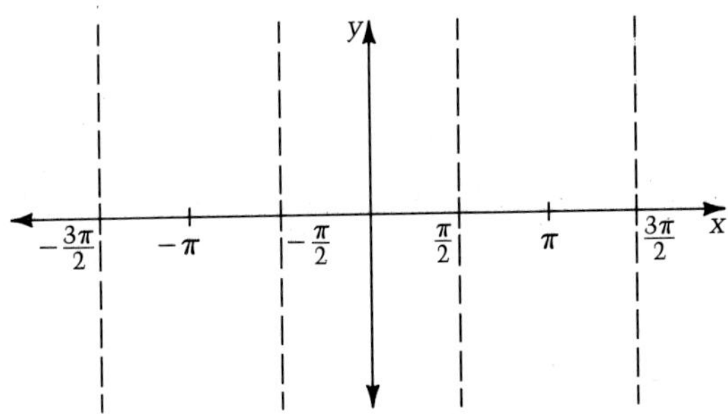


**Figure 3.17**

Notice that the tangent function is defined for all numbers between $-\frac{\pi}{2}$ and $\frac{\pi}{2}$. Let us consider what the graph will look like on that interval.

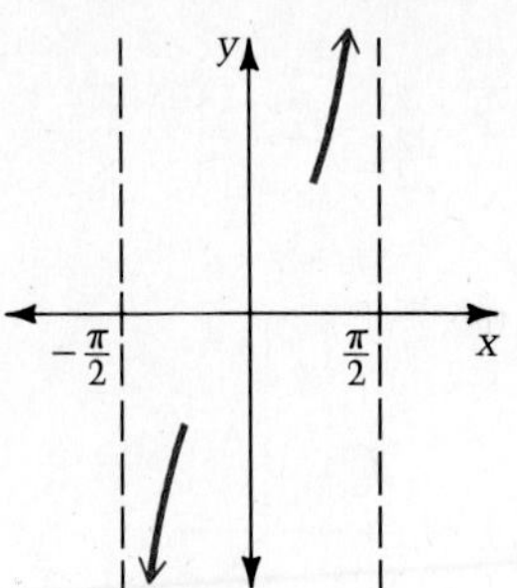


Figure 3.18

- As $x$ increases to $\frac{\pi}{2}$, $\sin x$ is nearly 1, and $\cos x$ is nearly 0, so that $\tan x$ $\left(\text{which equals } \frac{\sin x}{\cos x}\right)$ becomes a very large positive number.
- A similar argument establishes that as $x$ decreases towards $-\frac{\pi}{2}$, $\sin x$ is nearly $-1$, $\cos x$ is nearly 0, and so $\tan x$ takes on negative numbers with large absolute values.

To get a more complete graph of the tangent function for values between $-\frac{\pi}{2}$ and $\frac{\pi}{2}$, we construct a table of special values of $x$, plot the corresponding ordered pairs, and draw a smooth curve through the points. This curve is then repeated within each pair of adjacent asymptotes.

| $x$ | $-\frac{\pi}{2}$ | $-\frac{\pi}{3}$ | $-\frac{\pi}{4}$ | $-\frac{\pi}{6}$ | 0 | $\frac{\pi}{6}$ | $\frac{\pi}{4}$ | $\frac{\pi}{3}$ | $\frac{\pi}{2}$ |
|---|---|---|---|---|---|---|---|---|---|
| $\tan x$ | ⋆ | $-\sqrt{3}$ | $-1$ | $-\frac{\sqrt{3}}{3}$ | 0 | $\frac{\sqrt{3}}{3}$ | 1 | $\sqrt{3}$ | ⋆ |

⋆undefined

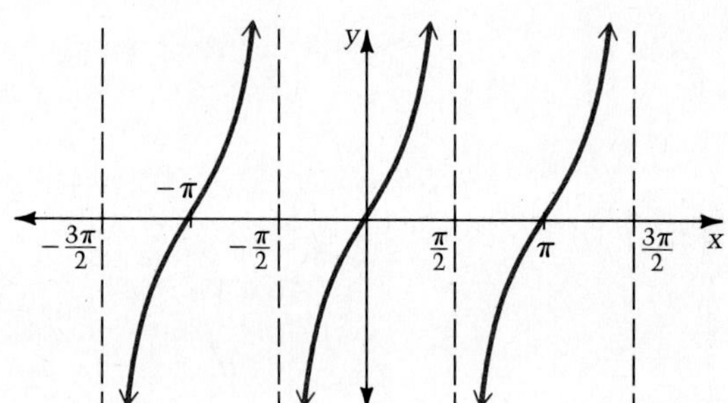


Figure 3.19 $y = \tan x$

**Properties of the Tangent Function**

**Domain** all real numbers *except* $x = \frac{\pi}{2} + k\pi$, for any integer $k$

**Range** all real numbers

**Period** $\pi$

**Symmetry** with respect to the origin; tangent is an "odd" function, thus $\tan(-x) = -\tan x$.

**Asymptotes** vertical asymptotes $x = \frac{\pi}{2} + k\pi$, for any integer $k$.

The graph of $y = \tan x$ is periodic, but unlike the sine and cosine functions, the period is $\pi$. Thus, the portion of the graph between $-\frac{\pi}{2}$ and $\frac{\pi}{2}$ repeats itself over and over as suggested in Figure 3.19.

The graph of the cotangent function can be found by using a method similar to that used to derive the graph of $y = \tan x$. Recall that

$$\cot x = \frac{1}{\tan x}.$$

This fact suggests that the graph of $y = \cot x$ becomes infinite (has a vertical asymptote) where $\tan x$ is 0, and is zero where the graph of $y = \tan x$ has vertical asymptotes. The complete graph is shown in Figure 3.20. Like the tangent function, the cotangent function has period $\pi$. In addition, the graph of the cotangent function, like that of the tangent function, is symmetric with respect to the origin.

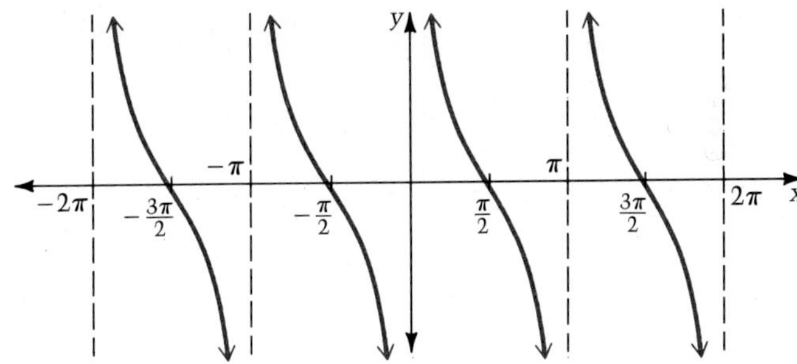


**Figure 3.20** $y = \cot x$

**Properties of the Cotangent Function**

| | |
|---|---|
| **Domain** | all real numbers *except* $x = k\pi$, for any integer $k$ |
| **Range** | all real numbers |
| **Period** | $\pi$ |
| **Symmetry** | with respect to the origin; cotangent is an "odd" function, thus $\cot(-x) = -\cot x$. |
| **Asymptotes** | vertical asymptotes $x = k\pi$, for any integer $k$. |

**Example 2** Verify that the graphs of the tangent and cotangent functions are symmetric with respect to the origin.

***Solution*** Recall that to verify that the graph of a function $y = f(x)$ is symmetric with respect to the origin, we must show that when $y$ is replaced with $-y$ and $x$ is replaced with $-x$, the resulting equation is equivalent to $y = f(x)$.

$$y = \tan x = \frac{\sin x}{\cos x}$$

$$-y = \frac{\sin(-x)}{\cos(-x)}$$ ■ $x$ is replaced with $-x$ and $y$ is replaced with $-y$.

$$-y = \frac{-\sin x}{\cos x}$$ ■ Recall that $\sin(-x) = -\sin x$ and $\cos(-x) = \cos x$.

$$y = \frac{\sin x}{\cos x}$$ ■ Both sides have been multiplied by $-1$.

$$y = \tan x$$

Thus, the graph of $y = \tan x$ is symmetric with respect to the origin. Symmetry with respect to the origin for the graph of $y = \cot x$ is established similarly.

The graph of the secant function can be easily produced by recalling that

$$\sec x = \frac{1}{\cos x}, \quad \text{for } \cos x \neq 0.$$

Since the secant function is not defined for values of $x$ where $\cos x$ is 0, its domain does not include any value in the set

$$\left\{\ldots, -\frac{5\pi}{2}, -\frac{3\pi}{2}, -\frac{\pi}{2}, \frac{\pi}{2}, \frac{3\pi}{2}, \frac{5\pi}{2}, \ldots\right\},$$

and its graph has vertical asymptotes at these values. Since the secant function can be thought of as the reciprocal of the cosine function, it is helpful to graph the cosine function as a dashed curve, and then generate the graph of the secant function by viewing the $y$-values of the secant function as reciprocals of the $y$-values of the cosine function.

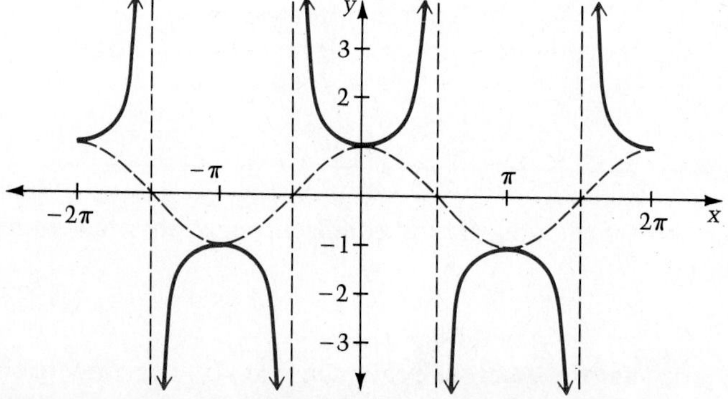


■ Notice that wherever the cosine function is 0, the secant function is undefined, and wherever the cosine function is 1, so is the secant function. Furthermore, whenever the cosine function is positive (negative), the secant function is positive (negative).

**Figure 3.21** $y = \sec x$ ($y = \cos x$ is shown as a dashed curve).

**Properties of the Secant Function**

**Domain** all real numbers *except* $x = \frac{\pi}{2} + k\pi$, for any integer $k$
**Range** all real numbers $y$ such that $y \leq -1$ or $y \geq 1$
**Period** $2\pi$
**Symmetry** symmetry with respect to the $y$-axis (an even function, like its reciprocal, cosine)
**Asymptotes** vertical asymptotes $x = \frac{\pi}{2} + k\pi$ (values where cosine is 0)

Note that the secant function has period $2\pi$, the same as that of its reciprocal, the cosine function. Moreoever, like the cosine function, the graph of the secant function is symmetric with respect to the $y$-axis.

The graph of the cosecant function is given in Figure 3.22. The cosecant function is the reciprocal of the sine function:

$$y = \frac{1}{\sin x}, \text{ for } \sin x \neq 0.$$

(The graph of the sine function is shown as a dashed curve.) Like the sine function, the cosecant function has period $2\pi$, and is symmetric with respect to the origin.

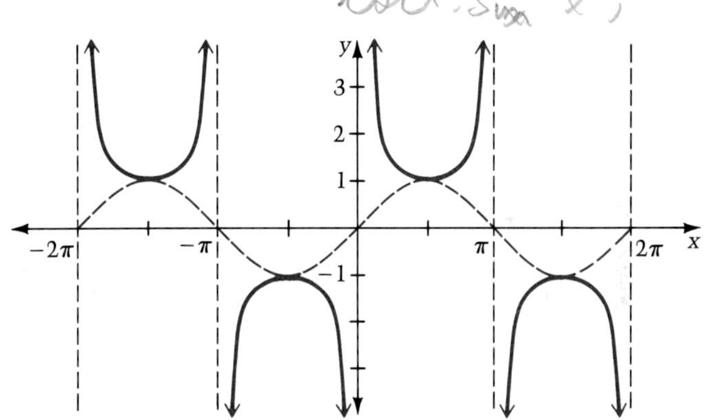


**Figure 3.22** $y = \csc x$ ($y = \sin x$ is shown as a dashed curve).

**Properties of the Cosecant Function**

**Domain** all real numbers *except* $x = k\pi$, for any integer $k$
**Range** all real numbers $y$ such that $y \leq -1$ or $y \geq 1$
**Period** $2\pi$
**Symmetry** symmetry with respect to the origin (an odd function, like its reciprocal, sine)
**Asymptotes** vertical asymptotes $x = k\pi$ (values where sine is 0)

**Example 3** Use the graphs of $y = \csc\left(x - \frac{\pi}{2}\right)$ and $y = -\sec x$ to determine whether $\csc\left(x - \frac{\pi}{2}\right) = -\sec x$ is an identity.

***Solution***

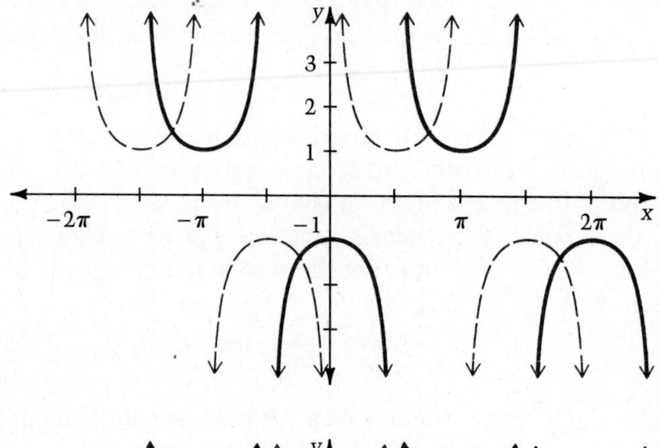


■ $y = \csc\left(x - \frac{\pi}{2}\right)$. The graph of $y = \csc x$ (dashed curve) is translated $\frac{\pi}{2}$ units to the right.

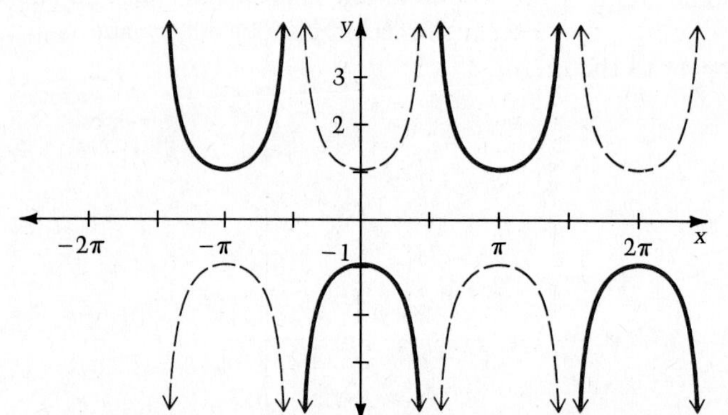


■ $y = -\sec x$. The graph of $y = \sec x$ (dashed curve) is reflected across the $x$-axis.

■ Since the graphs of $y = \csc\left(x - \frac{\pi}{2}\right)$ and $y = -\sec x$ are identical, the equation $\csc\left(x - \frac{\pi}{2}\right) = -\sec x$ is an identity.

**Example 4** Answer True or False.

(a) $\frac{7\pi}{2}$ is not in the domain of the tangent function.

(b) $\frac{3}{2}$ is in the range of the cosine function.

(c) The sine, cosine, and tangent functions all have the same period.

***Solution***

(a) True ■ $\tan x = \frac{\sin x}{\cos x}$, and $\cos \frac{7\pi}{2} = 0$.

(b) False ■ $-1 \le \cos x \le 1$ for all real $x$.

(c) False ■ Tangent has period $\pi$; sine and cosine have period $2\pi$.

## Exercise Set 3.3 ☰ Calculator Exercises in the Appendix

### Goal A

In exercises 1–12, graph each pair of functions on the same axes.

1. $y = \sin x, y = \sin(x + \pi)$
2. $y = \cos x, y = \cos(x - \pi)$
3. $y = \sin x, y = \sin\left(\frac{\pi}{2} + x\right)$
4. $y = \sin x, y = \sin\left(x - \frac{\pi}{2}\right)$
5. $y = \cos x, y = \cos(-x)$
6. $y = \sin x, y = -\sin x$
7. $y = \sin x, y = \sin\left(\frac{\pi}{4} - x\right)$
8. $y = \cos x, y = \cos\left(-x - \frac{\pi}{4}\right)$
9. $y = \cos(-x), y = -\cos(-x - \pi)$
10. $y = -\sin x, y = -\sin(-x)$
11. $y = \sin x, y = \sin(x - 2\pi)$
12. $y = \cos x, y = \cos(x + 4\pi)$

### Goal B

In exercises 13–22, graph each pair of functions on the same set of axes.

13. $y = \tan x, y = \tan\left(x - \frac{\pi}{2}\right)$
14. $y = \sec x, y = \sec\left(x + \frac{\pi}{3}\right)$
15. $y = \cot x, y = \cot\left(x - \frac{\pi}{6}\right)$
16. $y = \tan x, y = \tan(-x)$
17. $y = \csc x, y = -\csc x$
18. $y = \cot x, y = -\cot x$
19. $y = \sec x, y = \sec\left(\frac{\pi}{4} - x\right)$
20. $y = \tan x, y = \tan\left(\frac{\pi}{2} - x\right)$
21. $y = \csc(-x), y = -\csc(-x)$
22. $y = \csc(-x), y = \csc\left(-x - \frac{\pi}{2}\right)$

In exercises 23–30, determine whether the statement is True or False.

23. $-\frac{\pi}{2}$ is not in the domain of $y = \cot x$.
24. $7\pi$ is in the domain of $y = \csc x$.
25. $2\pi$ is in the range of $y = \cot x$.
26. $x = 2\pi$ is an asymptote of $y = \cot x$.
27. Secant and sine have the same period.
28. All trigonometric functions except for tangent and cotangent have period $2\pi$.
29. $\csc(-x) = \sin x$ is an identity.
30. $\cot(-x) = -\tan x$ is an identity.

### Superset

In exercises 31–38, sketch the graph of each function.

31. $y = |\sin x|$
32. $y = |\cot x|$
33. $y = \tan|x|$
34. $y = \sec|x|$
35. $y = \csc|x + \pi|$
36. $y = \sin|x - \pi|$
37. $y = \cos\left|\frac{\pi}{4} - x\right|$
38. $y = \csc\left|\frac{\pi}{3} + x\right|$

In exercises 39–42, sketch the graphs of each pair of equations on the same axes.

39. $y = \sin x, x = \sin y$
40. $y = \cos x, x = \cos y$
41. $y = \tan x, x = \tan y, -\frac{\pi}{2} < x < \frac{\pi}{2}$
42. $y = \cot x, x = \cot y, 0 < x < \pi$

## 3.4 Transformations of the Trigonometric Functions

**Goal A** *to graph functions of the form* $y = a \sin bx$ *and* $y = a \cos bx$

In this section we will rely heavily on our knowledge of transformations to develop an efficient way of graphing periodic functions. Essentially we will be concerned with translations, stretchings, and shrinkings of the basic trigonometric graphs.

For example, consider the function $y = 3 \sin x$. For this function, each $y$-coordinate is three times the corresponding $y$-coordinate of the function $y = \sin x$.

| $x$ | 0 | $\frac{\pi}{6}$ | $\frac{\pi}{4}$ | $\frac{\pi}{3}$ | $\frac{\pi}{2}$ | $\frac{2\pi}{3}$ | $\frac{3\pi}{4}$ | $\pi$ | $\frac{3\pi}{2}$ | $2\pi$ |
|---|---|---|---|---|---|---|---|---|---|---|
| $\sin x$ | 0 | $\frac{1}{2}$ | $\frac{\sqrt{2}}{2}$ | $\frac{\sqrt{3}}{2}$ | 1 | $\frac{\sqrt{3}}{2}$ | $\frac{\sqrt{2}}{2}$ | 0 | $-1$ | 0 |
| $3 \sin x$ | 0 | $\frac{3}{2}$ | $\frac{3\sqrt{2}}{2}$ | $\frac{3\sqrt{3}}{2}$ | 3 | $\frac{3\sqrt{3}}{2}$ | $\frac{3\sqrt{2}}{2}$ | 0 | $-3$ | 0 |

In Figure 3.23, the graph of $y = 3 \sin x$ is shown as the result of stretching the graph of $y = \sin x$ vertically (away from the $x$-axis). Recall that, in general, the graph of $y = k \cdot f(x)$ is found by vertically stretching or shrinking the graph of $y = f(x)$. If $k < 0$, the stretching or shrinking is accompanied by a reflection across the $x$-axis.

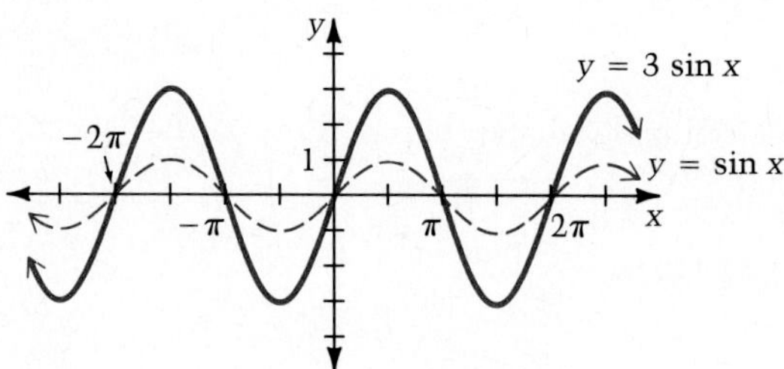


**Figure 3.23**

Notice that the range of the function $y = \sin x$ is the set of real numbers from $-1$ to 1 inclusive, whereas the range of $y = 3 \sin x$ is the set of real numbers from $-3$ to 3 inclusive. We call 3 the *amplitude* of $y = 3 \sin x$.

**Definition**

For functions of the form $y = a \sin x$ and $y = a \cos x$, the number $|a|$ is called the **amplitude.** The amplitude of a periodic function is one-half the difference between the maximum and minimum values of the function.

height is amplitude

y or V

**Example 1** Sketch the graph of

(a) $y = \frac{1}{2} \cos x$ (b) $y = -2 \sin x$

***Solution*** (a)

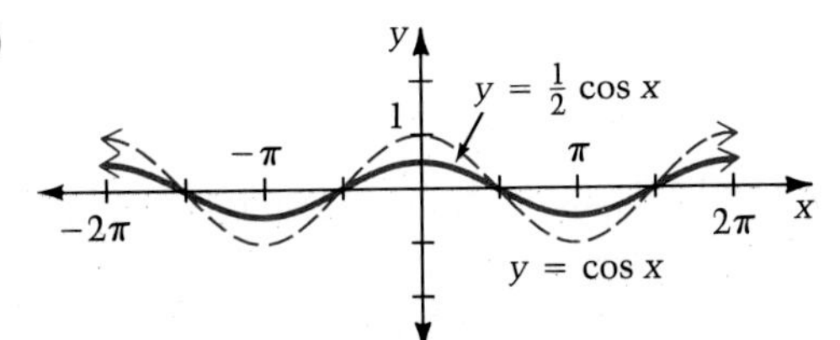


(b)

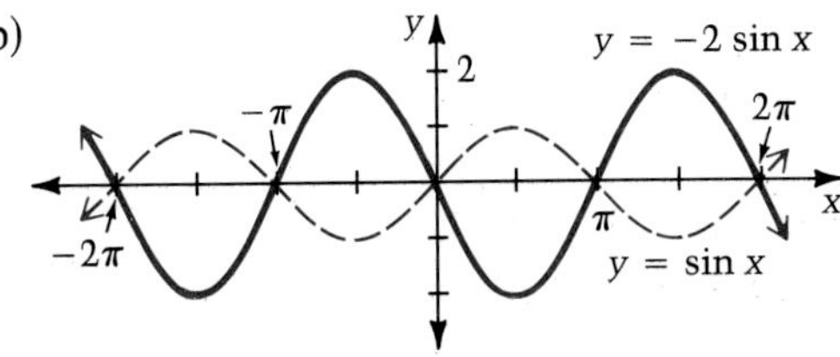


In (a), each $y$-coordinate for the graph of $y = \frac{1}{2} \cos x$ is $\frac{1}{2}$ the corresponding $y$-coordinate for the graph of $y = \cos x$. In (b), the graph of $y = -2 \sin x$ involves both a vertical stretching and a reflection across the $x$-axis.

Next we consider what happens to the graph of $y = \sin x$ when $x$ is multiplied by 3. That is, we wish to determine how the graph of $y = \sin x$ is transformed to produce the graph of $y = \sin 3x$. Consider the following table of values.

| $x$ | $0$ | $\frac{\pi}{6}$ | $\frac{\pi}{4}$ | $\frac{\pi}{3}$ | $\frac{\pi}{2}$ | $\frac{2\pi}{3}$ | $\frac{3\pi}{4}$ | $\pi$ | $\frac{3\pi}{2}$ | $2\pi$ |
|---|---|---|---|---|---|---|---|---|---|---|
| $3x$ | $0$ | $\frac{\pi}{2}$ | $\frac{3\pi}{4}$ | $\pi$ | $\frac{3\pi}{2}$ | $2\pi$ | $\frac{9\pi}{4}$ | $3\pi$ | $\frac{9\pi}{2}$ | $6\pi$ |
| $\sin 3x$ | $0$ | $1$ | $\frac{\sqrt{2}}{2}$ | $0$ | $-1$ | $0$ | $\frac{\sqrt{2}}{2}$ | $0$ | $1$ | $0$ |

Notice that as $x$ takes on values from 0 to $2\pi$, $3x$ takes on values from 0 to $6\pi$. As a result, the graph of $y = \sin 3x$ completes *three* full cycles as $x$ goes from 0 to $2\pi$, with one full cycle as $x$ goes from 0 to $\frac{2\pi}{3}$. Recall that the period is the length of the interval over which a periodic function makes one complete cycle. We therefore conclude that the period of $y = \sin 3x$ is $\frac{2\pi}{3}$, or one-third the period of $y = \sin x$. Graphs of

$$y = \sin 3x \quad \text{and} \quad y = \sin x$$

are shown below.

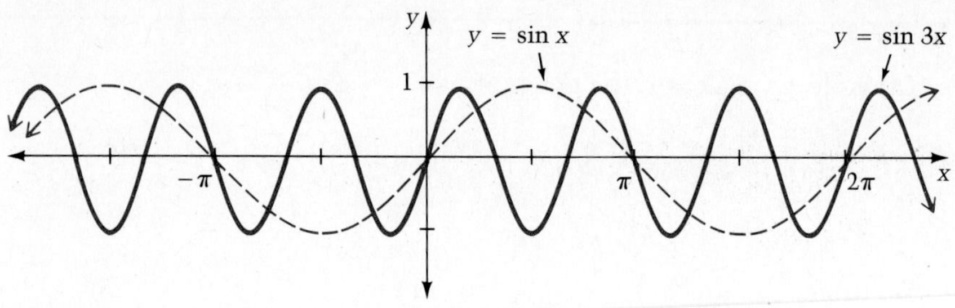


**Figure 3.24**

The graphs above suggest the following generalization.

**Fact**

Functions of the form $y = \sin bx$ and $y = \cos bx$ have period equal to $\left|\frac{2\pi}{b}\right|$.

**Example 2** Sketch the graph of $y = \cos 2x$.

***Solution***

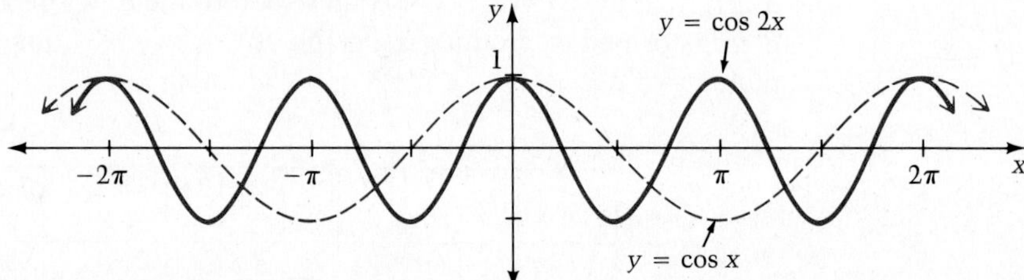


The period of $y = \cos 2x$ is $\left|\frac{2\pi}{2}\right| = \pi$, thus the given function completes one full cycle every $\pi$ units.

Let us summarize our results for functions of the following two forms:

$$y = a \sin bx \quad \text{or} \quad y = a \cos bx.$$

- The amplitude $|a|$ indicates a vertical stretching of the basic sine or cosine curve if $|a| > 1$, and a vertical shrinking if $0 < |a| < 1$.
- The value $|b|$ indicates a horizontal stretching of the basic sine or cosine curve if $0 < |b| < 1$, and a horizontal shrinking if $|b| > 1$.

**Example 3** Sketch the graph of (a) $y = -2 \sin \frac{1}{3}x$ (b) $y = 3 \sin(-4x)$.

***Solution*** (a) We begin by sketching the graph of $y = \sin \frac{1}{3}x$ as a dashed curve. It is then stretched vertically and reflected across the $x$-axis to produce the graph of $y = -2 \sin \frac{1}{3}x$.

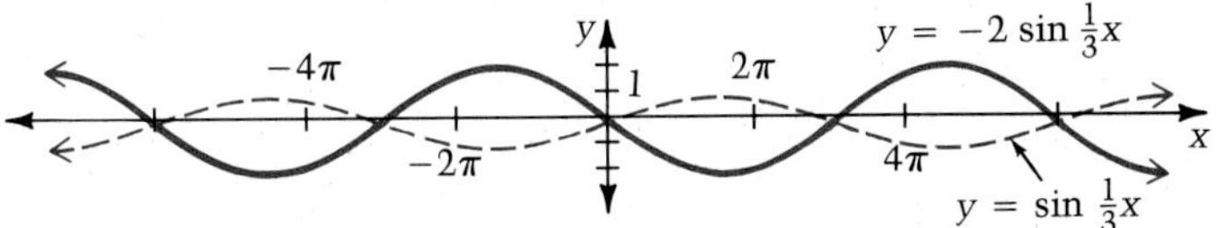


■ The period is $\left|\frac{2\pi}{\frac{1}{3}}\right| = 6\pi$.

(b) Recall that the graph of $y = f(-x)$ is the result of reflecting $y = f(x)$ across the $y$-axis. We begin by sketching the graph of $y = 3 \sin(4x)$ as a dashed curve, then reflect it across the $y$-axis to produce $y = 3 \sin(-4x)$.

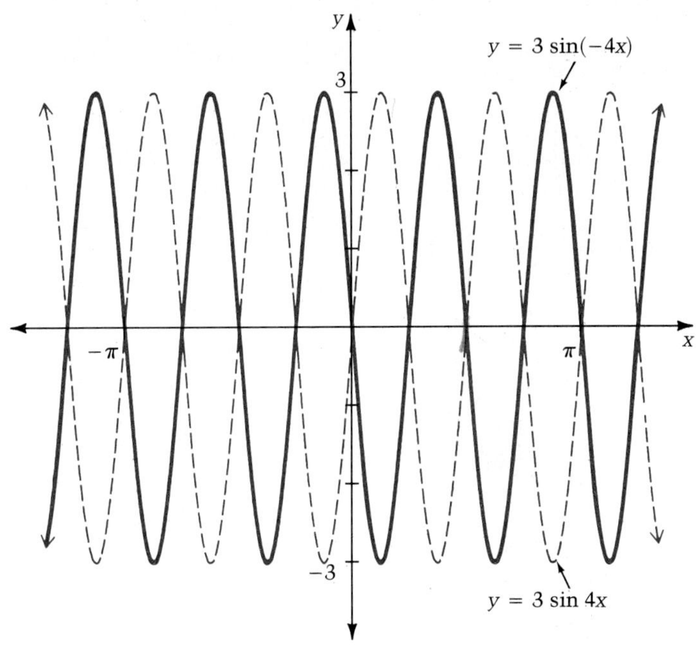


■ The period is $\left|\frac{2\pi}{-4}\right| = \frac{\pi}{2}$.

**Goal B** *to graph functions of the form* $y = k + a \sin b(x - h)$ *and* $y = k + a \cos b(x - h)$

Consider the function $y = 1 + 5 \sin 6(x - \frac{\pi}{4})$. If this were simply the function $y = 5 \sin 6x$, then we could use results already discussed in this section to conclude that the amplitude is equal to 5 and the period is equal to $|\frac{2\pi}{6}| = \frac{\pi}{3}$. As it turns out, the function $y = 1 + 5 \sin 6(x - \frac{\pi}{4})$ has the same period and amplitude as $y = 5 \sin 6x$.

**Fact**

The function $y = k + a \sin b(x - h)$ has the same amplitude $|a|$ and period $\left|\frac{2\pi}{b}\right|$ as the function $y = a \sin bx$.

The question then is how the graph of $y = 1 + 5 \sin 6(x - \frac{\pi}{4})$ differs from that of $y = 5 \sin 6x$. Recall that the graph of the function $y = k + f(x - h)$ can be produced by translating the graph of $y = f(x)$ horizontally $h$ units and vertically $k$ units.

**Fact**

The graph of $y = k + a \sin b(x - h)$ is produced by translating the graph of $y = a \sin bx$:

$$|h| \text{ units } \begin{cases} \text{to the right,} & \text{if } h > 0, \\ \text{to the left,} & \text{if } h < 0, \end{cases}$$

$$|k| \text{ units } \begin{cases} \text{upward,} & \text{if } k > 0, \\ \text{downward,} & \text{if } k < 0. \end{cases}$$

(Similar statements hold for the cosine function.)

The number $h$ that determines the extent of horizontal translation is called the **phase shift.** When determining the phase shift, be careful to express the periodic function in precisely the form stated above. Example 4 shows how the facts stated above may be applied to the cosine function.

**Example 4** For the function $y = 3 \cos\left(\frac{1}{2}x - \frac{\pi}{8}\right) - 1$, (a) determine the period, amplitude, and phase shift, and (b) sketch the graph.

***Solution*** (a) $y = -1 + 3 \cos \frac{1}{2}\left(x - \frac{\pi}{4}\right)$ ■ Begin by restating as $y = k + a \cos b(x - h)$.

The period is $\left|\frac{2\pi}{b}\right| = \frac{2\pi}{\frac{1}{2}} = 4\pi$, the amplitude is $|3| = 3$, and the phase shift is $\frac{\pi}{4}$.

(b) We begin by sketching the graph of $y = 3 \cos\left(\frac{1}{2}x\right)$ as a dashed curve, then translate this graph $\frac{\pi}{4}$ units to the right and one unit downward to produce the graph of

$y = 3 \cos \frac{1}{2}\left(x - \frac{\pi}{4}\right) - 1.$

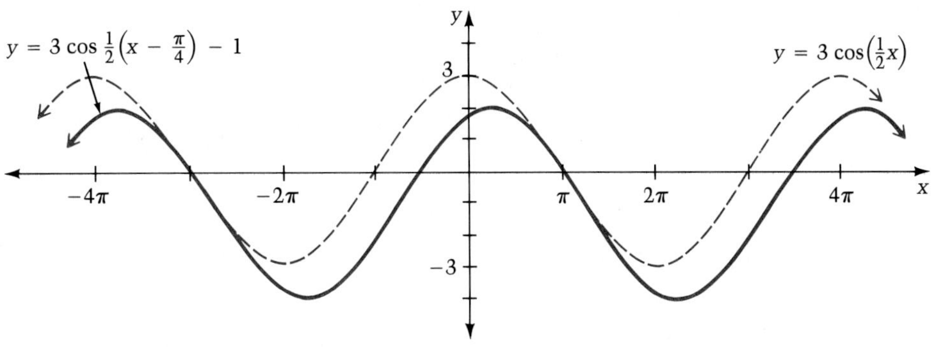


**Goal C** *to graph by the addition of ordinates*

Suppose we wish to sketch the graph of the function $f(x) = \frac{1}{2}x + \sin x$. Taken individually, the two component functions $y = \frac{1}{2}x$ and $y = \sin x$ are familiar to us (a straight line and the sine curve, respectively). One method for graphing the function $f(x) = \frac{1}{2}x + \sin x$ is to shift each point on the graph of $y = \sin x$ by an amount equal to $\frac{1}{2}x$. Example 5 uses this method, which is called graphing by **addition of ordinates.**

**Example 5** Sketch the graph of $y = \dfrac{1}{2}x + \sin x$.

***Solution***

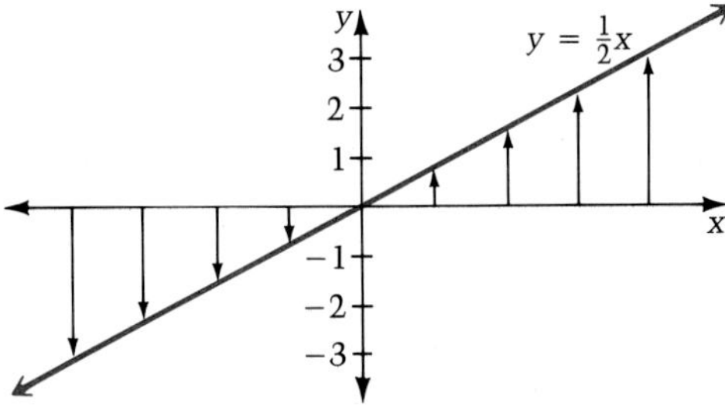


■ Sketch the graph of $y = \dfrac{1}{2}x$ and use arrows to represent $y$-coordinates for various values of $x$.

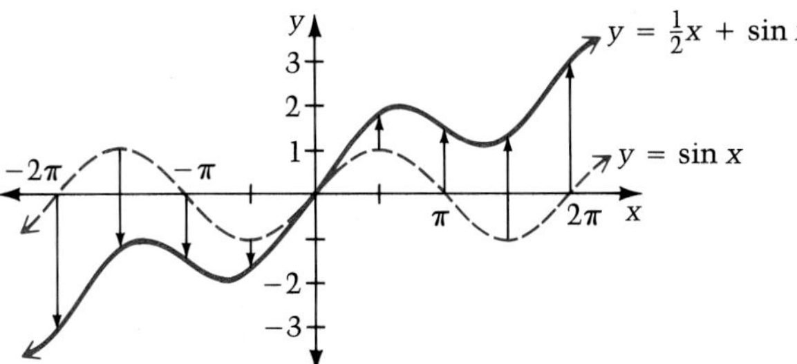


■ Sketch the graph of $y = \sin x$ as a dashed curve and use the arrows from the figure above to indicate the result of adding $\dfrac{1}{2}x$. Draw a smooth curve through the arrowheads.

## Exercise Set 3.4 ⊟ Calculator Exercises in the Appendix

### Goal A

In exercises 1–16, sketch the graph of the given function.

1. $y = \frac{1}{3}\cos x$
2. $y = 4 \sin x$
3. $y = -\frac{1}{2}\sin x$
4. $y = -\frac{3}{2}\cos x$
5. $y = -3 \sin x$
6. $y = \sqrt{2} \sin x$
7. $y = \cos \frac{1}{2}x$
8. $y = \sin(-\frac{1}{3}x)$
9. $y = \cos(-3x)$
10. $y = \sin \frac{3}{2}x$
11. $y = 2 \sin \frac{1}{2}x$
12. $y = \frac{1}{2}\cos 2x$
13. $y = -3 \sin \frac{1}{3}x$
14. $y = -2 \cos 3x$
15. $y = \sqrt{2}\cos(-2x)$
16. $y = \sqrt{3}\cos(-\frac{1}{2}x)$

### Goal B

In exercises 17–30, determine the amplitude, period, and phase shift, and sketch the graph of the given function.

17. $y = 2\cos\left(x + \frac{\pi}{2}\right)$
18. $y = 3\cos\left(x - \frac{\pi}{2}\right)$
19. $y = 2\cos\frac{1}{2}\left(x - \frac{\pi}{3}\right)$
20. $y = \frac{1}{2}\sin 2(x - \pi)$
21. $y = 2\sin\left(\frac{1}{2}x + \frac{\pi}{4}\right)$
22. $y = 2\cos(2x + \pi)$
23. $y = \sqrt{3}\sin\left(-\frac{x}{2} - \frac{\pi}{2}\right)$
24. $y = 3\sin\left(-x + \frac{\pi}{2}\right)$
25. $y = -3\sin(2x - \pi)$
26. $y = -2\cos\left(\frac{x}{2} + \frac{\pi}{2}\right)$
27. $y = 3\cos\left(\frac{1}{2}x + \frac{\pi}{2}\right) - 2$
28. $y = -3\sin\left(2x + \frac{\pi}{2}\right) + 1$
29. $y = 2\sin\left(x - \frac{\pi}{2}\right) + 1$
30. $y = -2\cos\left(\frac{x}{2} + \frac{\pi}{2}\right) - 1$

### Goal C

In exercises 31–38, sketch the graph of the given function.

31. $y = \sin x - 2x$
32. $y = 2x - \sin x$
33. $y = \frac{1}{2}x + \sin 2x$
34. $y = \frac{1}{2}x - \cos 2x$
35. $y = -3\cos 2x + x$
36. $y = -2\sin\frac{1}{2}x + 3x$
37. $y = -2\sin\frac{1}{2}\left(x - \frac{\pi}{2}\right) - 2x$
38. $y = -\frac{1}{2}\cos\left(x - \frac{\pi}{2}\right) + 4x$

### Superset

39. Find $b$ such that the period of the function $y = \frac{1}{2}\sin bx$ is $\frac{\pi}{4}$.

40. Find $b$ such that the period of the function $y = 3\cos b(x - \pi)$ is $4\pi$.

In exercises 41–48, sketch the graph of the given function.

41. $y = |\sin 2x|$
42. $y = |3 \sin x|$
43. $y = |-2\cos x|$
44. $y = \cos|-2x|$
45. $y = \sin x + \cos x$
46. $y = \sin x - \cos x$
47. $y = \sqrt{3}\sin x - \cos x$
48. $y = \frac{\sqrt{3}}{2}\sin x + \frac{1}{2}\cos x$

In exercises 49–52, determine whether the statement is True or False.

49. The function $y = \sin x + \cos x$ is periodic with period $4\pi$.
50. The maximum value of $4\sin\frac{1}{2}x$ is 2.
51. The maximum value of $-2\cos\frac{1}{2}x$ is 2.
52. The minimum value of $-\frac{3}{2}\cos\frac{2}{3}x$ is $\frac{3}{2}$.

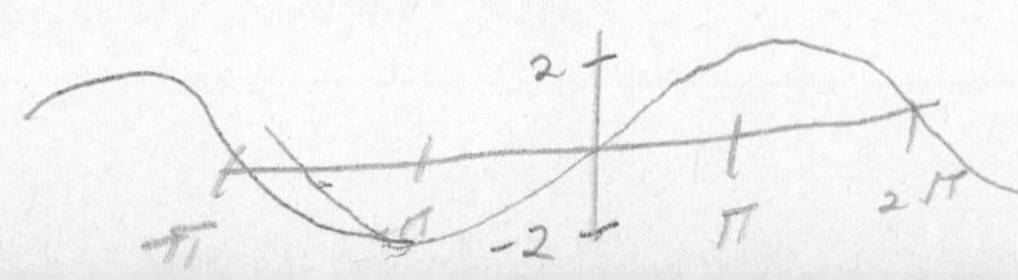

## 3.5 The Inverse Trigonometric Functions

**Goal A** *to use special angles to evaluate inverse trigonometric functions*

By the definition of a function, we are assured that to each value in the domain of a trigonometric function, there is assigned exactly one range value. For example, the sine function assigns to the domain value $\frac{\pi}{6}$ exactly one range value, namely $\frac{1}{2}$. Thus, if we know that $\sin\frac{\pi}{6} = y$, then $y$ must be $\frac{1}{2}$. However, it is not true that if we know a particular $y$-value, say $\frac{1}{2}$, the corresponding $x$-value is unique. For example, if $\sin x = \frac{1}{2}$, then $x$ can be any number in the set $\{\cdots, -\frac{11\pi}{6}, -\frac{7\pi}{6}, \frac{\pi}{6}, \frac{5\pi}{6}, \frac{13\pi}{6}, \cdots\}$.

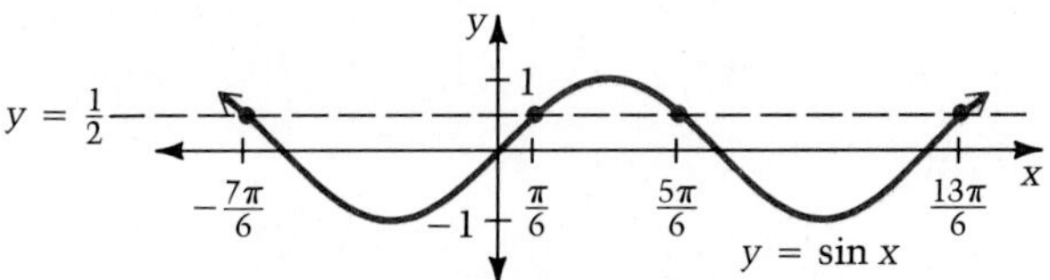


**Figure 3.25** There are an infinite number of values of $x$ such that $\sin x = \frac{1}{2}$.

In this section we would like to define an inverse sine function, that is a function which "undoes" what the sine function "does." To be a function, this "inverse sine" must take a value like $\frac{1}{2}$ and produce *exactly one number* whose sine is $\frac{1}{2}$, for example, $\frac{\pi}{6}$. (If it produced more than one, it would not be a function!) The problem is that the sine function is not one-to-one, and thus cannot have an inverse function. To resolve this problem, we look at the sine function and restrict its domain to some interval on which it *is* one-to-one. In Figure 3.26(a), we have graphed the sine function for domain values in the interval $[-\frac{\pi}{2}, \frac{\pi}{2}]$. For these values, the sine function is one-to-one and thus has an inverse function. Recall that the graph of an inverse can be found by reflecting the original graph across the line $y = x$. The graph of the inverse sine function is shown below. We describe this inverse with the equation $y = \sin^{-1} x$.

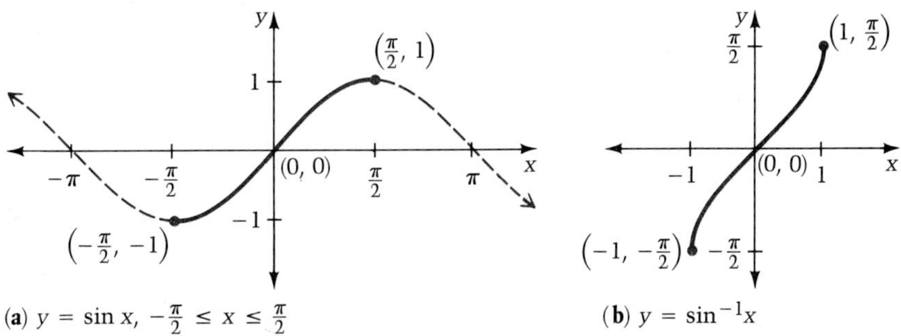


**Figure 3.26** (a) The restricted sine function. (b) The inverse sine function.

**Careful!** The symbol $\sin^{-1} x$ does not mean $\dfrac{1}{\sin x}$; as you know, $\dfrac{1}{\sin x} = \csc x$.)

**Definition**

The inverse sine function is defined as follows:

$$y = \sin^{-1} x \quad \text{with domain } -1 \leq x \leq 1$$

if and only if $\sin y = x$ and $-\dfrac{\pi}{2} \leq y \leq \dfrac{\pi}{2}$.

As a consequence of the definition of $y = \sin^{-1} x$, it is often useful to think of the expression $\sin^{-1} x$ as

the angle between $-\dfrac{\pi}{2}$ and $\dfrac{\pi}{2}$ inclusive whose sine is $x$.

The function name "arcsin" is sometimes used instead of $\sin^{-1}$ to refer to the inverse sine function. Thus, arcsin and $\sin^{-1}$ mean the same thing and can be used interchangeably.

---

**Example 1** Determine the value of each expression without using Table 4 or a calculator.

(a) $\sin^{-1}\left(\dfrac{1}{2}\right)$ (b) $\arcsin\left(\dfrac{\sqrt{2}}{2}\right)$ (c) $\sin^{-1}\left(-\dfrac{\sqrt{3}}{2}\right)$ (d) $\tan\left(\arcsin\left(-\dfrac{1}{2}\right)\right)$

(e) $\sin^{-1}\left(\sin \dfrac{5\pi}{4}\right)$

***Solution***

(a) Think of $\sin^{-1}\left(\dfrac{1}{2}\right)$ as the angle between $-\dfrac{\pi}{2}$ and $\dfrac{\pi}{2}$ whose sine is $\dfrac{1}{2}$. Since $\sin\left(\dfrac{\pi}{6}\right) = \dfrac{1}{2}$, and $\dfrac{\pi}{6}$ is between $-\dfrac{\pi}{2}$ and $\dfrac{\pi}{2}$, we conclude that $\sin^{-1}\left(\dfrac{1}{2}\right) = \dfrac{\pi}{6}$.

(b) $\arcsin\left(\dfrac{\sqrt{2}}{2}\right)$ is the angle between $-\dfrac{\pi}{2}$ and $\dfrac{\pi}{2}$ whose sine is $\dfrac{\sqrt{2}}{2}$. Since $\sin\left(\dfrac{\pi}{4}\right) = \dfrac{\sqrt{2}}{2}$, and $\dfrac{\pi}{4}$ is between $-\dfrac{\pi}{2}$ and $\dfrac{\pi}{2}$, $\arcsin\left(\dfrac{\sqrt{2}}{2}\right) = \dfrac{\pi}{4}$.

(c) $\sin^{-1}\left(-\dfrac{\sqrt{3}}{2}\right)$ is the angle between $-\dfrac{\pi}{2}$ and $\dfrac{\pi}{2}$ whose sine is $-\dfrac{\sqrt{3}}{2}$. Since $\sin\left(-\dfrac{\pi}{3}\right) = -\dfrac{\sqrt{3}}{2}$, and $-\dfrac{\pi}{3}$ is between $-\dfrac{\pi}{2}$ and $\dfrac{\pi}{2}$, $\sin^{-1}\left(-\dfrac{\sqrt{3}}{2}\right) = -\dfrac{\pi}{3}$.

(d) First determine $\arcsin\left(-\frac{1}{2}\right)$: $\arcsin\left(-\frac{1}{2}\right)$ is the angle between $-\frac{\pi}{2}$ and $\frac{\pi}{2}$ whose sine is $-\frac{1}{2}$. Since $\sin\left(-\frac{\pi}{6}\right) = -\frac{1}{2}$,

$$\arcsin\left(-\frac{1}{2}\right) = -\frac{\pi}{6}.$$

Thus,

$$\tan\left(\arcsin\left(-\frac{1}{2}\right)\right) = \tan\left(-\frac{\pi}{6}\right)$$

$$= -\tan\left(\frac{\pi}{6}\right)$$ ■ Tangent is an odd function.

$$= -\frac{\sqrt{3}}{3}$$

$$\tan\left(\arcsin\left(-\frac{1}{2}\right)\right) = -\frac{\sqrt{3}}{3}$$

(e) First determine $\sin\frac{5\pi}{4}$.

$$\sin\left(\frac{5\pi}{4}\right) = -\sin\left(\frac{\pi}{4}\right) = -\frac{\sqrt{2}}{2}$$ ■ Use the reference angle $\frac{\pi}{4}$.

Thus,

$$\sin^{-1}\left(\sin\frac{5\pi}{4}\right) = \sin^{-1}\left(-\frac{\sqrt{2}}{2}\right)$$

$$= -\frac{\pi}{4}$$ ■ $\sin^{-1}\left(-\frac{\sqrt{2}}{2}\right)$ is the angle between $-\frac{\pi}{2}$ and $\frac{\pi}{2}$ whose sine is $-\frac{\sqrt{2}}{2}$, namely, $-\frac{\pi}{4}$.

$$\sin^{-1}\left(\sin\frac{5\pi}{4}\right) = -\frac{\pi}{4}$$

Recall that if $f$ and $f^{-1}$ are inverses of one another, then

$f(f^{-1}(x)) = x$ for all $x$ in the domain of $f^{-1}$, and
$f^{-1}(f(x)) = x$ for all $x$ in the domain of $f$.

If we let $f(x) = \sin x$ and $f^{-1}(x) = \sin^{-1} x$, then those statements become

$\sin(\sin^{-1} x) = x$ for all $x$ such that $-1 \leq x \leq 1$, and

$\sin^{-1}(\sin x) = x$ for all $x$ such that $-\frac{\pi}{2} \leq x \leq \frac{\pi}{2}$.

By virtue of the last equation, it is *not* true that $\sin^{-1}(\sin x) = x$ for all $x$ in the domain of the sine function—Example 1(e) is a case in point. However, for all $x$ between $-\frac{\pi}{2}$ and $\frac{\pi}{2}$ (the domain of the restricted sine function), it is true that $\sin^{-1}(\sin x) = x$.

In a manner similar to the case of $\sin^{-1} x$, we can restrict the domains of the other trigonometric functions so that inverse functions can be defined. By restricting $y = \cos x$ to values of $x$ between 0 and $\pi$ inclusive, the function becomes one-to-one and has an inverse cosine function.

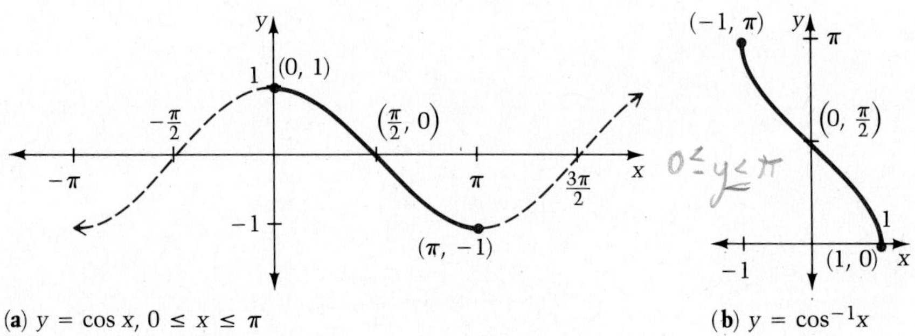


(a) $y = \cos x,\ 0 \le x \le \pi$ (b) $y = \cos^{-1}x$

**Figure 3.27** (a) The restricted cosine function. (b) The inverse cosine function.

**Definition**

The inverse cosine function is defined as follows:

$$y = \cos^{-1} x \quad \text{with domain } -1 \le x \le 1$$

if and only if $\cos y = x$ and $0 \le y \le \pi$.

By restricting $y = \tan x$ to values of $x$ between $-\dfrac{\pi}{2}$ and $\dfrac{\pi}{2}$, we can define the inverse function $y = \tan^{-1} x$.

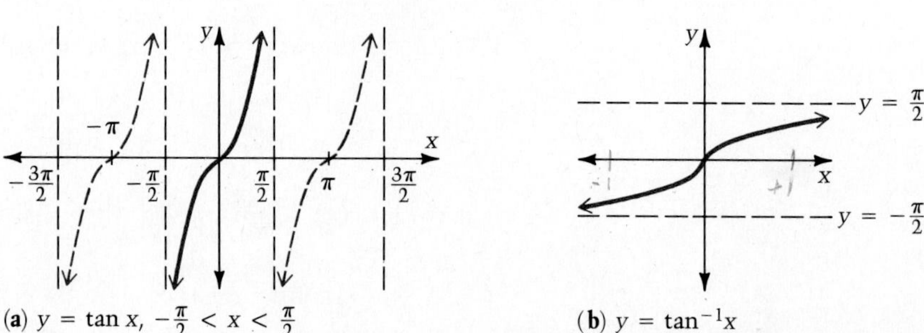


(a) $y = \tan x,\ -\frac{\pi}{2} < x < \frac{\pi}{2}$ (b) $y = \tan^{-1}x$

**Figure 3.28** (a) The restricted tangent function. (b) The inverse tangent function.

**Definition**

The inverse tangent function is defined as follows:

$$y = \tan^{-1} x \quad \text{with domain } -\infty < x < +\infty,$$

if and only if $\tan y = x$ and $-\frac{\pi}{2} < y < \frac{\pi}{2}$.

By virtue of the definitions of $\cos^{-1} x$ and $\tan^{-1} x$, we can think of $\cos^{-1} x$ as "the angle between 0 and $\pi$ whose cosine is equal to $x$," and $\tan^{-1} x$ as "the angle between $-\frac{\pi}{2}$ and $\frac{\pi}{2}$ whose tangent is equal to $x$." The expressions $\arccos x$ and $\arctan x$ are used interchangeably with $\cos^{-1} x$ and $\tan^{-1} x$.

**Example 2** Determine the value of each expression, without using Table 4.

(a) $\arctan(1)$ (b) $\cos^{-1}\left(-\frac{\sqrt{3}}{2}\right)$ (c) $\tan^{-1}\left(\tan\frac{2\pi}{3}\right)$

***Solution*** (a) $\arctan(1)$ is the angle between $-\frac{\pi}{2}$ and $\frac{\pi}{2}$ whose tangent is 1. Since $\tan\left(\frac{\pi}{4}\right) = 1$, and $\frac{\pi}{4}$ is between $-\frac{\pi}{2}$ and $\frac{\pi}{2}$, $\arctan(1) = \frac{\pi}{4}$.

(b) $\cos^{-1}\left(-\frac{\sqrt{3}}{2}\right)$ is the angle between 0 and $\pi$ inclusive whose cosine is $-\frac{\sqrt{3}}{2}$. Since $\cos\left(\frac{\pi}{6}\right)$ is $\frac{\sqrt{3}}{2}$, we are looking for an angle that has $\frac{\pi}{6}$ as a reference angle, and whose cosine is negative. Since the angle must be between 0 and $\pi$, our only choice is $\frac{5\pi}{6}$. Thus, $\cos^{-1}\left(-\frac{\sqrt{3}}{2}\right) = \frac{5\pi}{6}$.

(c) We begin by determining $\tan\left(\frac{2\pi}{3}\right)$:

$$\tan\left(\frac{2\pi}{3}\right) = -\tan\left(\frac{\pi}{3}\right) = -\sqrt{3}.$$

Now determine $\tan^{-1}(-\sqrt{3})$: $\tan^{-1}(-\sqrt{3})$ is the angle between $-\frac{\pi}{2}$ and $\frac{\pi}{2}$ whose tangent is $-\sqrt{3}$. Since the tangent is negative, the angle is between $-\frac{\pi}{2}$ and 0, and since the value of the tangent is $-\sqrt{3}$, the reference angle is $\frac{\pi}{3}$. Thus $\tan^{-1}(-\sqrt{3}) = -\frac{\pi}{3}$, and we have $\tan^{-1}\left(\tan\left(\frac{2\pi}{3}\right)\right) = -\frac{\pi}{3}$.

Inverses can be defined for the cotangent, secant, and cosecant functions by suitably restricting the domains of these three trigonometric functions. We leave these problems for the exercise set.

**Goal B** *to use right triangles to evaluate trigonometric and inverse trigonometric functions*

Sometimes it is necessary to evaluate expressions such as $\tan(\arcsin(-\frac{3}{5}))$ or $\cos(\tan^{-1}(\frac{5}{12}))$. Such composite expressions can be evaluated without resorting to calculators, tables, or facts about special angles. These problems require that you recall three things:

- The ranges of the inverse trigonometric functions:

$$-\frac{\pi}{2} \le \sin^{-1} x \le \frac{\pi}{2}, \quad 0 \le \cos^{-1} x \le \pi, \quad -\frac{\pi}{2} < \tan^{-1} x < \frac{\pi}{2}.$$

- The ASTC memory device for determining the quadrant in which an angle in standard position will terminate.
- The trigonometric function values of an angle in standard position can be determined by the coordinates of any point on the terminal side of the angle.

---

**Example 3** Determine the exact values of the following:

(a) $\tan\left(\arcsin\left(-\frac{3}{5}\right)\right)$ (b) $\cos\left(\tan^{-1}\left(\frac{5}{12}\right)\right)$

***Solution*** (a) Begin by letting $t = \arcsin\left(-\frac{3}{5}\right)$. Then $\sin t = \sin\left(\arcsin\left(-\frac{3}{5}\right)\right) = -\frac{3}{5}$. Since $\sin t = -\frac{3}{5}$, a negative number, $t$ must terminate in either the third or fourth quadrant (use ASTC). But, because $t$ is in the range of the arcsin function, $-\frac{\pi}{2} \le t \le \frac{\pi}{2}$, and thus $t$ cannot terminate in the third quadrant. Thus, $t$ must be a negative angle terminating in the fourth quadrant.

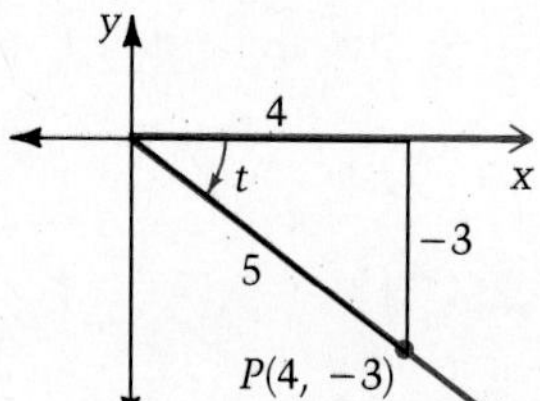


- Draw angle $t$ and label point $P$ on its terminal side.
- Since $\sin t = \frac{y}{r} = \frac{-3}{5}$, let $P$ have $y$-coordinate $-3$, and lie 5 units from the origin.

Find $x$ by applying the Pythagorean Theorem to the right triangle with legs $x$ and 3 and hypotenuse 5:

$$\begin{aligned} x^2 + y^2 &= r^2 \\ x^2 + 3^2 &= 5^2 \\ x^2 &= 16 \end{aligned}$$

Since $x > 0$, we have $x = 4$. Thus, $\tan\left(\arcsin\left(-\frac{3}{5}\right)\right) = \tan t = \frac{y}{x} = -\frac{3}{4}$.

(b) Begin by letting $t = \tan^{-1}\left(\frac{5}{12}\right)$. Then $\tan t = \tan\left(\tan^{-1}\left(\frac{5}{12}\right)\right) = \frac{5}{12}$. Since $\tan t$ is a positive number, $t$ must terminate in either the first or third quadrant. The range of the inverse tangent function requires that $-\frac{\pi}{2} < t < \frac{\pi}{2}$. Thus, $t$ is a first quadrant angle.

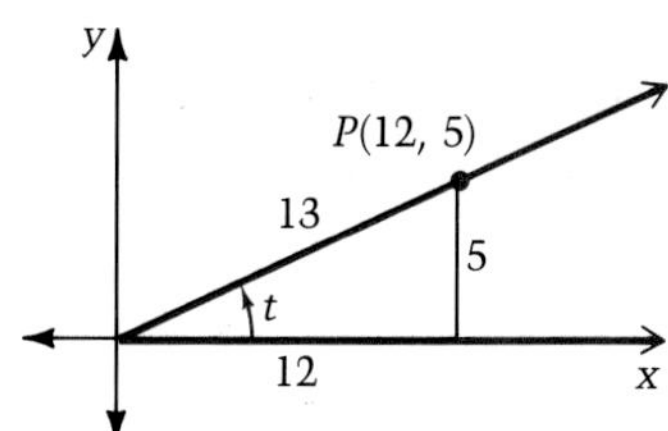


■ Draw $t$ and a point $P$ on its terminal side. Since $\tan t = \frac{y}{x} = \frac{5}{12}$, let $P$ have coordinates (12, 5). By the Pythagorean Theorem, we have

$$\begin{aligned} r^2 &= 12^2 + 5^2 = 169 \\ r &= 13. \end{aligned}$$

Thus, $\cos\left(\tan^{-1}\left(\frac{5}{12}\right)\right) = \cos t = \frac{x}{r} = \frac{12}{13}$.

**Example 4** Write $\cos^2(\tan^{-1} z)$, for $z \geq 0$, as an expression involving no trigonometric functions.

***Solution*** Let $t = \tan^{-1} z$. Then $\tan t = z$. Since $z \geq 0$, $t$ must be between 0 and $\frac{\pi}{2}$.

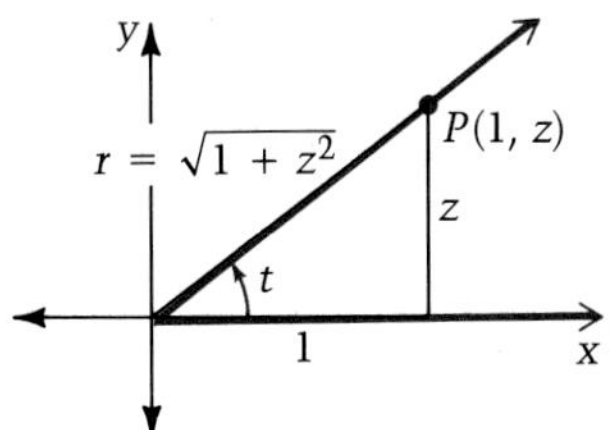


■ Since $\tan t = \frac{y}{x} = \frac{z}{1}$, draw angle $t$ with $P(1, z)$ on its terminal side. By the Pythagorean Theorem,

$$\begin{aligned} r^2 &= 1^2 + z^2 \\ r &= \sqrt{1 + z^2}. \end{aligned}$$

Thus, $\cos^2(\tan^{-1} z) = (\cos t)^2 = \left(\frac{x}{r}\right)^2 = \left(\frac{1}{\sqrt{1 + z^2}}\right)^2 = \frac{1}{1 + z^2}$.

## Exercise Set 3.5 Calculator Exercises in the Appendix

### Goal A

In exercises 1–24, evaluate the expression without using Table 4 or a calculator.

1. $\sin^{-1}\frac{\sqrt{3}}{2}$
2. $\sin^{-1}1$
3. $\cos^{-1}(-1)$
4. $\tan^{-1}(\sqrt{3})$
5. $\tan^{-1}\left(-\frac{\sqrt{3}}{3}\right)$
6. $\cos^{-1}\left(-\frac{\sqrt{2}}{2}\right)$
7. $\arctan(-1)$
8. $\tan^{-1}(-\sqrt{3})$
9. $\arcsin 0$
10. $\arccos 0$
11. $\sin\left(\cos^{-1}\left(\frac{\sqrt{3}}{2}\right)\right)$
12. $\tan\left(\cos^{-1}\left(\frac{\sqrt{3}}{2}\right)\right)$
13. $\tan\left(\cos^{-1}\left(-\frac{1}{2}\right)\right)$
14. $\sin\left(\cos^{-1}\left(-\frac{1}{2}\right)\right)$
15. $\cos(\sin^{-1}(-1))$
16. $\cos\left(\sin^{-1}\left(-\frac{\sqrt{2}}{2}\right)\right)$
17. $\sin(\tan^{-1}(-\sqrt{3}))$
18. $\cos(\tan^{-1}(-1))$
19. $\cos\left(\cos^{-1}\left(-\frac{\sqrt{2}}{2}\right)\right)$
20. $\cos\left(\sin^{-1}\left(-\frac{1}{2}\right)\right)$
21. $\tan^{-1}\left(\tan\frac{3\pi}{4}\right)$
22. $\cos^{-1}\left(\cos\frac{\pi}{3}\right)$
23. $\sin^{-1}\left(\cos\left(-\frac{\pi}{3}\right)\right)$
24. $\tan(\tan^{-1}(-\sqrt{3}))$

### Goal B

In exercises 25–34, use a right triangle to evaluate the expression.

25. $\sin\left(\sin^{-1}\frac{4}{5}\right)$
26. $\cos\left(\sin^{-1}\frac{4}{5}\right)$
27. $\cos\left(\sin^{-1}\left(-\frac{3}{5}\right)\right)$
28. $\tan\left(\arccos\frac{4}{5}\right)$
29. $\cos\left(\tan^{-1}\left(-\frac{3}{4}\right)\right)$
30. $\sin\left(\tan^{-1}\frac{7}{24}\right)$
31. $\sin\left(\arccos\left(-\frac{12}{13}\right)\right)$
32. $\tan\left(\sin^{-1}\left(\frac{5}{13}\right)\right)$
33. $\tan\left(\arcsin\left(-\frac{2}{7}\right)\right)$
34. $\sin(\arctan 2)$

In exercises 35–40, rewrite each expression without using trigonometric functions. Assume $z \geq 0$.

35. $\sin^2(\arctan z)$
36. $\cos^2(\tan^{-1} z)$
37. $\cos^2(\sin^{-1} z)$
38. $\tan^2(\sin^{-1} z)$
39. $\tan(\arcsin(-z))$
40. $\sin(\arcsin(-z))$

### Superset

Inverse cotangent, secant, and cosecant functions can be defined as follows:

$y = \cot^{-1} x$ if and only if $\cot y = x$, where $-\infty < x < +\infty$, and $0 < y < \pi$,

$y = \sec^{-1} x$ if and only if $\sec y = x$, where $x \leq -1$ or $x \geq 1$, and $0 \leq y \leq \pi$, $y \neq \frac{\pi}{2}$,

$y = \csc^{-1} x$ if and only if $\csc y = x$, where $x \leq -1$ or $x \geq 1$, and $-\frac{\pi}{2} \leq y \leq \frac{\pi}{2}$, $y \neq 0$.

Use these definitions in evaluating the expressions in exercises 41–56.

41. $\operatorname{arcsec}(-2)$
42. $\operatorname{arccsc}(-\sqrt{2})$
43. $\sin\left(\csc^{-1}\left(-\frac{13}{5}\right)\right)$
44. $\cos\left(\sec^{-1}\left(-\frac{13}{12}\right)\right)$
45. $\cos\left(\sec^{-1}\frac{3}{\sqrt{5}}\right)$
46. $\cot\left(\sec^{-1}\frac{7}{3}\right)$
47. $\cot\left(\sin^{-1}\frac{2}{3}\right)$
48. $\sec\left(\sin^{-1}\left(-\frac{1}{2}\right)\right)$
49. $\sec\left(\cot^{-1}\left(-\frac{\sqrt{11}}{2}\right)\right)$
50. $\cot\left(\tan^{-1}\left(\frac{\sqrt{6}}{4}\right)\right)$
51. $\arccos(0.9483)$
52. $\sin^{-1}(0.4226)$
53. $\tan^{-1}(7.115)$
54. $\operatorname{arccot}(-28.64)$
55. $\sin(\cot^{-1}(12.25))$
56. $\sin(\cos^{-1}(-0.2221))$

In exercises 57–60, sketch the graph.

57. $y = 3\cos^{-1} x$
58. $y = \cos^{-1}(\frac{1}{3}x)$
59. $y = \sin^{-1}(2x)$
60. $y = \frac{1}{2}\sin^{-1} x$

# Chapter Review & Test

## Chapter Review

### 3.1 Radian Measure and the Unit Circle (pp. 70–76)

The unit circle is the circle of radius 1 with center at the origin. A point $P(u, v)$ is on the unit circle if and only if $u^2 + v^2 = 1$ (p. 70) A *standard arc* $s$ is an arc on the unit circle which starts at (1, 0), has length $|s|$, and travels counterclockwise if $s$ is positive, and clockwise if $s$ is negative. (p. 71)

A *central angle* $\theta$ is an angle whose vertex is at the center of a circle. (p. 72) The *radian measure* of a positive central angle $\theta$ is defined as a ratio:

$$\theta = \frac{s}{r},$$

where $s$ is the arclength of the arc cut by $\theta$, and $r$ is the radius of the circle. (p. 73)

To convert the degree measure $d$ of an angle to its radian measure $\alpha$, and vice versa, use the following proportion. (p. 74)

$$\frac{d}{180°} = \frac{\alpha}{\pi}.$$

### 3.2 Trigonometric Functions of Real Numbers (pp. 77–84)

Every real number $x$ can be associated uniquely with a central angle $\theta$ that cuts standard arc $x$ on the unit circle. (p. 77) To define the trigonometric functions of any real number $x$, let $T(u, v)$ be the terminal point on the standard arc $x$. (p. 78)

$$\sin x = v \qquad \cos x = u \qquad \tan x = \frac{v}{u}$$

$$\csc x = \frac{1}{v} \qquad \sec x = \frac{1}{u} \qquad \cot x = \frac{u}{v}$$

To find the trigonometric function value of any real number $x$, we determine the sign by means of the ASTC chart, and then find the trigonometric function value of the reference value for $x$:

| $x$ between | 0 and $\frac{\pi}{2}$ | $\frac{\pi}{2}$ and $\pi$ | $\pi$ and $\frac{3\pi}{2}$ | $\frac{3\pi}{2}$ and $2\pi$ |
|---|---|---|---|---|
| reference value | $x$ | $\pi - x$ | $x - \pi$ | $2\pi - x$ |

If $x$ is outside the interval $[0, 2\pi)$, begin by adding an appropriate integer multiple of $2\pi$ to produce a value between 0 and $2\pi$, then use the procedure described above. (p. 81)

### 3.3 The Trigonometric Functions: Basic Graphs and Properties (pp. 85–95)

A function $f$ is called a *periodic function* if there is a positive real number $p$ such that $f(x) = f(x + p)$ for all $x$ in the domain of $f$. The smallest such positive number $p$ is called the *period* of the function. (p. 85)

### 3.4 Transformations of the Trigonometric Functions (pp. 96–102)

For functions described by equations of the form $y = k + a \sin b(x - h)$ or $y = k + a \cos b(x - h)$, the number $|a|$ is the *amplitude*, the number $\left|\frac{2\pi}{b}\right|$ is the *period*, and the number $h$ is the *phase shift*. (pp. 96–100)

To graph functions of the form $y = a \sin bx + g(x)$ or $y = a \cos bx + g(x)$, use the method of *addition of ordinates*. (p. 101)

### 3.5 The Inverse Trigonometric Functions (pp. 103–110)

If each of the trigonometric functions is restricted to a portion of its domain, the resulting function is one-to-one, and thus has an inverse function. The inverses of the sine, cosine, and tangent functions are defined as follows:

$$y = \sin^{-1} x \quad \text{with domain } -1 \le x \le 1 \text{ if and only if } \sin y = x \text{ and } -\frac{\pi}{2} \le y \le \frac{\pi}{2},$$

$$y = \cos^{-1} x \quad \text{with domain } -1 \le x \le 1 \text{ if and only if } \cos y = x \text{ and } 0 \le y \le \pi,$$

$$y = \tan^{-1} x \quad \text{with domain } -\infty < x < +\infty \text{ if and only if } \tan y = x \text{ and } -\frac{\pi}{2} < x < \frac{\pi}{2}.$$

## Chapter Test

3.1A Determine whether the given point is on the unit circle.

**1.** $\left(-\frac{5}{13}, \frac{12}{13}\right)$ **2.** $\left(\frac{\sqrt{2}}{6}, \frac{4}{6}\right)$

Represent the given real number as a standard arc.

**3.** $\frac{10\pi}{3}$ **4.** $\frac{7\pi}{2}$ **5.** $\frac{-25\pi}{4}$

**3.1B** Convert the measure of the given angle from degrees to radians.

**6.** $-135°$ **7.** $240°$ **8.** $-735°$

Convert the measure of the given angle from radians to degrees.

**9.** $\frac{9\pi}{2}$ **10.** $\frac{16\pi}{3}$

**3.1C** Determine the measure in radians and in degrees of the central angle which cuts an arc of length $s$ in a circle with radius $r$.

**11.** $s = 12\pi$ cm, $r = 8$ cm **12.** $s = \frac{3\pi}{2}$ cm, $r = 6$ cm

**13.** On a steeple clock the hour hand is 3 ft long and the minute hand is 4 ft long. How far does the tip of the hour hand move in 4 hr 30 min?

**3.2A** Determine the given trigonometric function value.

**14.** $\sin\left(\frac{5\pi}{4}\right)$ **15.** $\cos\left(-\frac{4\pi}{3}\right)$ **16.** $\sec\left(\frac{19\pi}{2}\right)$ **17.** $\tan\left(-\frac{17\pi}{3}\right)$

Use Table 4 or a calculator to determine the trigonometric function value.

**18.** $\sin 4.3$ **19.** $\cos 5.6$

**3.3A** Graph the following pair of functions on the same axes.

**20.** $y = \sin x,\ y = -\sin\left(x + \frac{\pi}{2}\right)$

**3.3B** Graph the following pair of functions on the same axes.

**21.** $y = \tan x,\ y = \tan\left(x - \frac{\pi}{4}\right)$

Determine whether each statement is true or false.

**22.** $\tan x$ is not defined for $x = -3\pi$.

**23.** $\cot x$ is not defined for $x = -\frac{3\pi}{2}$.

**24.** $\csc(-x) = -\frac{1}{\sin x}$ is an identity.

**25.** $\cos(-x) = -\cos x$ is an identity.

**3.4A** Determine the amplitude and period, and sketch the graph.

**26.** $y = 3 \cos \frac{1}{2}x$

**27.** $y = -3 \sin 2x$

**3.4B** Determine the amplitude, period, and phase shift, and sketch the graph.

**28.** $y = 2 \cos\left(2x - \frac{\pi}{2}\right)$

**29.** $y = -4 \sin(2x + \pi)$

**3.4C** Sketch the graph of each of the following trigonometric functions.

**30.** $y = -3 \sin \frac{1}{2}x + x$

**31.** $y = \frac{1}{2} \cos 2x - 2x$

**3.5A** Evaluate each expression without using Table 4 or a calculator.

**32.** $\arcsin\left(-\frac{1}{2}\right)$

**33.** $\cos^{-1}\left(\frac{\sqrt{3}}{2}\right)$

**34.** $\sin^{-1}\left(\cot \frac{3\pi}{4}\right)$

**3.5B** Use a right triangle to evaluate the expression.

**35.** $\cos\left(\tan^{-1} \frac{3}{4}\right)$

**36.** $\sin\left(\operatorname{arcsec} \frac{13}{5}\right)$

## Superset

In exercises 37–42, determine whether the statement is true or false.

**37.** $2 \sin \frac{\pi}{6} \cos \frac{\pi}{6} = \sin \frac{\pi}{3}$

**38.** $1 + \tan^2 \frac{\pi}{4} = \sec^2 \frac{\pi}{2}$

**39.** $\sin \frac{\pi}{6} = \sqrt{\dfrac{1 + \cos \frac{\pi}{3}}{2}}$

**40.** $\sin \frac{\pi}{4} = \sqrt{\dfrac{1 - \sin \frac{\pi}{2}}{2}}$

In exercises 41–44, find a value of $b$ such that the period $p$ of the function is the given value.

**41.** $y = \sin bx,\ p = \frac{\pi}{3}$

**42.** $y = \cos bx,\ p = \frac{\pi}{2}$

**43.** $y = \cot bx,\ p = 2\pi$

**44.** $y = \tan bx,\ p = \frac{\pi}{6}$

**45.** An equilateral triangle is inscribed in a circle with a 6-inch radius. Find the length of the arc cut by one side of the triangle.

**46.** An automobile has tires with a 30-inch diameter. If the tires are revolving at 264 rpm, find the speed of the car in miles per hour.

## Cumulative Test: *Chapters 1–3*

In exercises 1–2, determine the distance between the given points.

**1.** $(-7, 2), (-1, -6)$ **2.** $(-4, 3), (8, -2)$

In exercises 3–4, determine $f(3), f(-2), f(h^2), f(h - 4)$ for each function.

**3.** $f(x) = x^2 - 2x + 3$ **4.** $f(x) = 9 - x^2$

In exercises 5–6, determine whether each of the functions is one-to-one.

**5.** $f(x) = x^3 - 3$ **6.** $f(x) = -3x + 5$

In exercises 7–8, find an equation that describes $(f \circ g)(x)$ and state the domain of the composite function.

**7.** $f(x) = x^2 - 4;\ g(x) = 2x - 5$ **8.** $f(x) = \sqrt{x - 2};\ g(x) = 4x^2 + 1$

In exercises 9–10, the given function $y = f(x)$ is one-to-one. Find the equation describing the inverse function, sketch the inverse, and identify its domain and range.

**9.** $f(x) = 2x - 5;\ 0 \le x \le 3$ **10.** $f(x) = \sqrt{x - 1},\ 1 \le x \le 5$

In exercises 11–12, sketch the graph of the second equation by reflecting the graph of the first equation.

**11.** $y = |x - 1|;\quad y = -|x - 1|$ **12.** $y = \sqrt{x + 2};\quad y = \sqrt{2 - x}$

In exercises 13–14, use transformations to sketch the following.

**13.** $y = -(x - 2)^2 + 2$ **14.** $y = -2|x + 1|$

In exercises 15–16, determine an equation of the line satisfying the stated conditions. Write your answer in (a) point-slope form, (b) slope-intercept form, and (c) general form.

**15.** The line is perpendicular to the graph of $2x - 3y + 5 = 0$ with $y$-intercept $-2$.

**16.** The line is parallel to the graph of $2y - 6x + 3 = 0$ and passes through $(-3, 0)$.

In exercises 17–18, sketch the graph of each function. Label the vertex, the $x$-intercepts, the axis of symmetry, and two additional points.

**17.** $f(x) = x^2 - 4x + 4$ **18.** $f(x) = x^2 - 2x + 5$

In exercises 19–20, a right triangle $ABC$ is given with $c$ the length of the hypotenuse, and $a$ and $b$ the lengths of the legs. Find the length of the third side given the lengths of two sides.

**19.** $a = 8, b = 15$ **20.** $c = 25, b = 7$

In exercises 21–22, $\triangle ABC$ has right angle at $C$ and sides of length $a$, $b$, and $c$. Given the following information, find the indicated parts.

**21.** $m(\angle A) = 60°$, $a = 20$, find $c$

**22.** $m(\angle A) = 60°$, $c = 14$, find $b$.

In exercises 23–24, the given point is located on the terminal side of angle $\varphi$. Find $\sin\varphi$, $\cos\varphi$ and $\tan\varphi$.

**23.** $(-8, -20)$

**24.** $(22, -12)$

In exercises 25–28, evaluate each of the following.

**25.** $\sin 225°$

**26.** $\cos(-585°)$

**27.** $\tan(-480°)$

**28.** $\cot 330°$

In exercises 29–32, convert the measure of the given angle from degrees to radians or vice versa.

**29.** $-630°$

**30.** $-315°$

**31.** $\dfrac{8\pi}{3}$

**32.** $\dfrac{9\pi}{4}$

In exercises 33–34, use Table 4 or a calculator to determine the trigonometric function value.

**33.** $\cos(-3.8)$

**34.** $\sin(7.2)$

In exercises 35–36, determine the amplitude, period and phase shift, and sketch the graph.

**35.** $y = 3\sin(2x - \pi)$

**36.** $y = 2\cos\left(\frac{1}{2} + \pi\right)$

**Superset**

**37.** For the function $f(x) = 3x^2 + 2x$, find the following: $\dfrac{f(2+h) - f(2)}{h}$.

**38.** Sketch the graph of the following function: $f(x) = \begin{cases} x, & \text{if } x < -3, \\ x^2 + 1, & \text{if } x > 0. \end{cases}$

**39.** The value of one trigonometric function of $\theta$ is given. If the terminal side of $\theta$ lies in the given quadrant, find the values of the other five trigonometric functions.

(a) fourth quadrant, $\cos\theta = \dfrac{9}{41}$

(b) third quadrant, $\sin\theta = -\dfrac{7}{25}$

**40.** A jogger runs six laps in five minutes on a circular track that has a diameter of 60 yd. What is the jogger's average speed?

# 4 Trigonometric Identities and Equations

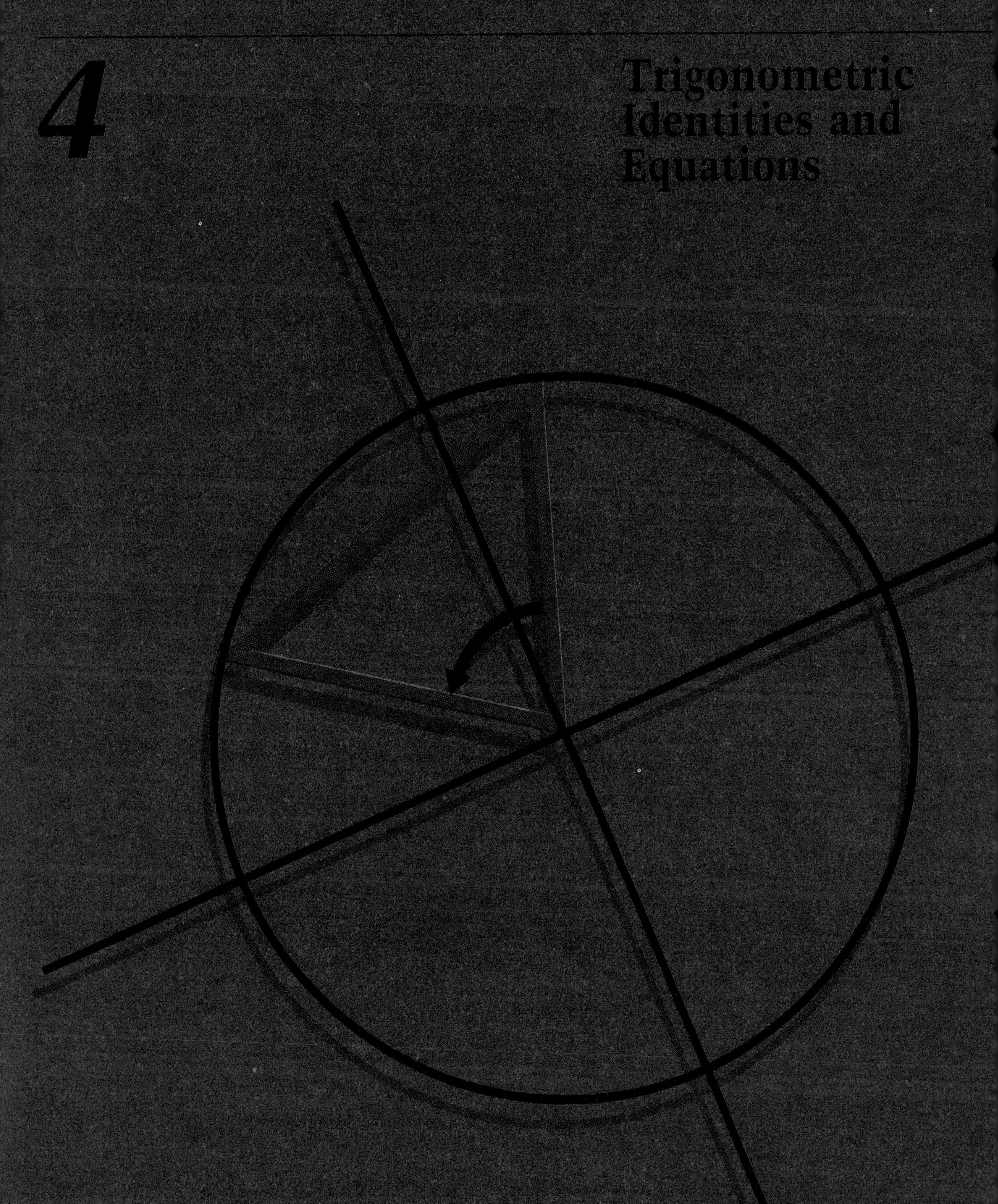

## 4.1 Basic Trigonometric Identities

**Goal A** *to simplify an expression by using identities*

We refer to equations such as

$$(x + 1)^2 = x^2 + 2x + 1 \quad \text{or} \quad \frac{x^2 - 1}{x - 1} = x + 1$$

as identities. An **identity** is an equation that is true for all values of $x$ for which the expressions are defined. The equation,

$$(x + 1)^2 = x^2 + 2x + 1,$$

is true for all real numbers $x$. The equation

$$\frac{x^2 - 1}{x - 1} = x + 1,$$

is true for $x \neq 1$. As long as $x \neq 1$, the expression on the left-hand side of the equation is defined, and may be simplified by factoring and canceling:

$$\frac{x^2 - 1}{x - 1} = \frac{(x - 1)(x + 1)}{x - 1} = x + 1.$$

We have already seen trigonometric identities such as

$$\tan\theta = \frac{\sin\theta}{\cos\theta}.$$

This identity is true for all $\theta$ such that $\cos\theta \neq 0$. It is one of the basic identities we derived earlier. Recall that we derived eleven basic identities using the definitions of the trigonometric functions and the properties of the unit circle. They are called the **basic identities** because only the trigonometric definitions are used to derive them.

We classified the eleven basic identities into three categories:

**Basic Trigonometric Identities**

I. Reciprocal Identities

(1) $$\sin\theta = \frac{1}{\csc\theta}$$

(2) $$\cos\theta = \frac{1}{\sec\theta}$$

(3) $$\tan\theta = \frac{1}{\cot\theta}$$

(4) $$\cot\theta = \frac{1}{\tan\theta}$$

(5) $$\sec\theta = \frac{1}{\cos\theta}$$

(6) $$\csc\theta = \frac{1}{\sin\theta}$$

II. Quotient Identities

(7) $$\frac{\sin\theta}{\cos\theta} = \tan\theta$$

(8) $$\frac{\cos\theta}{\sin\theta} = \cot\theta$$

III. Pythagorean Identities

(9) $$\sin^2\theta + \cos^2\theta = 1$$

(10) $$\tan^2\theta + 1 = \sec^2\theta$$

(11) $$1 + \cot^2\theta = \csc^2\theta$$

You should memorize the basic identities listed above. We now consider an example in which knowledge of the basic identities is essential in simplifying the given expression.

**Example 1** Show that the expression $\frac{1 + \tan^2 \theta}{\csc^2 \theta}$ may be simplified to $\tan^2 \theta$.

***Solution***

$$\frac{1 + \tan^2 \theta}{\csc^2 \theta} = \frac{\sec^2 \theta}{\csc^2 \theta}$$ ■ By identity (10), $\tan^2 \theta + 1 = \sec^2 \theta$.

$$= \frac{\frac{1}{\cos^2 \theta}}{\frac{1}{\sin^2 \theta}}$$ ■ By (5), $\sec^2 \theta = \frac{1}{\cos^2 \theta}$. ■ By (6), $\csc^2 \theta = \frac{1}{\sin^2 \theta}$.

$$= \frac{\sin^2 \theta}{\cos^2 \theta}$$ ■ Divide: multiply by the reciprocal of the denominator.

$$= \tan^2 \theta$$ ■ By (7), $\frac{\sin \theta}{\cos \theta} = \tan \theta$.

There are no standard steps to take to simplify a trigonometric expression. Simplifying trigonometric expressions is similar to factoring polynomials: by trial and error and by experience, you learn what will work in which situations. One useful technique is to begin by rewriting the entire expression in terms of sines and cosines.

**Example 2** Express $\left(1 - \frac{1}{\csc \theta}\right)^2 + \cos^2 \theta$ in terms of $\sin \theta$.

***Solution***

$$\left(1 - \frac{1}{\csc \theta}\right)^2 + \cos^2 \theta = (1 - \sin \theta)^2 + \cos^2 \theta$$ ■ By (1)

$$= 1 - 2 \sin \theta + \sin^2 \theta + \cos^2 \theta$$ ■ $(1 - \sin \theta)^2$ is expanded.

$$= 1 - 2 \sin \theta + 1$$ ■ By (9)

$$= 2 - 2 \sin \theta$$

**Goal B** *to verify simple identities*

We are now in a position to begin verifying new trigonometric identities. The primary strategy that we will use is to choose the expression on one side of the identity and to transform it so that it is identical to the expression on the other side. This "transformation" will come about by using the rules of algebra and the Basic Identities. Before we begin with some examples, let us offer a few words of advice.

1. Memorize the Basic Identities. Verifying a new identity often means that you must run through a mental checklist of options that might prove effective in rewriting one of the expressions.
2. It is often best to proceed by first writing down the more complicated side of the identity, and then transforming it in a step-by-step fashion until it looks exactly like the other side of the identity.
3. As an aid in rewriting the more complicated side of the identity, it is often useful to rewrite this expression in terms of sines and cosines only.

**Careful!** One approach *not* to be used is to write down the entire identity as your first step, and "do the same thing to both sides," as you would in solving an equation. Working on both sides of an equation in this way can be done only when the equation is assumed to be true. In verifying an identity, you are trying to prove that the equation is true. Thus, you may not assume that it is already true.

**Example 3** Verify the identity: $\dfrac{1 - \cos^2 \theta}{\tan \theta} = \sin \theta \cos \theta$.

***Solution***

$$\frac{1 - \cos^2 \theta}{\tan \theta} = \frac{\sin^2 \theta}{\tan \theta}$$

■ By (9), $1 - \cos^2 \theta = \sin^2 \theta$.

$$= \frac{\sin^2 \theta}{\dfrac{\sin \theta}{\cos \theta}}$$

■ By (7)

$$= \frac{\sin^2 \theta}{1} \cdot \frac{\cos \theta}{\sin \theta}$$

$$= \sin \theta \cos \theta$$

**Example 4** Verify the identity: $\dfrac{\tan x}{1 - \sec x} = -\dfrac{1 + \sec x}{\tan x}$

***Solution***

$$\frac{\tan x}{1 - \sec x} = \frac{\tan x}{1 - \sec x} \cdot \frac{1 + \sec x}{1 + \sec x}$$

■ Notice that multiplying by $\dfrac{1 + \sec x}{1 + \sec x}$ transforms the denominator into $1 - \sec^2 x$.

$$= \frac{(\tan x)(1 + \sec x)}{1 - \sec^2 x}$$

$$= \frac{(\tan x)(1 + \sec x)}{-\tan^2 x}$$

■ By (10), $1 - \sec^2 x = -\tan^2 x$.

$$= -\frac{1 + \sec x}{\tan x}$$

■ $\tan x$ has been canceled from the numerator and denominator.

## Exercise Set 4.1

### Goal A

In exercises 1–8, show that the first expression may be rewritten as the second by using the Reciprocal and Quotient Identities.

1. $\sin\alpha\sec\alpha;\ \tan\alpha$
2. $\dfrac{\csc\alpha}{\sec\alpha};\ \cot\alpha$
3. $\dfrac{\sin\varphi}{\csc\varphi};\ \sin^2\varphi$
4. $\dfrac{\cos\varphi}{\sec\varphi};\ \cos^2\varphi$
5. $\csc\beta;\ \sec\beta\cot\beta$
6. $\sin\beta;\ \dfrac{\tan\beta}{\sec\beta}$
7. $\csc\theta\cot\theta;\ \cos\theta\csc^2\theta$
8. $\csc^2\theta\tan\theta;\ \sec\theta\csc\theta$

In exercises 9–18, rewrite the expression in terms of only the sine or only the tangent.

9. $\dfrac{\csc\alpha}{1+\cot^2\alpha}$
10. $\dfrac{\tan\alpha\sec\alpha}{1+\tan^2\alpha}$
11. $\cot\beta(\sec^2\beta-1)$
12. $\sec^2\alpha-2$
13. $\csc\beta-\cos\beta\cot\beta$
14. $\dfrac{\sec\beta-\cos\beta}{\tan\beta}$
15. $(1+\tan^2\varphi)\sin^2\varphi$
16. $\dfrac{\sec^2\varphi-1}{\sec^2\varphi}$
17. $\dfrac{\sin\alpha+\cos\alpha}{\cos\alpha}$
18. $\dfrac{\csc\alpha+\sec\alpha}{\csc\alpha}$

### Goal B

In exercises 19–38, verify each identity by using the Basic Identities.

19. $\tan\alpha+\cot\alpha=\csc\alpha\sec\alpha$
20. $\dfrac{\sec\alpha}{\tan\alpha}-\dfrac{\tan\alpha}{\sec\alpha}=\cos\alpha\cot\alpha$
21. $(1+\cot\beta)^2-\csc^2\beta=2\cot\beta$
22. $(1+\csc^2\beta)-\cot^2\beta=2$
23. $\cos\alpha(\tan\alpha+\sec\alpha)=1+\sin\alpha$
24. $\sin\alpha(\cot\alpha+\csc\alpha)=\cos\alpha+1$
25. $\sec\beta+\csc\beta\cot\beta=\csc^2\beta\sec\beta$
26. $\sec\beta-\sin\beta\cot\beta=\tan\beta\sin\beta$
27. $\dfrac{\sin\gamma}{\cos\gamma+1}+\dfrac{\cos\gamma-1}{\sin\gamma}=0$
28. $\dfrac{\sin\gamma}{\csc\gamma}+\dfrac{\cos\gamma}{\sec\gamma}=1$
29. $\dfrac{(\sin\gamma+\cos\gamma)^2}{\sin\gamma\cos\gamma}=\csc\gamma\sec\gamma+2$
30. $\left(\dfrac{\sin\gamma}{\cos\gamma}+\dfrac{\cos\gamma}{\sin\gamma}\right)^2=\csc^2\gamma\sec^2\gamma$
31. $(\tan^2\alpha+1)(1+\cos^2\alpha)=\tan^2\alpha+2$
32. $(1+\cot^2\alpha)(1+\sin^2\alpha)=2+\cot^2\alpha$
33. $\csc x\sec x=\tan x+\cot x$
34. $\cos x=(\csc x-\sin x)\tan x$
35. $\dfrac{\tan\beta-\sin\beta}{\sin^3\beta}=\dfrac{\sec\beta}{1+\cos\beta}$
36. $\dfrac{\cot\beta-\cos\beta}{\cos^3\beta}=\dfrac{\csc\beta}{1+\sin\beta}$
37. $(\sin\alpha+\cos\alpha)^2-(\sin\beta+\cos\beta)^2$
$=2(\sin\alpha\cos\alpha-\sin\beta\cos\beta)$
38. $(\sin\alpha-\sin\beta)^2+(\cos\alpha-\cos\beta)^2$
$=2(1-\sin\alpha\sin\beta-\cos\alpha\cos\beta)$

### Superset

In exercises 39–43, the quadrantal angle is given. Express the first trigonometric function in terms of the second.

39. $\sin\alpha$ in terms of $\cot\alpha$; $\alpha$ in Quadrant I
40. $\cos\alpha$ in terms of $\cot\alpha$; $\alpha$ in Quadrant III
41. $\sec\beta$ in terms of $\sin\beta$; $\beta$ in Quadrant II
42. $\tan\varphi$ in terms of $\cos\varphi$; $\varphi$ in Quadrant IV
43. $\csc\gamma$ in terms of $\sec\gamma$; $\gamma$ in Quadrant III
44. Express the other 5 trigonometric functions in terms of (a) $\sin\alpha$ for $\alpha$ in Quadrant II; (b) $\cos\alpha$ for $\alpha$ in the Quadrant III.

## 4.2 Sum and Difference Identities

**Goal A** *to apply the sum and difference identities*

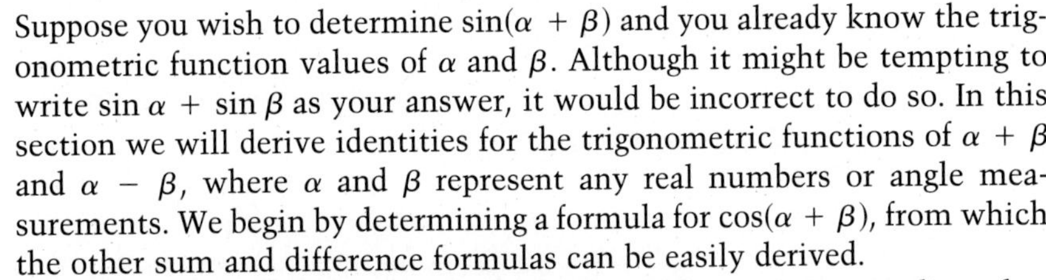

Suppose you wish to determine $\sin(\alpha + \beta)$ and you already know the trigonometric function values of $\alpha$ and $\beta$. Although it might be tempting to write $\sin \alpha + \sin \beta$ as your answer, it would be incorrect to do so. In this section we will derive identities for the trigonometric functions of $\alpha + \beta$ and $\alpha - \beta$, where $\alpha$ and $\beta$ represent any real numbers or angle measurements. We begin by determining a formula for $\cos(\alpha + \beta)$, from which the other sum and difference formulas can be easily derived.

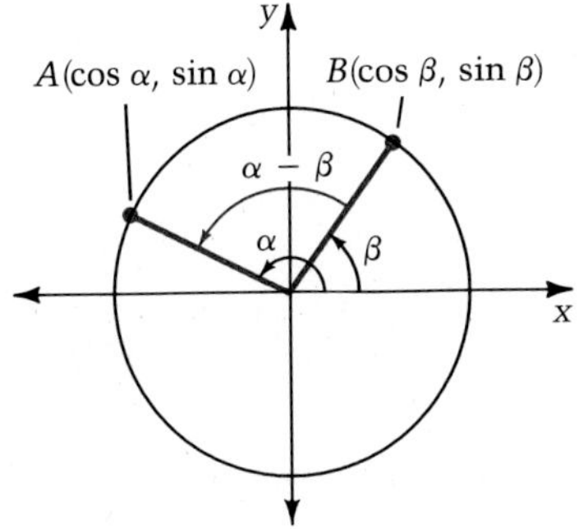


Figure 4.1

Suppose $\alpha$ and $\beta$ are positive real numbers and $\alpha > \beta$. To visualize what we are about to do, we can think of $\alpha$ and $\beta$ as the lengths of standard arcs, or the measures of angles in standard position on the unit circle. Interpreted this way, $\alpha$ and $\beta$ determine points $A$ and $B$ on the unit circle, shown in Figure 4.1. Note that $A$ has coordinates $(\cos \alpha, \sin \alpha)$, and $B$ has coordinates $(\cos \beta, \sin \beta)$.

In Figure 4.2(a) we show $\alpha - \beta$ and a line segment connecting the points $A$ and $B$. The line segment joining the endpoints of an arc is called a **chord.** We let $d$ be the length of chord $AB$. In Figure 4.2(b), the angle $\alpha - \beta$ has been redrawn in standard position, and the chord of length $d$ now has endpoints $(\cos(\alpha - \beta), \sin(\alpha - \beta))$ and $(1, 0)$, labeled $A'$ and $B'$ respectively.

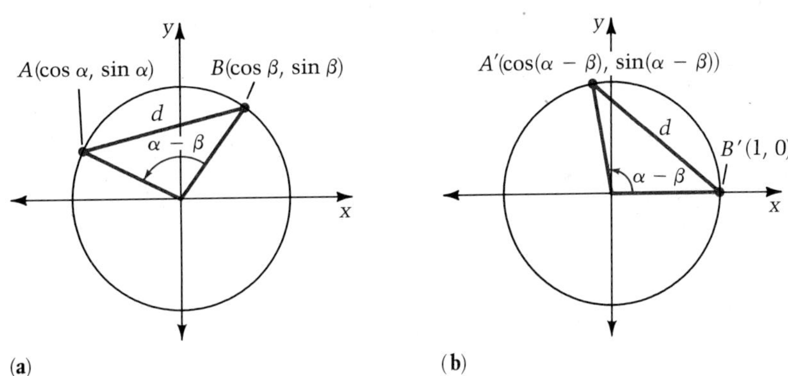


Figure 4.2

By means of the distance formula, we express $d^2$ in two different ways: first by using the coordinates of $A$ and $B$ (Figure 4.2(a)), and then by using the coordinates of $A'$ and $B'$ (Figure 4.2(b)). Using $A$ and $B$, we have

$$
\begin{aligned}
d^2 &= (\text{distance between } A \text{ and } B)^2 = (\cos \alpha - \cos \beta)^2 + (\sin \alpha - \sin \beta)^2 \\
&= (\cos^2 \alpha - 2 \cos \alpha \cos \beta + \cos^2 \beta) + (\sin^2 \alpha - 2 \sin \alpha \sin \beta + \sin^2 \beta) \\
&= (\sin^2 \alpha + \cos^2 \alpha) + (\sin^2 \beta + \cos^2 \beta) - 2 \cos \alpha \cos \beta - 2 \sin \alpha \sin \beta \\
&= \quad 1 \quad + \quad 1 \quad - 2(\cos \alpha \cos \beta + \sin \alpha \sin \beta) \\
&= 2 - 2(\cos \alpha \cos \beta + \sin \alpha \sin \beta)
\end{aligned}
$$

Using $A'$ and $B'$, we have

$$\begin{aligned} d^2 &= (\text{distance between } A' \text{ and } B')^2 \\ &= (\cos(\alpha - \beta) - 1)^2 + (\sin(\alpha - \beta) - 0)^2 \\ &= \cos^2(\alpha - \beta) - 2\cos(\alpha - \beta) + 1 + \sin^2(\alpha - \beta) \\ &= \underbrace{\sin^2(\alpha - \beta) + \cos^2(\alpha - \beta)}_{1} + 1 - 2\cos(\alpha - \beta) \\ &= 2 - 2\cos(\alpha - \beta) \end{aligned}$$

Equating the two expressions for $d^2$, we have

$$2 - 2\cos(\alpha - \beta) = 2 - 2(\cos\alpha\cos\beta + \sin\alpha\sin\beta).$$

Then, by adding $(-2)$ to both sides and dividing by $(-2)$, we conclude that

(12) $$\cos(\alpha - \beta) = \cos\alpha\cos\beta + \sin\alpha\sin\beta.$$

Although we have developed identity (12) for $\alpha$ and $\beta$ that are positive with $\alpha > \beta$, a similar argument establishes (12) for all $\alpha$ and $\beta$.

**Example 1** Determine cos 15° without using Table 4 or a calculator.

***Solution***

$$\begin{aligned} \cos 15^\circ &= \cos(45^\circ - 30^\circ) \\ &= \cos 45^\circ \cos 30^\circ + \sin 45^\circ \sin 30^\circ && \blacksquare \text{ By identity (12)} \\ &= \left(\frac{\sqrt{2}}{2}\right)\left(\frac{\sqrt{3}}{2}\right) + \left(\frac{\sqrt{2}}{2}\right)\left(\frac{1}{2}\right) \\ &= \frac{\sqrt{6}}{4} + \frac{\sqrt{2}}{4} \end{aligned}$$

Thus, $\cos 15^\circ = \dfrac{\sqrt{6} + \sqrt{2}}{4}$. = .9658

Recall that sine is an odd function and cosine is an even function. These two facts are represented as follows: for all real numbers $x$,

$$\sin(-x) = -\sin x \quad \text{and} \quad \cos(-x) = \cos x.$$

We use these facts to establish another identity:

$$\begin{aligned} \cos(\alpha + \beta) &= \cos(\alpha - (-\beta)) \\ &= \cos\alpha\cos(-\beta) + \sin\alpha \cdot (\sin(-\beta)) && \blacksquare \text{ By (12)} \\ &= \cos\alpha\cos\beta + \sin\alpha\,(-\sin\beta) && \blacksquare \text{ cosine is even; sine is odd} \\ &= \cos\alpha\cos\beta - \sin\alpha\sin\beta \end{aligned}$$

Thus,

(13) $$\cos(\alpha + \beta) = \cos \alpha \cos \beta - \sin \alpha \sin \beta$$

Recall that a trigonometric function value of an acute angle $\theta$ is equal to the co-trigonometric function value of the complement of $\theta$. For example, for any acute angle $\theta$,

$$\sin \theta = \cos\left(\frac{\pi}{2} - \theta\right).$$

By virtue of (12) we can now prove such a statement true for any $\theta$.

$$\begin{aligned}\cos\left(\frac{\pi}{2} - \theta\right) &= \cos \frac{\pi}{2} \cos \theta + \sin \frac{\pi}{2} \sin \theta \\ &= (0) \cdot \cos \theta + (1) \cdot \sin \theta \\ &= \sin \theta\end{aligned}$$

In a similar manner, we can show that $\sin\left(\frac{\pi}{2} - \theta\right) = \cos \theta$.

We will use these results in deriving an identity for $\sin(\alpha + \beta)$.

$$\begin{aligned}\sin(\alpha + \beta) &= \cos\left(\frac{\pi}{2} - (\alpha + \beta)\right) && \blacksquare\ \sin \theta = \cos\left(\frac{\pi}{2} - \theta\right). \\ &= \cos\left(\left(\frac{\pi}{2} - \alpha\right) - \beta\right) \\ &= \cos\left(\frac{\pi}{2} - \alpha\right) \cos \beta + \sin\left(\frac{\pi}{2} - \alpha\right) \sin \beta && \blacksquare\ \text{By (12)} \\ &= \sin \alpha \cos \beta + \cos \alpha \sin \beta\end{aligned}$$

Thus,

(14) $$\sin(\alpha + \beta) = \sin \alpha \cos \beta + \cos \alpha \sin \beta.$$

Replacing $\beta$ with $(-\beta)$ in (14), we have

$$\begin{aligned}\sin(\alpha - \beta) &= \sin(\alpha + (-\beta)) \\ &= \sin \alpha \cos(-\beta) + \cos \alpha \sin(-\beta) && \blacksquare\ \text{By (14)} \\ &= \sin \alpha \cos \beta + \cos \alpha (-\sin \beta) && \blacksquare\ \text{Cosine is even; sine is odd.} \\ &= \sin \alpha \cos \beta - \cos \alpha \sin \beta\end{aligned}$$

Thus,

(15) $$\sin(\alpha - \beta) = \sin \alpha \cos \beta - \cos \alpha \sin \beta.$$

**Example 2** Determine $\sin \frac{5\pi}{12}$ without using Table 4 or a calculator.

***Solution***

$$\begin{aligned} \sin \frac{5\pi}{12} &= \sin\left(\frac{\pi}{4} + \frac{\pi}{6}\right) \\ &= \sin \frac{\pi}{4} \cos \frac{\pi}{6} + \cos \frac{\pi}{4} \sin \frac{\pi}{6} \\ &= \frac{\sqrt{2}}{2} \cdot \frac{\sqrt{3}}{2} + \frac{\sqrt{2}}{2} \cdot \frac{1}{2} \end{aligned}$$

Thus,

$$\sin \frac{5\pi}{12} = \frac{\sqrt{6}}{4} + \frac{\sqrt{2}}{4} = \frac{\sqrt{6} + \sqrt{2}}{4}.$$

Notice that in Example 1 we found that $\cos 15° = \frac{\sqrt{6} + \sqrt{2}}{4}$, and in Example 2 we found that $\sin \frac{5\pi}{12}$ also equals $\frac{\sqrt{6} + \sqrt{2}}{4}$. Since $\frac{5\pi}{12} = 75°$, this result is expected because of the cofunction identity:

$$\cos 15° = \sin(90° - 15°) = \sin 75°.$$

**Example 3** If $\sin \alpha = \frac{12}{13}$ for some second quadrant angle $\alpha$, and $\sin \beta = \frac{3}{5}$ for some first quadrant angle $\beta$, determine (a) $\sin(\alpha - \beta)$ and (b) $\sin(\alpha + \beta)$.

***Solution*** We begin by using the Pythagorean Identity (9) to find $\cos \alpha$ and $\cos \beta$.

$$\cos \alpha = \pm\sqrt{1 - \sin^2 \alpha}$$

$$\cos \alpha = -\sqrt{1 - \left(\frac{12}{13}\right)^2}$$ ■ Since $\alpha$ is a second quadrant angle, $\cos \alpha$ is negative.

$$\cos \alpha = -\sqrt{1 - \frac{144}{169}}$$

$$\cos \alpha = -\sqrt{\frac{25}{169}} = -\frac{5}{13}$$

$$\cos \beta = \sqrt{1 - \left(\frac{3}{5}\right)^2}$$ ■ Since $\beta$ is a first quadrant angle, $\cos \beta$ is positive.

$$\cos \beta = \sqrt{\frac{16}{25}} = \frac{4}{5}$$

(a) $\sin(\alpha - \beta) = \sin\alpha\cos\beta - \cos\alpha\sin\beta = \left(\frac{12}{13}\right)\left(\frac{4}{5}\right) - \left(\frac{-5}{13}\right)\left(\frac{3}{5}\right) = \frac{63}{65}$

(b) $\sin(\alpha + \beta) = \sin\alpha\cos\beta + \cos\alpha\sin\beta = \left(\frac{12}{13}\right)\left(\frac{4}{5}\right) + \left(\frac{-5}{13}\right)\left(\frac{3}{5}\right) = \frac{33}{65}$

The identity for $\tan(\alpha + \beta)$ may be derived by using the quotient identity (7) and the identities for the sine and cosine of a sum.

$$\tan(\alpha + \beta) = \frac{\sin(\alpha + \beta)}{\cos(\alpha + \beta)}$$ ■ By (7)

$$= \frac{\sin\alpha\cos\beta + \cos\alpha\sin\beta}{\cos\alpha\cos\beta - \sin\alpha\sin\beta}$$ ■ By (14) and (13)

$$= \frac{\dfrac{\sin\alpha\cos\beta}{\cos\alpha\cos\beta} + \dfrac{\cos\alpha\sin\beta}{\cos\alpha\cos\beta}}{\dfrac{\cos\alpha\cos\beta}{\cos\alpha\cos\beta} - \dfrac{\sin\alpha\sin\beta}{\cos\alpha\cos\beta}}$$ ■ Divide each term by $\cos\alpha\cos\beta$.

$$= \frac{\tan\alpha + \tan\beta}{1 - \tan\alpha\tan\beta}$$ ■ (7) is used four times.

Therefore,

(16) $$\tan(\alpha + \beta) = \frac{\tan\alpha + \tan\beta}{1 - \tan\alpha\tan\beta}$$

In a similar manner,

(17) $$\tan(\alpha - \beta) = \frac{\tan\alpha - \tan\beta}{1 + \tan\alpha\tan\beta}$$

The derivation of identity (17) is left as an exercise.

**Example 4** Without using Table 4 or a calculator, find $\tan\frac{7\pi}{12}$.

***Solution***

$$\tan\left(\frac{7\pi}{12}\right) = \tan\left(\frac{\pi}{4} + \frac{\pi}{3}\right) = \frac{\tan\frac{\pi}{4} + \tan\frac{\pi}{3}}{1 - \tan\frac{\pi}{4}\tan\frac{\pi}{3}}$$ ■ By (16)

$$= \frac{1 + \sqrt{3}}{1 - \sqrt{3}}$$

When an answer has a radical in both the numerator and the denominator, it is often easier to approximate the answer if we rationalize the denominator first. In the case of Example 4, we have,

$$\frac{1+\sqrt{3}}{1-\sqrt{3}} = \frac{(1+\sqrt{3})(1+\sqrt{3})}{(1-\sqrt{3})(1+\sqrt{3})} = \frac{1+2\sqrt{3}+3}{1-3} = \frac{4+2\sqrt{3}}{-2} = -(2+\sqrt{3}).$$

Thus, we see that $\tan\frac{7\pi}{12}$ is approximately $-3.73$. ($\sqrt{3} \approx 1.73$.)

**Example 5** Simplify the following trigonometric expression as a single function of some angle:

$$\frac{\sin(\alpha+\beta)+\sin(\alpha-\beta)}{\cos(\alpha+\beta)+\cos(\alpha-\beta)}.$$

***Solution***

$$\begin{aligned}\frac{\sin(\alpha+\beta)+\sin(\alpha-\beta)}{\cos(\alpha+\beta)+\cos(\alpha-\beta)} &= \frac{(\sin\alpha\cos\beta+\cos\alpha\sin\beta)+(\sin\alpha\cos\beta-\cos\alpha\sin\beta)}{(\cos\alpha\cos\beta-\sin\alpha\sin\beta)+(\cos\alpha\cos\beta+\sin\alpha\sin\beta)}\\ &= \frac{2\sin\alpha\cos\beta}{2\cos\alpha\cos\beta}\\ &= \frac{\sin\alpha}{\cos\alpha} = \tan\alpha\end{aligned}$$

The sum and difference identities derived in this section are listed below for easy reference.

**Sum and Difference Identities**

(12) $\cos(\alpha-\beta) = \cos\alpha\cos\beta + \sin\alpha\sin\beta$

(13) $\cos(\alpha+\beta) = \cos\alpha\cos\beta - \sin\alpha\sin\beta$

(14) $\sin(\alpha+\beta) = \sin\alpha\cos\beta + \cos\alpha\sin\beta$

(15) $\sin(\alpha-\beta) = \sin\alpha\cos\beta - \cos\alpha\sin\beta$

(16) $\tan(\alpha+\beta) = \dfrac{\tan\alpha+\tan\beta}{1-\tan\alpha\tan\beta}$

(17) $\tan(\alpha-\beta) = \dfrac{\tan\alpha-\tan\beta}{1+\tan\alpha\tan\beta}$

## Exercise Set 4.2 ▤ Calculator Exercises in the Appendix

### Goal A

In exercises 1–12, without using Table 4 or a calculator, find the following trigonometric values by using the sum and difference identities.

1. sin 120°
2. cos 120°
3. sin 15°
4. cos 15°
5. cos 75°
6. sin 75°
7. sin 195°
8. cos 105°
9. cos(−45°)
10. sin(−45°)
11. sin(−135°)
12. cos(−135°)

In exercises 13–20, without using Table 4 or a calculator, find the following trigonometric values.

13. $\cos \frac{5\pi}{12}$
14. $\sin \frac{5\pi}{12}$
15. $\sin \frac{7\pi}{12}$
16. $\cos \frac{7\pi}{12}$
17. $\cos\left(-\frac{5\pi}{12}\right)$
18. $\sin\left(-\frac{5\pi}{12}\right)$
19. $\cos\left(-\frac{11\pi}{12}\right)$
20. $\sin\left(-\frac{11\pi}{12}\right)$

In exercises 21–26, find the following trigonometric values without using Table 4 or a calculator.

21. tan 75°
22. tan 15°
23. $\tan \frac{19\pi}{12}$
24. $\tan \frac{\pi}{12}$
25. $\tan\left(-\frac{\pi}{12}\right)$
26. $\tan\left(-\frac{5\pi}{12}\right)$

In exercises 27–30, use the given information to determine the following: (a) $\sin(\alpha + \beta)$, (b) $\cos(\alpha + \beta)$, and (c) $\tan(\alpha + \beta)$.

27. $\sin \alpha = \frac{4}{5}$, $\alpha$ is a first quadrant angle, $\beta$ is a third quadrant angle, and $\sin \beta = -\frac{12}{13}$.

28. $\sin \alpha = -\frac{4}{5}$, $\alpha$ is a fourth quadrant angle, $\beta$ is a second quadrant angle, and $\sin \beta = \frac{3}{5}$.

29. $\cos \alpha = \frac{5}{13}$, $\alpha$ is a first quadrant angle, $\beta$ is a second quadrant angle, and $\tan \beta = -\frac{3}{2}$.

30. $\cos \alpha = -\frac{5}{12}$, $\alpha$ is a second quadrant angle, $\beta$ is a third quadrant angle, and $\tan \beta = \frac{3}{2}$.

In exercises 31–34, given the information in exercises 27–30, find the following:
(a) $\sin(\alpha - \beta)$, (b) $\cos(\alpha - \beta)$, and (c) $\tan(\alpha - \beta)$.

In exercises 35–44, verify each identity.

35. $\sin(\alpha - \beta)\cos \beta + \cos(\alpha - \beta)\sin \beta = \sin \alpha$

36. $\sin(\alpha - \beta)\sin(\alpha + \beta) = \sin^2 \alpha - \sin^2 \beta$

37. $\cos(\alpha - \beta)\cos(\alpha + \beta) = \cos^2 \alpha - \sin^2 \beta$

38. $\cos(\alpha - \beta) + \cos(\alpha + \beta) = 2 \cos \alpha \cos \beta$

39. $\tan\left(\alpha + \frac{\pi}{4}\right) = \frac{\cos \alpha + \sin \alpha}{\cos \alpha - \sin \alpha}$

40. $\tan\left(\frac{\pi}{4} - \alpha\right) = \frac{\cos \alpha - \sin \alpha}{\cos \alpha + \sin \alpha}$

41. $\tan\left(\frac{\pi}{4} + \alpha\right)\tan\left(\frac{\pi}{4} - \alpha\right) = \tan \frac{\pi}{4}$

42. $\frac{\tan(\alpha - \beta)}{\tan(\beta - \alpha)} = -1$

43. $\frac{\sin(x + y)}{\sin(x - y)} = \frac{\tan x + \tan y}{\tan x - \tan y}$

44. $\frac{\sin(x + y)}{\sin x \sin y} = \frac{1}{\tan x} + \frac{1}{\tan y}$

### Superset

In exercises 45–52, simplify the given sum or difference.

45. $\cos\left(\alpha + \frac{\pi}{6}\right)$
46. $\sin\left(\alpha + \frac{\pi}{6}\right)$
47. $\tan\left(\alpha + \frac{\pi}{4}\right)$
48. $\cos\left(\beta + \frac{\pi}{4}\right)$
49. $\sin\left(\frac{\pi}{2} - \varphi\right)$
50. $\tan\left(\frac{\pi}{2} - \beta\right)$
51. $\sin\left(\frac{\pi}{4} - \alpha\right)$
52. $\cos\left(\frac{\pi}{4} - \beta\right)$

In exercises 53–58, derive an identity involving the given expression.

53. $\tan(\alpha - \beta)$
54. $\cot(\alpha + \beta)$
55. $\sec(\alpha + \beta)$
56. $\csc(\alpha + \beta)$
57. $\sin 2\alpha$ (Hint: $2\alpha = \alpha + \alpha$.)
58. $\cos 2\alpha$

## 4.3 The Double-Angle and Half-Angle Identities

**Goal A** *to apply the double-angle identities*

In this section we will derive identities for $\sin 2\theta$, $\cos 2\theta$, $\tan 2\theta$, $\sin \frac{\theta}{2}$, $\cos \frac{\theta}{2}$, and $\tan \frac{\theta}{2}$. The first three identities are called **double-angle identities;** the other three are called **half-angle identities.**

The double-angle identities follow easily from the sum identities. For example, consider the identity $\sin(\alpha + \beta) = \sin \alpha \cos \beta + \cos \alpha \sin \beta$. If $\alpha = \beta$, we have $\sin(2\alpha) = \sin(\alpha + \alpha) = \sin \alpha \cos \alpha + \cos \alpha \sin \alpha$. Thus, for any angle $\theta$

(18) $$\sin 2\theta = 2 \sin \theta \cos \theta.$$

Similarly, $\cos 2\theta = \cos(\theta + \theta) = \cos \theta \cos \theta - \sin \theta \sin \theta$, that is,

(19) $$\cos 2\theta = \cos^2 \theta - \sin^2 \theta.$$

Using the Pythagorean Identity, $\sin^2 \theta + \cos^2 \theta = 1$, we can use (19) to derive two other identities for $\cos 2\theta$:

$$\begin{aligned} \cos 2\theta &= \cos^2 \theta - \sin^2 \theta \\ &= (1 - \sin^2 \theta) - \sin^2 \theta \end{aligned}$$

(20) $$\cos 2\theta = 1 - 2 \sin^2 \theta.$$

Also,

$$\begin{aligned} \cos 2\theta &= \cos^2 \theta - \sin^2 \theta \\ &= \cos^2 \theta - (1 - \cos^2 \theta) \end{aligned}$$

(21) $$\cos 2\theta = 2 \cos^2 \theta - 1.$$

We leave as an exercise to show that

(22) $$\tan 2\theta = \frac{2 \tan \theta}{1 - \tan^2 \theta}.$$

**Example 1** If $\sin \theta = 0.6$ and $\cos \theta = 0.8$, find (a) $\sin 2\theta$, and (b) $\cos 2\theta$.

***Solution*** (a) $\sin 2\theta = 2 \sin \theta \cos \theta = 2(0.6)(0.8) = 0.96$ ■ By (18)

(b) $\cos 2\theta = \cos^2 \theta - \sin^2 \theta = (0.8)^2 - (0.6)^2 = 0.28$ ■ By (19)

**Example 2** Use double-angle identities and the values of sin 30°, cos 30°, and tan 30° from the Table of Special Angles to determine the following: (a) sin 60°, (b) cos 60°, (c) tan 60°.

***Solution***

(a) $\sin 60° = \sin 2(30°) = 2 \sin 30° \cos 30° = 2\left(\frac{1}{2}\right)\left(\frac{\sqrt{3}}{2}\right) = \frac{\sqrt{3}}{2}$

(b) $\cos 60° = \cos 2(30°) = \cos^2 30° - \sin^2 30° = \left(\frac{\sqrt{3}}{2}\right)^2 - \left(\frac{1}{2}\right)^2 = \frac{1}{2}$

(c) $\tan 60° = \tan 2(30°) = \dfrac{2 \tan 30°}{1 - \tan^2 30°} = \dfrac{2\left(\frac{\sqrt{3}}{3}\right)}{1 - \left(\frac{\sqrt{3}}{3}\right)^2}$

$$= \frac{\frac{2\sqrt{3}}{3}}{1 - \frac{3}{9}} = \frac{2\sqrt{3}}{3} \cdot \frac{9}{6} = \sqrt{3}$$

**Example 3** Given that $\sin \frac{\pi}{9} \approx 0.3420$, find $\tan \frac{2\pi}{9}$.

***Solution***

$$\tan \frac{2\pi}{9} = \frac{\sin \frac{2\pi}{9}}{\cos \frac{2\pi}{9}} = \frac{\sin 2\left(\frac{\pi}{9}\right)}{\cos 2\left(\frac{\pi}{9}\right)}$$ ■ By (7)

$$= \frac{2 \sin \frac{\pi}{9} \cos \frac{\pi}{9}}{1 - 2 \sin^2\left(\frac{\pi}{9}\right)}$$ ■ By (18) ■ By (20)

To proceed we need to determine $\cos \frac{\pi}{9}$.

$$\sin^2\left(\frac{\pi}{9}\right) + \cos^2\left(\frac{\pi}{9}\right) = 1$$ ■ By (9)

$$\cos \frac{\pi}{9} = \sqrt{1 - \sin^2\left(\frac{\pi}{9}\right)}$$ ■ Since $\frac{\pi}{9}$ is a first quadrant angle, $\cos \frac{\pi}{9}$ is positive.

$$\approx \sqrt{1 - (0.3420)^2} = 0.9397$$

Thus, $$\tan \frac{2\pi}{9} = \frac{2 \sin \frac{\pi}{9} \cos \frac{\pi}{9}}{1 - 2 \sin^2 \frac{\pi}{9}} \approx \frac{2(0.3420)(0.9397)}{1 - 2(0.3420)^2} \approx 0.8390.$$

**Example 4** Given $\sin\theta = -\frac{4}{5}$ and that $\theta$ is a third quadrant angle, determine (a) $\sin 2\theta$ (b) $\cos 2\theta$ (c) $\tan 2\theta$.

***Solution*** (a) To apply the identity $\sin 2\theta = 2\sin\theta\cos\theta$, we need to determine $\cos\theta$.

$$\cos\theta = -\sqrt{1-\sin^2\theta}$$

■ By (9). Since $\theta$ is a third quadrant angle, $\cos\theta$ is negative.

$$= -\sqrt{1-\left(-\frac{4}{5}\right)^2} = -\sqrt{1-\frac{16}{25}} = -\frac{3}{5}$$

$$\sin 2\theta = 2\left(-\frac{4}{5}\right)\left(-\frac{3}{5}\right) = \frac{24}{25}.$$

■ By (18)

(b) $$\cos 2\theta = 1 - 2\sin^2\theta$$

■ Identity (20)

$$= 1 - 2\left(-\frac{4}{5}\right)^2 = 1 - \frac{32}{25} = -\frac{7}{25}$$

(c) $$\tan 2\theta = \frac{\sin 2\theta}{\cos 2\theta} = \frac{\frac{24}{25}}{-\frac{7}{25}} = -\frac{24}{7}$$

In Example 4(c), we used the quotient identity (7) to determine $\tan 2\theta$, since we already knew the values of $\sin 2\theta$ and $\cos 2\theta$ from parts (a) and (b). We could also have used the double-angle identity (22),

$$\tan 2\theta = \frac{2\tan\theta}{1-\tan^2\theta}.$$

In that case, we would first need to find the value of $\tan\theta$ by computing $\frac{\sin\theta}{\cos\theta}$.

**Goal B** *to use the half-angle identities*

To derive the first half-angle identity, we solve the double-angle identity $\cos 2\theta = 1 - 2\sin^2\theta$ for $\sin^2\theta$:

$$\sin^2\theta = \frac{1-\cos 2\theta}{2}.$$

If we replace $\theta$ with $\frac{\theta}{2}$, we have

$$\sin^2\frac{\theta}{2} = \frac{1-\cos\theta}{2}.$$

Thus, we can state our first half-angle identity as follows:

(23) $$\sin\frac{\theta}{2} = \pm\sqrt{\frac{1-\cos\theta}{2}}$$

■ The positive or negative sign is used depending on the quadrant in which $\frac{\theta}{2}$ is located.

**Example 5** Use identity (23) to determine $\sin 15°$.

***Solution***

$$\sin 15° = \sin\frac{30°}{2} = \sqrt{\frac{1-\cos 30°}{2}}$$

$$= \sqrt{\frac{1-\frac{\sqrt{3}}{2}}{2}} = \sqrt{\frac{1}{2}-\frac{\sqrt{3}}{4}} = \sqrt{\frac{2}{4}-\frac{\sqrt{3}}{4}} = \frac{\sqrt{2-\sqrt{3}}}{2}$$

■ We choose the positive sign since 15° is a first quadrant angle.

To derive a half-angle identity for the cosine function, we solve the identity $\cos 2\theta = 2\cos^2\theta - 1$ for $\cos\theta$:

$$\cos\theta = \pm\sqrt{\frac{1+\cos 2\theta}{2}}.$$

Replacing $\theta$ with $\frac{\theta}{2}$, we conclude that

(24) $$\cos\frac{\theta}{2} = \pm\sqrt{\frac{1+\cos\theta}{2}}.$$

**Example 6** Use identity (24) to determine $\cos\frac{7\pi}{12}$.

***Solution***

$$\cos\frac{7\pi}{12} = \cos\left(\frac{1}{2}\left(\frac{7\pi}{6}\right)\right) = -\sqrt{\frac{1+\cos\frac{7\pi}{6}}{2}}$$

$$= -\sqrt{\frac{1-\frac{\sqrt{3}}{2}}{2}} = -\frac{\sqrt{2-\sqrt{3}}}{2}$$

■ We choose the negative sign since $\frac{7\pi}{12}$ is a second quadrant angle and cosine is negative there.

■ $\cos\frac{7\pi}{6} = -\cos\frac{\pi}{6} = -\frac{\sqrt{3}}{2}$.

Using identities (7), (23), and (24), we can establish the following:

(25) $$\tan\frac{\theta}{2} = \pm\sqrt{\frac{1-\cos\theta}{1+\cos\theta}}.$$

There are two other ways of describing $\tan\frac{\theta}{2}$ that are often useful.

$$\tan\frac{\theta}{2} = \frac{\sin\frac{\theta}{2}}{\cos\frac{\theta}{2}} = \frac{\sin\frac{\theta}{2}}{\cos\frac{\theta}{2}}\cdot\frac{2\cos\frac{\theta}{2}}{2\cos\frac{\theta}{2}} = \frac{2\sin\frac{\theta}{2}\cos\frac{\theta}{2}}{2\cos^2\left(\frac{\theta}{2}\right)}$$

$$= \frac{\sin\left(2\cdot\frac{\theta}{2}\right)}{1+\left(2\cos^2\left(\frac{\theta}{2}\right)-1\right)}$$ ■ By (18) ■ Add 1 and subtract 1.

$$= \frac{\sin\theta}{1+\cos\theta}$$ ■ By (21)

Thus,

(26) $$\tan\frac{\theta}{2} = \frac{\sin\theta}{1+\cos\theta}.$$

Multiplying the right-hand side of (26) by $\frac{1-\cos\theta}{1-\cos\theta}$, yields still another identity for $\tan\frac{\theta}{2}$.

(27) $$\tan\frac{\theta}{2} = \frac{1-\cos\theta}{\sin\theta}.$$

Identities (26) and (27) have an advantage over identity (25) in that they do not contain a radical sign.

**Example 7** Use a half-angle identity to determine $\tan(-202.5°)$.

***Solution*** Solving this problem requires that we recognize that $-202.5°$ is $\frac{1}{2}(-405°)$ and that $-405°$ is a fourth quadrant angle which has 45° as its reference angle.

$$\tan(-202.5°) = \tan\left(\frac{-405°}{2}\right) = \frac{\sin(-405°)}{1+\cos(-405°)}$$ ■ By (26)

$$= \frac{-\frac{\sqrt{2}}{2}}{1+\frac{\sqrt{2}}{2}} = -\frac{\sqrt{2}}{2+\sqrt{2}}$$ ■ $\sin(-405°) = -\sin 45° = -\frac{\sqrt{2}}{2}$ ■ $\cos(-405°) = +\cos 45° = \frac{\sqrt{2}}{2}$

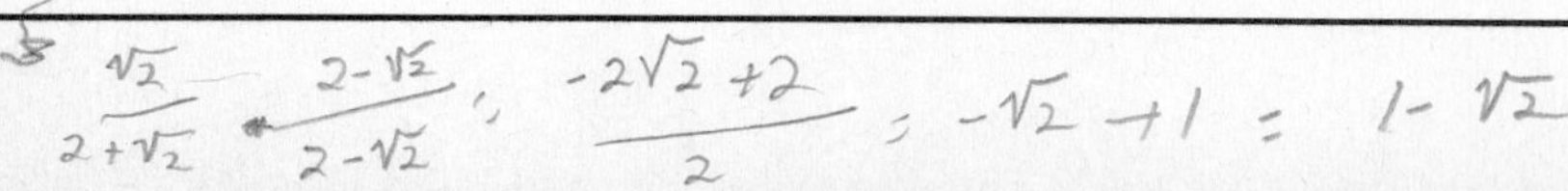

In Example 7 we can estimate the value of tan $(-202.5°)$ most easily if we rationalize the denominator first.

$$\frac{-\sqrt{2}}{2+\sqrt{2}} = \frac{-\sqrt{2}}{2+\sqrt{2}} \cdot \frac{2-\sqrt{2}}{2-\sqrt{2}} = \frac{-2\sqrt{2}+2}{2}$$
$$= -\sqrt{2} + 1 \approx 1 - 1.414 = -0.414$$

That is, $\tan(-202.5°)$ is approximately $-0.414$.

For easy reference, we summarize the identities presented in this section.

**Double-Angle Identities**

(18) $\sin 2\theta = 2 \sin \theta \cos \theta$

(19) $\cos 2\theta = \cos^2 \theta - \sin^2 \theta$

(20) $\cos 2\theta = 1 - 2 \sin^2 \theta$

(21) $\cos 2\theta = 2 \cos^2 \theta - 1$

(22) $\tan 2\theta = \dfrac{2 \tan \theta}{1 - \tan^2 \theta}$

**Half-Angle Identities**

(23) $\sin \dfrac{\theta}{2} = \pm \sqrt{\dfrac{1 - \cos \theta}{2}}$

(24) $\cos \dfrac{\theta}{2} = \pm \sqrt{\dfrac{1 + \cos \theta}{2}}$

(25) $\tan \dfrac{\theta}{2} = \pm \sqrt{\dfrac{1 - \cos \theta}{1 + \cos \theta}}$

(26) $\tan \dfrac{\theta}{2} = \dfrac{\sin \theta}{1 + \cos \theta}$

(27) $\tan \dfrac{\theta}{2} = \dfrac{1 - \cos \theta}{\sin \theta}$

## Exercise Set 4.3 Calculator Exercises in the Appendix

### Goal A

In exercises 1–18, given $\alpha$, find $\sin 2\alpha$, $\cos 2\alpha$, and $\tan 2\alpha$ using the double-angle identities.

**1.** 60° **2.** 45° **3.** 120°

**4.** 180° **5.** 135° **6.** 150°

**7.** $-300°$ **8.** $-225°$ **9.** 585°

**10.** 765° **11.** $\dfrac{5\pi}{6}$ **12.** $\dfrac{3\pi}{4}$

**13.** $-\dfrac{2\pi}{3}$ **14.** $-\dfrac{7\pi}{6}$ **15.** $-\dfrac{3\pi}{4}$

16. $-3\pi$

17. $\frac{22\pi}{8}$

18. $\frac{26\pi}{12}$

In exercises 19–26, from the given information, determine (a) $\sin 2\alpha$, (b) $\cos 2\alpha$, (c) $\tan 2\alpha$.

19. $\cos \alpha = -\frac{4}{5}$, $\alpha$ in Quadrant II

20. $\cos \alpha = \frac{4}{5}$, $\alpha$ in Quadrant IV

21. $\sin \alpha = -\frac{12}{13}$, $\alpha$ in Quadrant IV

22. $\tan \alpha = \frac{4}{3}$, $\alpha$ in Quadrant III

23. $\tan \alpha = \frac{9}{40}$, $\alpha$ in Quadrant I

24. $\sin \alpha = \frac{5}{13}$, $\alpha$ in Quadrant I

25. $\cot \alpha = 2$, $\alpha$ in Quadrant III

26. $\sec \alpha = -\frac{25}{24}$, $\alpha$ in Quadrant II

## Goal B

In exercises 27–46, given $\alpha$, find $\sin \frac{\alpha}{2}$, $\cos \frac{\alpha}{2}$, and $\tan \frac{\alpha}{2}$ using the half-angle identities.

27. $60°$

28. $180°$

29. $90°$

30. $120°$

31. $210°$

32. $315°$

33. $-225°$

34. $-495°$

35. $675°$

36. $855°$

37. $4\pi$

38. $\frac{2\pi}{3}$

39. $\frac{4\pi}{3}$

40. $\frac{5\pi}{2}$

41. $\frac{\pi}{6}$

42. $\frac{5\pi}{4}$

43. $-\frac{11\pi}{6}$

44. $-\frac{3\pi}{4}$

45. $\frac{19\pi}{6}$

46. $\frac{16\pi}{3}$

In exercises 47–54, use the information given in exercises 19–26 to determine (a) $\sin \frac{\alpha}{2}$, (b) $\cos \frac{\alpha}{2}$, and (c) $\tan \frac{\alpha}{2}$.

## Superset

55. Derive identity (22) for $\tan 2\alpha$.

56. Derive identity (25) for $\tan \frac{\alpha}{2}$.

57. Work out the details to show identity (27).

58. Express $\sin 3\alpha$ in terms of $\sin \alpha$ only.

59. Express $\cos 3\alpha$ in terms of $\cos \alpha$ only.

60. Express $\cos 4\alpha$ in terms of $\cos \alpha$ only.

61. Express $\sin 4\alpha$ in terms of $\sin \alpha$ and $\cos \alpha$.

62. Express $\tan 4\alpha$ in terms of $\tan \alpha$ only.

In exercises 63–72, express the first trigonometric function as specified.

63. $\cos 12\alpha$ in terms of $\cos 6\alpha$

64. $\cos 12\alpha$ in terms of $\sin 6\alpha$

65. $\sin 8\alpha$ in terms of $\sin 4\alpha$ and $\cos 4\alpha$

66. $\cos 8\alpha$ in terms of $\sin 4\alpha$ and $\cos 4\alpha$

67. $\tan 6\alpha$ in terms of $\tan 2\alpha$

68. $\cot 6\alpha$ in terms of $\tan 2\alpha$

69. $\cot 6\alpha$ in terms of $\cot 12\alpha$ and $\csc 12\alpha$

70. $\tan 6\alpha$ in terms of $\tan 12\alpha$ and $\sec 12\alpha$

71. $\sin 8\alpha$ in terms of $\sin 4\alpha$

72. $\sin 2\alpha$ in terms of $\cot 4\alpha$ and $\csc 4\alpha$

In exercises 73–84, simplify each of the trigonometric expressions.

73. $\frac{\sin 2\alpha}{2 \sin \alpha}$

74. $\frac{\sin 2\alpha}{2 \cos \alpha}$

75. $2 \sin \frac{\alpha}{2} \cos \frac{\alpha}{2}$

76. $2 \sin 2\alpha \cos 2\alpha$

77. $1 - 2 \sin^2 \frac{\alpha}{2}$

78. $2 \cos^2 \frac{\alpha}{2} - 1$

79. $(\sin \alpha + \cos \alpha)^2 - \sin 2\alpha$

80. $(\sin \alpha - \cos \alpha)^2 + \sin 2\alpha$

81. $\frac{\cos^2 \alpha}{1 + \sin \alpha}$

82. $\frac{\cos^3 \alpha - \sin^3 \alpha}{\cos \alpha - \sin \alpha}$

83. $\cos^4 \alpha - \sin^4 \alpha$

84. $\frac{\cos^4 \alpha - \sin^4 \alpha}{\sin^2 \alpha - \cos^2 \alpha}$

## 4.4 Identities Revisited

**Goal A** *to simplify a trigonometric expression using identities*

As we have already seen, the primary strategy used to verify a trigonometric identity is to transform the expression on one side of the identity so that it is identical to the expression on the other side. Frequently this means that you must recognize that a form present in the problem, is a part of a known identity.

**Example 1** Simplify the expression $\cos 2x \cos x + \sin 2x \sin x$.

***Solution*** Observe that $\cos 2x \cos x + \sin 2x \sin x$ is simply the right-hand side of identity (12): $\cos(\alpha - \beta) = \cos \alpha \cos \beta + \sin \alpha \sin \beta$, with $\alpha$ replaced by $2x$ and $\beta$ replaced by $x$. Thus,

$$\cos 2x \cos x + \sin 2x \sin x = \cos(2x - x) = \cos x.$$

**Example 2** Simplify the expression $\dfrac{\sin 2x}{1 + \cos 2x}$.

***Solution*** The given expression is the right-hand side of identity (26): $\tan \dfrac{\theta}{2} = \dfrac{\sin \theta}{1 + \cos \theta}$, with $\theta$ replaced by $2x$. Thus,

$$\frac{\sin 2x}{1 + \cos 2x} = \tan \frac{2x}{2} = \tan x.$$

**Example 3** Simplify the expression $1 - 2 \sin^2 3A$.

***Solution*** This expression is the right-hand side of identity (20): $\cos 2\theta = 1 - 2 \sin^2 \theta$, with $\theta$ replaced by $3A$. Thus

$$1 - 2 \sin^2 3A = \cos 2(3A) = \cos 6A.$$

**Goal B** *to verify trigonometric identities*

In section 1 of this chapter, we introduced techniques for verifying identities, but had only the Basic Identities (1)–(11) at our disposal. We now have, in addition, identities (12)–(27) to assist us in verifying new identities. We now consider several examples that draw upon the knowledge of all the identities established thus far.

**Example 4** Verify the identity: $\dfrac{\sin 2\theta}{1 + \cos 2\theta} = \tan \theta$.

***Solution***

$$\frac{\sin 2\theta}{1 + \cos 2\theta} = \frac{2 \sin \theta \cos \theta}{1 + (2 \cos^2 \theta - 1)}$$

■ (18) is used in the numerator. (21) is used in the denominator to produce an expression that will have only one term.

$$= \frac{2 \sin \theta \cos \theta}{2 \cos^2 \theta}$$

■ Denominator is simplified.

$$= \frac{\sin \theta}{\cos \theta} = \tan \theta$$

■ Common factors are canceled.

**Example 5** Verify the identity: $\dfrac{\tan \theta + \sin \theta}{2 \tan \theta} = \cos^2 \dfrac{\theta}{2}$.

***Solution***

$$\frac{\tan \theta + \sin \theta}{2 \tan \theta} = \frac{\tan \theta}{2 \tan \theta} + \frac{\sin \theta}{2 \tan \theta}$$

■ We work with the left-hand side of the identity since it is more complicated and offers the best chance for simplification.

$$= \frac{\tan \theta}{2 \tan \theta} + \frac{\sin \theta}{2} \cdot \frac{\cos \theta}{\sin \theta}$$

■ Since $\tan \theta = \dfrac{\sin \theta}{\cos \theta}$, we can divide by $\tan \theta$ by multiplying by its reciprocal $\dfrac{\cos \theta}{\sin \theta}$.

$$= \frac{1}{2} + \frac{\cos \theta}{2} = \frac{1 + \cos \theta}{2}$$

$$= \cos^2 \frac{\theta}{2}$$

■ By (24)

**Example 6** Verify the identity: $\sin(\alpha + \beta) \sin(\alpha - \beta) = \sin^2 \alpha - \sin^2 \beta$.

***Solution***

$$\sin(\alpha + \beta) \sin(\alpha - \beta)$$

$$= (\sin \alpha \cos \beta + \cos \alpha \sin \beta)(\sin \alpha \cos \beta - \cos \alpha \sin \beta)$$

■ By (14) and (15)

$$= (\sin \alpha \cos \beta)^2 - (\cos \alpha \sin \beta)^2$$

■ $(A + B)(A - B) = A^2 - B^2$.

$$= \sin^2 \alpha \cos^2 \beta - \cos^2 \alpha \sin^2 \beta$$

$$= \sin^2 \alpha (1 - \sin^2 \beta) - (1 - \sin^2 \alpha)(\sin^2 \beta)$$

■ We want an expression that involves only $\sin \alpha$ and $\sin \beta$.

$$= \sin^2 \alpha - \sin^2 \alpha \sin^2 \beta - \sin^2 \beta + \sin^2 \alpha \sin^2 \beta$$

$$= \sin^2 \alpha - \sin^2 \beta$$

**Example 7** Verify the identity: $\dfrac{\sec \beta}{1 + \cos \beta} = \dfrac{\tan \beta - \sin \beta}{\sin^3 \beta}$.

***Solution***

$$\frac{\sec \beta}{1 + \cos \beta} = \frac{\sec \beta}{1 + \cos \beta} \cdot \frac{1 - \cos \beta}{1 - \cos \beta}$$

■ We will work with the left-hand side of the identity. The denominator is multiplied by $1 - \cos \beta$ in order to get an expression involving only $\sin \beta$.

$$= \frac{\sec\beta(1 - \cos\beta)}{1 - \cos^2\beta}$$

$$= \frac{\sec\beta - \sec\beta(\cos\beta)}{\sin^2\beta}$$

■ By (9)

$$= \frac{\sec\beta - 1}{\sin^2\beta}$$

■ $\sec\beta(\cos\beta) = \sec\beta\left(\frac{1}{\sec\beta}\right) = 1.$

$$= \frac{\sec\beta - 1}{\sin^2\beta} \cdot \frac{\sin\beta}{\sin\beta}$$

■ We know that we want $\sin^3\beta$ in the denominator, so we multiply by $\frac{\sin\beta}{\sin\beta}$.

$$= \frac{\sec\beta\sin\beta - \sin\beta}{\sin^3\beta}$$

$$= \frac{\tan\beta - \sin\beta}{\sin^3\beta}$$

■ $\sec\beta\sin\beta = \frac{1}{\cos\beta}\sin\beta = \tan\beta.$

## Goal C *to use conversion identities*

We now consider a group of identities that allow us to restate a product of two trigonometric function values as a sum, and a sum of two trigonometric function values as a product. These new identities are derived from the sum and difference identities for the sine and cosine (identities (12)–(15)). For example, based on identities (14) and (15), we have

$$\begin{aligned}\sin(\alpha + \beta) + \sin(\alpha - \beta) &= (\sin\alpha\cos\beta + \cos\alpha\sin\beta) \\ &\quad + (\sin\alpha\cos\beta - \cos\alpha\sin\beta) \\ &= 2\sin\alpha\cos\beta.\end{aligned}$$

Thus, $\sin(\alpha + \beta) + \sin(\alpha - \beta) = 2\sin\alpha\cos\beta$, or

(28) $$\sin\alpha\cos\beta = \frac{1}{2}[\sin(\alpha + \beta) + \sin(\alpha - \beta)]$$

**Example 8** Determine sin 15° cos 75° by using identity (28).

***Solution***

$$\begin{aligned}\sin 15^\circ\cos 75^\circ &= \frac{1}{2}[\sin(15^\circ + 75^\circ) + \sin(15^\circ - 75^\circ)] \\ &= \frac{1}{2}[\sin 90^\circ + \sin(-60^\circ)] \\ &= \frac{1}{2}[\sin 90^\circ - \sin 60^\circ] \\ &= \frac{1}{2}\left(1 - \frac{\sqrt{3}}{2}\right) = \frac{2 - \sqrt{3}}{4}\end{aligned}$$

■ Recall that $\sin(-\theta) = -\sin\theta$ because sine is an odd function.

In identity (28), let us replace $\alpha + \beta$ with $x$ and $\alpha - \beta$ with $y$. Since

$$x + y = (\alpha + \beta) + (\alpha - \beta) = 2\alpha, \quad \text{and}$$
$$x - y = (\alpha + \beta) - (\alpha - \beta) = 2\beta,$$

we have $\alpha = \dfrac{x + y}{2}$, and $\beta = \dfrac{x - y}{2}$. Thus identity (28) can be rewritten as

$$\sin\left(\frac{x + y}{2}\right) \cos\left(\frac{x - y}{2}\right) = \frac{1}{2}(\sin x + \sin y),$$

or equivalently as

(31) $$\sin x + \sin y = 2\left[\sin\left(\frac{x + y}{2}\right) \cos\left(\frac{x - y}{2}\right)\right].$$

Notice that identity (31) expresses the sum of two trigonometric function values as a product of trigonometric function values.

Employing arguments similar to those used to derive identities (28) and (31), we can develop five other identities. These identities are called **conversion identities** because they allow us to convert products of trigonometric function values into sums, and vice versa.

**Conversion Identities**

(28) $$\sin \alpha \cos \beta = \frac{1}{2}[\sin(\alpha + \beta) + \sin(\alpha - \beta)]$$

(29) $$\sin \alpha \sin \beta = \frac{1}{2}[\cos(\alpha - \beta) - \cos(\alpha + \beta)]$$

(30) $$\cos \alpha \cos \beta = \frac{1}{2}[\cos(\alpha + \beta) + \cos(\alpha - \beta)]$$

(31) $$\sin x + \sin y = 2 \sin\left(\frac{x + y}{2}\right) \cos\left(\frac{x - y}{2}\right)$$

(32) $$\sin x - \sin y = 2 \cos\left(\frac{x + y}{2}\right) \sin\left(\frac{x - y}{2}\right)$$

(33) $$\cos x + \cos y = 2 \cos\left(\frac{x + y}{2}\right) \cos\left(\frac{x - y}{2}\right)$$

(34) $$\cos x - \cos y = -2 \sin\left(\frac{x + y}{2}\right) \sin\left(\frac{x - y}{2}\right)$$

**Example 9** Express (a) $\cos 8\alpha + \cos 6\alpha$ as a product (b) $\cos 2\theta \sin \theta$ as a sum or difference.

***Solution*** (a) $\cos 8\alpha + \cos 6\alpha = 2\cos\left(\frac{8\alpha + 6\alpha}{2}\right)\cos\left(\frac{8\alpha - 6\alpha}{2}\right)$

$= 2\cos 7\alpha \cos \alpha.$

■ Identity (33) is used with $x = 8\alpha$ and $y = 6\alpha$.

(b) $\cos 2\theta \sin \theta = \frac{1}{2}[\sin(3\theta) + \sin(-\theta)]$

$= \frac{1}{2}[\sin 3\theta - \sin \theta].$

■ Identity (28) is used with $\alpha = \theta$ and $\beta = 2\theta$.

**Example 10** Verify the following identity: $\dfrac{\sin 4x - \sin 3x}{\cos 4x + \cos 3x} = \dfrac{1 - \cos x}{\sin x}$.

***Solution***

$$\frac{\sin 4x - \sin 3x}{\cos 4x + \cos 3x} = \frac{2\cos\left(\frac{4x + 3x}{2}\right)\sin\left(\frac{4x - 3x}{2}\right)}{2\cos\left(\frac{4x + 3x}{2}\right)\cos\left(\frac{4x - 3x}{2}\right)}$$

■ By (32)

■ By (33)

$$= \frac{\sin\frac{x}{2}}{\cos\frac{x}{2}} = \tan\frac{x}{2} = \frac{1 - \cos x}{\sin x}$$

■ By (27)

## Exercise Set 4.4

### Goal A

In exercises 1–8, simplify the expression.

1. $\cos 4\alpha \cos \alpha + \sin \alpha \sin 4\alpha$
2. $\cos 4\alpha \cos \alpha - \sin \alpha \sin 4\alpha$
3. $\sin 4\alpha \cos \alpha - \sin \alpha \cos 4\alpha$
4. $\sin 4\alpha \cos \alpha + \sin \alpha \cos 4\alpha$
5. $\dfrac{2\tan 2\varphi}{1 - \tan^2 2\varphi}$
6. $\dfrac{\sin 2\alpha}{1 + \cos 2\alpha}$
7. $\dfrac{\sec \alpha - 1}{\sec \alpha}$
8. $\dfrac{\tan \beta}{\sec^2 \beta - 2}$

### Goal B

In exercises 9–20, verify the identity.

9. $\tan x = \dfrac{2\tan\frac{1}{2}x}{1 - \tan^2\frac{1}{2}x}$
10. $\cot x = \dfrac{\cot^2\frac{1}{2}x - 1}{2\cot\frac{1}{2}x}$
11. $\dfrac{\cos(x + y)}{\cos x \sin y} = \cot y - \tan x$
12. $\dfrac{\sin(x + y)}{\sin(x - y)} = \dfrac{\tan x + \tan y}{\tan x - \tan y}$

13. $\dfrac{1 - \cos 2\alpha}{1 + \cos 2\alpha} = \tan^2 \alpha$

14. $\dfrac{\sin(\alpha + \beta)}{\sin \alpha \sin \beta} = \cot \alpha + \cot \beta$

15. $\dfrac{\sin \alpha + \sin \beta}{\cos \alpha - \cos \beta} = -\cot \frac{1}{2}(\alpha - \beta)$

16. $\dfrac{\cos(\alpha + \beta)}{\cos(\alpha - \beta)} = -\dfrac{\tan \beta - \cot \alpha}{\tan \beta + \cot \alpha}$

17. $\csc x - \sin x = \sin x \cot^2 x$

18. $\cos x \csc x \tan x = 1$

19. $\dfrac{1 + \cos \varphi}{\sin \varphi} + \dfrac{\sin \varphi}{1 + \cos \varphi} = 2 \csc \varphi$

20. $\dfrac{\sin \varphi}{1 - \sin \varphi} + \dfrac{1 + \sin \varphi}{\sin \varphi} = \dfrac{\csc \varphi}{1 - \sin \varphi}$

In exercises 21–26, determine whether the given equation is or is not an identity.

21. $\tan s = \dfrac{\tan s + 1}{\cot s + 1}$

22. $\dfrac{\sin s}{1 + \cos s} = \dfrac{1 - \cos s}{\sin s}$

23. $\tan 3\alpha = \dfrac{1 + \cos 2\alpha}{\sin 2\alpha}$

24. $\dfrac{\sin 3\alpha}{\sin \alpha} + \dfrac{\cos 3\alpha}{\cos \alpha} = 1$

25. $\dfrac{1 + \sin \theta - \cos \theta}{1 + \sin \theta + \cos \theta} = \tan \frac{1}{2}\theta$

26. $\dfrac{1 + \sin \theta}{1 - \sin \theta} - \dfrac{1 - \sin \theta}{1 + \sin \theta} = \dfrac{4 \tan \theta}{\cos \theta}$

## Goal C

In exercises 27–34, evaluate by using the conversion identities.

27. $\sin 75° + \sin 15°$

28. $\cos 75° - \cos 15°$

29. $\cos 105° + \cos 15°$

30. $\sin 105° - \sin 15°$

31. $\sin \dfrac{\pi}{12} - \sin \dfrac{5\pi}{12}$

32. $\cos \dfrac{11\pi}{12} + \cos \dfrac{5\pi}{12}$

33. $\cos \dfrac{7\pi}{12} - \cos \dfrac{\pi}{12}$

34. $\sin \dfrac{13\pi}{12} + \sin \dfrac{7\pi}{12}$

For exercises 35–38, derive the identities (29), (30), (32), and (33).

In exercises 39–44, express the trigonometric expression as a product.

39. $\sin 3x + \sin x$

40. $\sin 3x - \sin x$

41. $\sin 6x + \sin 2x$

42. $\sin 6x - \sin 2x$

43. $\cos 6x - \cos 2x$

44. $\cos 6x + \cos 2x$

In exercises 45–48, write the expression as a sum or difference involving sines and cosines.

45. $\sin 2x \cos 3x$

46. $\sin 3x \cos 2x$

47. $\sin 3x \sin 2x$

48. $\cos 3x \cos 2x$

## Superset

49. Find an expression for $\tan 2x + \sec 2x$ in terms only of $\sin x$ and $\cos x$.

50. If $0 < \alpha < \frac{\pi}{2}$ and $\beta = \frac{\pi}{2} - \alpha$, show that
(a) $\sin \alpha \cos \beta + \cos \alpha \sin \beta = 1$
(b) $\cos \alpha \cos \beta - \sin \alpha \sin \beta = 0$
(c) $\sin \alpha + \sin \beta = \cos \alpha + \cos \beta$

51. Verify each of the following.
(a) $2 \arctan(\frac{1}{3}) + \arctan(\frac{1}{7}) = \frac{\pi}{4}$
(b) $\arctan(\frac{1}{2}) + \arctan(\frac{1}{3}) = \frac{\pi}{4}$
(c) $\tan(\arctan(a) - \arctan(b)) = \dfrac{a - b}{1 + ab}$

52. For each of the following, find two nonzero numbers $M$ and $N$ such that
(a) $M + N \cos^2 x = \sin^4 x - \cos^4 x$
(b) $\cos^4 x - \sin^4 x = M - N \sin^2 x$
(c) $M + N \cos x = \dfrac{\sin^2 x}{1 - \cos x}$
(d) $\tan x \csc x \cos x + \cot x \sec x \sin x = M$
(e) $(\tan x + \cot x)^2 \sin^2 x = M + N \tan^2 x$

## 4.5 Trigonometric Equations

**Goal A** *to solve trigonometric equations using knowledge of the special angles*

Much of algebra is concerned with techniques for solving equations like

$$2x + 1 = 0, \quad \text{or} \quad 2x^2 - x = 1,$$

where $x$ represents a real number. We now wish to consider equations that involve trigonometric functions, such as

$$2 \cos \alpha + 1 = 0, \quad \text{or} \quad 2 \cos^2 \alpha - \cos \alpha = 1.$$

To solve such equations, we first solve for $\cos \alpha$ using algebraic techniques, and then we use our knowledge of trigonometry to solve for $\alpha$. For example, to solve the trigonometric equation

$$2 \cos \alpha + 1 = 0,$$

we begin by solving for $\cos \alpha$. This requires the same algebraic steps as solving the equation $2x + 1 = 0$ for $x$.

**Example 1** Solve $2 \cos \alpha + 1 = 0$ for $\alpha$. Express the solution in radians.

***Solution***

$2 \cos \alpha + 1 = 0$ ■ Begin by solving for $\cos \alpha$.

$2 \cos \alpha = -1$

$\cos \alpha = -\dfrac{1}{2}$ ■ Having solved for $\cos \alpha$, we now solve for $\alpha$.

Recall that $\cos \frac{\pi}{3} = \frac{1}{2}$, and that cosine is negative for second and third quadrant angles. Thus, $\alpha$ must be a second or third quadrant angle whose reference angle is $\frac{\pi}{3}$, namely $\frac{2\pi}{3}$ or $\frac{4\pi}{3}$. Since adding any multiple of $2\pi$ to either of these values produces an angle whose cosine is also $-\frac{1}{2}$, we conclude

$$\alpha = \frac{2\pi}{3} + 2n\pi \quad \text{or} \quad \alpha = \frac{4\pi}{3} + 2n\pi \quad \text{for any integer } n.$$

When solving trigonometric equations, we often restrict the solutions to values in the interval $[0, 2\pi)$ or angles in the interval $[0°, 360°)$.

**Example 2** Determine all solutions of $2\sin^2\alpha - \sin\alpha = 1$ in the interval $[0, 2\pi)$.

***Solution***

$$2\sin^2\alpha - \sin\alpha = 1$$

■ Begin by solving for $\sin\alpha$.

$$2\sin^2\alpha - \sin\alpha - 1 = 0$$

■ Now factor the left side; treat $\sin\alpha$ as the variable.

$$(2\sin\alpha + 1)(\sin\alpha - 1) = 0$$

| $2\sin\alpha + 1 = 0$ | $\sin\alpha - 1 = 0$ |
|---|---|
| $\sin\alpha = -\dfrac{1}{2}$ | $\sin\alpha = 1$ |

■ Use the Principle of Zero Products.

To solve $\sin\alpha = -\frac{1}{2}$, recall that $\sin\frac{\pi}{6} = \frac{1}{2}$ and sine is negative in the third and fourth quadrants. Thus, $\alpha$ is a third or fourth quadrant angle whose reference angle is $\frac{\pi}{6}$. So $\alpha = \frac{7\pi}{6}$ or $\frac{11\pi}{6}$. To solve $\sin\alpha = 1$, recall that the only value in the interval $[0, 2\pi)$ whose sine is 1 is $\alpha = \frac{\pi}{2}$.

The solutions of the given equation in the interval $[0, 2\pi)$ are $\frac{\pi}{2}$, $\frac{7\pi}{6}$, and $\frac{11\pi}{6}$.

If the domain in the Example 2 had not been restricted to the interval $[0, 2\pi)$, the solutions would be $\frac{7\pi}{6} + 2n\pi$, $\frac{11\pi}{6} + 2n\pi$, and $\frac{\pi}{2} + 2n\pi$, for any integer $n$.

When solving trigonometric equations, it is important to remember the domains and ranges of the trigonometric functions. Example 3 illustrates how knowledge of the range can be crucial.

**Example 3** Determine the solutions of $\cos^2\varphi = 2 - \cos\varphi$ in the interval $[0°, 360°)$.

***Solution***

$$\cos^2\varphi = 2 - \cos\varphi$$

■ First solve for $\cos\varphi$.

$$\cos^2\varphi + \cos\varphi - 2 = 0$$

$$(\cos\varphi + 2)(\cos\varphi - 1) = 0$$

■ Now use the Principle of Zero Products.

| $\cos\varphi + 2 = 0$ | $\cos\varphi - 1 = 0$ |
|---|---|
| $\cos\varphi = -2$ | $\cos\varphi = 1$ |

There is no solution to the equation $\cos\varphi = -2$ because the range of the cosine function is the interval $[-1, 1]$, therefore there is no angle having a cosine of $-2$. To solve $\cos\varphi = 1$, recall that the only value in $[0°, 360°)$ whose cosine is 1 is $\varphi = 0°$. The only solution of the given equation in the interval $[0°, 360°)$ is 0°.

In the next two examples, we solve equations involving two different trigonometric functions. Given such an equation, you should try to use algebraic techniques and trigonometric identities to rewrite the equation in a form involving only one trigonometric function.

**Example 4** Determine all solutions of the equation $\sin x + 1 = \cos x$ in the interval $[0, 2\pi)$.

***Solution***

$$\begin{aligned} \sin x + 1 &= \cos x \\ (\sin x + 1)^2 &= (\cos x)^2 \\ \sin^2 x + 2\sin x + 1 &= \cos^2 x \\ \sin^2 x + 2\sin x + 1 &= 1 - \sin^2 x \\ \sin^2 x + 2\sin x &= -\sin^2 x \\ 2\sin^2 x + 2\sin x &= 0 \\ 2\sin x(\sin x + 1) &= 0 \end{aligned}$$

■ Square both sides of the equation to produce $\cos^2 x$ on the right, which can be rewritten in terms of $\sin x$.

■ $\cos^2 x$ is replaced with $1 - \sin^2 x$ (identity (9)).

| $2\sin x = 0$ | $\sin x + 1 = 0$ |
|---|---|
| $\sin x = 0$ | $\sin x = -1$ |
| $x = 0$ or $\pi$ | $x = \dfrac{3\pi}{2}$ |

■ Because we squared both sides of the original equation, we must check for extraneous solutions.

**Must check:**

| $x = 0$ | $x = \pi$ | $x = \dfrac{3\pi}{2}$ |
|---|---|---|
| $\sin 0 + 1 = \cos 0$ | $\sin \pi + 1 = \cos \pi$ | $\sin \dfrac{3\pi}{2} + 1 = \cos \dfrac{3\pi}{2}$ |
| $0 + 1 = 1$ ■ True | $0 + 1 = -1$ ■ False | $-1 + 1 = 0$ ■ True |

The solutions of the given equation in the interval $[0, 2\pi)$ are 0 and $\dfrac{3\pi}{2}$.

**Example 5** Solve the equation $\cos 2x + \sin x = 0$ for $x$. Express solutions in radians.

***Solution***

$$\begin{aligned} \cos 2x + \sin x &= 0 \\ 1 - 2\sin^2 x + \sin x &= 0 \\ 2\sin^2 x - \sin x - 1 &= 0 \end{aligned}$$

■ Use identity (20) to rewrite $\cos 2x$ in terms of $\sin x$.

■ The equation was multiplied by $-1$.

This equation was solved over the interval $[0, 2\pi)$ in Example 2. The solutions were found to be $\frac{\pi}{2}$, $\frac{7\pi}{6}$, and $\frac{11\pi}{6}$. Thus the solutions of the given equation are

$$\frac{\pi}{2} + 2n\pi, \quad \frac{7\pi}{6} + 2n\pi, \quad \text{and} \quad \frac{11\pi}{6} + 2n\pi \quad \text{for any integer } n.$$

**Example 6** Solve the equation $\sin\theta + 2\sin\frac{\theta}{2} = \cos\frac{\theta}{2} + 1$ over the interval $[0°, 360°)$.

***Solution*** We begin by making the substitution $\alpha = \frac{\theta}{2}$ in order to simplify the equation.

$$\sin 2\alpha + 2\sin\alpha = \cos\alpha + 1$$
■ Since $\alpha = \frac{\theta}{2}$, $\theta = 2\alpha$.

$$2\sin\alpha\cos\alpha + 2\sin\alpha = \cos\alpha + 1$$
■ By identity (18)

$$2\sin\alpha\cos\alpha + 2\sin\alpha - \cos\alpha - 1 = 0$$

$$2\sin\alpha(\cos\alpha + 1) - 1(\cos\alpha + 1) = 0$$
■ Factor.

$$(2\sin\alpha - 1)(\cos\alpha + 1) = 0$$
■ Factor again, and use the Principle of Zero Products.

| $2\sin\alpha - 1 = 0$ | $\cos\alpha + 1 = 0$ |
|---|---|
| $\sin\alpha = \frac{1}{2}$ | $\cos\alpha = -1$ |
| $\alpha = 30°$ or $150°$ | $\alpha = 180°$ |

By virtue of our substitution, $\theta = 2\alpha$, and so the values of $\theta$ corresponding to $\alpha = 30°$, $\alpha = 150°$, and $\alpha = 180°$, are $\theta = 60°$, $\theta = 300°$, and $\theta = 360°$. Since the last value is not in the interval $[0°, 360°)$, we conclude that the solutions of the original equation in the given interval are

$$\theta = 60° \quad\text{and}\quad \theta = 300°.$$

**Example 7** Solve the equation $\sin 3x + \sin x + \cos x = 0$ for $x$. Express solutions in degrees.

***Solution***

$$\sin 3x + \sin x + \cos x = 0$$

$$\sin(2x + x) + \sin(2x - x) + \cos x = 0$$
■ $3x$ and $x$ are rewritten to allow the use of identity (28).

$$2\sin 2x\cos x + \cos x = 0$$
■ By (28) with $\alpha = 2x$ and $\beta = x$: $2\sin\alpha\cos\beta = \sin(\alpha + \beta) + \sin(\alpha - \beta)$.

$$\cos x(2\sin 2x + 1) = 0$$

| $\cos x = 0$ | $2\sin 2x + 1 = 0$ |
|---|---|
| $x = 90°$ or $270°$ | $\sin 2x = -\frac{1}{2}$ |
| | $2x = 210°$ or $330°$ |
| | $x = 105°$ or $165°$ |

The solutions of the original equation are all angles of the form

$90° + n\cdot 360°$, $105° + n\cdot 360°$, $165° + n\cdot 360°$, and $270° + n\cdot 360°$ for any integer $n$.

**Goal B** *to solve trigonometric equations by using tables or a calculator*

In the previous section, the solutions of the trigonometric equations involved special angles ($0, \frac{\pi}{6}, \frac{\pi}{4}$, etc). An equation such as $2 - 7\cos x = 0$ has a solution which is not a special angle. We must use a table or a calculator to determine such a solution.

**Example 8** Solve the equation $2 - 7\cos x = 0$ over the interval $[0°, 360°)$.

***Solution***

$$2 - 7\cos x = 0$$
$$\cos x = \frac{2}{7} \approx 0.2857$$

Using a calculator or Table 4 and rounding to the nearest ten minutes, we find that $x \approx 73°20'$ (the interpolated value is $73°24'$). Since we are given that $\cos x$ is positive, $x$ could be a first or fourth quadrant angle. The fourth quadrant angle having $73°20'$ as a reference angle is $286°40'$.

The solutions rounded to the nearest ten minutes are $73°20'$ and $286°40'$.

**Example 9** Solve the equation $\sin^2\theta + 2\sin\theta - 1 = 0$ over the interval $[0°, 360°)$.

***Solution***

$$\sin^2\theta + 2\sin\theta - 1 = 0$$
$$x^2 + 2x - 1 = 0$$

■ Replace $\sin\theta$ with $x$, and solve the quadratic equation for $x$.

$$x = \frac{-2 \pm \sqrt{2^2 - 4(1)(-1)}}{2(1)}$$

■ Use the quadratic formula with $a = 1$, $b = 2$, and $c = -1$.

$$x = \frac{-2 \pm \sqrt{8}}{2} = \frac{-2 \pm 2\sqrt{2}}{2} = -1 \pm \sqrt{2}$$

Using $\sqrt{2} \approx 1.41$, and recalling that $x$ represents $\sin\theta$, we have

$$\sin\theta \approx -1 + 1.41 = 0.41 \quad \text{or} \quad \sin\theta \approx -1 - 1.41 = -2.41.$$

Since the range of the sine function is the interval $[-1, 1]$, there is no $\theta$ such that $\sin\theta = -2.41$. Thus, our only solutions result from the equation $\sin\theta = 0.41$. Using a calculator or Table 4 and rounding to the nearest ten minutes, we obtain $\theta \approx 24°10'$ (the interpolated value is $24°12'$). Since 0.41 is positive, $\theta$ could also be the second quadrant angle with $24°10'$ as its reference angle, namely $155°50'$. Thus, the solutions to the nearest ten minutes are

$$\theta = 24°10' \quad \text{and} \quad \theta = 155°50'.$$

**Goal C** *to solve equations involving inverse trigonometric functions*

The following examples illustrate techniques for solving equations that involve inverse trigonometric functions.

**Example 10** Solve $\sin^{-1}\left(\frac{5}{13}\right) + \cos^{-1}\left(\frac{3}{5}\right) = \sin^{-1} x$ for $x$.

***Solution*** By the definition of $\sin^{-1}(x)$, we know that $\sin(\sin^{-1}(x)) = x$. Taking the sine of both sides of the given equation we get

$$\sin\left(\sin^{-1}\left(\frac{5}{13}\right) + \cos^{-1}\left(\frac{3}{5}\right)\right) = x.$$

Let $\alpha = \sin^{-1}\left(\frac{5}{13}\right)$, and $\beta = \cos^{-1}\left(\frac{3}{5}\right)$. Draw $\alpha$ and $\beta$ as acute angles in two different right triangles.

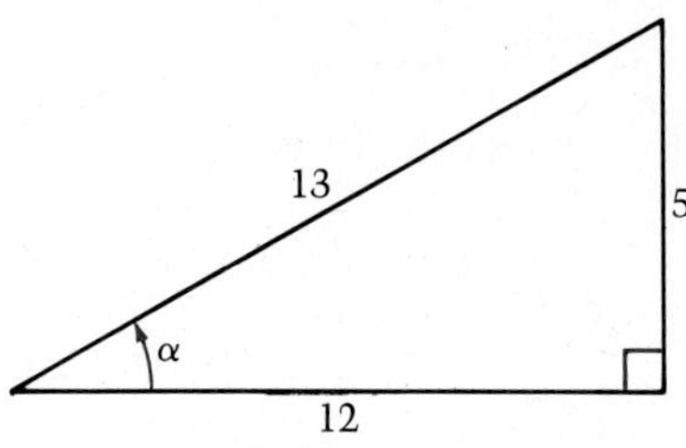


■ Since $\alpha = \sin^{-1}(\frac{5}{13})$, $\sin\alpha = \sin(\sin^{-1}(\frac{5}{13})) = \frac{5}{13}$. We label the sides of the triangle accordingly.

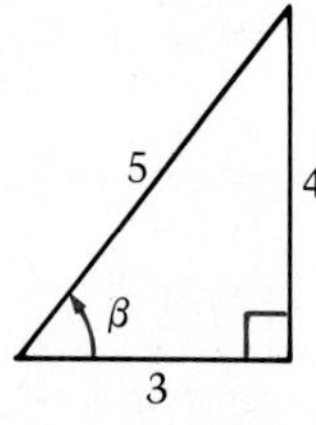


■ Since $\beta = \cos^{-1}(\frac{3}{5})$, $\cos\beta = \cos(\cos^{-1}(\frac{3}{5})) = \frac{3}{5}$. We label the triangle accordingly.

$$x = \sin\left(\sin^{-1}\left(\frac{5}{13}\right) + \cos^{-1}\left(\frac{3}{5}\right)\right)$$

$$= \sin(\alpha + \beta) = \sin\alpha\cos\beta + \cos\alpha\sin\beta$$ ■ By identity (14).

$$= \frac{5}{13}\cdot\frac{3}{5} + \frac{12}{13}\cdot\frac{4}{5} = \frac{15}{65} + \frac{48}{65} = \frac{63}{65}$$ ■ From the two triangles.

**Example 11** Find the smallest positive angle $x$, in degrees, such that

$$\tan^{-1}\left(\frac{1}{2}\right) + \tan^{-1}\left(\frac{1}{3}\right) = x.$$

***Solution***

$$\tan^{-1}\left(\frac{1}{2}\right) + \tan^{-1}\left(\frac{1}{3}\right) = x$$

Let $\alpha = \tan^{-1}(\frac{1}{2})$ and $\beta = \tan^{-1}(\frac{1}{3})$; the equation becomes $\alpha + \beta = x$. Then, taking the tangent of both sides of this equation, we have

$$\tan x = \tan(\alpha + \beta) = \frac{\tan\alpha + \tan\beta}{1 - \tan\alpha\tan\beta}$$ ■ Identity (16)

Therefore, $\tan x = \dfrac{\frac{1}{2} + \frac{1}{3}}{1 - \left(\frac{1}{2}\right)\left(\frac{1}{3}\right)} = \dfrac{\frac{5}{6}}{\frac{5}{6}} = 1$ ■ $\tan\alpha = \tan(\tan^{-1}(\frac{1}{2})) = \frac{1}{2}$. $\tan\beta = \tan(\tan^{-1}(\frac{1}{3})) = \frac{1}{3}$.

The smallest positive angle having a tangent function value of 1 is 45°.

## Exercise Set 4.5 ☰ Calculator Exercises in the Appendix

### Goal A

In exercises 1–10, solve the equation for $x$. Express your answer in radians.

1. $2\sin x + \sqrt{3} = 0$
2. $3\cot x - \sqrt{3} = 0$
3. $2\sin^2 x - \sin x = 0$
4. $2\cos^2 x - \cos x - 1 = 0$
5. $4\cos^2 x - 3 = 0$
6. $\tan^2 x - 1 = 0$
7. $2\cos^2 x + 5\cos x - 3 = 0$
8. $\sin x\cos x - \cos x = 0$
9. $\sin x\cos x - \sin x + \cos x - 1 = 0$
10. $2\sin^2 x - \sqrt{2}\sin x - 4\sin x + 2\sqrt{2} = 0$

In exercises 11–18, determine all the solutions of the equation in the interval $[0, 2\pi)$.

11. $2\sin^2 x + \sin x - 1 = 0$
12. $2\cos^2 x - \cos x - 1 = 0$
13. $2\sin^2 x + 3\sin x - 2 = 0$
14. $\sqrt{2}\cos^2 x + (1 - \sqrt{2})\cos x - 1 = 0$
15. $2\cos^2 x + 3\sin x - 2 = 0$
16. $3\cot^2 x - 1 = 0$
17. $2\sin^2 3\theta - 4\sin 3\theta - 6 = 0$
18. $2\sin^2\frac{1}{2}\theta + 3\sin\frac{1}{2}\theta - 2 = 0$

In exercises 19–32, solve the equation for $x$ in the indicated domain.

**19.** $\sin 2x = \cos 2x$; $x$ in $[0, 2\pi)$

**20.** $\cos x \tan x = 0$; $x$ in $[0, 360°)$

**21.** $3 \cos^2 x = \sin^2 x$; $x$ in $[-90°, 90°]$

**22.** $\sin x \cos x = \cos x$; $x$ in $[0, 180°]$

**23.** $\dfrac{\sqrt{2} \sin^2 x}{\cos x} - \tan x = 0$; $x$ in $[0, 2\pi)$

**24.** $\cos^2 x - \sin x = 1$; $x$ in $[0, \pi]$

**25.** $2 \sin x - \tan x = 0$; $x$ in $[0°, 180°]$

**26.** $3 \sin^2 x - \cos x = 3$; $x$ in $[0, \pi]$

**27.** $\cos 5x + \cos x = 0$; $x$ in $[0°, 360°)$

**28.** $\cos \frac{7}{2}x + \cos \frac{1}{2}x = 0$; $x$ in $[0, \pi)$

**29.** $\sin 2x + \cos x = 0$; $x$ in $[0, 2\pi)$

**30.** $\cos 2x + \cos x + 1 = 0$; $x$ in $[0, 2\pi)$

**31.** $\cos 2x + \sin x = 1$; $x$ in $[0, 2\pi)$

**32.** $\cos 2x - 1 - \tan x = 0$; $x$ in $[0, 2\pi)$

## Goal B

In exercises 33–44, solve the equation for $x$ in $[0°, 360°)$ using Table 4 or a calculator.

**33.** $4 \sin^2 x - 3 \sin x = 0$

**34.** $5 \cos^2 x - 2 \cos x = 0$

**35.** $3 \sin^2 x + 2 \sin x - 1 = 0$

**36.** $3 \cos^2 x - \cos x - 2 = 0$

**37.** $6 \sin^2 x - \sin x - 2 = 0$

**38.** $2 \sin^2 x + \frac{1}{2} \sin x - 2 = 0$

**39.** $(4 \sin x + 1)^2 - 3 = 0$

**40.** $(3 \cos x - 5)^2 - 8 = 0$

**41.** $2 \tan^2 x + 2 \tan x + 3 = 0$

**42.** $3 \tan^2 x + 4 \tan x - 4 = 0$

**43.** $3 \sin^2 2x - 2 = 0$

**44.** $(2 \cos^2 2x + 3)^2 = 1$

## Goal C

In exercises 45–56, solve the equation for $x$.

**45.** $\arcsin\left(-\frac{1}{2}\right) = x$

**46.** $\arccos\left(-\frac{1}{2}\right) = x$

**47.** $\tan^{-1}\sqrt{3} = x$

**48.** $\frac{1}{2}\pi - \tan^{-1}(-1) = x$

**49.** $\sin\left(2 \arcsin \frac{3}{5}\right) = x$

**50.** $\sin\left(2 \arccos \frac{3}{5}\right) = x$

**51.** $\cos^{-1} \frac{3}{5} + \sin^{-1} \frac{4}{5} = \cos^{-1} x$

**52.** $\cos^{-1} \frac{3}{5} - \sin^{-1} \frac{4}{5} = \cos^{-1} x$

**53.** $\arcsin \frac{3}{5} - \arccos \frac{5}{13} = \arcsin x$

**54.** $\arcsin \frac{3}{5} + \arccos \frac{12}{13} = \arccos x$

**55.** $\tan^{-1} \frac{4}{3} + \tan^{-1} \frac{1}{7} = x$

**56.** $\arctan\left(-\frac{1}{3}\right) - \arctan\left(\frac{2}{3}\right) = \arctan x$

## Superset

**57.** Solve for $x$: $\arctan x + \arctan \frac{1}{3} = \frac{\pi}{4}$.

**58.** Solve the following system of equations:

$$\begin{cases} \sin x + \cos x = 1 \\ 2 \sin x = \sin 2x \end{cases}$$

**59.** Solve simultaneously:

$$\begin{cases} r = 1 - \cos \theta \\ r = \cos \theta \end{cases}$$

where $r > 0$ and $0 \le \theta \le 2\pi$

**60.** Solve for $x$: $\tan x + \sqrt{\tan x + 5} = 7$.

**61.** Solve for $x$: $\sqrt{2 \cot x - 1} + \sqrt{\cot x - 4} = 4$.

# Chapter Review and Test

## Chapter Review

### 4.1 Basic Trigonometric Identities (pp. 118–122)

Reciprocal Identities

(1) $\sin\theta = \dfrac{1}{\csc\theta}$

(2) $\cos\theta = \dfrac{1}{\sec\theta}$

(3) $\tan\theta = \dfrac{1}{\cot\theta}$

(4) $\cot\theta = \dfrac{1}{\tan\theta}$

(5) $\sec\theta = \dfrac{1}{\cos\theta}$

(6) $\csc\theta = \dfrac{1}{\sin\theta}$

Quotient Identities

(7) $\tan\theta = \dfrac{\sin\theta}{\cos\theta}$

(8) $\cot\theta = \dfrac{\cos\theta}{\sin\theta}$

Pythagorean Identities

(9) $\sin^2\theta + \cos^2\theta = 1$

(10) $\tan^2\theta + 1 = \sec^2\theta$

(11) $1 + \cot^2\theta = \csc^2\theta$

### 4.2 Sum and Difference Identities (pp. 123–129)

(12) $\cos(\alpha - \beta) = \cos\alpha\cos\beta + \sin\alpha\sin\beta$

(13) $\cos(\alpha + \beta) = \cos\alpha\cos\beta - \sin\alpha\sin\beta$

(14) $\sin(\alpha + \beta) = \sin\alpha\cos\beta + \cos\alpha\sin\beta$

(15) $\sin(\alpha - \beta) = \sin\alpha\cos\beta - \cos\alpha\sin\beta$

(16) $\tan(\alpha + \beta) = \dfrac{\tan\alpha + \tan\beta}{1 - \tan\alpha\tan\beta}$

(17) $\tan(\alpha - \beta) = \dfrac{\tan\alpha - \tan\beta}{1 + \tan\alpha\tan\beta}$

### 4.3 The Double-Angle and Half-Angle Identities (pp. 130–136)

(18) $\sin 2\theta = 2\sin\theta\cos\theta$

(19) $\cos 2\theta = \cos^2\theta - \sin^2\theta$

(20) $\cos 2\theta = 1 - 2\sin^2\theta$

(21) $\cos 2\theta = 2\cos^2\theta - 1$

(22) $\tan 2\theta = \dfrac{2\tan\theta}{1 - \tan^2\theta}$

(23) $\sin\dfrac{\theta}{2} = \pm\sqrt{\dfrac{1 - \cos\theta}{2}}$

(24) $\cos\dfrac{\theta}{2} = \pm\sqrt{\dfrac{1 + \cos\theta}{2}}$

(25) $\tan\dfrac{\theta}{2} = \pm\sqrt{\dfrac{1 - \cos\theta}{1 + \cos\theta}}$

(26) $$\tan\frac{\theta}{2} = \frac{\sin\theta}{1+\cos\theta}$$ (27) $$\tan\frac{\theta}{2} = \frac{1-\cos\theta}{\sin\theta}$$

**4.4 The Conversion Identities (pp. 137–142)**

(28) $$\sin\alpha\cos\beta = \frac{1}{2}[\sin(\alpha+\beta)+\sin(\alpha-\beta)]$$

(29) $$\sin\alpha\sin\beta = \frac{1}{2}[\cos(\alpha-\beta)-\cos(\alpha+\beta)]$$

(30) $$\cos\alpha\cos\beta = \frac{1}{2}[\cos(\alpha+\beta)+\cos(\alpha-\beta)]$$

(31) $$\sin x+\sin y = 2\sin\left(\frac{x+y}{2}\right)\cos\left(\frac{x-y}{2}\right)$$

(32) $$\sin x-\sin y = 2\cos\left(\frac{x+y}{2}\right)\sin\left(\frac{x-y}{2}\right)$$

(33) $$\cos x+\cos y = 2\cos\left(\frac{x+y}{2}\right)\cos\left(\frac{x-y}{2}\right)$$

(34) $$\cos x-\cos y = -2\sin\left(\frac{x+y}{2}\right)\sin\left(\frac{x-y}{2}\right)$$

**4.5 Trigonometric Equations (pp. 143–150)**

A trigonometric equation is an equation that involves trigonometric functions. For example, $2\cos^2 x - \cos x = 1$ is a trigonometric equation. To solve such equations usually involves a two step process, first solve for $\cos x$ and then, using trigonometry, solve for $x$. We often restrict the solutions to real numbers in the interval $[0, 2\pi)$ or angles in the interval $[0°, 360°)$. p. 143

## Chapter Test

**4.1A** Show that the first expression may be rewritten as the second by using the Basic Trigonometric Identities.

**1.** $\dfrac{1-\tan x}{\sec x} + \sin x$; $\cos x$ **2.** $\dfrac{1}{2}\left(\dfrac{1+\cos x}{\sin x} + \dfrac{\sin x}{1+\cos x}\right)$; $\csc x$

**4.1B** Verify the identity.

**3.** $\dfrac{\tan^2 x+1}{\tan^2 x+2} = \dfrac{1}{1+\cos^2 x}$ **4.** $\dfrac{1+\cot^2 x}{2+\cot^2 x} = \dfrac{1}{1+\sin^2 x}$

**4.2A** Find the following trigonometric values without using Table 4 or a calculator.

**5.** $\cot \dfrac{5\pi}{12}$

**6.** $\tan\left(-\dfrac{11\pi}{12}\right)$

Verify each identity.

**7.** $\dfrac{\sin(a + b)}{\cos a \cos b} = \tan a + \tan b$

**8.** $\dfrac{\tan a - \tan b}{\tan a + \tan b} = \dfrac{\sin(a - b)}{\sin(a + b)}$

**4.3A** Given $\alpha$, find $\sin 2\alpha$, $\cos 2\alpha$, and $\tan 2\alpha$ using the double-angle identities.

**9.** $315°$

**10.** $-\dfrac{7\pi}{4}$

**4.3B** From the given information, determine $\cos \dfrac{\alpha}{2}$.

**11.** $\sin \alpha = -\dfrac{2}{3}$, $\alpha$ in Quadrant III

**12.** $\tan \alpha = -\dfrac{3}{2}$, $\alpha$ in Quadrant II

**13.** $\tan \alpha = \dfrac{3}{4}$, $\alpha$ in Quadrant III

**14.** $\cot \alpha = -\dfrac{3}{2}$, $\alpha$ in Quadrant II

**4.4A** Simplify each expression.

**15.** $\cos x(\tan x + \cot x)$

**16.** $\dfrac{1 + \cot x}{\sin x + \cos x}$

**17.** $\dfrac{\tan x + 1}{\sec x + \csc x}$

**18.** $\dfrac{\sec^2 \theta - \tan^2 \theta}{\csc \theta}$

**19.** $\dfrac{\cot \varphi - \tan \varphi}{\cot \varphi + \tan \varphi}$

**20.** $\dfrac{\sin 2\alpha}{1 - \cos 2\alpha}$

**4.4B** Verify each identity.

**21.** $\cot \theta \sin^2 \theta = \tan \theta \cos^2 \theta$

**22.** $\dfrac{1}{\sin x} - \sin x = \dfrac{\sin x}{\tan^2 x}$

**23.** $\sin\left(\dfrac{\pi}{4} - \alpha\right) = \cos\left(\alpha + \dfrac{\pi}{4}\right)$

**24.** $\sin \alpha \cos \beta \cot \alpha \tan \beta = \cos \alpha \sin \beta$

**4.4C** Evaluate by using the conversion identities.

**25.** $\cos \frac{11\pi}{12} + \cos \frac{5\pi}{12}$

**26.** $\cos \frac{13\pi}{12} - \cos \frac{5\pi}{12}$

Write each expression as a sum or difference.

**27.** $\sin 4x \cos 3x$

**28.** $\cos 4x \cos 2x$

**4.5A** Solve for $x$. Express the answer in radians.

**29.** $2 \cos^2 x - \cos x = 1$

**30.** $2 \sin^2 x + \sin x - 1 = 0$

**31.** $\sin^2 x + 3 \sin x = 4$

**32.** $\cos^2 x + \cos x - 2 = 0$

**4.5B** Solve each equation for $x$ in $[0°, 360°)$ using tables or a calculator.

**33.** $3 \sin^2 x - 2 \sin x - 1 = 0$

**34.** $\tan^2 x - 2 \tan x - 3 = 0$

**35.** $7 \cos^2 x - 6 \cos x - 1 = 0$

**36.** $12 \sin^2 x + 12 \sin x - 6 = 0$

**4.5C** Solve each equation for $x$.

**37.** $\arctan(1) = x - \frac{\pi}{3}$

**38.** $\arccos\left(-\frac{\sqrt{3}}{2}\right) = x$

**39.** $\arcsin\left(\frac{3}{5}\right) - \arccos\left(\frac{4}{5}\right) = \sin^{-1} x$

**40.** $\tan^{-1}(1) + \tan^{-1}\left(\frac{12}{5}\right) = x$

## Superset

**41.** Express $\sin \alpha$ in terms of $\cot \alpha$ for $\alpha$ in Quadrant II.

**42.** Express $\sec \alpha$ in terms of $\sin \alpha$ for $\alpha$ in Quadrant I.

**43.** Simplify each expression.

(a) $\cos\left(\frac{3\pi}{2} - \alpha\right)$

(b) $\sin(\pi + \beta)$

**44.** Simplify each expression.

(a) $\frac{\sin 2x}{\sin x} - \frac{\cos 2x}{\cos x}$

(b) $\frac{1}{\csc \theta - \cot \theta} - \frac{1}{\csc \theta + \cot \theta}$

**45.** Express $\tan 6x$ in terms of $\tan 2x$.

# 5 Applications in Trigonometry

## 5.1 Applications Involving Right Triangles

**Goal A** *to solve a right triangle*

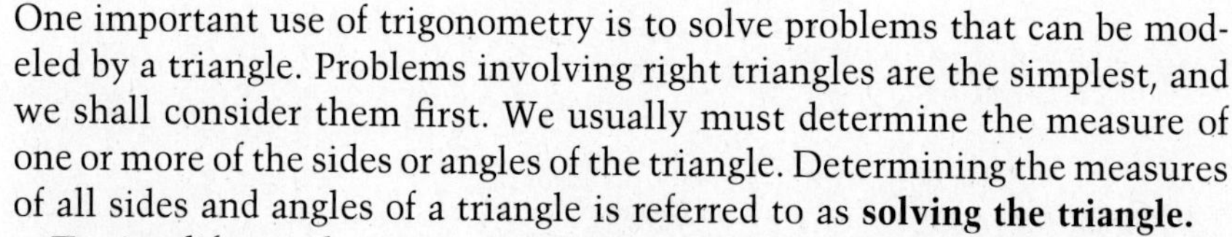

One important use of trigonometry is to solve problems that can be modeled by a triangle. Problems involving right triangles are the simplest, and we shall consider them first. We usually must determine the measure of one or more of the sides or angles of the triangle. Determining the measures of all sides and angles of a triangle is referred to as **solving the triangle.**

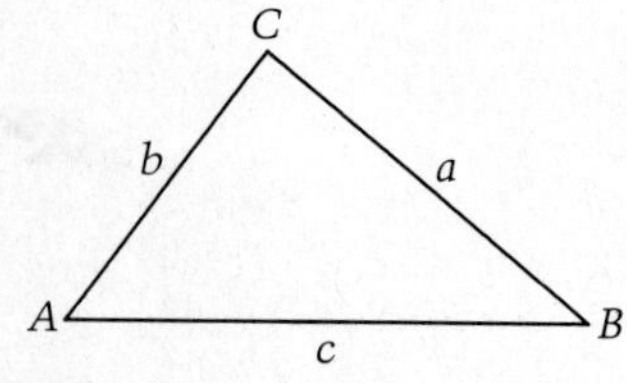

**Figure 5.1**

To simplify our discussion, we will agree that in $\triangle ABC$, the vertices are $A$, $B$, and $C$, and the sides opposite these vertices are $a$, $b$, and $c$, respectively (Figure 5.1). Also, we will agree that $A = 42°$ will mean "the measure of the angle at vertex $A$ is 42°."

**Example 1** Solve the right triangle $ABC$, given that $C = 90°$, $B = 40°$, and $a = 10.25$.

***Solution***

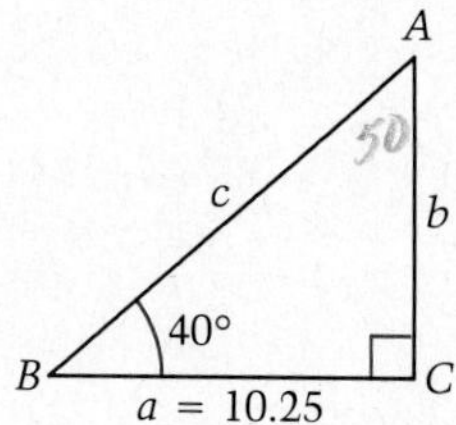


■ Begin by drawing a triangle and labeling the parts.

**Find $c$:** $\cos 40° = \dfrac{10.25}{c}$ ■ $\cos B = \frac{a}{c}$.

$$c = \frac{10.25}{\cos 40°} = \frac{10.25}{0.7660} \approx 13.38$$

We chose $\cos B = \frac{a}{c}$ because this equation involved the unknown $c$ that we wished to find and two variables, $B$ and $a$, that were known. Trying $\sin B = \frac{b}{c}$ would have been pointless because two variables in that equation were then unknown.

**Find $b$:** $\tan 40° = \dfrac{b}{10.25}$ ■ $\tan B = \frac{b}{a}$ is used to find $b$ because $B$ and $a$ are known.

$$b = (10.25)(\tan 40°)$$
$$b = (10.25)(0.8391) \approx 8.60$$

**Find $a$:** Since $A$ and $B$ are complementary angles, $A = 50°$.

We must digress for a moment to discuss accuracy of answers to applied problems. The trigonometric function values in Table 4 are approximations, and measuring instruments are subject to some measurement error and thus produce approximations. Thus, when we solve problems involving lengths and angles, our solutions are approximations. The question then is how should such solutions be "rounded"? To answer this we must discuss the notion of *significant digits.*

The length 10.25 in Example 1 is said to have four significant digits. The number 0.061 has two significant digits, and the number 72,000, if it has been rounded to the nearest thousand, also has two significant digits. We apply the following rule of thumb:

> If a number $N$ can be written in scientific notation as $N = A \cdot 10^n$ where $|A|$ is greater than or equal to 1 and less than 10, then the number of **significant digits** in $N$ is the number of digits in $A$.

When solving problems involving angles and sides of triangles, you should round your calculations according to these standard rules.

| Number of significant digits of the length | Measure of angle in degrees should be rounded to |
|---|---|
| 1 | the nearest multiple of 10° |
| 2 | the nearest degree |
| 3 | the nearest multiple of 10′ |
| 4 | the nearest minute |

When measurements are added or subtracted, the answer is rounded to *the least number of decimal places* of any of the measurements. When measurements are multiplied or divided, the answer is rounded to the *least number of significant digits* of any of the measurements. The power or root of a measurement is rounded to the same number of significant digits as is the measurement itself.

**Example 2** Solve $\triangle ABC$ given that it is a right triangle with $C = 90°$, $a = 16.5$, and $c = 30.2$.

***Solution***

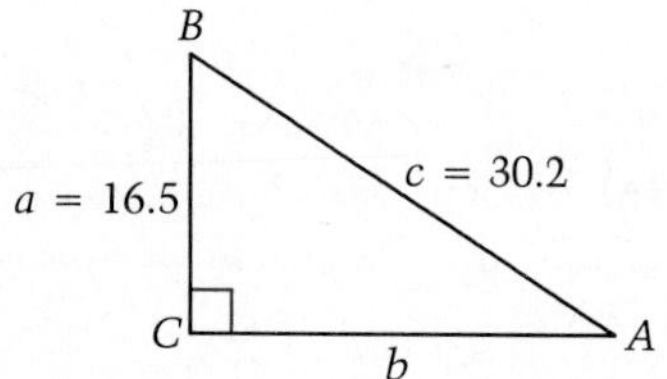


■ Draw a triangle and label the parts.

**Find $b$:** $(30.2)^2 = (16.5)^2 + b^2$

$b^2 \approx 912 - 272 = 640$

$b \approx 25.3$

**Find $A$:** $\sin A = \dfrac{16.5}{30.2} \approx 0.546$

$A \approx 33°10'$

**Find $B$:** $B \approx 56°50'$

Note that $A$ and $B$ are rounded to the nearest 10′.

## Goal B *to solve applied problems by using right triangles*

It should be clear from Examples 1 and 2 that a right triangle can be solved provided either one side and one of the acute angles are known, or two sides are known. Suppose you are given both acute angles. Is that enough information to solve the triangle? (The answer is "No.")

Some practical applications involve angles formed by the horizontal and the line of sight of an observer. If the line of sight is above the horizontal, the angle is called an **angle of elevation.** If the line of sight is below the horizontal, the angle is called an **angle of depression.**

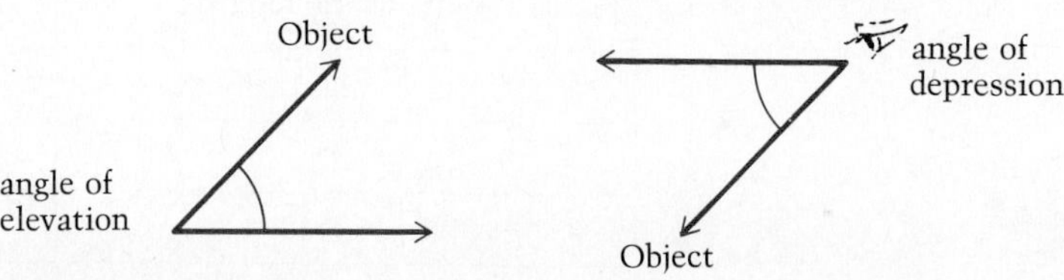


**Figure 5.2**

**Example 3** A hot air balloon is rising vertically in still air. An observer is standing on level ground, 100 feet away from the point of launch. At one instant the observer measures the angle of elevation of the balloon as 30°00′. One minute later, the angle of elevation is 76°10′. How far did the balloon travel during that minute?

***Solution***

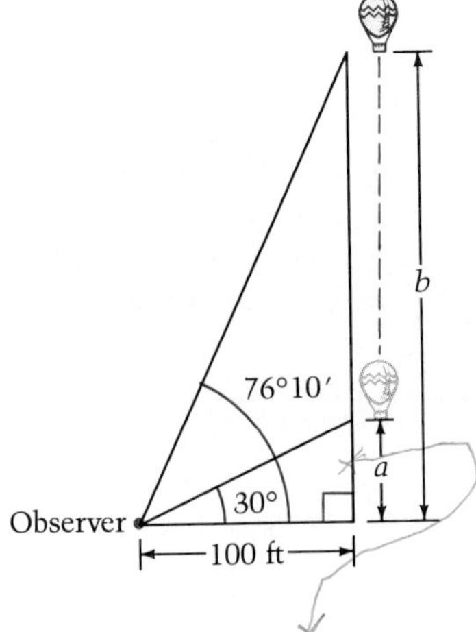


■ Draw a triangle that models the problem. We wish to determine $b - a$, the distance traveled during the one-minute period.

**Find *a*:**
$$\tan 30° = \frac{a}{100}$$
$$a = 100(\tan 30°)$$
$$= 100\left(\frac{\sqrt{3}}{3}\right) \approx 57.7$$

**Find *b*:**
$$\tan 76°10' = \frac{b}{100}$$
$$b = 100(\tan 76°10')$$
$$= 100(4.061) \approx 406$$

The balloon traveled $406 - 57.7 \approx 348$ feet during the minute.

Surveyors and navigators measure angles in terms of the north-south line. Two methods of measure are used: azimuth and bearing. The **azimuth** is the measure of an angle from due north in the clockwise direction.

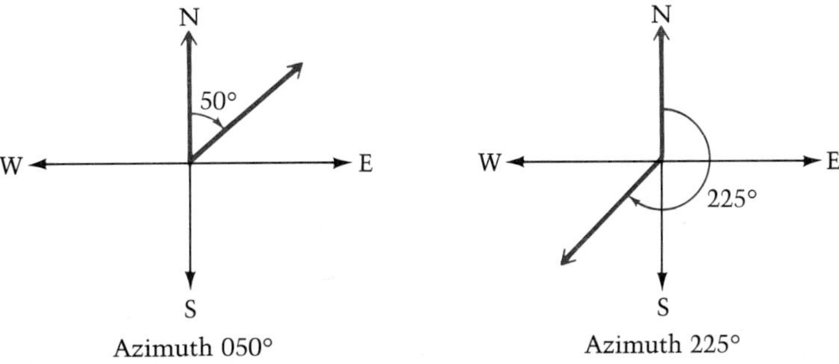


**Figure 5.3**

The **bearing** is the measure of the acute angle between the north-south line and the line representing the direction. It is described in terms of compass directions.

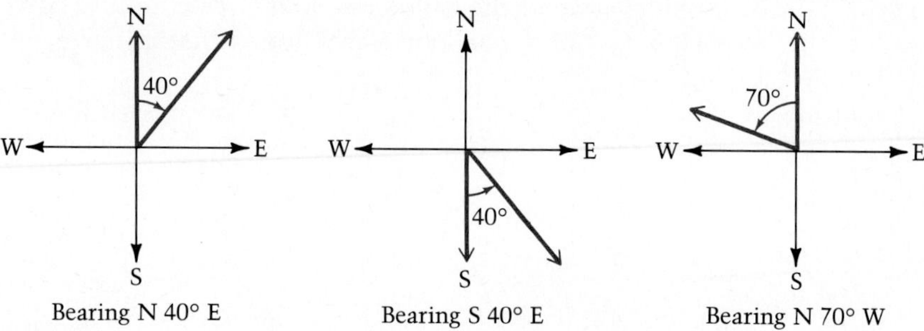


**Figure 5.4**

**Example 4** A plane flies from County Airport on a bearing of N 45° E for three hours, and then flies on a bearing of S 45° E for four hours. If the speed of the plane is 400 mph, and we ignore the effects of wind, what is the plane's distance and azimuth from the County Airport after the seven hours?

***Solution***

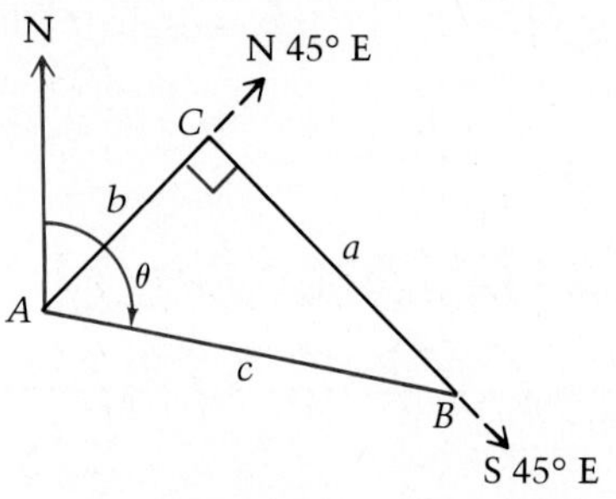


■ We begin by drawing a figure. Note that $C = 90°$ so $\triangle ABC$ is a right triangle. We wish to determine side $c$ and angle (azimuth) $\theta$. Note that $\theta = A + 45°$.

$$b = \text{(speed)(time on N 45° E bearing)} = (400 \text{ mph})(3 \text{ hr}) = 1200 \text{ mi}$$

$$a = \text{(speed)(time on S 45° E bearing)} = (400)(4 \text{ hr}) = 1600 \text{ mi}$$

$$c = \sqrt{1200^2 + 1600^2} = 2000$$

$$\tan A = \frac{\text{opposite}}{\text{adjacent}} = \frac{a}{b} = \frac{1600}{1200} = \frac{4}{3}$$

$$A \approx 53°$$

■ $A$ is rounded to the nearest degree.

Since $\theta = 45° + A = 45° + 53° = 98°$, we conclude that the azimuth of the plane is approximately 098°. The plane is at a distance of 2000 mi from County Airport.

## Exercise Set 5.1 ⊟ Calculator Exercises in the Appendix

### Goal A

In exercises 1–10, some information about the right triangle $ABC$ is given. Use only the Table of Special Angles to solve $\triangle ABC$.

1. $B = 45°$, $c = 64$
2. $A = 30°$, $c = 48$
3. $A = 60°$, $b = 44$
4. $B = 60°$, $b = 12$
5. $a = 15$, $b = 15\sqrt{3}$
6. $b = 40$, $c = 80$
7. $B = 30°$, $a = 24$
8. $A = 45°$, $a = 44$
9. $a = 25\sqrt{2}$, $c = 50$
10. $b = 25$, $a = 25$

In exercises 11–20, use the given information to solve the right triangle $ABC$. Be sure to adjust the number of significant digits in the answer.

11. $A = 36°$, $a = 70$
12. $B = 72°$, $a = 40$
13. $B = 12°$, $a = 9$
14. $A = 53°$, $c = 20$
15. $c = 12.3$, $A = 38°$
16. $a = 42.5$, $A = 67°$
17. $b = 0.244$, $A = 28°20'$
18. $A = 51°40'$, $b = 7.85$
19. $c = 142.5$, $B = 71°18'$
20. $c = 6.782$, $B = 23°37'$

### Goal B

21. A 24 ft ladder is leaning against a building. If the ladder makes an angle of 60° with the ground, how far up the building does the ladder reach?
22. If an 18 ft ladder is placed against a building so that it reaches a window sill 9 ft off the ground, what is the acute angle between the ladder and the ground?
23. A jet is flying at an altitude of 2000 ft. The angle of depression to an aircraft carrier is 20°. How far is the jet from the carrier?
24. When the angle of elevation of the sun is 60°, a certain flagpole casts a shadow 30 ft long. How tall is this flagpole?
25. A 60 ft long ramp is inclined at an angle of 5° with the level ground. How high does the ramp rise above the ground?
26. A ramp for wheelchairs is to be built beside the main steps of the library. The total vertical rise of the steps is 3 ft and the ramp will be inclined at an angle of 12°. How long a ramp is needed?
27. A television crew 2600 ft from a launch pad is filming the launch of a space shuttle. What is the angle of elevation of the camera when the shuttle is 4000 ft directly above the pad?
28. From an 80 ft lighthouse on the coast, an overturned sailboat is sighted. If the angle of depression is 9°, how far is the boat from the lighthouse?
29. A 50 ft tall flagpole casts a shadow on level ground. What is the angle of elevation of the sun when the shadow is (a) 29 ft long? (b) 60.0 ft long? (c) 12.2 ft long? (d) 125.00 ft long?
30. The Charleston Light in Charleston, SC is one of the most powerful lighthouses in the Western Hemisphere. It is 163 ft high, and its light can be seen 19 mi out at sea. What is the distance of a small boat from the foot of the tower if the angle of depression of the boat from the tower is (a) 30°? (b) 9°32′?

### Superset

31. From level ground, the angle of elevation to a distant cliff is 30°. By walking a distance of 2000 ft directly toward the foot of the cliff, the angle of elevation becomes 45°. What is the height of the cliff?
32. A 20 ft flagpole is mounted on the edge of the roof of a building. A person standing level with the base of the building measures the angle of elevation to the top of the flagpole to be 65°. From the same spot, the angle of elevation of the foot of the flagpole is 60°. What is the height of the building?
33. A video camera is to be installed in a bank to monitor the bank teller's counter, which is 4 ft high. The camera will be mounted on the wall at a height of 10 ft. The counter is 20 ft from the wall on which the camera is to be mounted. To aim the camera, what should the angle of depression of the camera be?

## 5.2 Law of Sines

**Goal A** *to solve a triangle, given one side and two angles*

In the last section we used trigonometric functions to solve right triangles. In this section and the next, we shall derive the Law of Sines and the Law of Cosines, which allow us to solve triangles which are *not* right triangles. Triangles that are not right triangles are called **oblique.**

First, we shall consider the Law of Sines. We begin by considering any oblique triangle $ABC$. There are two possibilities. **Case 1:** $\triangle ABC$ is an *acute* triangle (all angles are less than 90°) as shown in Figure 5.5(a). **Case 2:** $\triangle ABC$ is an *obtuse* triangle (one of its angles is greater than 90°) as shown in Figure 5.5(b). In each case, we have drawn the altitude $h$ from vertex $B$ to side $b$, or the line containing side $b$ (in Case 2).

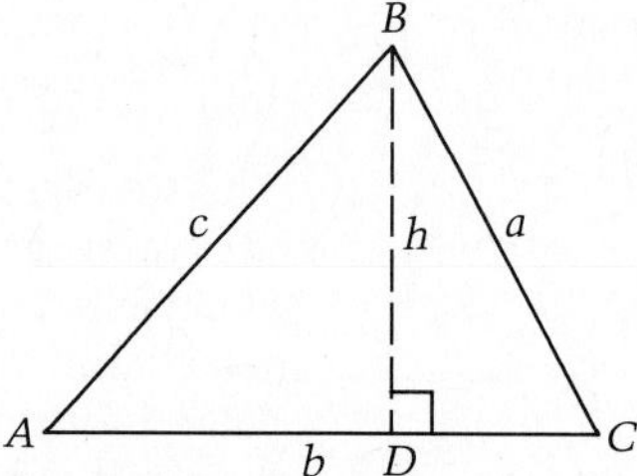


(**a**) Case 1: acute triangle

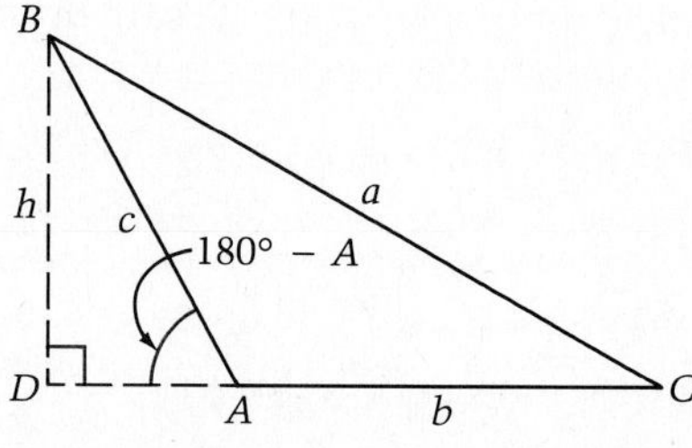


(**b**) Case 2: obtuse triangle

**Figure 5.5**

Note that in Case 2 we have

$$\begin{aligned} \sin(180° - A) &= \sin 180° \cos A - \cos 180° \sin A \\ &= (0)(\cos A) - (-1)(\sin A) \\ &= \sin A. \end{aligned}$$

In both cases,

$$h = a \sin C, \quad \text{and} \quad h = c \sin A.$$

Equating the two expressions for $h$, we get $a \sin C = c \sin A$, or

$$\frac{\sin A}{a} = \frac{\sin C}{c}.$$

Had we drawn the altitude from vertex $C$ to side $c$, we would have concluded

$$\frac{\sin A}{a} = \frac{\sin B}{b}.$$

We summarize our results as follows:

**Law of Sines**

In any triangle with angles $A$, $B$, and $C$, and opposite sides $a$, $b$, and $c$, respectively,

$$\frac{\sin A}{a} = \frac{\sin B}{b} = \frac{\sin C}{c}.$$

**Example 1** Given $\triangle ABC$ with $A = 50°10'$, $B = 70°40'$, and $c = 10.5$, solve the triangle.

***Solution***

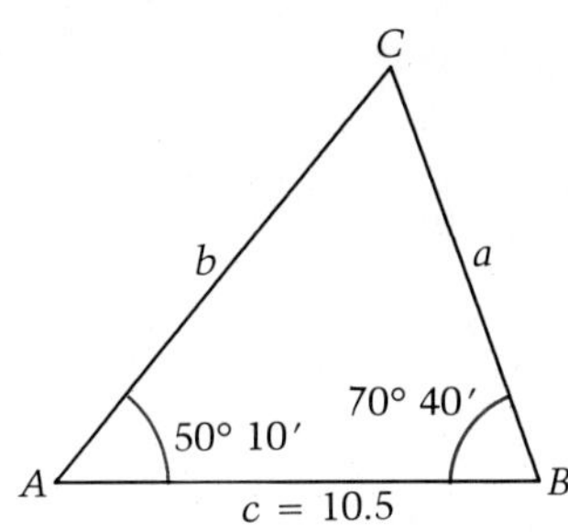


■ Begin by drawing a triangle and labeling the parts.

**Find $C$:** $C = 180° - 50°10' - 70°40' = 59°10'$

**Find $a$:** $\dfrac{\sin 50°10'}{a} = \dfrac{\sin 59°10'}{10.5}$ ■ $\dfrac{\sin A}{a} = \dfrac{\sin C}{c}$.

$$a = \frac{(10.5)(\sin 50°10')}{\sin 59°10'}$$

$$= \frac{(10.5)(0.7679)}{0.8587} \approx 9.39$$

■ Round to three significant digits.

**Find $b$:** $\dfrac{\sin 70°40'}{b} = \dfrac{\sin 59°10'}{10.5}$ ■ $\dfrac{\sin B}{b} = \dfrac{\sin C}{c}$.

$$b = \frac{(10.5)(0.9436)}{0.8587} \approx 11.5$$

■ Round to three significant digits.

**Goal B** *to solve a triangle given two sides and an angle opposite one of them*

We use the Law of Sines to solve oblique triangles when we are given (1) one side and two angles, or (2) two sides and the angle opposite one of them. The first type of problem was treated in Example 1. The second type is more complicated.

Consider what might happen if we are given two sides, say $a$ and $b$, and acute angle $A$. There are four situations that can occur. (In the figures below, $h$ is the altitude from vertex $C$.)

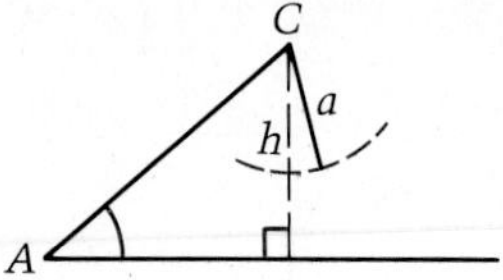


(**a**) $a < h$: no triangle possible

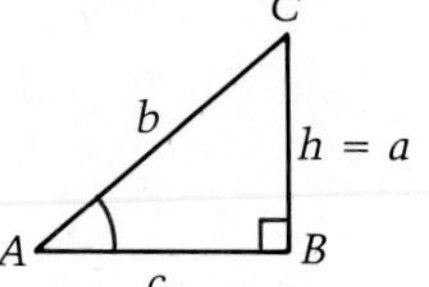


(**b**) $a = h$: one triangle possible

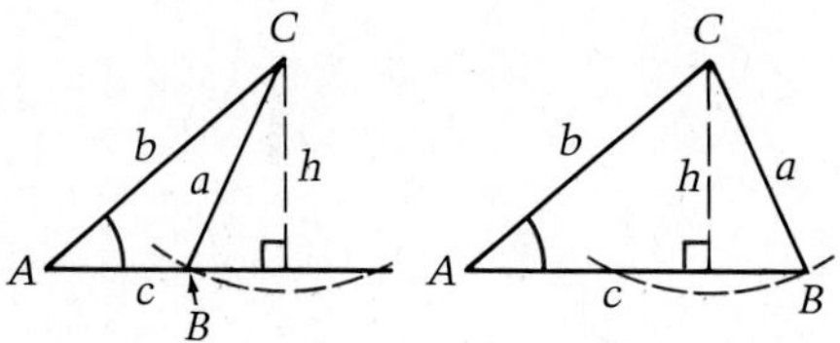


(**c**) $h < a < b$: two triangles possible

(**d**) $a \geq b$: one triangle possible

**Figure 5.6**

Example 2 illustrates case (c) above, namely the case where two triangles are possible. This is generally called *the ambiguous case.*

**Example 2** Solve $\triangle ABC$, given that $A = 24°30'$, $a = 6.00$, and $b = 12.2$.

***Solution*** Before we can draw the triangle, we must determine another angle.

**Find $B$:** $\dfrac{\sin B}{12.2} = \dfrac{\sin 24°30'}{6.00}$ ■ $\dfrac{\sin B}{b} = \dfrac{\sin A}{a}$.

$$\sin B = \frac{0.4147}{6.00}(12.2)$$

Recall that the sine function is positive for angles in the first or second quadrant. Since $\sin 57°30' \approx 0.8432$, the second quadrant angle having the same sine function value is $180° - 57°30' = 122°30'$. Thus, we have two cases:

$$B = 57°30' \quad \text{or} \quad 122°30'.$$

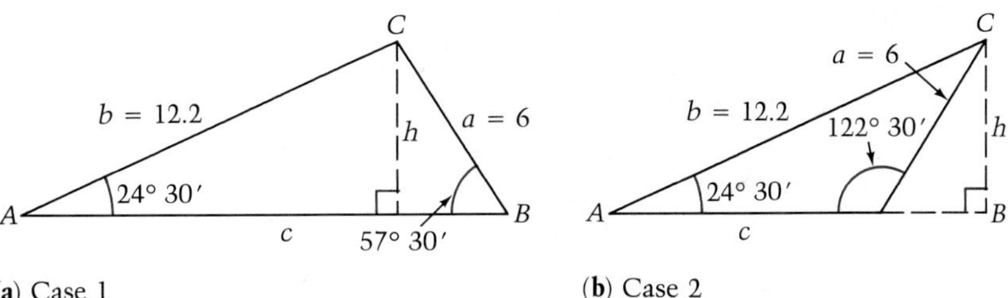

(**a**) Case 1 (**b**) Case 2

Case 1: $B = 57°30'$

**Find $C$:** $C = 180° - 24°30' - 57°30' = 98°00'$

**Find $c$:** $\dfrac{\sin 98°00'}{c} = \dfrac{\sin 24°30'}{6.00}$

■ $\dfrac{\sin C}{c} = \dfrac{\sin A}{a}$.

$$\frac{0.9903}{c} = \frac{0.4147}{6.00}$$

■ $\sin 98°00' = \sin(180° - 98°00')$
$= \sin 82°00' = 0.9903$

$$c = \left(\frac{6.00}{0.4147}\right)(0.9903) \approx 14.3$$

Case 2: $B = 122°30'$

**Find $C$:** $C = 180° - 24°30' - 122°30' = 33°00'$

**Find $c$:** $\dfrac{\sin 33°00'}{c} = \dfrac{0.4147}{6.00}$

■ $\dfrac{\sin C}{c} = \dfrac{\sin A}{a}$.

$$c = \frac{6.00}{0.4147}(0.5446) \approx 7.9$$

In discussing the case where two sides and an opposite angle are given, we have not considered the case where the given angle is obtuse.

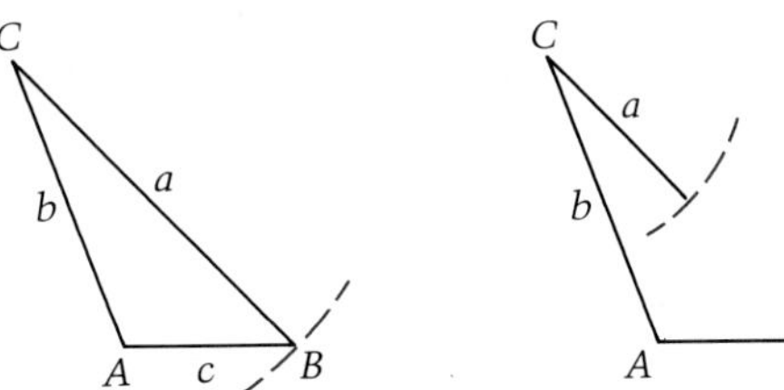

**Figure 5.7**

Clearly in this situation, a triangle can be formed only if $a > b$.

In practice, applying the Law of Sines will supply you with the information you need in order to decide which of the many possible cases can occur for the given data. For example, in the case of "no possible triangle," one step in the computation might yield that the sine of an unknown angle is greater than 1, which is impossible.

**Goal C** *to solve applied problems by using the Law of Sines*

**Example 3** A forest fire is spotted by observers in two fire towers 12 miles apart. Tower $B$ is on a bearing of S 12°10′ E from Tower $A$. If the bearing of the fire from Tower $A$ is S 45°40′ W and from Tower $B$ is N 75°20′ W, how far is the fire from Tower $B$.

***Solution***

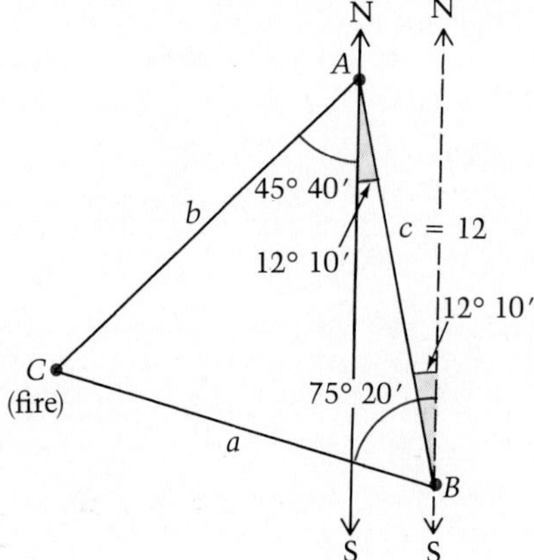


- Draw a figure to represent the given data.
- To determine $B$, use the fact that when parallel lines are cut by a transversal, alternate interior angles (shaded angles) are equal. Thus $B = 75°20' - 12°10' = 63°10'$.
- Note that $A = 45°40' + 12°10' = 57°50'$.

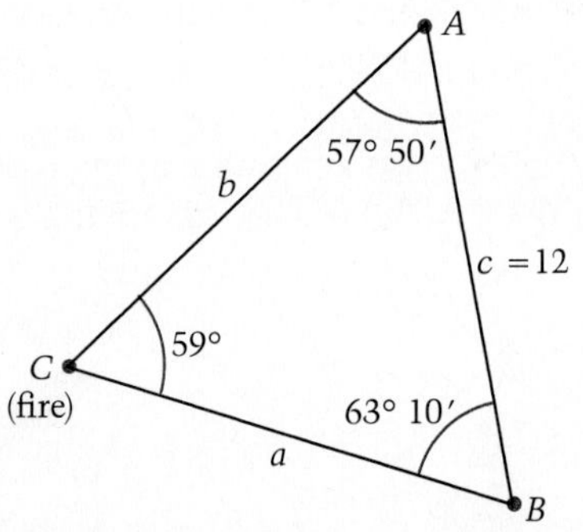


- Let us now simplify our diagram, showing the known parts of the triangle.
- $C = 180° - A - B$
  $= 180° - 57°50' - 63°10'$
  $= 59°00'$.

- $\dfrac{\sin A}{a} = \dfrac{\sin C}{c}$.

**Find *a*:** $\dfrac{\sin 57°50'}{a} = \dfrac{\sin 59°00'}{12.0}$

$$a = \frac{12.0}{0.8572}(0.8465) \approx 11.9 \text{ mi}$$

The fire is approximately 11.9 mi from Tower B.

## Exercise Set 5.2 Calculator Exercises in the Appendix

### Goal A

In exercises 1–10, the measures of the two angles and one side of $\triangle ABC$ are given. Solve the triangle.

1. $A = 30°$, $B = 45°$, $a = 10\sqrt{2}$
2. $A = 60°$, $B = 45°$, $b = 8\sqrt{6}$
3. $B = 30°$, $C = 135°$, $b = 4\sqrt{2}$
4. $B = 45°$, $C = 120°$, $c = 4\sqrt{6}$
5. $a = 24.0$, $A = 100°$, $B = 24°$
6. $b = 54.0$, $A = 76°$, $B = 41°$
7. $c = 42.0$, $B = 36°$, $C = 64°$
8. $a = 96.0$, $A = 105°$, $C = 21°$
9. $a = 0.7280$, $B = 71°36'$, $C = 66°54'$
10. $b = 0.4980$, $A = 37°18'$, $C = 92°06'$

### Goal B

In exercises 11–18, you are given measures for $A$, $a$, and $b$. State whether it is possible to have $\triangle ABC$. If it is, solve the triangle. (Determine both solutions if there is more than one.)

11. $A = 30°$, $a = 12$, $b = 24$
12. $A = 60°$, $a = 36$, $b = 24$
13. $A = 120°$, $a = 18$, $b = 24$
14. $A = 135°$, $a = 24$, $b = 24\sqrt{2}$
15. $A = 120°$, $a = 54.3$, $b = 48.8$
16. $A = 135°$, $b = 36.4$, $a = 44.7$
17. $a = 32.2$, $b = 36.4$, $A = 52°30'$
18. $b = 24.8$, $a = 20.6$, $A = 38°30'$

### Goal C

19. Points $A$ and $B$ are on opposite sides of the Grand Canyon. Point $C$ is 200 yd from $A$, $m(\angle BAC) = 87°30'$, and $m(\angle ACB) = 67°12'$. What is the distance between $A$ and $B$?
20. Two observers standing on shore $\frac{1}{2}$ mi apart at points $A$ and $B$ measure the angle to a sailboat at point $C$ at the same time. If $m(\angle CAB) = 63°24'$ and $m(\angle CBA) = 56°36'$, find the distance from each observer to the boat.
21. Ship A is 485 m due east of ship B. A lighthouse 1600 m from ship A is on a bearing of N 17°18′ E from ship B. What is the bearing of the lighthouse from ship A?
22. Two observers 2 mi apart on level ground are in line with a spot directly below a hot air balloon. If the angle of elevation of the balloon for one observer is 68°54′ and, at the same time, 26°24′ for the other, what is the altitude of the balloon to the nearest tenth of a mile?
23. An observer on a ship spots a liferaft at a bearing of N 75°24′ E while, at the same instant, a second observer on another ship takes a bearing of the liferaft of N 15°54′ E. If the second ship is at a distance of 5.5 miles and a bearing of S 27°54′ E from the first ship, find the distance from each ship to the liferaft.
24. An observer on a ship spots a liferaft at a bearing of N 35°18′ W while, at the same instant, a second observer on another ship takes a bearing of the liferaft of N 10°36′ E. If the second ship is 4.8 mi and on a bearing of S 55°42′ W from the first ship, how far is each ship from the liferaft.

### Superset

25. Express $x$ in terms of $\theta$, $\varphi$ and $m$.

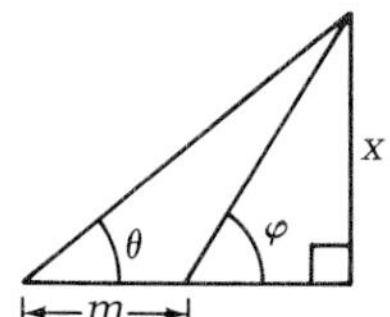


In exercises 26–27, prove each of the following.

26. $\dfrac{a+b}{b} = \dfrac{\sin A + \sin B}{\sin B}$

27. $\dfrac{a-b}{a+b} = \dfrac{\sin A - \sin B}{\sin A + \sin B}$

## 5.3 Law of Cosines

**Goal A** *to solve a triangle given two sides and the included angle, or three sides*

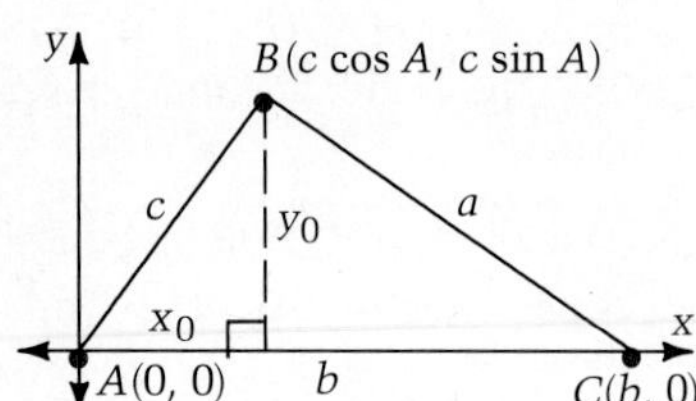


**Figure 5.8** $B$ has coordinates $x_0 = c \cos A$, $y_0 = c \sin A$.

In the last section, we considered the Law of Sines and how it is used to solve triangles when we are given (1) one side and two angles, or (2) two sides and an angle opposite one of them. We now derive a formula, called the Law of Cosines, which will be useful if we are given (3) two sides and the included angle, or (4) three sides. Suppose we have $\triangle ABC$ positioned so that side $b$ lies on the $x$-axis. Then

$$x_0 = c \cos A, \quad \text{and} \quad y_0 = c \sin A.$$

■ $\cos A = \frac{x_0}{c}$, and $\sin A = \frac{y_0}{c}$.

By the distance formula, we have

$$\begin{aligned} a^2 &= (c \cos A - b)^2 + (c \sin A - 0)^2 \\ &= (c \cos A)^2 - 2(c \cos A)(b) + b^2 + (c \sin A)^2 \\ &= c^2 \cos^2 A - 2bc \cos A + b^2 + c^2 \sin^2 A \\ &= b^2 + c^2(\sin^2 A + \cos^2 A) - 2bc \cos A \\ &= b^2 + c^2 - 2bc \cos A \end{aligned}$$

■ $\sin^2 A + \cos^2 A = 1$.

Note that the formula gives us a means of determining the square of one side of a triangle, given the other two sides and their included angle. There are three forms of this formula, as stated below.

**Law of Cosines**

In any triangle with angles $A$, $B$, and $C$, and opposite sides $a$, $b$, and $c$, respectively,

$$\begin{aligned} c^2 &= a^2 + b^2 - 2ab \cos C, \\ a^2 &= b^2 + c^2 - 2bc \cos A, \\ b^2 &= a^2 + c^2 - 2ac \cos B. \end{aligned}$$

**Figure 5.9**

**Example 1** Given $\triangle ABC$ with $a = 4$, $b = 6$, and $C = 120°$, find $c$.

***Solution***

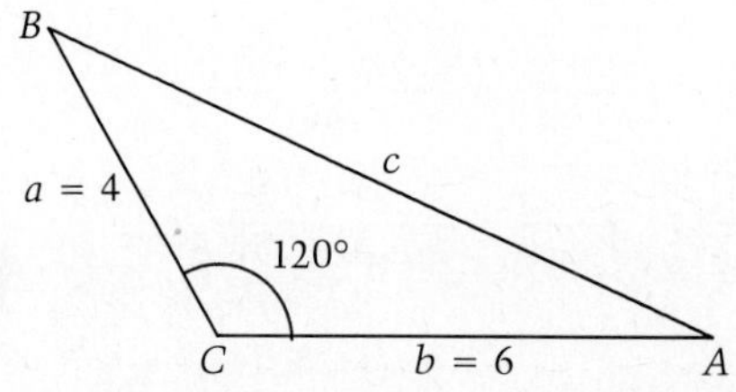


■ Begin by drawing a triangle that represents the given data.

$$c^2 = a^2 + b^2 - 2ab \cos C$$
$$c^2 = 4^2 + 6^2 - 2(4)(6) \cos 120°$$
$$c^2 = 16 + 36 - 48\left(-\frac{1}{2}\right)$$

■ $\cos 120° = -\cos 60° = -\frac{1}{2}$.

$$c^2 = 52 + 24 = 76$$
$$c = \sqrt{76} \approx 8.7$$

**Example 2** Given $\triangle ABC$ with $a = 6.00$, $b = 12.0$, and $c = 7.00$, find $B$.

***Solution***

$$b^2 = a^2 + c^2 - 2ac \cos B$$
$$12^2 = 6^2 + 7^2 - 2(6)(7) \cos B$$

■ The form of the Law of Cosines that involves angle $B$ is used.

$$\cos B = \frac{144 - 36 - 49}{-2(6)(7)} \approx -0.7024$$

■ Since $\cos B$ is negative, $B$ is the second quadrant angle whose reference angle has cosine function value 0.7024.

Since $\cos 45°20' \approx 0.7024$, $B \approx 180° - 45°20' = 134°40'$.

**Example 3** Highway 102 runs east-west and is intersected by Route 66, in a direction 20° north of due east. Car A is traveling along Highway 102 and is 4 miles east of the intersection. Car B is traveling eastbound on Route 66 and is 18 miles past the intersection. What is the distance between the two cars?

***Solution***

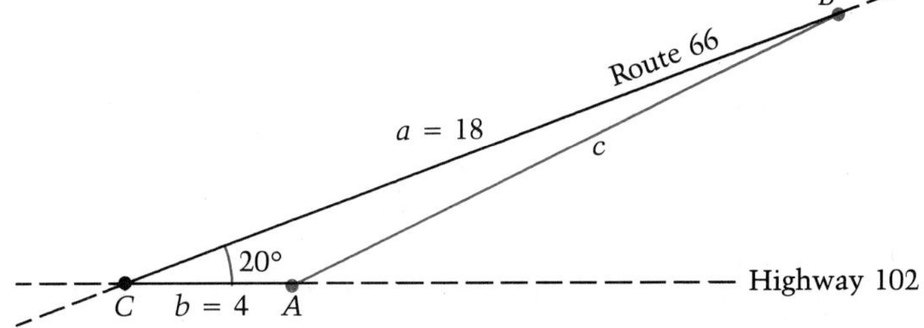


We wish to determine the distance $c$ shown above.

$$c^2 = a^2 + b^2 - 2ab \cos C$$
$$c^2 = 18^2 + 4^2 - 2(4)(18) \cos 20°$$
$$c^2 = 324 + 16 - 144(0.9397) \approx 204.7$$
$$c \approx 14.31$$

The distance between the two cars is approximately 14 mi.

**Example 4** Points $B$ and $C$ are on opposite sides of a reservoir. If the distance from point $A$ to $B$ is known to be 1.25 mi, and from point $A$ to $C$ is 1.15 mi, what is the distance between $B$ and $C$ if $\angle BAC = 55°10'$?

***Solution***

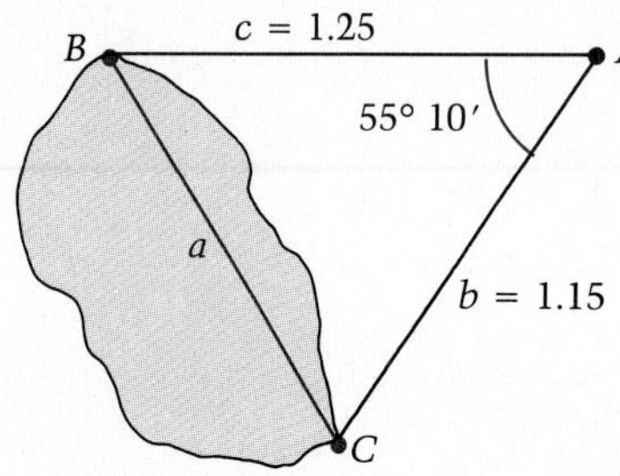


■ Draw a figure that represents the data.

$$a^2 = (1.15)^2 + (1.25)^2 - 2(1.15)(1.25)(\cos 55°10') \approx 1.24$$

■ Apply Law of Cosines to $\triangle ABC$. The desired distance is $a$.

$$a \approx 1.11$$

■ The distance between $B$ and $C$ is approximately 1.11 mi.

**Goal B** *to solve problems involving the area of a triangle*

We now return to the figure we used to motivate the derivation of the Law of Cosines (see Figure 5.10). According to the formula for the area of a triangle, we can describe the area $\mathbb{A}$ of $\triangle ABC$ as

$$\mathbb{A} = \frac{1}{2}(\text{altitude})(\text{base}) = \frac{1}{2}(c \sin A)(b) = \frac{1}{2}bc \sin A.$$

**Figure 5.10**

In general, the area of a triangle is half the product of any two sides times the sine of the included angle. For $\triangle ABC$, $\mathbb{A}$ can be expressed three ways:

$$\mathbb{A} = \frac{1}{2}ab \sin C, \qquad \mathbb{A} = \frac{1}{2}ac \sin B, \qquad \mathbb{A} = \frac{1}{2}bc \sin A.$$

**Example 5** Find the area of $\triangle ABC$ if $a = 13$ cm, $b = 10$ cm, and $C = 30°$.

***Solution***

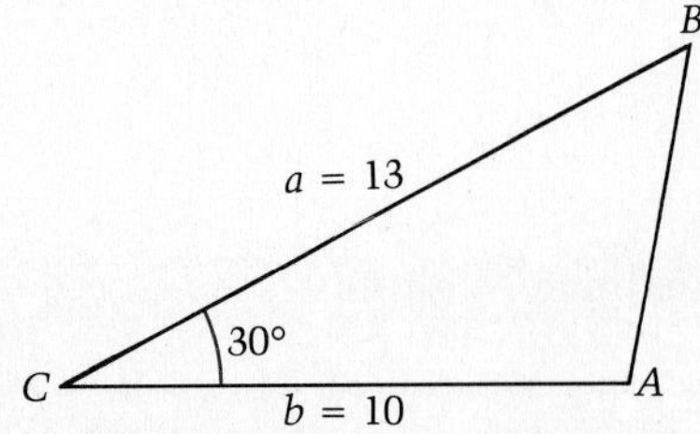


■ Draw a triangle that represents the data.

$$\begin{aligned}\text{Area} &= \frac{1}{2}(\text{product of two sides}) \cdot \sin(\text{included angle}) \\ &= \frac{1}{2}(ab)(\sin 30°) \\ &= \frac{1}{2}(13 \cdot 10)\left(\frac{1}{2}\right) = 32.5\end{aligned}$$

The area of the triangle is 32.5 cm$^2$.

**Example 6** Find the area of $\triangle ABC$ if $a = 22$ in, $b = 16$ in, and $c = 18$ in.

***Solution***

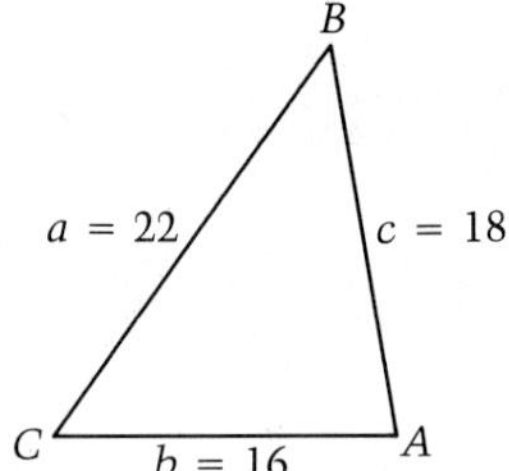


■ Draw a triangle that represents the data.

Since we are not given the lengths of two sides *and* the measure of an included angle, we first apply the Law of Cosines to determine one of the angles. We will find $C$.

$$c^2 = a^2 + b^2 - 2ab \cos C$$

$$\cos C = \frac{c^2 - a^2 - b^2}{-2ab} = \frac{a^2 + b^2 - c^2}{2ab} = \frac{22^2 + 16^2 - 18^2}{2(22)(16)} \approx 0.5909$$

$$C \approx 54°$$

Now that we have $a$ and $b$ and angle $C$, we can determine the area.

$$\text{Area} = \frac{1}{2}(ab) \sin 54° = \frac{1}{2}(22 \cdot 16)(0.8090) \approx 140 \text{ in}^2 \quad \text{(two significant digits)}$$

In the last example, it did not matter which of the three angles we chose in order to apply the formula for area; the result would have been the same. In addition, once we found that $\cos C \approx 0.5909$, we could have noted that since we need $\sin C$ in the formula for area, and since

$$\sin C = \sqrt{1 - \cos^2 C}$$

■ The Pythagorean Identity

we could have determined $\sin C$ (necessary for computing the area) without ever determining the measure of $C$.

## Exercise Set 5.3 ⊟ Calculator Exercises in the Appendix

### Goal A

In exercises 1–8, solve $\triangle ABC$, given two sides and the included angle.

1. $a = 12$, $b = 10$, $C = 60°$
2. $a = 12$, $b = 10$, $C = 30°$
3. $b = 20$, $c = 16$, $A = 120°$
4. $b = 24$, $c = 30$, $A = 135°$
5. $a = 112$, $c = 96$, $B = 54°$
6. $a = 78$, $c = 125$, $B = 128°$
7. $A = 147°36'$, $b = 12.6$, $c = 16.3$
8. $B = 34°54'$, $a = 22.3$, $c = 18.2$

In exercises 9–16, solve $\triangle ABC$, given $a$, $b$, and $c$.

9. $a = 3.0$, $b = 4.0$, $c = 5.0$
10. $a = 5.0$, $b = 12.0$, $c = 13.0$
11. $a = 2.8$, $b = 2.7$, $c = 2.4$
12. $a = 3.6$, $b = 5.2$, $c = 4.8$
13. $a = 3.2$, $b = 4.8$, $c = 6.4$
14. $a = 9.0$, $b = 6.3$, $c = 5.4$
15. $a = 45.0$, $b = 30.0$, $c = 50.0$
16. $a = 30.0$, $b = 20.0$, $c = 35.0$
17. Points $A$ and $B$ are sighted from point $C$. If $C = 98°$, $AC = 128$ m, and $BC = 96$ m, how far apart are the points $A$ and $B$?
18. Points $A$ and $B$ are sighted from point $C$. If $C = 36°$, $AC = 118$ ft, and $BC = 105$ ft, how far apart are the points $A$ and $B$?
19. Two sides and the included angle of a parallelogram have measures 3.2, 4.8, and 54°24′ respectively. Find the lengths of the diagonals.
20. The lengths of two sides of a parallelogram are 24.6 in and 38.2 in. The angle at one vertex has measure 108°42′. Find the lengths of the diagonals.
21. A bridge is supported by triangular braces. If the sides of each brace have lengths 63 ft, 46 ft, and 40 ft, find the measure of the angle opposite the 46 ft side.
22. The measures of two sides of a parallelogram are 28 in and 42 in. If the longer diagonal has measure 58 in, find the measures of the angles at the vertices.

### Goal B

In exercises 23–30, determine the area of $\triangle ABC$ using the given information.

23. $A = 60°$, $b = 12.6$, $c = 18.3$
24. $A = 45°$, $c = 23.7$, $b = 16.4$
25. $B = 37°12'$, $a = 10.9$, $c = 15.8$
26. $B = 24°54'$, $c = 10.5$, $a = 14.6$
27. $C = 112°$, $b = 44.6$, $a = 32.5$
28. $C = 118°$, $a = 18.7$, $b = 30.6$
29. $A = 13°30'$, $b = 254$, $c = 261$
30. $A = 66°24'$, $c = 0.231$, $b = 0.176$

### Superset

31. The lengths of two sides of a triangle are 12 in and 16 in, and the area is 87.36 in$^2$. Solve the triangle.
32. Given $\triangle ABC$ with $a = 12$, $b = 16$, and $c = 10$, find the length of the altitude from vertex $B$.
33. Find the area of a triangle with vertices at the points with coordinates $(-5, 0)$, $(6, 3)$, and $(8, -5)$.

In exercises 34–36, assume you are given an isosceles triangle $ABC$ with $a = c$ with $B = \theta$.

34. Find the length of the base $b$ in terms of $a$ and $\theta$.
35. Show that $b = 2a \sin \frac{1}{2}\theta$.
36. Prove that $b^2 = 2a^2(1 - \cos \theta)$.

## 5.4 Vectors in the Plane

**Goal A** *to determine the magnitude and direction of a vector*

The statement "The distance between points $A$ and $B$ is 12 miles" tells us only the distance between $A$ and $B$. The statement "An object undergoes a displacement of 12 miles due east from point $A$ to point $B$" tells us both the distance between $A$ and $B$ and the direction in which the object travels. Quantities like distance, which indicate a magnitude (size) but no direction, are called **scalar quantities.** Quantities like displacement, which indicate both magnitude and direction, are called **vector quantities.**

Since a vector quantity contains information about magnitude and direction, it is convenient to represent it by an arrow (a directed line segment) in the $xy$-plane. The magnitude of the vector quantity is given by the length of the arrow. The direction of the vector quantity is given by the angle that the arrow makes with the horizontal in the direction of the positive $x$-axis.

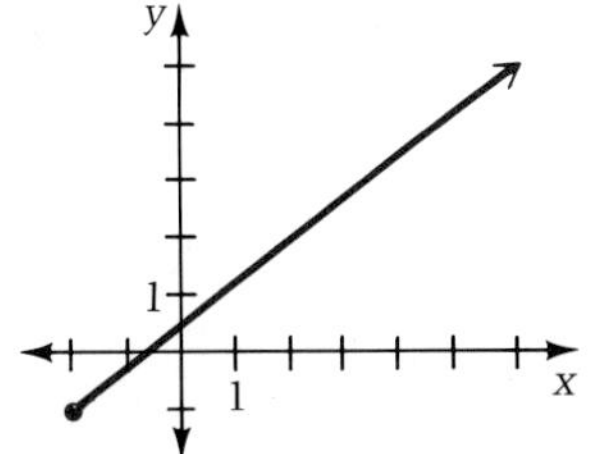


Figure 5.11

**Example 1** Determine the magnitude and direction of the vector represented in Figure 5.11.

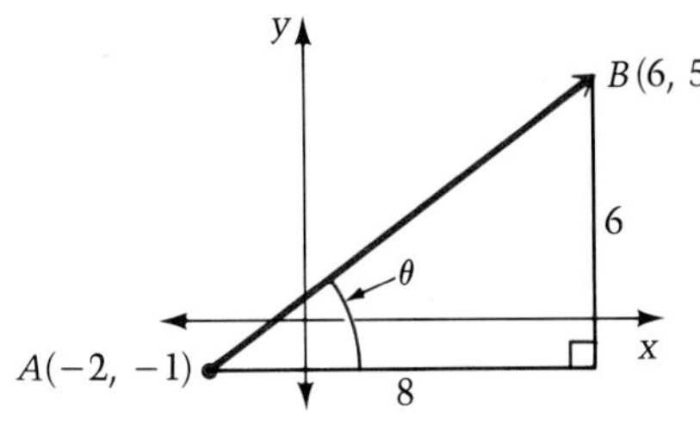


■ Begin by drawing a right triangle with the arrow as its hypotenuse. Label the coordinates of the tail $A$ and head $B$ of the vector.

$$\begin{aligned} d &= \sqrt{(x_2 - x_1)^2 + (y_2 - y_1)^2} \\ &= \sqrt{(6 - (-2))^2 + (5 - (-1))^2} \\ &= \sqrt{64 + 36} = 10 \end{aligned}$$

■ The magnitude of the vector can be found by applying the distance formula to the coordinates of the tail $A(-2, -1)$ and head $B(6, 5)$ of the vector.

$$\tan \theta = \frac{\text{opposite}}{\text{adjacent}} = \frac{6}{8} = 0.7500$$

$$\theta \approx 37°$$

■ The direction can be found by determining angle $\theta$ in the triangle.

The magnitude of the vector is 10, and it makes an angle of approximately 37° with the horizontal.

We usually use boldface letters to represent vector quantities.

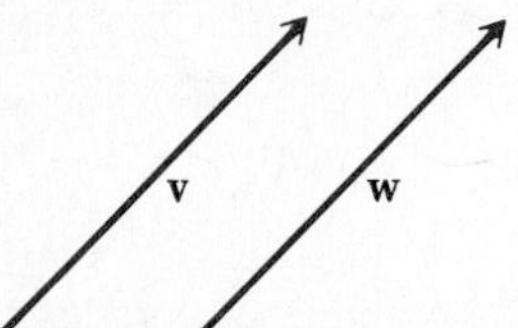


Vectors $\mathbf{v}$ and $\mathbf{w}$ are equal since they have the same magnitude and direction.

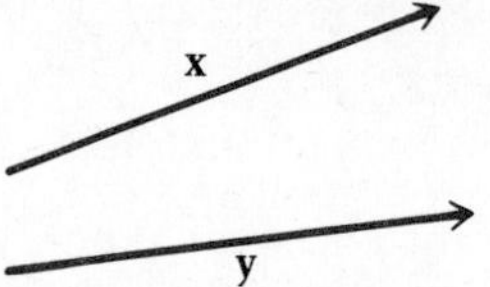


Vectors $\mathbf{x}$ and $\mathbf{y}$ have the same magnitude, but different directions.

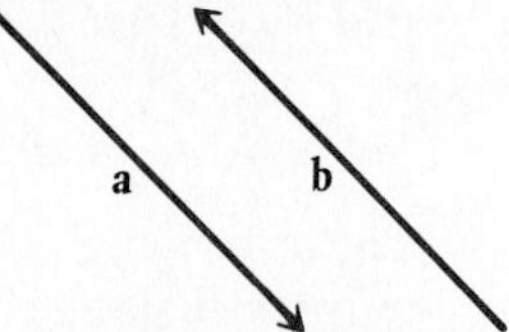


Vectors $\mathbf{a}$ and $\mathbf{b}$ are said to have the same magnitude but *opposite* directions.

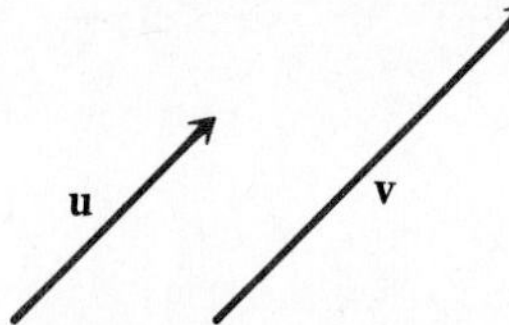


Vectors $\mathbf{u}$ and $\mathbf{v}$ have the same direction, but different magnitudes.

Figure 5.12

We will need to determine the sum of two vectors, and the product of a scalar (a real number) and a vector. The sum $\mathbf{a} + \mathbf{b}$ of two vectors $\mathbf{a}$ and $\mathbf{b}$ is found by first moving $\mathbf{b}$ without changing its magnitude or direction, so that the tail of $\mathbf{b}$ is placed at the head of $\mathbf{a}$. Then draw an arrow from the tail of $\mathbf{a}$ to the head of $\mathbf{b}$.

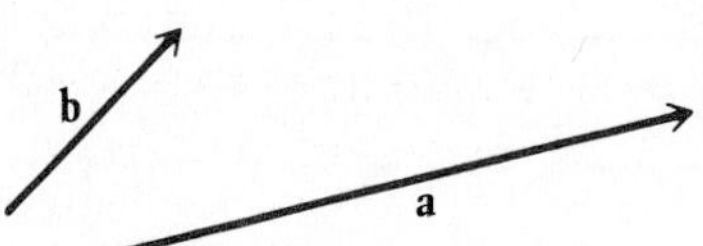


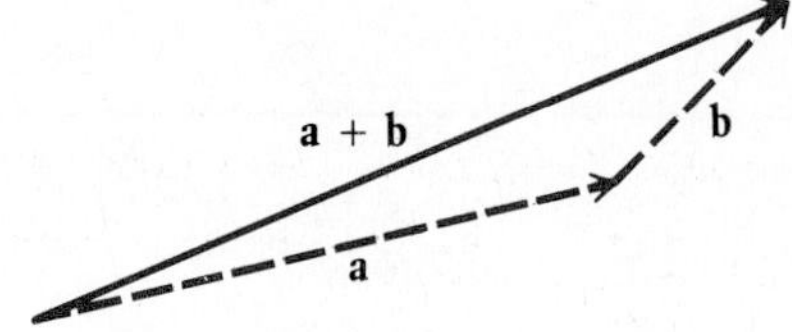


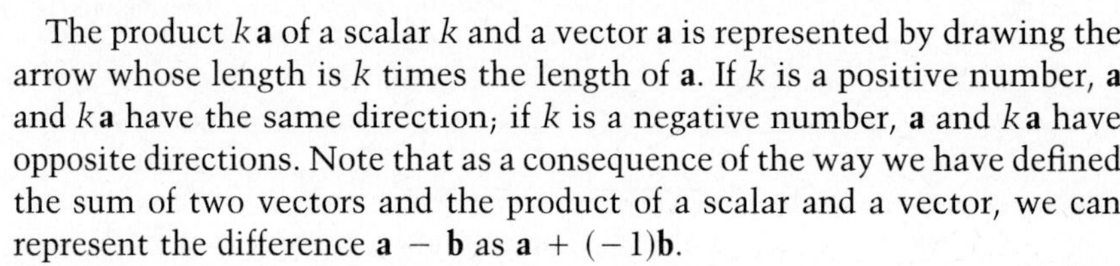

Figure 5.13

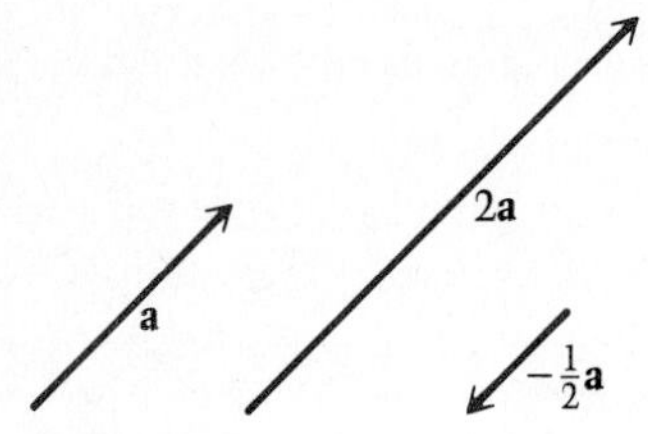


Figure 5.14

The product $k\mathbf{a}$ of a scalar $k$ and a vector $\mathbf{a}$ is represented by drawing the arrow whose length is $k$ times the length of $\mathbf{a}$. If $k$ is a positive number, $\mathbf{a}$ and $k\mathbf{a}$ have the same direction; if $k$ is a negative number, $\mathbf{a}$ and $k\mathbf{a}$ have opposite directions. Note that as a consequence of the way we have defined the sum of two vectors and the product of a scalar and a vector, we can represent the difference $\mathbf{a} - \mathbf{b}$ as $\mathbf{a} + (-1)\mathbf{b}$.

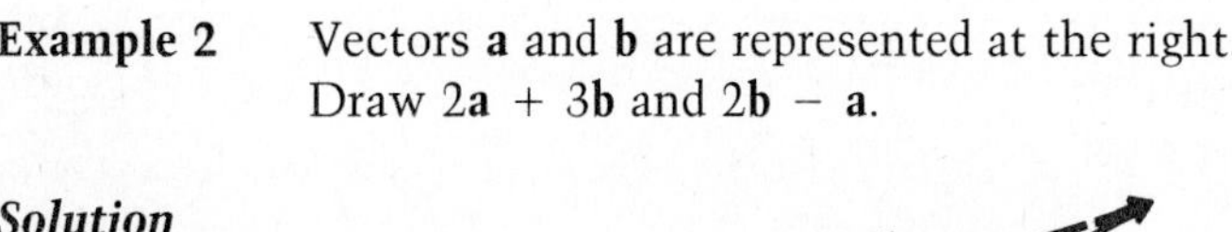

**Example 2** Vectors $\mathbf{a}$ and $\mathbf{b}$ are represented at the right. Draw $2\mathbf{a} + 3\mathbf{b}$ and $2\mathbf{b} - \mathbf{a}$.

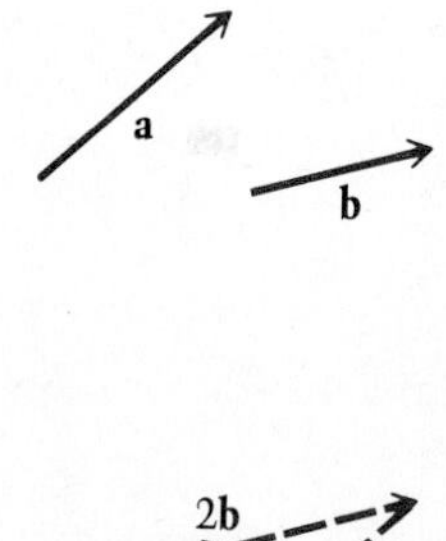


***Solution***

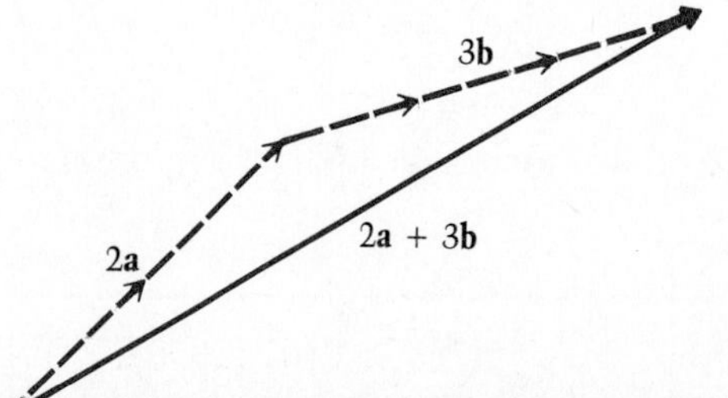


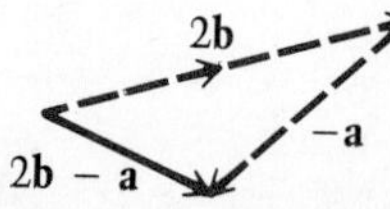

We have illustrated a method for adding (or subtracting) vectors by placing them in a tail-to-head arrangement. We can also add (or subtract) two vectors by placing them in a tail-to-tail arrangement. When positioned this way, the two vectors determine a parallelogram as shown in Figure 5.15. The sum **a** + **b** is found by drawing a vector from the point where the tails meet, along the diagonal of the parallelogram to the opposite vertex. The difference **a** − **b** is found by drawing a vector from the head of **b** to the head of **a**.

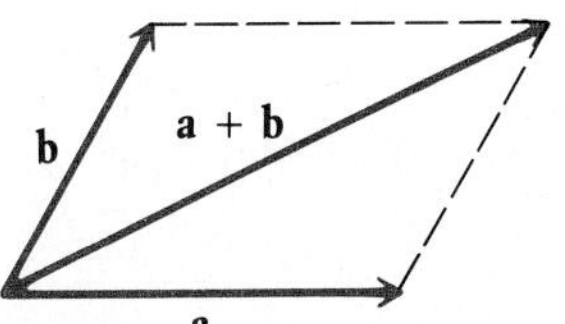


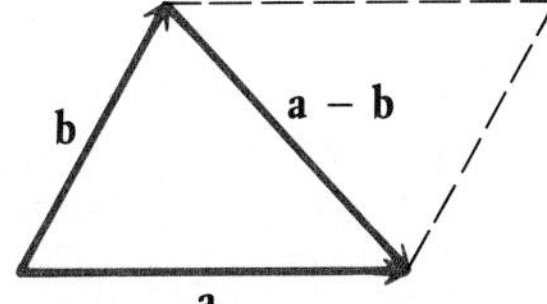


**Figure 5.15**

We have frequently referred to the tail and head of a vector. These points are alternately called the **initial point** and the **terminal point** of the vector, respectively. It is sometimes convenient to use these points when naming the vector. For example, if we know that vector **v** in Figure 5.16 has initial point $A$ and terminal point $B$, we may refer to it as $\overrightarrow{AB}$.

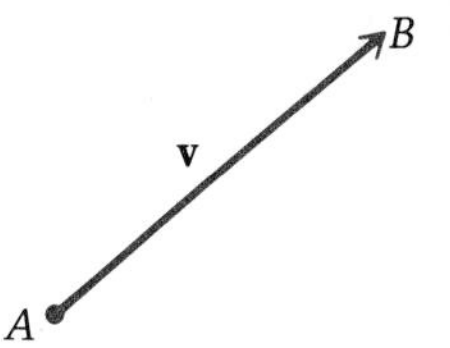


**Figure 5.16** $\mathbf{v} = \overrightarrow{AB}$

**Goal B** *to describe vectors as ordered pairs*

One of the most common ways of naming a vector is with an ordered pair of numbers. The first number of the ordered pair represents the change in $x$ from the initial point to the terminal point of the vector, and the second number represents the change in $y$ from the initial point to the terminal point. To distinguish ordered pairs representing vectors from ordered pairs representing points, we will use angular brackets $\langle\, , \rangle$ when referring to vectors. For example, in Figure 5.17, we denote the vector $\overrightarrow{PQ}$ by $\langle 3, -2\rangle$. This indicates that the change in $x$ from $P$ to $Q$ is 3 units in the positive $x$-direction and the change in $y$ is 2 units in the negative $y$-direction.

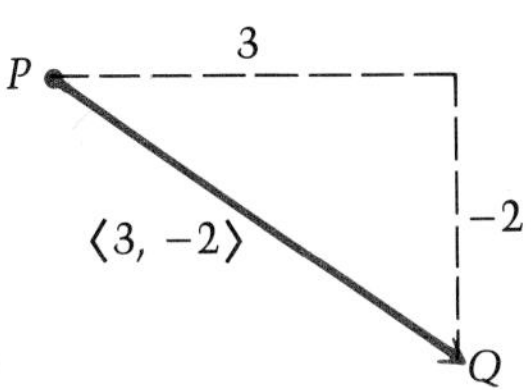


**Figure 5.17**

In a natural way, the $x$-coordinates and $y$-coordinates of two points $P_1$ and $P_2$ can be used to determine the vector $\overrightarrow{P_1P_2}$.

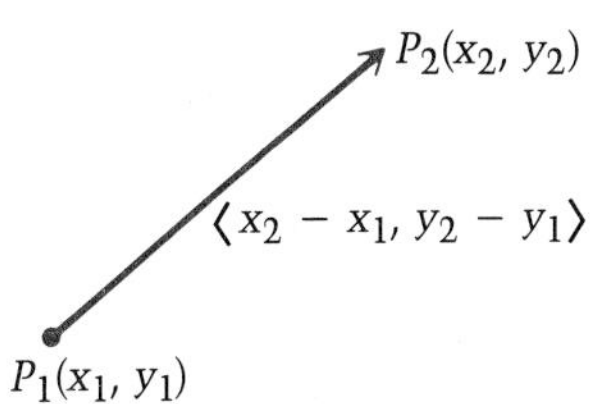


**Figure 5.18**

**Definition**

Given two points, $P_1(x_1, y_1)$ and $P_2(x_2, y_2)$, the vector with initial point $P_1$ and terminal point $P_2$ is given by the ordered pair

$$\langle x_2 - x_1, y_2 - y_1\rangle.$$

The numbers $x_2 - x_1$ and $y_2 - y_1$ are called the **scalar components** of the vector.

**Example 3** (a) Describe the vector from $A(-3, 2)$ to $B(9, -7)$ as an ordered pair.
(b) If $\mathbf{v} = \langle 3, -5 \rangle$ is positioned in the plane so that its initial point is at (1, 2), what are the coordinates of its terminal point?

***Solution*** (a) $\overrightarrow{AB} = \langle 9 - (-3), -7 - 2 \rangle = \langle 12, -9 \rangle$.

(Note: $\overrightarrow{BA} = \langle -3 - 9, 2 - (-7) \rangle = \langle -12, 9 \rangle$, thus, the vector $\overrightarrow{BA}$ is the same as the vector $-\overrightarrow{AB}$.

(b) The terminal point is found by adding 3 to the $x$-coordinate of the initial point and $-5$ to the $y$-coordinate of the initial point. Thus, the terminal point has coordinates $(1 + 3, 2 - 5) = (4, -3)$.

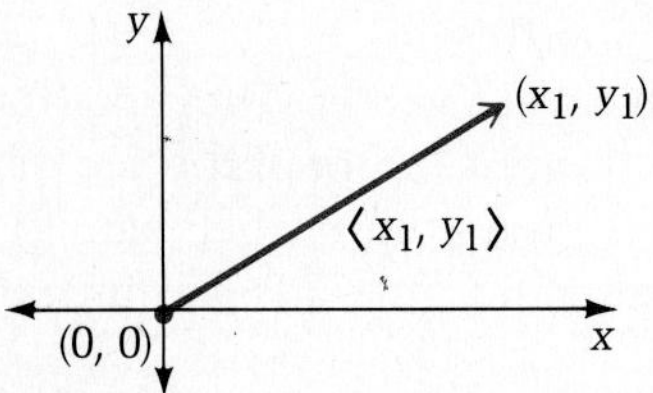


Figure 5.19

A vector that is positioned with its initial point at the origin (0, 0) is called a **position vector,** or **radius vector.** In this case, if the vector is given by $\langle x_1, y_1 \rangle$, then the terminal point is $(x_1, y_1)$. That is, the scalar components of a position vector are the same as the coordinates of the terminal point. The vector $\langle 0, 0 \rangle = \mathbf{0}$ is called the **zero vector.** We now formally define the notions of equality, addition, and scalar multiplication of vectors.

**Definition**

Given two vectors $\mathbf{a} = \langle a_1, a_2 \rangle$ and $\mathbf{b} = \langle b_1, b_2 \rangle$, and a scalar $k$, we have the following definitions.

| | |
|---|---|
| **Equality of vectors** | $\mathbf{a} = \mathbf{b}$ if and only if $a_1 = b_1$ and $a_2 = b_2$. |
| **Addition of vectors** | $\mathbf{a} + \mathbf{b} = \langle a_1 + b_1, a_2 + b_2 \rangle$ |
| **Scalar Multiplication of a vector** | $k\mathbf{a} = \langle ka_1, ka_2 \rangle$ |

**Example 4** Suppose $\mathbf{v} = \langle 2, 3 \rangle$, $\mathbf{w} = \langle 5, -4 \rangle$ and $\mathbf{j} = \langle 0, 1 \rangle$. Determine (a) $2\mathbf{v} + 3\mathbf{w}$ (b) $\mathbf{w} - 6\mathbf{j}$

***Solution***

(a)
$$\begin{aligned} 2\mathbf{v} + 3\mathbf{w} &= 2\langle 2, 3 \rangle + 3\langle 5, -4 \rangle \\ &= \langle 4, 6 \rangle + \langle 15, -12 \rangle \\ &= \langle 4 + 15, 6 - 12 \rangle \\ &= \langle 19, -6 \rangle \end{aligned}$$

(b)
$$\begin{aligned} \mathbf{w} - 6\mathbf{j} &= \langle 5, -4 \rangle + (-6)\langle 0, 1 \rangle \\ &= \langle 5, -4 \rangle + \langle 0, -6 \rangle \\ &= \langle 5 + 0, -4 - 6 \rangle \\ &= \langle 5, -10 \rangle \end{aligned}$$

The length of a vector $\mathbf{v}$ is called the **norm** of $\mathbf{v}$ and is denoted $\|\mathbf{v}\|$. Using the ordered pair definition of a vector and the Pythagorean Theorem, we have the following definition.

**Definition**

If $\mathbf{a} = \langle a_1, a_2 \rangle$, then $\|\mathbf{a}\|$ is the **norm** of $\mathbf{a}$, defined $\|\mathbf{a}\| = \sqrt{a_1^2 + a_2^2}$.

Note that the norm of a vector is a scalar quantity—it measures the length of the vector. A vector is called a **unit vector** if it has length equal to 1. Given a vector $\mathbf{a}$, we can determine a **unit vector in the direction of a,** denoted $\mathbf{u}_\mathbf{a}$, by multiplying vector $\mathbf{a}$ by the scalar $\frac{1}{\|\mathbf{a}\|}$. That is,

$$\mathbf{u}_\mathbf{a} = \frac{\mathbf{a}}{\|\mathbf{a}\|}$$

is the unit vector in the direction of $\mathbf{a}$.

**Example 5** Suppose $\mathbf{v} = \langle 3, -4 \rangle$. Determine the following.

(a) $\|\mathbf{v}\|$ (b) $\mathbf{u}_\mathbf{v}$ (c) a vector of length 2 in the direction of $\mathbf{v}$

***Solution*** (a) $\|\mathbf{v}\| = \sqrt{3^2 + (-4)^2} = \sqrt{9 + 16} = 5$

(b) $\mathbf{u}_\mathbf{v} = \frac{1}{\|\mathbf{v}\|}\mathbf{v} = \frac{1}{5}\langle 3, -4 \rangle = \left\langle \frac{3}{5}, -\frac{4}{5} \right\rangle$

Thus the unit vector in the direction of $\mathbf{v}$ is $\left\langle \frac{3}{5}, -\frac{4}{5} \right\rangle$.

(c) To determine a vector of length 2 in the direction of $\mathbf{v}$, multiply the unit vector (in that direction) by 2.

$$2\mathbf{u}_\mathbf{v} = 2\left\langle \frac{3}{5}, -\frac{4}{5} \right\rangle = \left\langle \frac{6}{5}, -\frac{8}{5} \right\rangle$$

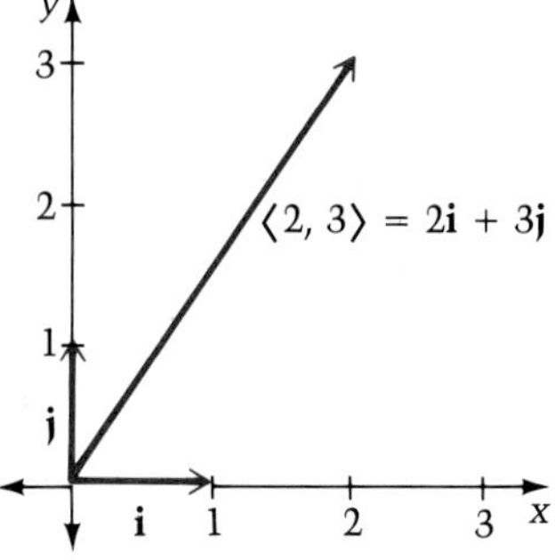


**Figure 5.20**

The unit vectors

$$\langle 1, 0 \rangle \quad \text{and} \quad \langle 0, 1 \rangle$$

occur so frequently that we give them the special names **i** and **j**, respectively. Thus, **i** is a vector of length 1 in the positive $x$-direction, and **j** is a vector of length 1 in the positive $y$-direction. These two vectors are referred to as the **unit coordinate vectors.** For any real numbers $a_1$ and $a_2$,

$$a_1\mathbf{i} + a_2\mathbf{j} = a_1\langle 1, 0 \rangle + a_2\langle 0, 1 \rangle = \langle a_1, 0 \rangle + \langle 0, a_2 \rangle = \langle a_1, a_2 \rangle.$$

Thus, any vector $\mathbf{a} = \langle a_1, a_2 \rangle$ can be written as $\mathbf{a} = a_1\mathbf{i} + a_2\mathbf{j}$, which is referred to as a **linear combination** of **i** and **j**.

**Example 6** Suppose $\mathbf{v} = 2\mathbf{i} + 3\mathbf{j}$ and $\mathbf{w} = \mathbf{i} - 4\mathbf{j}$. Determine the following.
(a) $5\mathbf{v} - \mathbf{w}$ (b) $\|\mathbf{v}\|$ (c) a unit vector in the direction of **v**

***Solution*** (a) $5\mathbf{v} - \mathbf{w} = 5(2\mathbf{i} + 3\mathbf{j}) - (\mathbf{i} - 4\mathbf{j}) = 10\mathbf{i} + 15\mathbf{j} - \mathbf{i} + 4\mathbf{j} = 9\mathbf{i} + 19\mathbf{j}$

(b) $\|\mathbf{v}\| = \|2\mathbf{i} + 3\mathbf{j}\| = \sqrt{2^2 + 3^2} = \sqrt{4 + 9} = \sqrt{13}$

(c) $\mathbf{u}_\mathbf{v} = \dfrac{1}{\|\mathbf{v}\|}\mathbf{v} = \dfrac{1}{\sqrt{13}}(2\mathbf{i} + 3\mathbf{j}) = \dfrac{2}{\sqrt{13}}\mathbf{i} + \dfrac{3}{\sqrt{13}}\mathbf{j}$

## Exercise Set 5.4 ⊟ Calculator Exercises in the Appendix

### Goal A

In exercises 1–8, determine the magnitude and direction of each vector.

1.
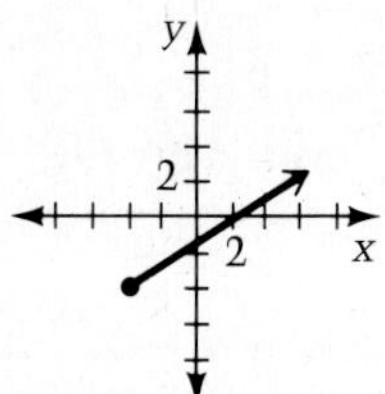

2.
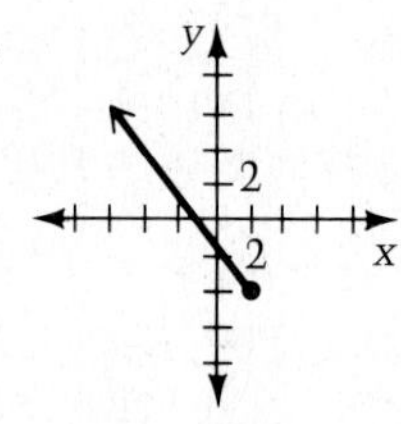

3.
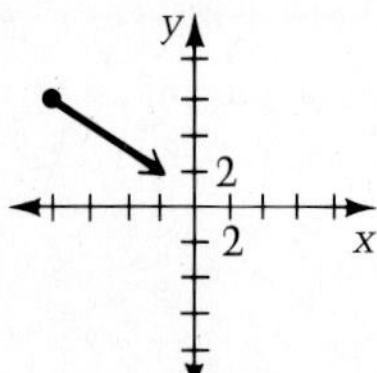

4.
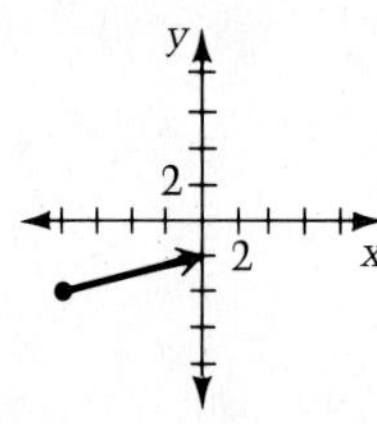

5.
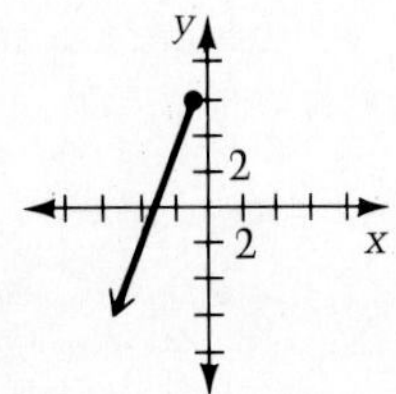

6.
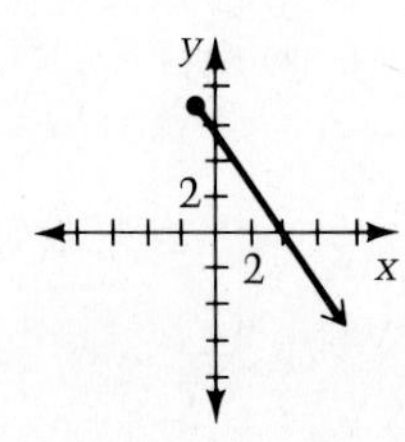

7.
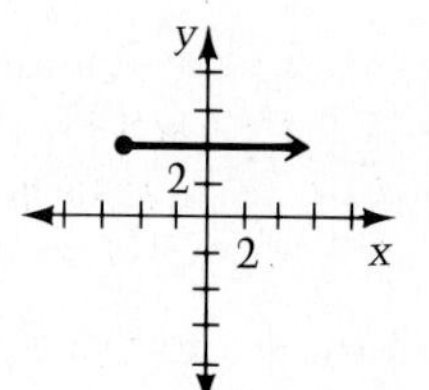

8.
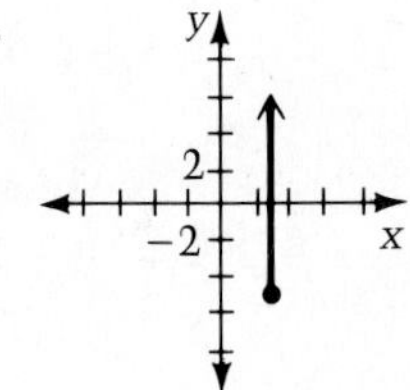

In exercises 9–20, using the given vectors, draw the vector sum or difference.

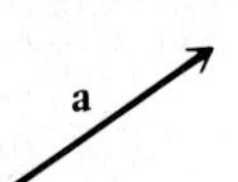


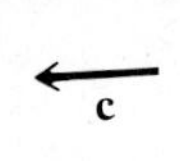


d

9. $\mathbf{a} + \mathbf{d}$
10. $\mathbf{b} + \mathbf{c}$
11. $\mathbf{a} - \mathbf{d}$
12. $\mathbf{b} - \mathbf{c}$
13. $2\mathbf{a} - 3\mathbf{b}$
14. $3\mathbf{d} - 4\mathbf{b}$
15. $\frac{1}{2}\mathbf{a} + \frac{1}{2}\mathbf{d}$
16. $2\mathbf{b} + 2\mathbf{c}$
17. $-2\mathbf{b}$
18. $-3\mathbf{c}$
19. $\mathbf{a} + 2\mathbf{b} - \mathbf{c}$
20. $2\mathbf{a} - \mathbf{c} - 2\mathbf{d}$

## Goal B

In exercises 21–26, points $A$ and $B$ are given. Describe the vector from $A$ to $B$ as an ordered pair.

**21.** $A(-1, -3)$, $B(4, 9)$ **22.** $A(2, -2)$, $B(-4, 6)$

**23.** $A(-5, 9)$, $B(-2, 5)$ **24.** $A(-10, -7)$, $B(-2, -1)$

**25.** $A(1, 4)$, $B(2, -4)$ **26.** $A\left(-2, 3\frac{1}{2}\right)$, $B(-5, 4)$

In exercises 27–32, given the vector **v** and the coordinates of its initial point, find the coordinates of its terminal point.

**27.** $\mathbf{v} = \langle 3, -2\rangle$, $(-1, 4)$

**28.** $\mathbf{v} = \langle -2, 3\rangle$, $(-1, 4)$

**29.** $\mathbf{v} = \langle -4, -5\rangle$, $\left(1\frac{1}{2}, 3\frac{1}{2}\right)$

**30.** $\mathbf{v} = \langle -5, 2\rangle$, $\left(1\frac{1}{2}, 3\frac{1}{2}\right)$

**31.** $\mathbf{v} = \left\langle \frac{13}{2}, \frac{7}{2}\right\rangle$, $(-4, -1)$

**32.** $\mathbf{v} = \left\langle -\frac{3}{2}, -\frac{7}{2}\right\rangle$, $(1, -2)$

In exercises 33–40, for the vectors $\mathbf{u} = \langle 3, 6\rangle$, $\mathbf{v} = \langle -8, 2\rangle$, and $\mathbf{w} = \langle 2, -1\rangle$, determine the following.

**33.** $3\mathbf{v} + 2\mathbf{w}$ **34.** $2\mathbf{v} + \mathbf{u}$

**35.** $\frac{1}{2}\mathbf{v} + 3\mathbf{u}$ **36.** $2\mathbf{w} + 3\mathbf{u}$

**37.** $-\frac{1}{2}\mathbf{v} + 3\mathbf{u}$ **38.** $-4\mathbf{w} + \frac{1}{3}\mathbf{u}$

**39.** $3\mathbf{v} + 2\mathbf{w} - \frac{1}{3}\mathbf{u}$ **40.** $\frac{1}{2}\mathbf{v} - 3\mathbf{w} + \frac{2}{3}\mathbf{u}$

In exercises 41–48, for each vector **v**, determine $\|\mathbf{v}\|$, $\mathbf{u}_\mathbf{v}$, and $-2\mathbf{v}$.

**41.** $\langle -6, 8\rangle$ **42.** $\langle -4, -3\rangle$ **43.** $\langle 0, -4\rangle$

**44.** $\langle -6, 0\rangle$ **45.** $\langle 2\sqrt{2}, 2\rangle$ **46.** $\langle 3, \sqrt{7}\rangle$

**47.** $\left\langle \frac{8}{3}, -2\right\rangle$ **48.** $\left\langle -\frac{5}{2}, 6\right\rangle$

In exercises 49–54, given vectors **v** and **w**, determine $2\mathbf{v} - 3\mathbf{w}$ as a linear combination of the unit vectors **i** and **j**. Determine $\|\mathbf{v}\|$ and $\mathbf{u}_\mathbf{v}$.

**49.** $\mathbf{v} = \mathbf{i} + \mathbf{j}$, $\mathbf{w} = \mathbf{i} - 2\mathbf{j}$

**50.** $\mathbf{v} = 2\mathbf{i} - \mathbf{j}$, $\mathbf{w} = \mathbf{i} - 2\mathbf{j}$

**51.** $\mathbf{v} = 3\mathbf{i} - 2\mathbf{j}$, $\mathbf{w} = \mathbf{i} + 4\mathbf{j}$

**52.** $\mathbf{v} = 2\mathbf{i} - 3\mathbf{j}$, $\mathbf{w} = -\mathbf{i} + 4\mathbf{j}$

**53.** $\mathbf{v} = \frac{1}{2}\mathbf{i} + \frac{5}{2}\mathbf{j}$, $\mathbf{w} = \frac{7}{2}\mathbf{i} - \frac{1}{2}\mathbf{j}$

**54.** $\mathbf{v} = \frac{2}{3}\mathbf{i} + 4\mathbf{j}$, $\mathbf{w} = \frac{4}{3}\mathbf{i} + \frac{1}{2}\mathbf{j}$

## Superset

In exercises 55–60, describe each radius vector as an ordered pair.

**55.**

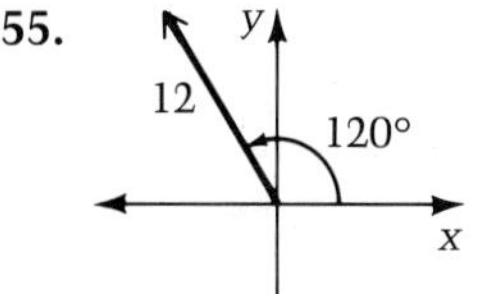


**56.**

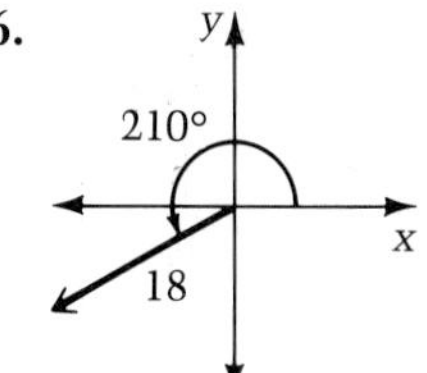


**57.**

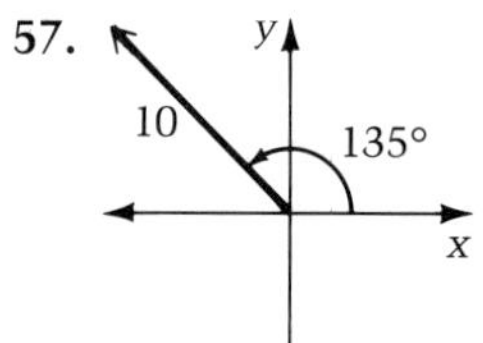


**58.**

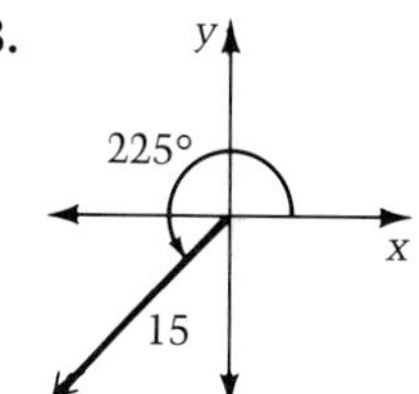


**59.**

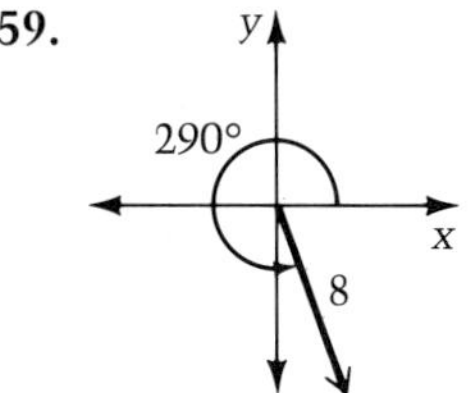


**60.**

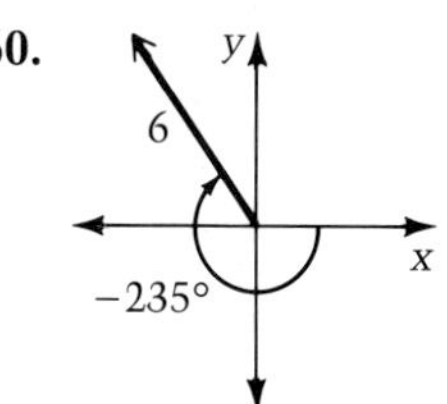


**61.** If $\mathbf{a} = 2\mathbf{i} + 3\mathbf{j}$, $\mathbf{b} = \mathbf{i} - 2\mathbf{j}$, and $\mathbf{c} = 4\mathbf{i} - \mathbf{j}$, find scalars $r$ and $s$ such that $\mathbf{c} = r\mathbf{a} - s\mathbf{b}$.

**62.** Find the angle between the radius vectors $\langle -2, 1\rangle$ and $\langle 3, 4\rangle$.

## 5.5 Vector Applications and the Dot Product

**Goal A** *to solve applied problems by using vector methods*

Often the quantities involved in applied problems are vectors, that is, quantities having both magnitude and direction. For example, the **velocity v** of an object is a vector quantity whose magnitude, $\|\mathbf{v}\|$, is the **speed** of the object and whose **direction** is the direction in which the object is moving.

In aviation, one frequently uses the concepts of *air speed* and *ground speed.* The **air speed** is the speed at which an airplane would fly in still air; the **ground speed** is the airplane's speed relative to the ground, after the effect of the wind has been accounted for. Thus, the (true) ground speed and true direction of an airplane are determined by forming the vector sum of the airplane's velocity vector and the wind's velocity vector.

**Example 1** An airplane's air speed is set at 300 km/h and its bearing is set at N 90° E (i.e., due east). A 50 km/h wind is blowing with a bearing S 60° E (i.e., in a direction 60° east of due south). Determine the ground speed and true direction of the airplane.

***Solution***

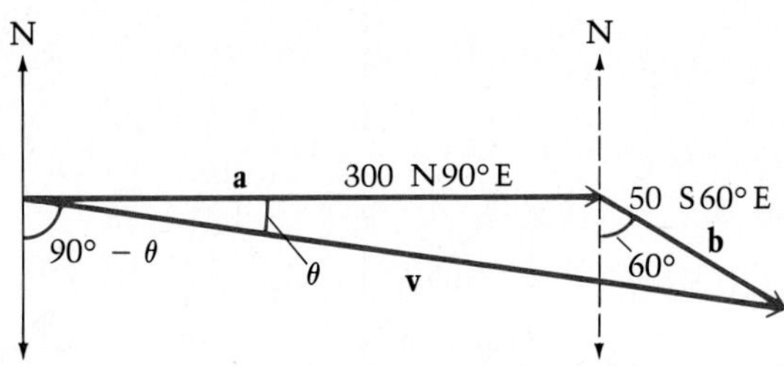


■ Draw a diagram. Let **a** be the airplane's velocity, and let **b** be the wind velocity. The ground velocity **v** is the vector sum of vectors **a** and **b**.

$$\|\mathbf{v}\|^2 = 300^2 + 50^2 - 2(300)(50)\cos 150^\circ$$
$$= 90{,}000 + 2500 - 30{,}000\left(-\frac{\sqrt{3}}{2}\right) \approx 118{,}481$$
$$\|\mathbf{v}\| \approx \sqrt{118{,}481} \approx 344 \text{ km/h}$$

■ The Law of Cosines is used to find $\|\mathbf{v}\|$, the ground speed.

$$\frac{\sin\theta}{50} = \frac{\sin 150^\circ}{344}$$
$$\sin\theta = \frac{50}{344}(\sin 150^\circ) = \frac{50}{344}\left(\frac{1}{2}\right) \approx 0.0727$$
$$\theta \approx 4^\circ 10'$$

■ The Law of Sines is used to find the true direction which is S $(90^\circ - \theta)$ E. Note the angle formed by **a** and **b** is $90^\circ + 60^\circ = 150^\circ$.

The airplane's approximate ground speed is 344 km/h; its true bearing is S 85°50′ E.

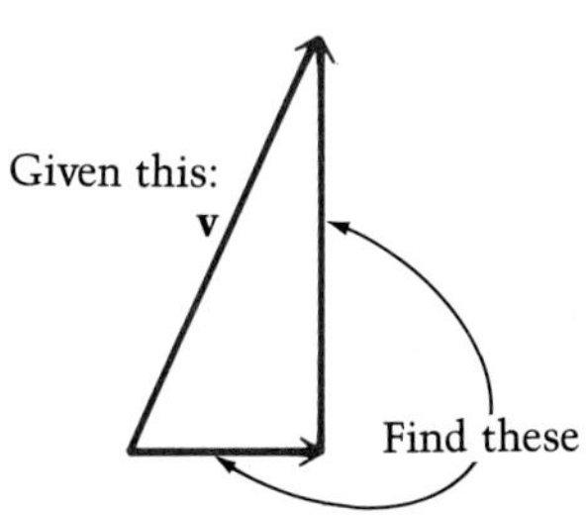


Figure 5.21

Our next application is a common problem in physics and engineering: given a vector **v**, find two other vectors whose vector sum is **v**. The two "other" vectors are usually perpendicular to each other, and are called **vector components of v.** The process of determining these two component vectors is referred to as **resolving v** into perpendicular components.

Of special importance is the resolution of a vector **v** into horizontal and vertical components. (We will refer to these vector components as $\mathbf{v}_x$ and $\mathbf{v}_y$, respectively, as shown in Figure 5.22.) To do this, we position **v** with its tail at the origin of the $xy$-plane, and draw perpendiculars from the head of **v** to the $x$-axis and $y$-axis.

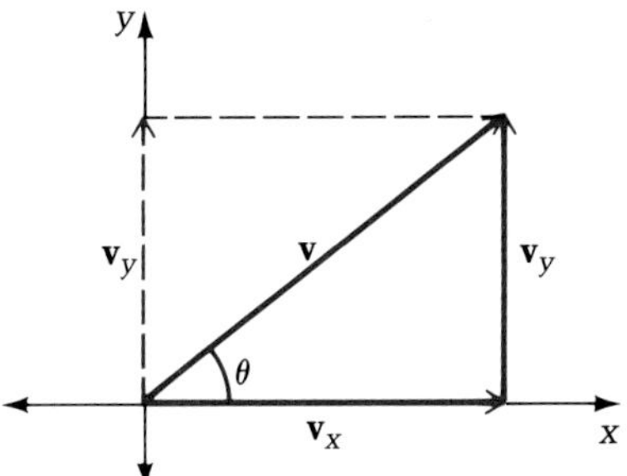


Figure 5.22

■ We move $\mathbf{v}_y$ to the position on the right in order to visualize it as one side in a right triangle with $\mathbf{v}_x$.

From the figure above, it is clear that

$$\cos\theta = \frac{\text{adjacent}}{\text{hypotenuse}} = \frac{\|\mathbf{v}_x\|}{\|\mathbf{v}\|}$$

and

$$\sin\theta = \frac{\text{opposite}}{\text{hypotenuse}} = \frac{\|\mathbf{v}_y\|}{\|\mathbf{v}\|}.$$

Thus,

$$\|\mathbf{v}_x\| = \|\mathbf{v}\|\cos\theta \quad \text{and} \quad \|\mathbf{v}_y\| = \|\mathbf{v}\|\sin\theta.$$

Since $\mathbf{v}_x$ is a vector with magnitude $\|\mathbf{v}\|\cos\theta$ in the direction of **i**, and $\mathbf{v}_y$ is a vector with magnitude $\|\mathbf{v}\|\sin\theta$ in the direction of **j**, we write

$$\mathbf{v}_x = (\|\mathbf{v}\|\cos\theta)\mathbf{i} \quad \text{and} \quad \mathbf{v}_y = (\|\mathbf{v}\|\sin\theta)\mathbf{j}$$

where $\theta$ is the angle formed by **v** and the positive $x$-axis. Note that

$$\mathbf{v} = \mathbf{v}_x + \mathbf{v}_y = (\|\mathbf{v}\|\cos\theta)\mathbf{i} + (\|\mathbf{v}\|\sin\theta)\mathbf{j},$$

and so $\mathbf{v}_x$ and $\mathbf{v}_y$ are truly vector components of **v**.

**Example 2** Vector **v** has magnitude 8. Resolve **v** into horizontal and vertical components if
(a) it makes an angle of 60° with the positive $x$-axis;
(b) it makes an angle of 135° with the positive $x$-axis.

***Solution*** (a)

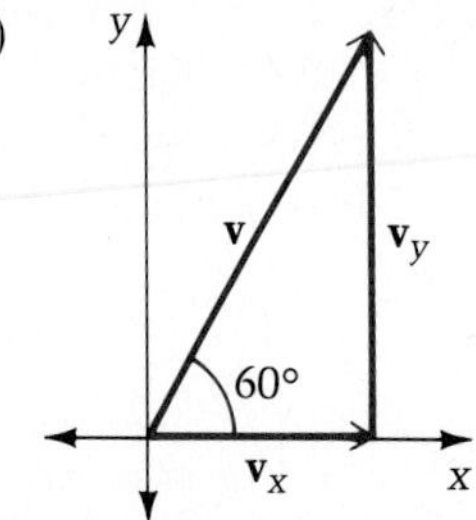


$$\mathbf{v_x} = (\|\mathbf{v}\| \cos\theta)\mathbf{i} = (8 \cdot \cos 60°)\mathbf{i} = (8)\left(\frac{1}{2}\right)\mathbf{i} = 4\mathbf{i}$$

$$\mathbf{v_y} = (\|\mathbf{v}\| \sin\theta)\mathbf{j} = (8 \cdot \sin 60°)\mathbf{j} = (8)\left(\frac{\sqrt{3}}{2}\right)\mathbf{j} = 4\sqrt{3}\mathbf{j}$$

$$\mathbf{v} = \mathbf{v_x} + \mathbf{v_y} = 4\mathbf{i} + 4\sqrt{3}\mathbf{j}.$$

(b)

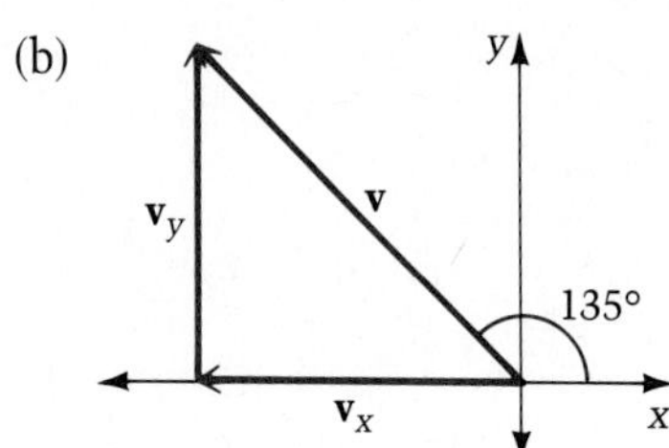


$$\mathbf{v_x} = (\|\mathbf{v}\| \cos\theta)\mathbf{i} = (8 \cdot \cos 135°)\mathbf{i} = \left[8 \cdot \left(-\frac{\sqrt{2}}{2}\right)\right]\mathbf{i} = -4\sqrt{2}\mathbf{i}$$

$$\mathbf{v_y} = (\|\mathbf{v}\| \sin\theta)\mathbf{j} = (8 \cdot \sin 135°)\mathbf{j} = \left(8 \cdot \frac{\sqrt{2}}{2}\right)\mathbf{j} = 4\sqrt{2}\mathbf{j}$$

$$\mathbf{v} = \mathbf{v_x} + \mathbf{v_y} = -4\sqrt{2}\mathbf{i} + 4\sqrt{2}\mathbf{j}.$$

To study an object under the influence of a force, it is useful to represent the force as a vector. We say that a force of one newton (N) is required to accelerate a mass of 1 kg at a rate of 1 m/s$^2$. One of the most common forces is **g**, the force that gravity exerts on an object (also called weight). If an object has a mass of $M$ kg, then the magnitude of **g** is $9.8 \cdot M$ newtons.

When an object is not accelerating (either it is moving at a constant velocity or it is at rest), we say that **forces are in equilibrium.** This means that the vector sum of all forces is zero.

**Example 3** A 400 kg piano is being rolled down a ramp. The ramp makes an angle of 30° with the level ground below. If we neglect friction, what is the force required to hold the piano stationary on the ramp?

***Solution***

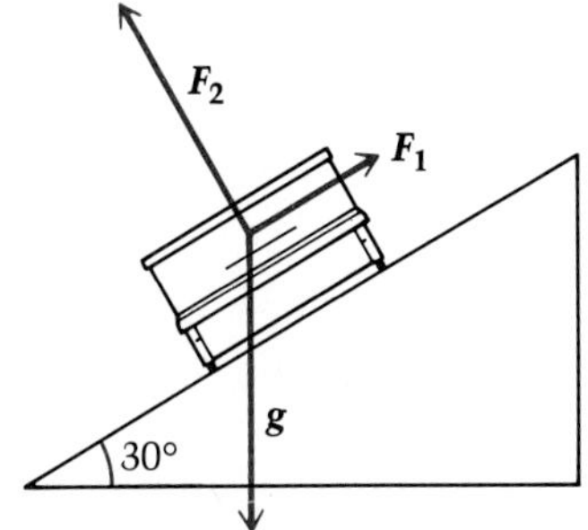


■ Begin by drawing a figure. Three forces act on the object: **g**, the force of gravity; $\mathbf{F}_1$, the force needed to hold the piano stationary ($\mathbf{F}_1$ is parallel to the ramp); $\mathbf{F}_2$, a force perpendicular to the surface of the ramp that keeps the piano from crashing through the ramp.

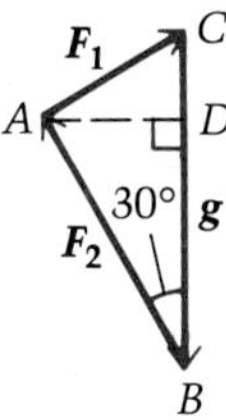


■ We draw a triangle representing the forces. Since there is no motion, the total vector sum is **0**. That is, $\mathbf{F}_1 + \mathbf{g} + \mathbf{F}_2 = \mathbf{0}$. Notice that $\mathbf{F}_1 + \mathbf{g} = -\mathbf{F}_2$.

The question asks us to determine $\mathbf{F}_1$. To determine $\angle B$ in the triangle above, we have drawn a perpendicular from $A$ that meets side $BC$ at point $D$. Given $\angle CAD = 30°$, and since $\angle CAB = 90°$, we conclude that $\angle DAB = 60°$. Thus, in right triangle $ADB$ we have $\angle B = 30°$.

$$\sin 30° = \frac{\text{opposite}}{\text{hypotenuse}} = \frac{\|\mathbf{F}_1\|}{\|\mathbf{g}\|} = \frac{\|\mathbf{F}_1\|}{(9.8)(400)}$$

■ To determine $\mathbf{F}_1$, consider $\sin B$ in right triangle $ABC$.

$$\|\mathbf{F}_1\| = (\sin 30°)(9.8)(400) = \left(\frac{1}{2}\right)(3920) = 1960 \text{ N}$$

The force required to hold the piano on the ramp is 1960 N (roughly 441 lb).

**Goal B** *to compute the dot product of two vectors*

Up to this point, the only type of vector multiplication that we have discussed is the product of a scalar $k$ and a vector **v** (the result $k\mathbf{v}$ is a vector). We now define a product of two vectors in such a way that the product is a scalar, not a vector.

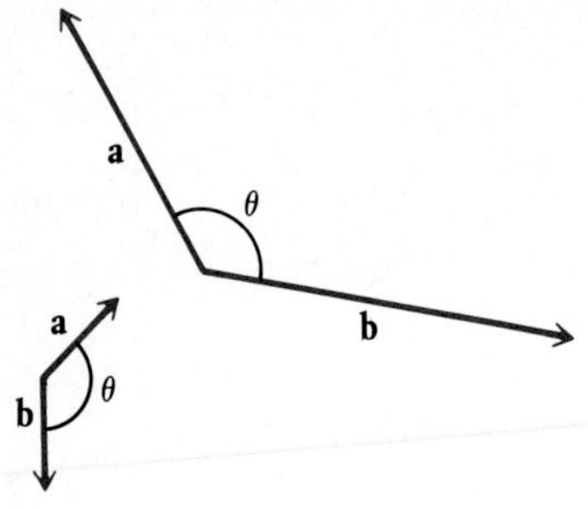

Figure 5.23

**Definition**

Let **a** and **b** be two nonzero vectors. The **dot product** of **a** and **b**, denoted $\mathbf{a} \cdot \mathbf{b}$, is defined as

$$\mathbf{a} \cdot \mathbf{b} = \|\mathbf{a}\| \, \|\mathbf{b}\| \cos \theta$$

where $\theta$ is the angle between **a** and **b** such that $0° \leq \theta \leq 180°$.

There is an alternate definition of the dot product which involves the components of the two vectors **a** and **b**.

**Alternate Definition**

Let $\mathbf{a} = \langle a_1, a_2 \rangle$ and $\mathbf{b} = \langle b_1, b_2 \rangle$. Then

$$\mathbf{a} \cdot \mathbf{b} = a_1 b_1 + a_2 b_2.$$

**Careful!** The dot product of two vectors is a scalar, not a vector.

Taken together, the two definitions of dot product provide a way to determine the angle between two vectors. The technique is illustrated in Example 4. Note that when the angle between the two vectors is 90°, $\cos \theta = 0$, and so the dot product is 0. In addition, if the dot product of two nonzero vectors is 0, then $\theta = 90°$. Thus, if **a** and **b** are nonzero vectors,

$\mathbf{a} \cdot \mathbf{b} = 0$ if and only if **a** and **b** are perpendicular.

**Example 4** Find $\mathbf{v} \cdot \mathbf{w}$ if $\mathbf{v} = \langle 3, 2 \rangle$ and $\mathbf{w} = \langle 4, -5 \rangle$. Determine the angle between **v** and **w**.

***Solution***

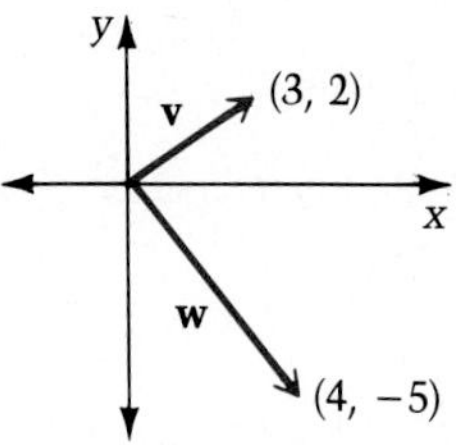


■ Although a figure is not necessary, it does help in visualizing the problem.

$$\mathbf{v} \cdot \mathbf{w} = \langle 3, 2 \rangle \cdot \langle 4, -5 \rangle = (3)(4) + (2)(-5) = 12 - 10 = 2$$

$$\mathbf{v} \cdot \mathbf{w} = \|\mathbf{v}\| \, \|\mathbf{w}\| \cos \theta$$

$$2 = \|\langle 3, 2 \rangle\| \, \|\langle 4, -5 \rangle\| \cos \theta$$

$$2 = [\sqrt{3^2 + 2^2} \cdot \sqrt{4^2 + (-5)^2}\,] \cos \theta$$

$$\cos \theta = \frac{2}{\sqrt{13} \cdot \sqrt{41}} = \frac{2}{\sqrt{533}} \approx 0.0866$$

$$\theta \approx 85°$$

■ Use $\mathbf{v} \cdot \mathbf{w}$ to approximate $\theta$, the angle between **v** and **w**.

## Exercise Set 5.5 ⊟ Calculator Exercises in the Appendix

### Goal A

In exercises 1–4, use that fact that eastbound airplanes "hitch a ride" on the jet stream.

1. A jet's air speed is 450 mph and its bearing is N 80° E. If the jet stream is blowing due east at 120 mph, what is the jet's ground speed and true bearing?
2. A jet's air speed is 450 mph and its bearing is S 80° W. If the jet stream is blowing due east at 120 mph, what is the jet's ground speed and true bearing?
3. A jet's ground speed is 410 mph and its true bearing is due north. If the jet stream is blowing due east at 120 mph, what is the jet's air speed and bearing?
4. A jet's ground speed is 510 mph and its true bearing is N 40° E. If the jet stream is blowing due east at 120 mph, what is the jet's air speed and bearing?
5. A 62 kg weight is on a ramp which is inclined at 40°. What is the force needed to hold the weight stationary on the ramp?
6. A crate containing a 300 kg wood stove is being unloaded from a delivery truck by sliding it down an inclined plank which is at an angle of 32° with the horizontal. What is the force needed to hold the crate at rest?
7. A 500 lb boat is being lowered into the water down an inclined ramp which is at an angle of 25° with the horizontal. What is the force needed to hold the boat at rest?
8. A force of 42 lb is required to hold a 250 lb block of granite from sliding down an inclined ramp. At what angle is the ramp inclined?

### Goal B

In exercises 9–16, determine the dot product of the given vectors. Find the angle between the vectors to the nearest degree.

9. $\langle 3, 1\rangle, \langle -2, 4\rangle$
10. $\langle 1, 1\rangle, \langle 3, -1\rangle$
11. $\langle 2, -1\rangle, \langle 2, -2\rangle$
12. $\langle -1, 2\rangle, \langle 4, 6\rangle$
13. $\langle 1, 9\rangle, \langle 9, -1\rangle$
14. $\langle 4, 3\rangle, \langle 3, -4\rangle$
15. $\langle -5, -3\rangle, \langle 4, -3\rangle$
16. $\langle -10, 2\rangle, \langle -3, -6\rangle$

### Superset

In exercises 17–22, two vectors are given. Determine whether they are parallel, perpendicular, or neither.

17. $\langle 2, -4\rangle, \langle 2, 1\rangle$
18. $\langle 1, -3\rangle, \langle 3, -1\rangle$
19. $\langle 1, 9\rangle, \left\langle \frac{3}{2}, \frac{1}{6}\right\rangle$
20. $\langle 8, 6\rangle, \left\langle \frac{1}{2}, -\frac{2}{3}\right\rangle$
21. $\langle -9, 6\rangle, \left\langle 2, -\frac{4}{3}\right\rangle$
22. $\left\langle -\frac{1}{3}, \frac{2}{3}\right\rangle, \langle 4, -8\rangle$

In physics, **work** is said to be done when a force applied to an object causes the object to move. In particular, if a constant force $\mathbf{F}$ is applied in the direction of motion, then the work $W$ done by force $\mathbf{F}$ in moving the object a distance $d$ is the scalar quantity $W = \|\mathbf{F}\| \cdot d$. However, if a constant force $\mathbf{F}$ is applied at an angle $\theta$ to the direction of the motion, then the work done by $\mathbf{F}$ in moving the object a distance $d$ is

$$W = \begin{pmatrix}\text{Component of force in}\\ \text{direction of motion}\end{pmatrix} \cdot (\text{Distance})$$
$$= (\|\mathbf{F}\| \cos \theta) \cdot d$$

In exercises 23–26, solve each work problem.

23. A wagon loaded with 180 lb of patio bricks is pulled 100 yd over level ground by a handle which makes an angle of 43° with the horizontal. Find the work done if a force of 22 lb is exerted in pulling the wagon.
24. A large crate is pushed across a level floor. Find the work done in moving the crate 20 ft if a force of 35 lb is applied at an angle of 18° with the horizontal.
25. A wagon is used to haul three small children a distance of a half mile over level ground. The handle used to pull the wagon makes an angle of 35° with the horizontal. Find the work done if a force of 25 lb is exerted on the handle.
26. A box is pulled by exerting a force of 16 lb on a rope attached to the box that makes an angle of 40° with the horizontal. Determine the work done in pulling the box 46 ft.

## 5.6 Simple Harmonic Motion

**Goal A** *to solve problems involving simple harmonic motion*

We now use our knowledge of trigonometric functions to describe motion that repeats itself periodically, such as the up-and-down bobbing motion of a buoy in the ocean, or the back-and-forth motion of a simple pendulum.

As a model for this type of motion, we consider a mass $m$ attached to a spring suspended from the ceiling (Figure 5.24). Initially the mass is at rest (Figure 5.24(a)). At that time, the mass is said to be in the **equilibrium position.** Next, the mass is pulled downward to a position $A$ units below equilibrium, and then released (Figure 5.24(b)). The spring then causes the mass to move upward, through the equilibrium position until it reaches a point $A$ units above equilibrium (Figures 5.24(c) and (d)). Then, the mass begins to move back downward towards the point from which it was released.

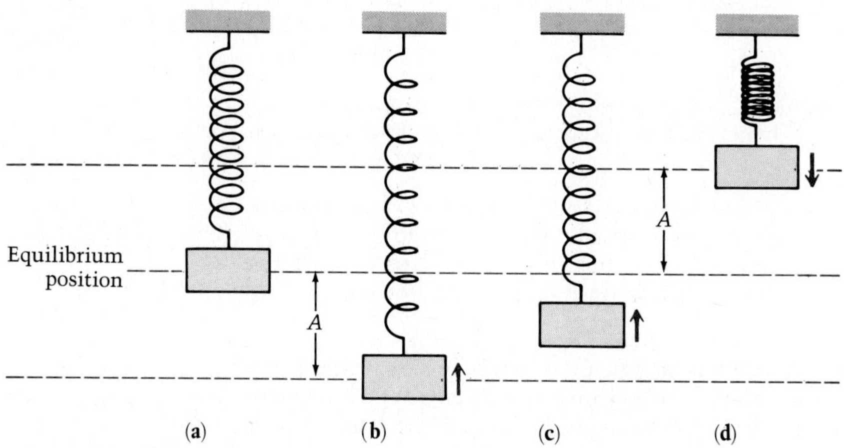


**Figure 5.24**

If the effects of friction in the spring are ignored, the up-and-down motion just described would continue forever. This "idealized" motion of the mass $m$ is called *simple harmonic motion.* It is the basic model for a variety of physical phenomena, including the vibration of a guitar string, the oscillation of atoms in a molecule, the propagation in air of sound waves, as well as various types of electromagnetic waves, such as radio waves and television signals.

The mathematical model for simple harmonic motion describes an object's displacement $x$, measured from the equilibrium position, as a function of time $t$. (The equilibrium position corresponds to the point where $x = 0$.) The sine and cosine functions are the building blocks of this model.

**Simple Harmonic Motion**

If an object exhibits simple harmonic motion, its displacement $x$ from the equilibrium position is a function of time $t$. This function can be described by an equation of the form

$$x = A\sin(\omega t - D),$$

or alternatively,

$$x = A\cos(\omega t - D).$$

Suppose that we are considering simple harmonic motion described by the equation $x = A\sin(\omega t - D)$ where $\omega$ and $D$ are positive numbers. Note that the equation can be rewritten as

$$x = A\sin\omega\left(t - \frac{D}{\omega}\right).$$

Recall that this form was extremely useful when we graphed transformations of the trigonometric functions. In particular, it follows from our earlier definitions that

the amplitude is $|A|$, and the period is $\left|\frac{2\pi}{\omega}\right| = \frac{2\pi}{\omega}$.

The **frequency** of the motion, denoted $f$, is the number of periods that are completed in one unit of time, and is described as the reciprocal of the period

$$f = \frac{1}{\text{period}} = \frac{1}{\frac{2\pi}{\omega}} = \frac{\omega}{2\pi}.$$

The graph of $x = A\sin(\omega t - D)$ is given below.

■ Note that this graph is a translation $\frac{D}{\omega}$ units to the right of the graph of $x = A\sin\omega t$.

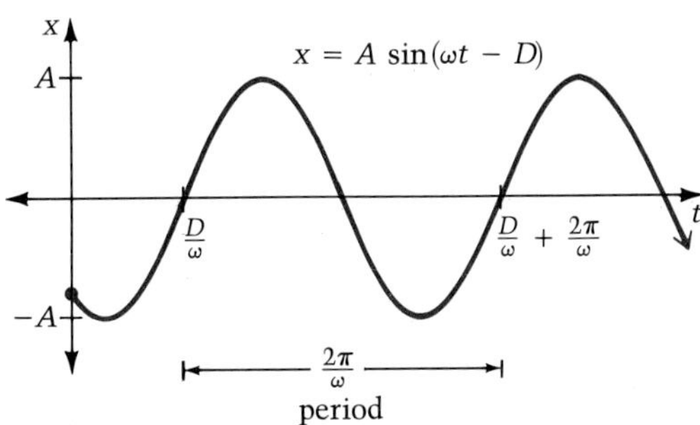


Figure 5.25

**Example 1** An object suspended on a spring is pulled down 4 units from the equilibrium point. When released, the object is in simple harmonic motion, and makes one complete up-and-down oscillation every 2 sec.

(a) Determine the frequency of the motion;
(b) Write an equation for the object's displacement ($x$) as a function of time ($t$). Take displacements below the equilibrium point as positive, and those above as negative.

***Solution*** (a) $f = \dfrac{1}{\text{period}} = \dfrac{1}{2}$ ■ Frequency is the reciprocal of the period.

Note that this means the object makes $\frac{1}{2}$ of an oscillation each second.

(b) We need to determine values for $A$, $\omega$, and $D$ in the general formula

$$x = A\sin(\omega t - D).$$

$A = 4$ ■ Initial displacement is 4 units below equilibrium point, so $A$ is (positive) 4. Remember that $|A|$ is the amplitude.

$f = \dfrac{\omega}{2\pi} = \dfrac{1}{2}$ ■ To find $\omega$, we use the definition of $f$ and the result from (a) that $f = \dfrac{1}{2}$.

$2\omega = 2\pi$

$\omega = \pi$

Now that we have found $A$ and $\omega$, we can determine $D$.

$x = 4\sin(\pi t - D)$

$4 = 4\sin(0 - D)$ ■ At release, $t = 0$ and $x = 4$. These values are substituted.

$1 = \sin(-D)$

$1 = -\sin D$ ■ Sine is odd: $\sin(-x) = -\sin x$.

$-1 = \sin D$

$D = \dfrac{3\pi}{2}$ ■ We have selected the smallest positive value of $D$ for which $\sin D = -1$.

An equation for the object's displacement is $x = 4\sin\left(\pi t - \dfrac{3\pi}{2}\right)$.

**Example 2** Show that the equation $x = 3\sin 2t - 4\cos 2t$, where $x$ is measured in inches and $t$ in seconds, describes simple harmonic motion. Determine the amplitude, period, and frequency.

***Solution*** We rewrite the equation in the form

$$x = A\sin(\omega t - D).$$

Recall: $\sin(\omega t - D) = (\sin \omega t)(\cos D) - (\cos \omega t)(\sin D).$

Thus, 
$$\begin{aligned} A\sin(\omega t - D) &= A[(\sin \omega t)(\cos D) - (\cos \omega t)(\sin D)] \\ &= (A\cos D)(\sin \omega t) - (A\sin D)(\cos \omega t) \end{aligned}$$

Compare: $x = (\quad 3 \quad)(\sin\ 2t) - (\quad 4 \quad)(\cos\ 2t)$

Therefore,

$$A\cos D = 3 \qquad A\sin D = 4 \qquad \text{and} \qquad \omega = 2$$

$$\cos D = \frac{3}{A} \qquad \sin D = \frac{4}{A}$$

Thus, since $1 = \cos^2 D + \sin^2 D = \frac{3^2}{A^2} + \frac{4^2}{A^2} = \frac{25}{A^2}$, we have $A = 5$.

Since $A = 5$, and $\omega = 2$, we have $x = 5\sin(2t - D)$.

To determine $D$, notice that

$$\tan D = \frac{\sin D}{\cos D} = \frac{4A}{3A} = \frac{4}{3}, \text{ and so } D \approx 0.93 \text{ radians}$$ ■ Table 4

Thus, $x = 5\sin(2t - 0.93)$, which is an equation for simple harmonic motion. The amplitude is 5, the period $= \frac{2\pi}{2} = \pi$, and the frequency $= \frac{1}{\pi}$.

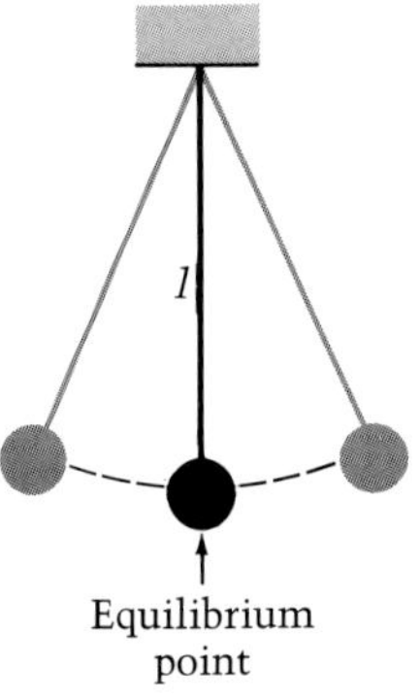

**Figure 5.26**

A pendulum in its simplest form consists of a point mass suspended by a "weightless" string of length $l$, such that the mass moves back and forth along a small arc. For small arcs the motion of the mass can be considered approximately straight-line motion. In this case the displacement $x$ from the equilibrium point is given by the equation

$$x = A\cos \omega t,$$

where $A$ is the initial displacement, and $\omega = \sqrt{\frac{g}{l}}$, with $g$ the acceleration due to gravity: 32 ft/sec$^2$, or 9.8 m/sec$^2$.

**Example 3** A pendulum in a grandfather clock is 2 ft long. It is released on an arc with initial displacement of 0.25 ft. (a) What is the period of the pendulum? (Assume that the acceleration due to gravity is 32 ft/sec$^2$.) (b) Write the equation for the motion of the pendulum and sketch the graph.

***Solution*** (a) period $= \dfrac{2\pi}{\omega}$

period $= \dfrac{2\pi}{4} = \dfrac{\pi}{2}$ ■ Since $\omega = \sqrt{\frac{g}{l}}$, we have $\omega = \sqrt{\frac{32}{2}} = \sqrt{16} = 4$.

This means that one back-and-forth oscillation takes approximately $\frac{\pi}{2} \approx 1.6$ sec.

(b) $x = A \cos \omega t$

$x = 0.25 \cos 4t$ ■ 0.25 substituted for $A$; 4 for $\omega$.

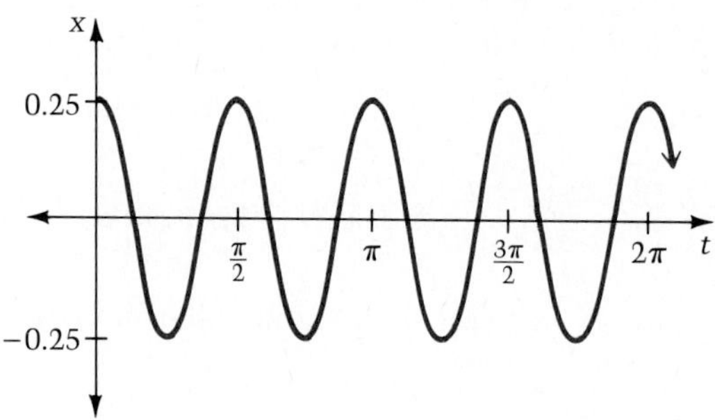


## Exercise Set 5.6 Calculator Exercises in the Appendix

### Goal A

In exercises 1–6, a mass suspended from a spring is in simple harmonic motion, described by the given equation, where $x$ is measured in inches and $t$ in seconds. Use the graph of the motion to determine at each of the following times where the object is and whether it is moving up or down.
(a) $t = 0$, (b) $t = \frac{1}{2}$, (c) $t = 1$, (d) $t = 2$, (e) $t = 3$, (f) $t = 4$, (g) $t = 6$. (Note: take displacement below the equilibrium point as positive and above as negative.)

**1.** $x = \frac{1}{2} \sin\left(\frac{1}{2}\pi t\right)$

**2.** $x = \frac{3}{2} \cos\left(\frac{1}{2}\pi t\right)$

**3.** $x = 3 \cos 2\pi\left(t + \frac{1}{2}\right)$

**4.** $x = 3 \sin 2\pi\left(t - \frac{1}{2}\right)$

**5.** $x = 0.20 \sin 4\pi t$

**6.** $x = 0.33 \cos 6\pi t$

In exercises 7–10, each equation represents simple harmonic motion. Give the amplitude, period, and frequency. Sketch the graph.

**7.** $x = 6 \sin 2t$

**8.** $x = 4 \cos \dfrac{t}{2}$

**9.** $x = 3 \cos\left(\frac{t}{2} + 2\pi\right)$

**10.** $x = 5 \sin(2t - \pi)$

In exercises 11–16, show that each equation, where $x$ is measured in centimeters and $t$ in seconds, describes simple harmonic motion. Determine the amplitude, period, and frequency.

**11.** $x = \frac{\sqrt{3}}{2} \sin t + \frac{1}{2} \cos t$

**12.** $x = \sqrt{3} \sin t - \cos t$

**13.** $x = \sin t - \cos t$

**14.** $x = \sin t + \cos t$

**15.** $x = \frac{5}{2} \sin 2t - 6 \cos 2t$

**16.** $x = 4 \sin 2t + \frac{15}{2} \cos 2t$

In exercises 17–24, a situation describing simple harmonic motion is given. (a) Write an equation for the object's displacement ($x$) as a function of time ($t$). (b) Determine the frequency of the motion.

**17.** An object suspended on a spring is pulled down 6 in below the equilibrium point. When released, the object is in simple harmonic motion and makes one complete oscillation every 3 sec.

**18.** An object attached to a spring is released from a compressed position 4 cm above its position of equilibrium. When released, the object is in simple harmonic motion and makes one complete oscillation every second.

**19.** A weight of 50 g is attached to a spring. The weight is pulled down 6 cm below the equilibrium point and then released. The object is then in simple harmonic motion and has a period of 1.6 sec.

**20.** The weight in exercise 19 is pulled down 3 cm and released. It makes one complete oscillation every 0.8 sec.

**21.** A pendulum is 8 ft long and is released with an initial displacement of 0.60 ft.

**22.** A pendulum is 6 ft long and is released with an initial displacement of 0.62 ft.

**23.** A pendulum 1 m long is released with an initial displacement of 2 cm.

**24.** A pendulum 1.6 m long is released with an initial displacement of 6 cm.

### Superset

In exercises 25–28, write an equation for simple harmonic motion, given the amplitude and period.

**25.** amplitude $= 3$, period $= \frac{4\pi}{3}$

**26.** amplitude $= \frac{3}{5}$, period $= \pi$

**27.** amplitude $= \frac{3}{2}$, period $= 3\pi$

**28.** amplitude $= 4$, period $= \frac{3\pi}{2}$

**29.** If the length of a pendulum is doubled, how does the period change?

**30.** How does the period of a pendulum compare with that of another pendulum which is one-fourth as long?

**31.** A guitar string is plucked so that a point on the string makes one complete oscillation every $\frac{1}{200}$ sec. If the string is plucked by lifting the point 0.01 cm and then releasing it, write an equation for the simple harmonic motion of the point.

**32.** Show that the motion of a particle which moves on a line according to the equation $x = 4 \sin 3t \cos 2t$ is the "sum" of two simple harmonic motions.

**33.** Is the motion represented by the equation $x = \sin^2 t$ simple harmonic motion?

**34.** A simple pendulum about 9.8 in long has a period of 1 sec at sea level. A pendulum 4 times as long has a period of 2 sec. One that is 9 times as long has a period of 3 sec, and so on. Verify each of these statements.

## 5.7 Polar Coordinates

**Goal A** *to plot points given their polar coordinates*

Up to this point, we have located points in a plane by specifying rectangular coordinates, such as $P(3, 4)$ or $Q(-1, 5)$. We now consider an alternate method for identifying points in a plane. The method depends on describing each point in terms of two numbers $r$ and $\theta$, known as polar coordinates. To simplify our work, we make the following definition:

**Definition**

A **$\theta$-ray** is a ray which has its initial point at (0, 0) and which makes an angle of $\theta$ with the positive $x$-axis. The ray in the direction opposite to that of a $\theta$-ray is called the **opposite of the $\theta$-ray.** Note that the opposite of the $\theta$-ray is the $(\theta + \pi)$-ray.

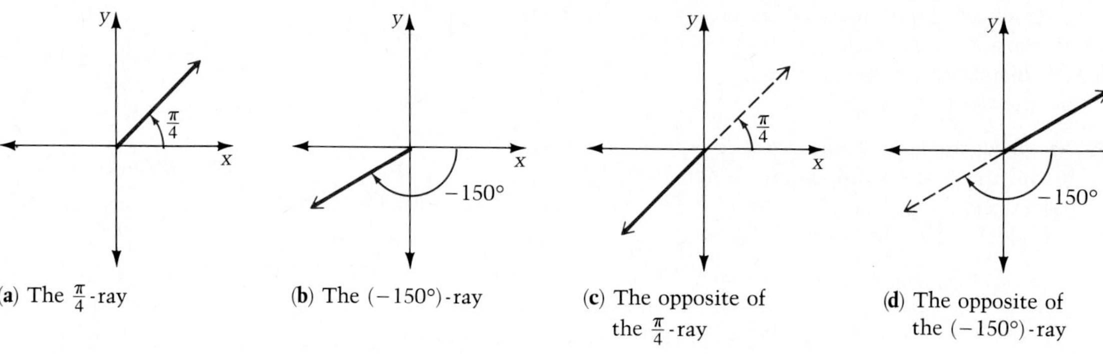


(**a**) The $\frac{\pi}{4}$-ray (**b**) The (−150°)-ray (**c**) The opposite of the $\frac{\pi}{4}$-ray (**d**) The opposite of the (−150°)-ray

**Figure 5.27**

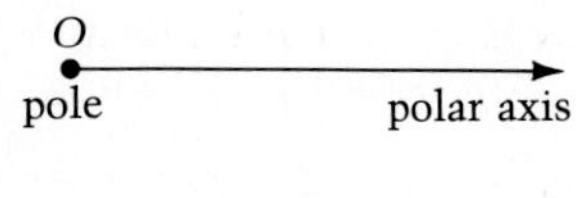


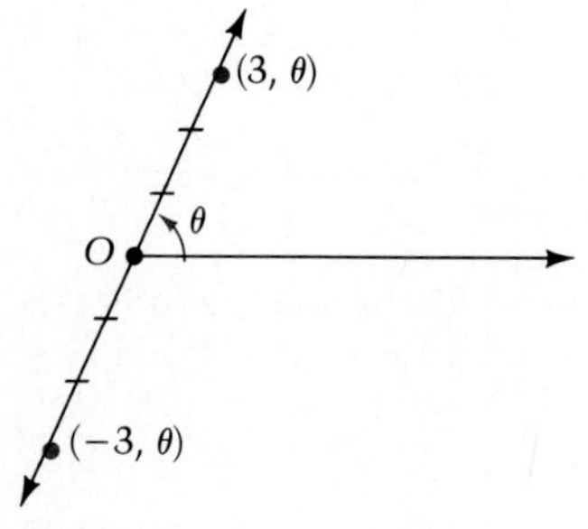


**Figure 5.28**

We construct what is called the *polar coordinate system* in the following way. We designate one point in the plane as the **origin** or **pole** (labeled $O$), and one ray emanating from $O$ as the **polar axis.** Points in the plane can then be described in polar form as follows.

**Definition**

The **polar coordinates** $(r, \theta)$ specify a point that lies at a distance $|r|$ from O. If $r > 0$, the point lies on the $\theta$-ray. If $r < 0$, the point lies on the opposite of the $\theta$-ray. If $r = 0$, the point lies at $O$, regardless of the value of $\theta$.

When plotting points given by polar coordinates, it is a common practice to use graph paper which displays concentric circles (centered at $O$) and rays emanating from $O$.

**Example 1** Graph the following points in the polar plane.

(a) $A(3, 135°)$ (b) $B(-4, 60°)$ (c) $C(-3, 315°)$ (d) $D\left(2, -\frac{3\pi}{2}\right)$

(e) $E(-5, \pi)$ (f) $F\left(0, \frac{\pi}{12}\right)$

***Solution***

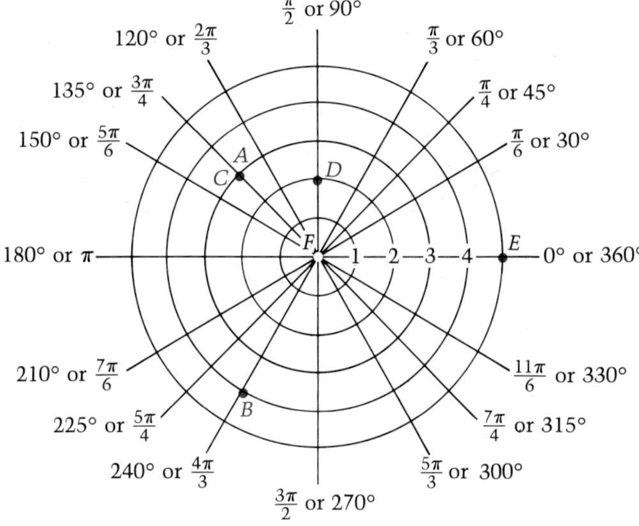


(a) Find the 135°-ray. Since 3 is positive, move 3 units away from the origin along that ray.
(b) Find the 60°-ray. Since $-4$ is negative, move 4 units away from the origin along the opposite of the 60°-ray.
(c) Find the 315°-ray. Since $-3$ is negative, move 3 units away from the origin along the opposite of the 315°-ray. Notice that this point coincides with (3, 135°).
(d) Find the $(-\frac{3\pi}{2})$-ray (it is the same as the $\frac{\pi}{2}$-ray). Since 2 is positive, move 2 units away from the origin on that ray.
(e) Find the $\pi$-ray. Since $-5$ is negative, move 5 units away from the origin on the opposite of the $\pi$-ray.
(f) Since $r = 0$, this point is at the origin, regardless of the value of $\theta$.

The description of a point in polar coordinates is not unique. See, for example, points $A$ and $C$ in Example 1. The coordinates (3, 135°) and $(-3, 315°)$ describe the same point. In addition, point $A(3, 135°)$ could alternately be described by adding any multiple of 360° to 135°, and keeping $r = 3$. In fact, for all integers $n$, any point $(r, \theta)$ can be described as

$$(r, \theta + n \cdot 360°), \quad \text{or} \quad (-r, (\theta + 180°) + n \cdot 360°).$$

**Goal B** *to translate ordered pairs and equations from rectangular form to polar form and vice versa*

When we superimpose the polar plane on the $xy$-plane, we discover some relationships that are useful in translating from one system to the other.

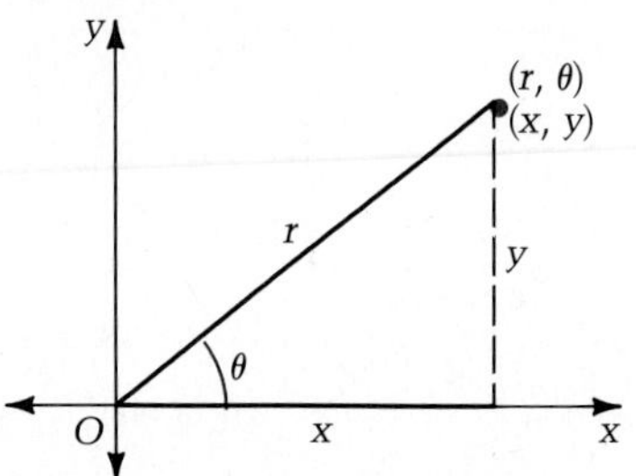


Figure 5.28

■ Since $\sin\theta = \frac{y}{r}$, and $\cos\theta = \frac{x}{r}$, we have
$y = r\sin\theta$,
$x = r\cos\theta$.

The following four relationships are used to solve translation problems.

**Rectangular ↔ Polar Relationships**

If point $P$ has rectangular coordinates $(x, y)$ and polar coordinates $(r, \theta)$, then

$$x = r\cos\theta, \qquad \tan\theta = \frac{y}{x}, \quad \text{provided } x \neq 0,$$
$$y = r\sin\theta, \qquad x^2 + y^2 = r^2.$$

**Example 2** Determine the $xy$-coordinates of the point having polar coordinates $\left(-1, \frac{\pi}{3}\right)$.

***Solution*** $x = r\cos\theta = (-1)\left(\cos\frac{\pi}{3}\right) = -\frac{1}{2} \qquad y = r\sin\theta = (-1)\left(\sin\frac{\pi}{3}\right) = -\frac{\sqrt{3}}{2}$

The rectangular coordinates of the polar point $\left(-1, \frac{\pi}{3}\right)$ are $\left(-\frac{1}{2}, -\frac{\sqrt{3}}{2}\right)$.

**Example 3** Determine all polar coordinates of the point having rectangular coordinates $(-1, 2)$.

***Solution***

$$r^2 = x^2 + y^2 = (-1)^2 + (2)^2 = 5$$
$$r = \pm\sqrt{5} \quad \text{(two cases)}$$

■ First we find $r$.

Case 1. Suppose $r = \sqrt{5}$:

$$x = r\cos\theta \qquad y = r\sin\theta$$
$$-1 = \sqrt{5}\cos\theta \qquad 2 = \sqrt{5}\sin\theta$$

$$\cos\theta = -\frac{1}{\sqrt{5}} \approx -0.4472 \qquad \sin\theta = \frac{2}{\sqrt{5}} \approx 0.8944$$

Thus $\theta$ has $63°30'$ as its reference angle (Table 4). Since $\cos\theta$ is negative and $\sin\theta$ is positive, $\theta$ is a second quadrant angle. Thus, $\theta = 180° - 63°30' = 116°30'$.

Case 2. If $r = -\sqrt{5}$, then $\theta = 116°30' + 180° = 296°30'$.

Thus, polar coordinates for the given point are of the form

$$(\sqrt{5}, 116°30' + n \cdot 360°) \quad \text{or} \quad (-\sqrt{5}, 296°30' + n \cdot 360), \quad \text{for all integers } n.$$

Certain graphs are represented by very simple equations in polar form. For example, the polar equation $r = c$ (where $c$ is a constant) represents a circle of radius $|c|$, centered at the origin. Also, the polar equation $\theta = \alpha$ (where $\alpha$ is a constant) is the line formed by joining the $\alpha$-ray and the opposite of the $\alpha$-ray. The slope of such a line is $\tan\alpha$.

**Example 4** Convert the following equations to rectangular form: (a) $r = 5$ (b) $r = 6\sin\theta$

***Solution*** (a)

$$r = 5$$

$$r^2 = 25 \qquad \blacksquare \text{ Both sides are squared.}$$

$$x^2 + y^2 = 25 \qquad \blacksquare\ r^2 \text{ is replaced by } x^2 + y^2.$$

Thus $r = 5$ is a circle with center at (0, 0) having radius 5.

(b)

$$r = 6\sin\theta$$

$$r^2 = 6(r\sin\theta) \qquad \blacksquare \text{ Both sides are multiplied by } r.$$

$$x^2 + y^2 = 6y \qquad \blacksquare\ r\sin\theta \text{ is replaced by } y;\ r^2 \text{ is replaced by } x^2 + y^2.$$

$$x^2 + (y^2 - 6y + \square) = \square \qquad \blacksquare \text{ We prepare to complete the square in } y.$$

$$x^2 + (y^2 - 6y + 9) = 9$$

$$x^2 + (y - 3)^2 = 9$$

Thus $r = 6\sin\theta$ is the polar equation of a circle of radius 3 centered at (0, 3).

Note that, in each problem in Example 4, we began by transforming the given equation so that one or more of the Rectangular ↔ Polar Relationships could be used. This is the key to handling such translation problems.

**Example 5** Convert the following rectangular equations to polar form.

(a) $y = 5x$ (b) $x^2 - y^2 = 3$

***Solution*** (a)

$$y = 5x$$
$$\frac{y}{x} = 5$$
$$\tan\theta = 5$$

(b)

$$x^2 - y^2 = 3$$
$$(r\cos\theta)^2 - (r\sin\theta)^2 = 3$$
$$r^2\cos^2\theta - r^2\sin^2\theta = 3$$
$$r^2(\cos^2\theta - \sin^2\theta) = 3$$
$$r^2\cos 2\theta = 3$$

■ $\cos^2\theta - \sin^2\theta = \cos 2\theta$.

**Goal C** *to graph a polar equation*

We end this section with the problem of graphing polar equations. In Example 6, we graph a polar equation by plotting a sufficient number of points to get a reasonable idea about the shape of the curve.

**Example 6** Sketch the graph of the polar equation $r = 1 + 2\cos\theta$.

***Solution*** We will provide 4 charts showing values of $r$ for selected values of $\theta$, and will use these charts to sketch the graph in stages. Values of $\cos\theta$ and $r$ are approximations.

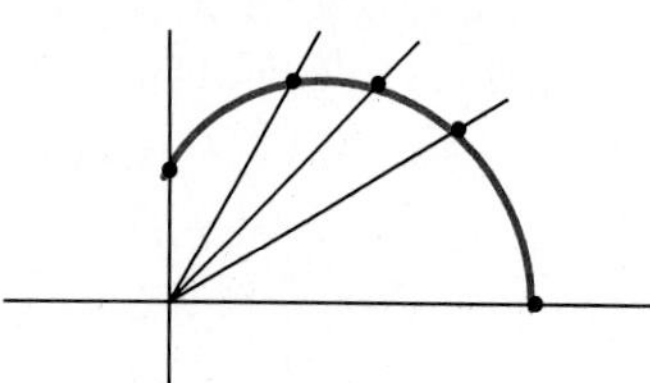

| $\theta$ | $0$ | $\frac{\pi}{6}$ | $\frac{\pi}{4}$ | $\frac{\pi}{3}$ | $\frac{\pi}{2}$ |
|---|---|---|---|---|---|
| $r$ | 3 | 2.7 | 2.4 | 2 | 1 |

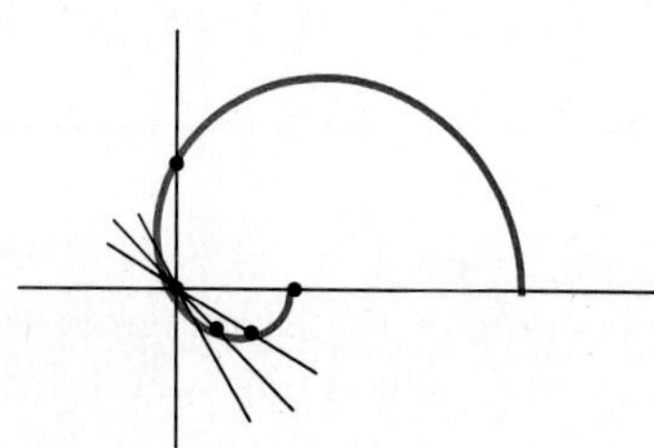

| $\theta$ | $\frac{\pi}{2}$ | $\frac{2\pi}{3}$ | $\frac{3\pi}{4}$ | $\frac{5\pi}{6}$ | $\pi$ |
|---|---|---|---|---|---|
| $r$ | 1 | 0 | −0.4 | −0.7 | −1 |

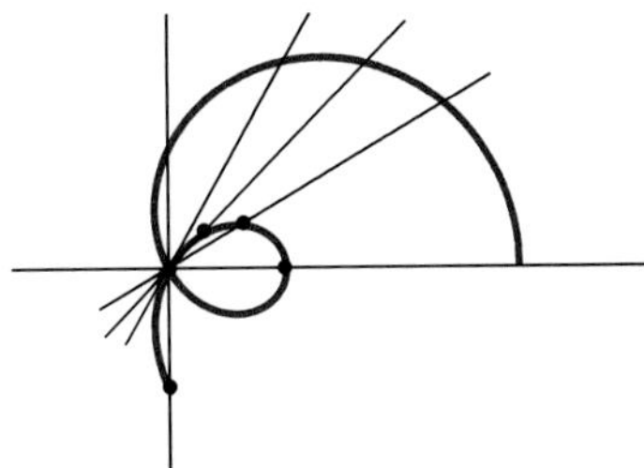

| $\theta$ | $\pi$ | $\frac{7\pi}{6}$ | $\frac{5\pi}{4}$ | $\frac{4\pi}{3}$ | $\frac{3\pi}{2}$ |
|---|---|---|---|---|---|
| $r$ | $-1$ | $-0.7$ | $-0.4$ | 0 | 1 |

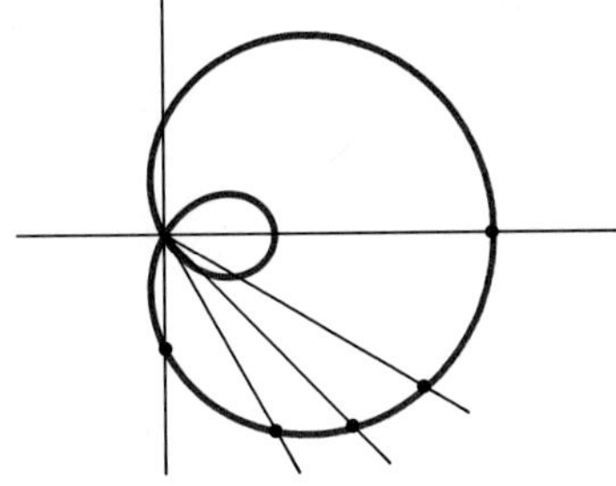

| $\theta$ | $\frac{3\pi}{2}$ | $\frac{5\pi}{3}$ | $\frac{7\pi}{4}$ | $\frac{11\pi}{6}$ | $2\pi$ |
|---|---|---|---|---|---|
| $r$ | 1 | 2 | 2.4 | 2.7 | 3 |

We now present a complete graph.

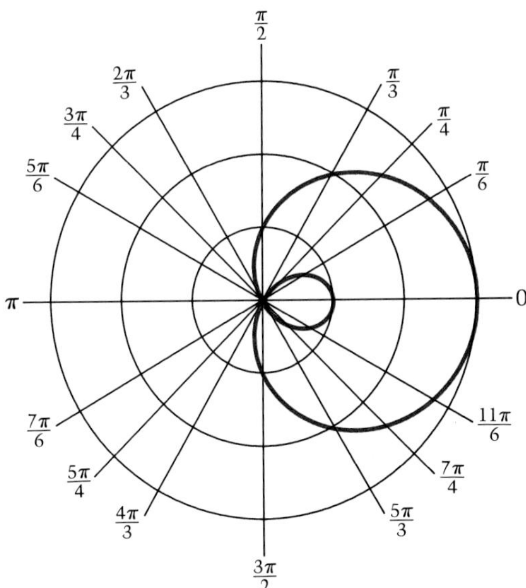

## Exercise Set 5.7

### Goal A

In exercises 1–16, graph the given point in the polar plane.

**1.** $(2, 120°)$ **2.** $(5, 30°)$ **3.** $(3, 135°)$

**4.** $(4, 210°)$ **5.** $(2, -120°)$ **6.** $(-3, 135°)$

**7.** $(-1, -270°)$ **8.** $(-5, -45°)$ **9.** $(-2, -420°)$

**10.** $(4, -570°)$ **11.** $(3, \pi)$ **12.** $\left(4, \frac{\pi}{4}\right)$

**13.** $\left(-2, \frac{5\pi}{6}\right)$ **14.** $\left(-5, -\frac{3\pi}{2}\right)$

**15.** $\left(-4, -\frac{\pi}{2}\right)$ **16.** $\left(-2, \frac{3\pi}{4}\right)$

### Goal B

For exercises 17–32, determine the $x$- and $y$-coordinates of the polar points in exercises 1–16.

In exercises 33–44, the rectangular coordinates of the following points are given. Determine two polar representations $(r, \theta)$, with $0 \le \theta < 2\pi$, of each point.

**33.** $(0, 4)$ **34.** $(-2, 0)$ **35.** $(-1, -1)$

**36.** $(1, -1)$ **37.** $(\sqrt{3}, -1)$ **38.** $(-1, -\sqrt{3})$

**39.** $(-\sqrt{2}, \sqrt{2})$ **40.** $(\sqrt{2}, -\sqrt{2})$ **41.** $(-3, 4)$

**42.** $(-5, -12)$ **43.** $(4, 6)$ **44.** $(5, -10)$

In exercises 45–54, convert the given polar equation to rectangular form.

**45.** $r = 3$ **46.** $2r = 5$

**47.** $\theta = \frac{3\pi}{4}$ **48.** $\theta = \frac{\pi}{3}$

**49.** $r(\cos\theta - 2\sin\theta) = 5$

**50.** $r(3\sin\theta - 4\cos\theta) = 5$

**51.** $r(1 + \cos\theta) = 1$ **52.** $r(1 + \sin\theta) = 2$

**53.** $r(1 - 2\sin\theta) = 2$ **54.** $r(2 - \cos\theta) = 2$

In exercises 55–66, convert the given rectangular equation to polar form.

**55.** $x = 0$ **56.** $y = -5$

**57.** $x + y = 1$ **58.** $x + 4 = 4y$

**59.** $x^2 - 2x + y^2 = 0$ **60.** $x^2 - 6y - y^2 = 0$

**61.** $x^2 = 6y + 9$ **62.** $y^2 = 8x + 16$

**63.** $2xy = 1$ **64.** $xy = 1$

**65.** $x^2 + y^2 = 16y$ **66.** $x^2 + y^2 = 4x$

### Goal C

In exercises 67–74, sketch the graph of the polar equation.

**67.** $r = -3$ **68.** $r^2 = 4$

**69.** $\theta^2 = \frac{\pi^2}{16}$ **70.** $\theta = \frac{5\pi}{4}$

**71.** $r = 3\cos\theta$ **72.** $r = 4\sin\theta$

**73.** $r = 2 + \sin\theta$ **74.** $r = 1 - \cos\theta$

### Superset

**75.** Determine the polar equation of the line through the points $(-2, 0)$ and $(1, 2)$.

**76.** Determine the polar equation of the line through the points that have polar coordinates $(1, \frac{\pi}{2})$ and $(-2, \pi)$.

**77.** (a) Find the distance between the points $(-2, \frac{\pi}{3})$ and $(4, \frac{3\pi}{4})$. (Hint: first convert the points to rectangular coordinates.)
(b) Sketch the points on the polar plane. Use the Law of Cosines to determine the distance between the two points.

In exercises 78–82, sketch the graph of each of the following polar equations for $0 \le \theta < \pi$.

**78.** $r = \theta$ **79.** $r = \frac{\theta}{2}$ **80.** $r(1 - \cos\theta) = 2$

**81.** $r(1 + \sin\theta) = 2$ **82.** $r = 4\sin 2\theta$

# Chapter Review & Test

## Chapter Review

### 5.1 Applications Involving Right Triangles (pp. 156–161)

The process of determining the measures of all the sides and all the angles in a triangle is referred to as *solving the triangle.* A right triangle can be solved whenever one side and one of the acute angles are known, or whenever two sides are known. (p. 158)

### 5.2 Law of Sines (pp. 162–167)

*Law of Sines:* In any triangle with angles $A$, $B$, $C$, and opposite sides $a$, $b$, and $c$ respectively,

$$\frac{\sin A}{a} = \frac{\sin B}{b} = \frac{\sin C}{c}.$$

The Law of Sines is used to solve oblique triangles when given (1) one side and two angles, or (2) two sides and the angle opposite one of them.

### 5.3 Law of Cosines (pp. 168–172)

*Law of Cosines:* In any triangle with angles $A$, $B$, and $C$, and opposite sides $a$, $b$, and $c$ respectively,

$$a^2 = b^2 + c^2 - 2bc \cos A,$$
$$b^2 = a^2 + c^2 - 2ac \cos B,$$
$$c^2 = a^2 + b^2 - 2ab \cos C.$$

### 5.4 Vectors in the Plane (pp. 173–179)

A vector with initial point $P_1(x_1, y_1)$ and terminal point $P_2(x_2, y_2)$ is given by the ordered pair $\langle x_2 - x_1, y_2 - y_1 \rangle$. The numbers $x_2 - x_1$ and $y_2 - y_1$ are called the *scalar components* of the vector. (p. 175)

The length of a vector $\mathbf{v}$ is called the *norm* of $\mathbf{v}$ and is denoted $\|\mathbf{v}\|$. If $\mathbf{a} = \langle a_1, a_2 \rangle$, then $\|\mathbf{a}\| = \sqrt{a_1^2 + a_2^2}$. (p. 177)

### 5.5 Vector Applications and the Dot Product (pp. 180–185)

For a given vector $\mathbf{a}$, the process of determining two component vectors $\mathbf{b}$ and $\mathbf{c}$ such that $\mathbf{b}$ is perpendicular to $\mathbf{c}$ and $\mathbf{b} + \mathbf{c} = \mathbf{a}$ is referred to as *resolving* $\mathbf{a}$ into perpendicular components. For a given vector $\mathbf{v}$, the horizontal component vector is denoted $\mathbf{v_x}$ and the vertical component vector is denoted $\mathbf{v_y}$. (p. 181)

The *dot product* of two nonzero vectors $\mathbf{a} = \langle a_1, a_2 \rangle$ and $\mathbf{b} = \langle b_1, b_2 \rangle$ is denoted $\mathbf{a} \cdot \mathbf{b}$, and it can be defined in two equivalent ways: (1) $\mathbf{a} \cdot \mathbf{b} = \|\mathbf{a}\| \|\mathbf{b}\| \cos \theta$, where $\theta$ is the angle between $\mathbf{a}$ and $\mathbf{b}$ such that $0° \leq \theta \leq 180°$; (2) $\mathbf{a} \cdot \mathbf{b} = a_1b_1 + a_2b_2$. (p. 184)

### 5.6 Simple Harmonic Motion (pp. 186–191)

If an object exhibits *simple harmonic motion*, its displacement $x$ from the equilibrium position is a function of $t$. This function can be described by either of the equations

$$x = A \sin(\omega t - D), \quad \text{or} \quad x = A \cos(\omega t - D).$$

### 5.7 Polar Coordinates (pp. 192–198)

The *polar coordinates* $(r, \theta)$ specify a point that lies at a distance $|r|$ from the origin. If $r > 0$, the point lies on the $\theta$-ray. If $r < 0$, the point lies on the opposite of the $\theta$-ray. If $r = 0$, the point lies at the origin, regardless of the value of $\theta$. (p. 192) If point $P$ has rectangular coordinates $(x, y)$ and polar coordinates $(r, \theta)$, then $x = r \cos \theta$, $y = r \sin \theta$, $\tan \theta = \frac{y}{x}$ (provided $x \neq 0$) and $x^2 + y^2 = r^2$. Using these facts, equations in rectangular form may be converted to polar form, and vice versa. (p. 194)

## Chapter Test

**5.1A** Solve the right triangle $ABC$ using the given information about the triangle.

**1.** $B = 60°$, $b = 12$

**2.** $A = 45°$, $c = 18$

**3.** $A = 33°42'$, $b = 96.8$

**4.** $B = 55°12'$, $a = 78.4$

**5.** $a = 36.3$, $b = 46.9$

**6.** $a = 52.6$, $c = 134.1$

**5.1B** Solve the following applied problems.

**7.** What is the angle of elevation of the sun at the time that a 68 ft tall flagpole casts a shadow 25 ft long?

**8.** A 25 ft long ladder is leaning against a building. If the foot of the ladder is 6 ft from the base of the building (on level ground), what acute angle does the ladder make with the ground?

**5.2A** Solve $\triangle ABC$ using the given information about the triangle.

**9.** $A = 41°$, $B = 83°$, $c = 44.6$

**10.** $A = 76°$, $C = 34°$, $b = 67.8$

**5.2B** Given the measure of one angle and two sides, state whether it is possible to have $\triangle ABC$. If it is, solve the triangle. (Determine both solutions if there is more than one.)

**11.** $A = 24°18'$, $c = 9.4$, $a = 6.2$

**12.** $B = 46°54'$, $a = 12.4$, $b = 10.6$

**5.2C** Solve the following applied problems.

**13.** A boat $B$ is observed simultaneously by two Coast Guard stations $S$ and $T$ which are 550 yd apart. Angles $STB$ and $TSB$ are observed to be $132°48'$ and $21°42'$, respectively. Find the distance from the boat to station $T$.

**14.** Points $A$ and $B$ are 126 ft apart and on the same side of a river. A tree at point $T$, on the other side of the river, is observed, with angles $ABT$ and $BAT$ equal to $74°12'$ and $42°36'$, respectively. Find the distance from point $T$ to point $B$.

**5.3A** Solve $\triangle ABC$, given the following.

**15.** $a = 1.52$, $b = 2.31$, $C = 119°$ **16.** $b = 67.2$, $c = 34.9$, $A = 41°$

**17.** $a = 26.7$, $b = 34.2$, $c = 41.8$ **18.** $a = 1.82$, $b = 2.63$, $c = 31.4$

**5.3B** Determine the area of $\triangle ABC$ using the given information.

**19.** $A = 30°$, $b = 18.2$, $c = 10.8$ **20.** $a = 18$, $b = 24$, $c = 36$

**5.4A** The vector **a** has initial point (0, 0) and terminal point (−3, 3). The vector **b** has initial point (0, 0) and terminal point (1, 1).

**21.** (a) Determine the magnitude and direction of vector **a**.
(b) Draw $\mathbf{a} - 2\mathbf{b}$.

**22.** (a) Determine the magnitude and direction of vector **b**.
(b) Draw $-\mathbf{a} + 2\mathbf{b}$.

**5.4B** Points $A$ and $B$ are given. Describe the vector $\overrightarrow{AB}$ as an ordered pair.

**23.** $A(-3, -2)$, $B(2, 7)$ **24.** $A(6, -3)$, $B(-1, 3)$

Use the vectors $\mathbf{a} = \left\langle -\frac{1}{2}, \frac{7}{2} \right\rangle$, $\mathbf{b} = \langle 6, -1 \rangle$ to determine the following.

**25.** $\|\mathbf{a}\|$ and $2\mathbf{a} - \frac{1}{2}\mathbf{b}$ **26.** $\|\mathbf{b}\|$ and $\mathbf{b} - 4\mathbf{a}$

**5.5A** The magnitude $\|\mathbf{v}\|$ and the angle $\theta$ between the vector $\mathbf{v}$ and the positive $x$-axis are given. Find the horizontal and vertical components of $\mathbf{v}$.

**27.** $\|\mathbf{v}\| = 8.6,\ \theta = 148°$

**28.** $\|\mathbf{v}\| = 10.8,\ \theta = 206°$

**29.** A plane's air speed is 500 mph and its bearing is N 75° E. An 8 mph wind is blowing in the direction N 70° W. Determine the ground speed and true bearing of the plane.

**30.** Find the force required to hold a 120 lb crate stationary on a ramp inclined at 25°.

**5.5B** Determine the dot product of the given vectors. Find the angle between the vectors to the nearest degree.

**31.** $\langle 3, -1\rangle, \langle -2, -2\rangle$

**32.** $\langle 4, 6\rangle, \langle -3, 2\rangle$

**5.6A** The equation represents simple harmonic motion. Determine the amplitude, period, and frequency. Sketch the graph.

**33.** $x = 4\sin\left(\frac{t}{2} - 2\pi\right)$

**34.** $x = 6\cos(2t + \pi)$

**5.7A** Graph the polar point and determine the $x$- and $y$-coordinates.

**35.** $(4, -135°)$

**36.** $(-3, 210°)$

**5.7B** Convert the polar equation to rectangular form.

**37.** $r = 3\sin\theta$

**38.** $r = -4\cos\theta$

**5.7C** Sketch the graph of the polar equation.

**39.** $r = 3 - 2\cos\theta$

**40.** $r = 5 - 4\sin\theta$

## Superset

Find the distance between the two polar points.

**41.** $(2, -120°), (4, -570°)$

**42.** $\left(6, \frac{\pi}{3}\right), \left(-2, \frac{3\pi}{2}\right)$

**43.** Let $\mathbf{a} = \langle -3, 1\rangle$, $\mathbf{b} = \langle 4, -3\rangle$ and $\mathbf{c} = \langle -6, 7\rangle$. Express $\mathbf{c}$ as a linear combination of $\mathbf{a}$ and $\mathbf{b}$.

**44.** Write the polar coordinates for the terminal point of the position vector $\mathbf{a}$ if $\mathbf{a} = -3\mathbf{i} + \sqrt{3}\mathbf{j}$.

**45.** Find the angle formed by an internal diagonal of a cube and one of its edges.

## Cumulative Test: *Chapters 4 and 5*

In exercises 1–4, find the following trigonometric values without using Table 4 or a calculator.

**1.** $\sin 165°$ **2.** $\tan 285°$ **3.** $\tan\left(-\frac{\pi}{12}\right)$ **4.** $\cos\left(-\frac{5\pi}{12}\right)$

In exercises 5–8, given $\alpha$, find $\sin 2\alpha$, $\cos 2\alpha$ and $\tan 2\alpha$ using the double-angle identities.

**5.** $-315°$ **6.** $405°$ **7.** $\frac{5\pi}{4}$ **8.** $-\frac{11\pi}{6}$

In exercises 9–10, from the given information determine $\sin \frac{\alpha}{2}$ and $\cos \frac{\alpha}{2}$.

**9.** $\sin \alpha = \frac{8}{15}$, $\alpha$ in Quadrant II **10.** $\cos \alpha = -\frac{7}{24}$, $\alpha$ in Quadrant III.

In exercises 11–12, evaluate by using the conversion identities.

**11.** $\cos 52° + \cos 16°$ **12.** $\sin 39° + \sin 11°$

In exercises 13–16, verify the identities.

**13.** $\frac{\cos 2x + \cos x}{\cos x + 1} = 2 \cos x - 1$ **14.** $\frac{1 - \cos 2x}{2 \cos x \sin x} = \tan x$

**15.** $\frac{(1 + \tan x)^2}{1 + \sin 2x} = 1 + \tan^2 x$ **16.** $\cot x + \tan x = \frac{2}{\sin 2x}$

In exercises 17–20, determine all the solutions of the equation in the interval $[0, \pi]$.

**17.** $\sin^2 x + \sin x = 2$ **18.** $2 \cos^2 x - 3 \cos x = 2$

**19.** $\sin 2x \cos x + \cos 2x \sin x = \frac{1}{2}$

**20.** $\cos 2x \cos x - \sin 2x \sin x = 0.$

In exercises 21–24, solve for $x$.

**21.** $\cos\left(\tan^{-1} \frac{5}{12}\right) = x$ **22.** $\tan\left(\arcsin \frac{8}{17}\right) = x$

**23.** $\cos^{-1} \frac{1}{2} + 2 \sin^{-1} x = \pi$ **24.** $\tan^{-1} x + 2 \tan^{-1} 1 = \frac{3\pi}{4}$

In exercises 25–28, solve the triangle for the part indicated.

**25.** $B = 45°$, $C = 90°$, $b = 24.0$, find $a$.

**26.** $A = 60°$, $C = 90°$, $c = 18.0$, find $b$.

**27.** $A = 48°12'$, $B = 67°42'$, $c = 32.4$, find $a$.

**28.** $A = 56°48'$, $C = 25°18'$, $a = 76.8$, find $b$.

In exercises 29–30, solve each applied problem.

**29.** A surveyor measures a triangular lot. The three sides have lengths of 73 yd, 106 yd, and 145 yd. Find the area of the lot.

**30.** A painter is painting a large triangular sign with area of 18 $ft^2$. If one side is 6 ft long and another side 8 ft long, what is the length of the third side?

In exercises 31–32, $\mathbf{a} = \langle -1, 2\rangle$, $\mathbf{b} = \langle 3, -1\rangle$. Determine $\|\mathbf{c}\|$, $\mathbf{u}_\mathbf{c}$, and $-\frac{1}{2}\mathbf{c}$.

**31.** $\mathbf{c} = 2\mathbf{a} + 3\mathbf{b}$

**32.** $\mathbf{c} = -\mathbf{a} + 4\mathbf{b}$

In exercises 33–34, solve each applied problem.

**33.** A ball is thrown due east at 15 yd/s from a train that is travelling due north at the rate of 20 yd/s. Find the speed of the ball and the direction of its path.

**34.** An airplane is headed due north at a speed of 420 mph. If the wind is blowing due east with a velocity of 32.0 mph, find the true bearing and ground speed of the plane.

In exercises 35–36, determine the dot product for the given vectors. Find the angle between the vectors to the nearest degree.

**35.** $\langle 2, -3\rangle, \langle -9, -6\rangle$

**36.** $\langle 1, -5\rangle, \langle 4, -2\rangle$

In exercises 37–38, the equation represents simple harmonic motion. Give the amplitude, period, and frequency, and state the position of the point at $t = \frac{\pi}{3}$.

**37.** $x = 4\cos\left(\frac{t}{2} + 2\pi\right)$

**38.** $x = 6\sin(2t - \pi)$

In exercises 39–40, the rectangular coordinates of a point are given. Determine a polar representation $(r, \theta)$, with $0 < \theta < \pi$.

**39.** $(-\sqrt{3}, -1)$

**40.** $(3\sqrt{3}, -3)$

## Superset

**41.** If $\tan 2\theta = -\frac{24}{7}$, find $\sin\theta$ and $\cos\theta$, if $\theta$ is an acute angle.

**42.** Express $\sin x + \sin 3x + \sin 5x + \sin 7x$ as a product.

**43.** The diagonals of a parallelogram intersect at an angle of 40°24′. If the diagonals are 16.8 in and 22.4 in long, find the perimeter of the parallelogram.

**44.** A force of 6 N is applied in dragging an object. What is the work done in moving the object 10 m if the force is applied at an angle of 60° with the ground.

# 6 Complex Numbers

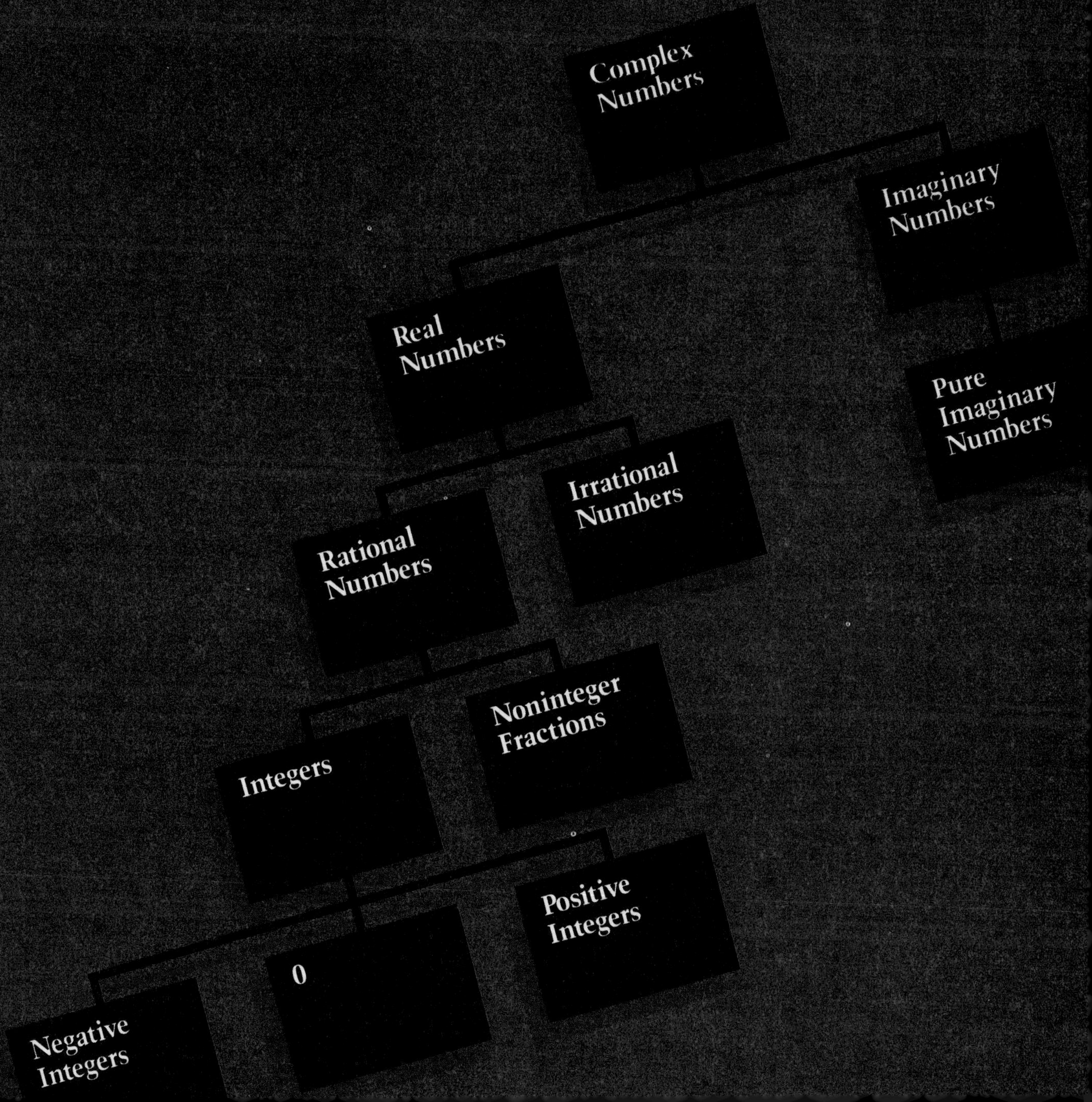

Complex Numbers

Real Numbers

Imaginary Numbers

Rational Numbers

Irrational Numbers

Pure Imaginary Numbers

Integers

Noninteger Fractions

Negative Integers

0

Positive Integers

## 6.1 Introduction to Complex Numbers

**Goal A** *to identify and simplify complex numbers*

When discussing sets of numbers in Chapter 1, we started with the set of positive integers and quickly enlarged that set to the set of all integers (positive, negative, and zero). The set of integers was then used to define the set of rational numbers, and the set of rational numbers was in turn joined to the set of irrational numbers to produce the set of all real numbers. At each step along the way, a certain set was enlarged to include more numbers. This process is called **extending the set.**

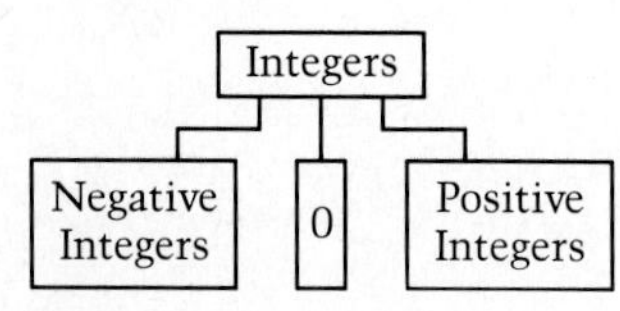

Figure 6.1

The need for extending sets becomes clear as we search for solutions to equations. For example, the equation

$$x + 5 = 1$$

cannot be solved over the set of positive integers. If the replacement set for $x$ is extended to the set of all integers (thereby including zero and the negative integers, as shown in Figure 6.1), the equation then has a solution, $-4$.

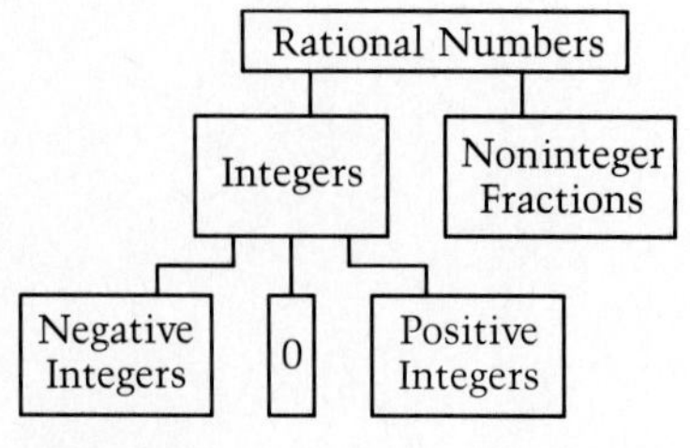

Figure 6.2

Similarly, the equation $5x = 1$ cannot be solved over the set of integers. If the replacement set for $x$ is extended to the set of rational numbers (thereby including noninteger fractions, as shown in Figure 6.2), the equation then has a solution, $\frac{1}{5}$.

Finally, the equation $x^2 = 2$ cannot be solved over the set of rational numbers. If the replacement set for $x$ is extended to the set of real numbers (thereby including irrational numbers, as shown in Figure 6.3), the equation then has solutions, $\sqrt{2}$ and $-\sqrt{2}$.

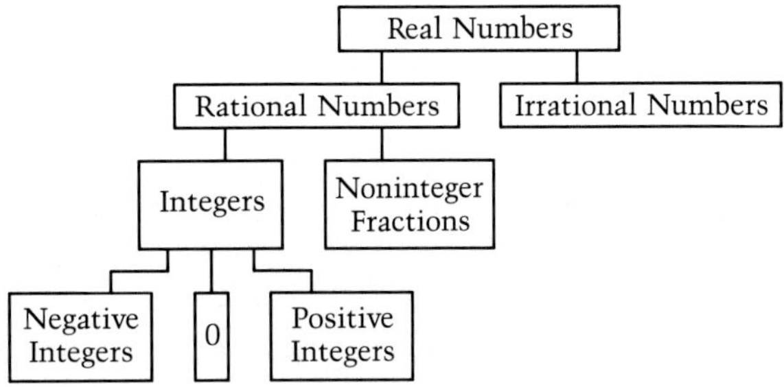


**Figure 6.3**

A quadratic equation such as

$$x^2 = -1,$$

however, has no real number solutions, for there is no real number whose square is $-1$. In order to solve such an equation, it will be necessary, once again, to extend the replacement set for $x$. To do this, we begin by defining a number that is a solution of the above equation.

**Definition**

The number $i$ has the property that

$$i^2 = -1.$$

For this reason, $i$ can be considered a square root of $-1$. Thus, we write

$$i = \sqrt{-1}.$$

Since $i^2 = -1$, we have invented a number $i$ that is a solution of $x^2 = -1$. With a definition for $\sqrt{-1}$, we can define square roots of other negative numbers. Observe that

$$-2 = (-1)(2) = i^2(\sqrt{2})^2 = (i \cdot \sqrt{2})^2.$$

Thus, $i\sqrt{2}$ is a square root of $-2$. (Note that when $i$ is multiplied by a radical, $i$ is generally written first to avoid confusing a number like $\sqrt{2}i$ with $\sqrt{2i}$). In fact, we can use the number $i$ to form square roots of any negative number.

**Definition**

If $p$ is a positive real number, then

$$\sqrt{-p} = i\sqrt{p}.$$

As for any positive real number, $-p$ has two square roots. These roots are $i\sqrt{p}$ and $-i\sqrt{p}$. We refer to $i\sqrt{p}$ as the **principal square root of $-p$.**

Using this definition, we can rewrite expressions containing square roots of negative numbers. This is illustrated in the next example.

**Example 1** Use the number $i$ to rewrite each of the following.

(a) $\sqrt{-5}$ (b) $\sqrt{-16}$ (c) $\sqrt{-8}$

***Solution*** (a) $\sqrt{-5} = i\sqrt{5}$ (b) $\sqrt{-16} = i\sqrt{16} = 4i$ (c) $\sqrt{-8} = i\sqrt{8} = 2i\sqrt{2}$

The equation $x^2 + 2x + 3 = 0$ has no real number solutions. Using the quadratic formula together with the definition of $i$, however, we can determine the solutions. Substituting $a = 1$, $b = 2$, and $c = 3$, we obtain the following (here we assume that the operations on real numbers also apply to complex numbers):

$$\begin{aligned} x &= \frac{-b \pm \sqrt{b^2 - 4ac}}{2a} \\ &= \frac{-2 \pm \sqrt{-8}}{2} \\ &= \frac{-2 \pm i\sqrt{8}}{2} \\ &= \frac{-2 \pm 2i\sqrt{2}}{2} \\ &= -1 \pm i\sqrt{2} \end{aligned}$$

Thus, the solutions are $-1 + i\sqrt{2}$ and $-1 - i\sqrt{2}$. The form of these two numbers motivates us to extend the set of real numbers to a larger set, called the *complex numbers.*

**Definition**

If $a$ and $b$ are real numbers, then any number of the form

$$a + bi$$

is called a **complex number.** The number $a$ is called the **real part** of the complex number, and $b$ is called the **imaginary part.**

We now consider three examples of complex numbers. The first is the real number 5. Since 5 can be written in the form

$$5 + 0i,$$

it is a complex number. Similarly, every real number $a$ can be written in the form $a + 0i$. Thus, the set of real numbers is contained in the set of complex numbers.

The number $2 - 4i$ is also a complex number since it can be written as $2 + (-4)i$. Any complex number of the form $a + bi$, where the imaginary part $b$ is not zero, is called an **imaginary number.** (Note that a complex number whose imaginary part is negative is usually written in the form $a - bi$.)

Finally, since $7i$ can be written

$$0 + 7i,$$

it is also a complex number. Since its imaginary part is not zero, it is also an imaginary number. Any number of the form $bi$, where $b \neq 0$, is called a **pure imaginary number.**

**Example 2** Determine the real part and the imaginary part of each complex number.

(a) $4 + 7i$ (b) $\sqrt{3}$ (c) $-2 + i$ (d) $8 - 6i$ (e) $5i$ (f) $0$

***Solution***

(a) The real part is 4; the imaginary part is 7.
(b) Since $\sqrt{3}$ can be written $\sqrt{3} + 0i$, the real part is $\sqrt{3}$ and the imaginary part is 0.
(c) Since $-2 + i$ can be written $-2 + 1i$, the real part is $-2$ and the imaginary part is 1.
(d) Since $8 - 6i$ can be written $8 + (-6)i$, the real part is 8 and the imaginary part is $-6$.
(e) Since $5i$ can be written $0 + 5i$, the real part is 0 and the imaginary part is 5.
(f) Since 0 can be written $0 + 0i$, the real part is 0 and the imaginary part is 0.

Thus, the set of complex numbers is formed by joining the set of real numbers and the set of imaginary numbers as shown in Figure 6.4.

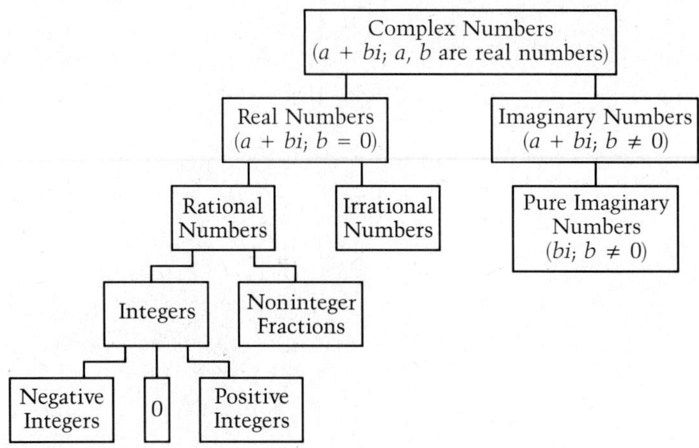


**Figure 6.4**

**Goal B** *to add, subtract, and multiply complex numbers*

The arithmetic of complex numbers follows patterns that we have already observed when working with binomials. In the case of complex numbers, however, we simplify results by using the fact that $i^2 = -1$.

Notice the following pattern:

$$i^1 = i \qquad i^5 = i^4(i) = (1)(i) = i$$
$$i^2 = -1 \qquad i^6 = i^4(i^2) = (1)(-1) = -1$$
$$i^3 = i^2(i) = -i \qquad i^7 = i^4(i^3) = (1)(-i) = -i$$
$$i^4 = (i^2)(i^2) = 1 \qquad i^8 = i^4(i^4) = (1)(1) = 1$$

It follows that

$$i^4, i^8, i^{12}, i^{16}, \ldots$$

are all equal to 1. In general, for any positive integer $n$, $i^{4n} = 1$. With this information, we can simplify a number like $i^{31}$ very easily. Since 28 is the highest multiple of 4 less than 31, we can write the following:

$$i^{31} = i^{28}i^3 = (1)(-i) = -i.$$

**Careful!** When simplifying an expression containing the square root of a negative number, you should always begin by rewriting the expression in terms of $i$. For example,

$$\sqrt{-4} \cdot \sqrt{-9} = (2i)(3i) = i^2 \cdot 6 = -6.$$

Failure to do this may lead to serious errors. Note that the radical expression $\sqrt{-4} \cdot \sqrt{-9} \neq \sqrt{(-4)(-9)}$. If we assumed that this statement were true, then we would obtain the following incorrect statement:

$$\sqrt{-4} \cdot \sqrt{-9} = \sqrt{(-4)(-9)} = \sqrt{36} = 6.$$

Thus, $\sqrt{a} \cdot \sqrt{b} = \sqrt{ab}$ only if $a$ and $b$ are nonnegative.

**Example 3** Simplify each expression.

(a) $i^{64}$ (b) $i^{10}$ (c) $\sqrt{-4} \cdot \sqrt{-8}$ (d) $i^{37}$ (e) $i^{127}$

(f) $\sqrt{-9} \cdot \sqrt{-18} \cdot \sqrt{-2}$

***Solution***

(a) $i^{64} = 1$ ■ Since 64 is a multiple of 4, $i^{64} = 1$.

(b) $i^{10} = i^8 i^2 = (1)(-1) = -1$

(c)
$$\begin{aligned} \sqrt{-4} \cdot \sqrt{-8} &= i\sqrt{4} \cdot i\sqrt{8} \\ &= 2i \cdot 2i\sqrt{2} \\ &= 4i^2\sqrt{2} \\ &= -4\sqrt{2} \end{aligned}$$

(d) $i^{37} = i^{36}i = (1)(i) = i$

(e) $i^{127} = i^{124}i^3 = (1)(-i) = -i$

(f)
$$\begin{aligned} \sqrt{-9} \cdot \sqrt{-18} \cdot \sqrt{-2} &= i\sqrt{9} \cdot i\sqrt{18} \cdot i\sqrt{2} \\ &= 3i \cdot 3i\sqrt{2} \cdot i\sqrt{2} \\ &= 9i^3(\sqrt{2})^2 \\ &= 18i^3 \\ &= -18i \end{aligned}$$

**Definition**

Two complex numbers $a + bi$ and $c + di$ are equal if and only if $a = c$ and $b = d$.

The previous definition states that two complex numbers are equal if and only if their real parts are equal and their imaginary parts are equal. Thus, if we are given that

$$a + 3i = -2 + (x - 5)i,$$

then equating the real parts, we have

$$a = -2,$$

and equating the imaginary parts, we have

$$3 = x - 5$$
$$x = 8.$$

Hence, the complex number represented by the expressions $a + 3i$ and $-2 + (x - 5)i$ is

$$-2 + 3i.$$

The sum of two complex numbers is a complex number whose real part is the sum of the real parts of the two numbers, and whose imaginary part is the sum of their imaginary parts. Their difference is found in a similar way.

**Definition**
If $a + bi$ and $c + di$ are complex numbers, then

$$(a + bi) + (c + di) = (a + c) + (b + d)i,$$

and

$$(a + bi) - (c + di) = (a - c) + (b - d)i.$$

We find the product of two complex numbers the same way we find the product of two binomials; we use the FOIL method.

$$\begin{aligned}(a + bi)(c + di) &= ac + adi + bci + bidi \\ &= ac + bdi^2 + (ad + bc)i \\ &= ac - bd + (ad + bc)i\end{aligned}$$

■ Since $i^2 = -1$.

**Definition**
If $a + bi$ and $c + di$ are complex numbers, then

$$(a + bi)(c + di) = (ac - bd) + (ad + bc)i$$

In the following examples, we shall practice the arithmetic of complex numbers.

**Example 4** Simplify each expression.

(a) $(3 + 3i) + (7 - 5i)$ (b) $(6 - 2i) + 2i$ (c) $(10 + 7i) - (10 - 4i)$

(d) $(8 + i) - 17i$

***Solution*** (a) $(3 + 3i) + (7 - 5i) = (3 + 7) + (3i - 5i)$ ■ Group like terms and combine.

$= 10 + (-2i)$

$= 10 - 2i$

(b) $(6 - 2i) + 2i = 6 + (-2i + 2i) = 6$

(c) $(10 + 7i) - (10 - 4i) = (10 - 10) + (7i + 4i)$

$= 11i$

(d) $(8 + i) - 17i = 8 + (i - 17i) = 8 - 16i$

**Example 5** Simplify each expression.

(a) $5(2 - 7i)$ (b) $-6i(4 + 9i)$ (c) $(2 + i)(3 - 4i)$ (d) $(3 - 5i)^2$

***Solution*** (a) $5(2 - 7i) = 5 \cdot 2 - 5 \cdot 7i$

$= 10 - 35i$

(b) $-6i(4 + 9i) = -24i - 54i^2$ ■ Replace $i^2$ with $-1$.

$= -24i - 54(-1)$

$= 54 - 24i$

(c) $(2 + i)(3 - 4i) = 2 \cdot 3 + 2(-4i) + 3i + i(-4i)$

$= 6 - 8i + 3i - 4i^2$ ■ Replace $i^2$ with $-1$.

$= 6 + 4 + [(-8i) + 3i]$

$= 10 - 5i$

(d) $(3 - 5i)^2 = (3 - 5i)(3 - 5i)$

$= 9 - 15i - 15i + 25i^2$

$= 9 - 25 - (15i + 15i)$

$= -16 - 30i$

## Exercise Set 6.1

### Goal A

In exercises 1–18, use the number $i$ to rewrite each expression.

**1.** $\sqrt{-13}$ **2.** $\sqrt{-6}$ **3.** $\sqrt{-15}$

**4.** $\sqrt{-21}$ **5.** $\sqrt{-4}$ **6.** $\sqrt{-25}$

**7.** $\sqrt{-100}$ **8.** $\sqrt{-81}$ **9.** $\sqrt{-24}$

**10.** $\sqrt{-27}$ **11.** $\sqrt{-98}$ **12.** $\sqrt{-75}$

**13.** $-\sqrt{-9}$ **14.** $-\sqrt{-36}$ **15.** $-\sqrt{-20}$

**16.** $-\sqrt{-18}$ **17.** $-\sqrt{-63}$ **18.** $-\sqrt{-32}$

In exercises 19–32, determine the real part and the imaginary part of each complex number.

**19.** $2 + 5i$ **20.** $7 + i$ **21.** $-3 + 4i$

**22.** $5 - 8i$ **23.** $-1 - 2i$ **24.** $-5 - 10i$

**25.** $\frac{1}{2} + 6i$ **26.** $-7 - \frac{3}{5}i$ **27.** $6i$

**28.** $14$ **29.** $-\sqrt{3}$ **30.** $i$

**31.** $\frac{1}{4} - i\sqrt{10}$ **32.** $-\sqrt{6} + \frac{2}{7}i$

### Goal B

In exercises 33–94, simplify each expression.

**33.** $i^{32}$ **34.** $i^{18}$

**35.** $i^{51}$ **36.** $i^{65}$

**37.** $(-i)^{10}$ **38.** $(-i)^{43}$

**39.** $(-i)^{25}$ **40.** $(-i)^{52}$

**41.** $(2i)^3 \cdot i^{12}$ **42.** $(4i)^2 \cdot i^{50}$

**43.** $(3i)^4 \cdot i^{34}$ **44.** $(5i)^2 \cdot i^{15}$

**45.** $\sqrt{-7} \cdot \sqrt{-8}$ **46.** $\sqrt{-12} \cdot \sqrt{-3}$

**47.** $\sqrt{8} \cdot \sqrt{-2}$ **48.** $\sqrt{-5} \cdot \sqrt{10}$

**49.** $\sqrt{-2} \cdot \sqrt{-3} \cdot \sqrt{12}$

**50.** $\sqrt{-5} \cdot \sqrt{-15} \cdot \sqrt{-3}$

**51.** $(6 + 4i) + (2 + 5i)$

**52.** $(2 + 2i) + (7 + 3i)$

**53.** $(2 + 3i) + (-4 + 5i)$

**54.** $(5 - 3i) + (8 + 2i)$

**55.** $(7 - 9i) + (3 - 6i)$

**56.** $(4 - 4i) + (-5 + i)$

**57.** $(3 - 4i) + (-5 - 6i)$

**58.** $(-1 - i) + (10 - 9i)$

**59.** $(-10 - 5i) + (7 + 2i)$

**60.** $(1 + 6i) + (-1 + 6i)$

**61.** $(9 + 2i) - (4 + 3i)$

**62.** $(3 + i) - (8 + 4i)$

**63.** $(8 + 7i) - (6 - 7i)$

**64.** $(9 - 2i) - (3 + 5i)$

**65.** $(-11 + 7i) - (11 - 7i)$

**66.** $(-12 - 4i) - (6 - i)$

**67.** $7i + (2 + 8i)$ **68.** $9 + (1 - 5i)$

**69.** $2i - (12 - 2i)$ **70.** $(7 - 4i) - 10$

**71.** $4(-6 + 8i)$ **72.** $-2(5 - 9i)$

**73.** $3i(2 - i)$ **74.** $-4i(1 + 8i)$

**75.** $(5 + 2i)(3 + i)$

**76.** $(1 + 4i)(1 + 2i)$

**77.** $(-7 + 2i)(2 + 5i)$

**78.** $(9 - 3i)(-2 + i)$

**79.** $(-1 - i)(8 + 2i)$

**80.** $(2 + 6i)(-3 - 3i)$

**81.** $(5 + 4i)(5 - 4i)$

**82.** $(2 + 3i)(2 - 3i)$

**83.** $(1 + i)(1 - i)$

**84.** $(\sqrt{7} - i)(\sqrt{7} + i)$

**85.** $(2 + 3i)^2$

**86.** $(3 + 4i)^2$

**87.** $(-5 + i)^2$

**88.** $(6 - i)^2$

**89.** $(-1 - 2i)^2 \cdot i$

**90.** $(2i)(2 + 5i)^2$

**91.** $(3i)^3(-2 - 2i)$

**92.** $(-8 + 6i)(-3i^5)$

**93.** $(4 - 2i)^2(1 + i)$

**94.** $(-1 + 3i)^3$

## Superset

In exercises 95–102, find real numbers $x$ and $y$ that satisfy each equation.

**95.** $2 + 3yi = -x + 9i$

**96.** $-4x + 7i = -8 - 7y$

**97.** $8x - 4i = 6 + 2yi$

**98.** $9 + 5yi = 6x - 7i$

**99.** $-7 + i = (x - y) + (2x + y)i$

**100.** $(2x + y) - 8i = 5 + (x - 3y)i$

**101.** $(x - y) + 4i = 8 - (x - 2y)i$

**102.** $3 - 2i = (x + y) - (3x - y)i$

In exercises 103–106, simplify each expression.

**103.** $i + i^2 + i^3 + \cdots + i^{48}$

**104.** $i + i^2 + i^3 + \cdots + i^{18}$

**105.** $i + i^2 + i^3 + \cdots + i^{29}$

**106.** $i + i^2 + i^3 + \cdots + i^{75}$

In exercises 107–114, determine whether each statement is true or false.

**107.** $3i$ is a solution of $x^2 + 9 = 0$.

**108.** $-3i$ is a solution of $x^2 + 9 = 0$.

**109.** $5 + i$ is a solution of $x^2 - 10x + 24 = 0$.

**110.** $5 - i$ is a solution of $x^2 - 10x + 24 = 0$.

**111.** $-3 + 2i$ is a solution of $x^2 + 6x + 13 = 0$.

**112.** $-3 - 2i$ is a solution of $x^2 + 6x + 13 = 0$.

**113.** Both $2 + 7i$ and $2 - 7i$ are solutions of the quadratic equation $x^2 - 4x + 53 = 0$.

**114.** Both $\frac{1}{2} + 3i$ and $\frac{1}{2} - 3i$ are solutions of the quadratic equation $4x^2 - 4x + 37 = 0$.

A complex number $z$ is called an ***n*th root of one** or an ***n*th root of unity** if $z^n = 1$.

**115.** Show that $1$, $i$, $-1$, and $-i$ are fourth roots of unity.

**116.** Show that $1$, $-\frac{1}{2} + \frac{\sqrt{3}}{2}i$, and $-\frac{1}{2} - \frac{\sqrt{3}}{2}i$ are cube roots of unity.

**117.** Show that $1$, $\frac{1}{2} + \frac{\sqrt{3}}{2}i$, $-\frac{1}{2} + \frac{\sqrt{3}}{2}i$, $-1$, $-\frac{1}{2} - \frac{\sqrt{3}}{2}i$, and $\frac{1}{2} - \frac{\sqrt{3}}{2}i$ are sixth roots of unity.

**118.** Show that the values $1$, $\frac{\sqrt{2}}{2} + \frac{\sqrt{2}}{2}i$, $i$, $-\frac{\sqrt{2}}{2} + \frac{\sqrt{2}}{2}i$, $-1$, $-\frac{\sqrt{2}}{2} - \frac{\sqrt{2}}{2}i$, $-i$, and $\frac{\sqrt{2}}{2} - \frac{\sqrt{2}}{2}i$ are eighth roots of unity.

In exercises 119–124, determine whether each statement is true or false. If false, illustrate this with an example.

**119.** If $a$ is a positive real number and $b$ is a negative real number, then $\sqrt{a} \cdot \sqrt{b} = \sqrt{ab}$.

**120.** If $a$ and $b$ are real numbers, then $\sqrt{a} \cdot \sqrt{b} = \sqrt{ab}$.

**121.** The product of two imaginary numbers is always imaginary.

**122.** The product of two imaginary numbers is always a real number.

**123.** If $a$ and $b$ are real numbers, then $(a + bi)(a - bi)$ is always a real number.

**124.** If $a$ and $b$ are real numbers, then $(a + bi)^2$ is always imaginary.

## 6.2 Properties of Complex Numbers

**Goal A** *to write the quotient of two complex numbers in the form $a + bi$*

By now you should have no difficulty solving linear equations like

$$3 + 4x = 2.$$

The last step in solving this equation involves multiplying both sides by the reciprocal of 4, namely $\frac{1}{4}$.

If we replaced the coefficient of $x$ in the above equation with an irrational number, say $1 - \sqrt{5}$, the equation would then be

$$3 + (1 - \sqrt{5})x = 2,$$

and the last step in solving this equation would require multiplying both sides by the reciprocal of $1 - \sqrt{5}$. Recall that we can simplify by rationalizing the denominator, i.e., by multiplying the numerator and denominator by $1 + \sqrt{5}$.

$$\begin{aligned}\frac{1}{1-\sqrt{5}} &= \frac{1}{1-\sqrt{5}} \cdot \frac{1+\sqrt{5}}{1+\sqrt{5}} \\ &= \frac{1+\sqrt{5}}{1-5} \\ &= -\frac{1+\sqrt{5}}{4}\end{aligned}$$

Suppose instead, that the coefficient of $x$ in the equation above were

$$3 - 2i.$$

Then the last step in solving the equation would involve multiplying both sides by the reciprocal of $3 - 2i$, $\frac{1}{3-2i}$. The reciprocal is a complex number and can be written in the form $a + bi$. To produce this form, we simply multiply the numerator and denominator by $3 + 2i$.

$$\frac{1}{3-2i} = \frac{1}{3-2i} \cdot \frac{3+2i}{3+2i} = \frac{3+2i}{9-4i^2} = \frac{3+2i}{9+4} = \frac{3}{13} + \frac{2i}{13}$$

The complex number $3 + 2i$ is called the **complex conjugate** of $3 - 2i$. In general, the complex conjugate of $a + bi$ is $a - bi$, and the complex conjugate of $a - bi$ is $a + bi$. Notice that since

$$(a + bi)(a - bi) = a^2 - b^2i^2 = a^2 + b^2,$$

the product of a complex number and its conjugate is a real number.

**Example 1** Write the reciprocal of each of the following in the form $a + bi$.

(a) $3 - 4i$ (b) $5 + 3i$

***Solution*** (a) The reciprocal of $3 - 4i$ is $\frac{1}{3 - 4i}$.

$$\begin{aligned} \frac{1}{3 - 4i} &= \frac{1}{3 - 4i} \cdot \frac{3 + 4i}{3 + 4i} \\ &= \frac{3 + 4i}{9 - 16i^2} \\ &= \frac{3 + 4i}{9 + 16} \\ &= \frac{3}{25} + \frac{4}{25}i \end{aligned}$$

■ Multiply the numerator and denominator by the conjugate of the denominator $3 - 4i$.

(b) $\frac{1}{5 + 3i} = \frac{1}{5 + 3i} \cdot \frac{5 - 3i}{5 - 3i} = \frac{5 - 3i}{25 - 9i^2} = \frac{5 - 3i}{25 + 9} = \frac{5}{34} - \frac{3}{34}i$

We can use the complex conjugate in a similar way when we divide complex numbers.

**Example 2** Simplify the quotient $\frac{2 - 3i}{-3 + i}$.

***Solution***

$$\begin{aligned} \frac{2 - 3i}{-3 + i} &= \frac{2 - 3i}{-3 + i} \cdot \frac{-3 - i}{-3 - i} \\ &= \frac{-6 - 2i + 9i + 3i^2}{9 - i^2} \\ &= \frac{-6 - 2i + 9i - 3}{9 - (-1)} \\ &= -\frac{9}{10} + \frac{7}{10}i \end{aligned}$$

■ Multiply the numerator and denominator by the conjugate of the denominator $-3 + i$.

■ We used the FOIL method to find the product in the numerator.

■ We used the fact that $i^2 = -1$.

**Goal B** *to use the symbol $\bar{z}$ to represent the conjugate of a complex number*

If we let the variable $z$ represent a complex number $a + bi$, it is customary to let $\bar{z}$ represent its complex conjugate $a - bi$; that is,

$$\overline{a + bi} = a - bi.$$

We can establish several facts about complex numbers and their conjugates by using the rules of arithmetic for complex numbers.

For example, the conjugate of the sum of two complex numbers is the sum of the conjugates of the two numbers:

$$\overline{z + w} = \bar{z} + \bar{w}.$$

To establish this fact, let $z = a + bi$ and $w = c + di$.

$$\overline{z + w} = \overline{(a + bi) + (c + di)}$$

■ Substitute $a + bi$ for $z$ and $c + di$ for $w$. Then combine like terms.

$$= \overline{(a + c) + (b + d)i}$$

■ Now use the definition of a complex conjugate.

$$= (a + c) - (b + d)i$$

■ Rearrange the terms.

$$= a - bi + c - di$$

■ The expression is rewritten as the sum of two complex numbers.

$$= \overline{a + bi} + \overline{c + di}$$

■ The complex numbers are rewritten as conjugates.

$$= \bar{z} + \bar{w}$$

Using the arithmetic of complex numbers, we can also show that the conjugate of the product of two complex numbers is the product of the conjugate of the two numbers, i.e., $\overline{z \cdot w} = \bar{z} \cdot \bar{w}$.

■ Again, let $z = a + bi$ and $w = c + di$.

$$\begin{aligned}
\overline{z \cdot w} &= \overline{(a + bi)(c + di)} \\
&= \overline{(ac - bd) + (ad + bc)i} \\
&= (ac - bd) - (ad + bc)i \\
&= ac - bd - adi - bci \\
&= ac - adi - bci + bdi^2 \\
&= a(c - di) - bi(c - di) \\
&= (a - bi)(c - di) \\
&= \overline{(a + bi)}\,\overline{(c + di)} \\
&= \bar{z} \cdot \bar{w}
\end{aligned}$$

■ $-bd$ is rewritten as $bdi^2$.

We now summarize some of the more useful properties of complex conjugates.

**Facts**

If $\overline{z}$ and $\overline{w}$ are complex conjugates of the complex numbers $z$ and $w$, respectively, then

1. $\overline{z + w} = \overline{z} + \overline{w}$ 2. $\overline{z - w} = \overline{z} - \overline{w}$
3. $\overline{z \cdot w} = \overline{z} \cdot \overline{w}$ 4. $\overline{\left(\frac{z}{w}\right)} = \frac{\overline{z}}{\overline{w}}$, for $w \neq 0$
5. $\overline{z^n} = (\overline{z})^n$, for every positive integer $n$
6. $\overline{z} = z$, if $z$ is a real number

**Example 3** If $z = 2 - 5i$ and $w = 3 + i$, verify that (a) $\overline{z + w} = \overline{z} + \overline{w}$ and (b) $\overline{z \cdot w} = \overline{z} \cdot \overline{w}$.

***Solution***

(a)
$$\begin{aligned}\overline{z + w} &= \overline{(2 - 5i) + (3 + i)} \\ &= \overline{(2 + 3) + (1 - 5)i} \\ &= \overline{5 - 4i} \\ &= 5 + 4i\end{aligned}$$

■ Begin by substituting values for $z$ and $w$. Then combine like terms.

$$\begin{aligned}\overline{z} + \overline{w} &= \overline{2 - 5i} + \overline{3 + i} \\ &= 2 + 5i + 3 - i \\ &= (2 + 3) + (5 - 1)i \\ &= 5 + 4i\end{aligned}$$

■ Begin by substituting values for $z$ and $w$. Then use the definition of conjugate.

Thus, since $\overline{z + w} = 5 + 4i$ and $\overline{z} + \overline{w} = 5 + 4i$, the two expressions $\overline{z + w}$ and $\overline{z} + \overline{w}$ are equal.

(b)
$$\begin{aligned}\overline{z \cdot w} &= \overline{(2 - 5i)(3 + i)} \\ &= \overline{6 + 2i - 15i - 5i^2} \\ &= \overline{6 + 5 + 2i - 15i} \\ &= \overline{11 - 13i} \\ &= 11 + 13i\end{aligned}$$

■ Begin by substituting values for $z$ and $w$.

■ Rearrange terms; let $i^2 = -1$.

$$\begin{aligned}\overline{z} \cdot \overline{w} &= \overline{2 - 5i} \cdot \overline{3 + i} \\ &= (2 + 5i)(3 - i) \\ &= 6 - 2i + 15i - 5i^2 \\ &= 6 + 5 + 13i \\ &= 11 + 13i\end{aligned}$$

■ Begin by substituting values for $z$ and $w$. Then use the definition of conjugate.

■ Rearrange terms; let $i^2 = -1$.

Thus, since $\overline{z \cdot w} = 11 + 13i$ and $\overline{z} \cdot \overline{w} = 11 + 13i$, we conclude $\overline{z \cdot w} = \overline{z} \cdot \overline{w}$.

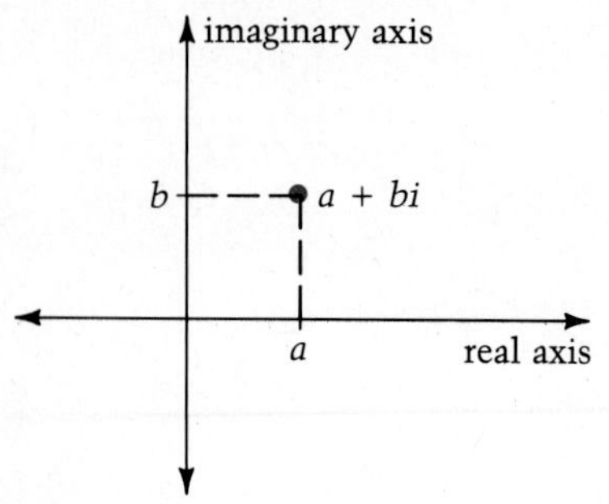


Figure 6.5 The Complex Plane.

**Goal C** *to represent a complex number as a point in the complex plane*

Recall that in order to visualize the real numbers, we associated each number with a point on a line, called the real line. Complex numbers can be visualized as points in a plane called the **complex plane,** shown in Figure 6.5.

Every complex number $a + bi$ can be associated with an ordered pair $(a, b)$ in the complex plane, where $a$ is the real part of the complex number and $b$ is the imaginary part. The horizontal axis in the complex plane is referred to as the real axis, since real numbers are graphed there. The vertical axis is called the imaginary axis. Pure imaginary numbers (numbers of the form $bi$) are graphed on the imaginary axis.

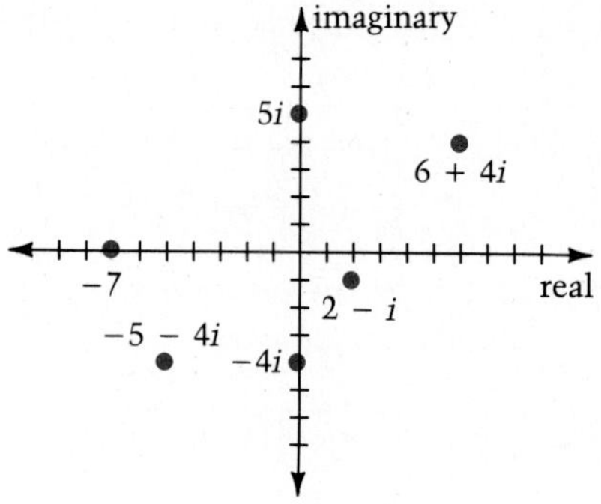


**Figure 6.6**

We will define the absolute value of a complex number $a + bi$ as the distance between the point corresponding to $a + bi$ and the origin. For example, the absolute value of $4 + 3i$ is the distance between the point $4 + 3i$ and the origin, and can be found by the Pythagorean Theorem:

$$\sqrt{4^2 + 3^2} = \sqrt{25} = 5.$$

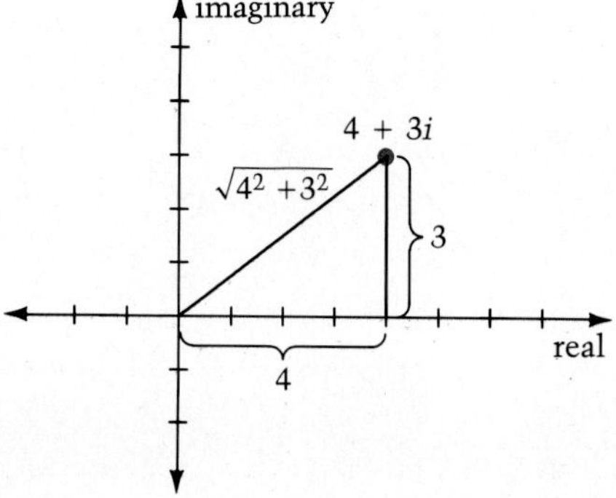


**Figure 6.7**

**Example 4** Represent each of the following complex numbers as a point in the complex plane.
(a) $4 - 3i$ (b) $-3 + 4i$ (c) $-3 - 4i$ (d) $\overline{-2 - 3i}$

***Solution***

(a)

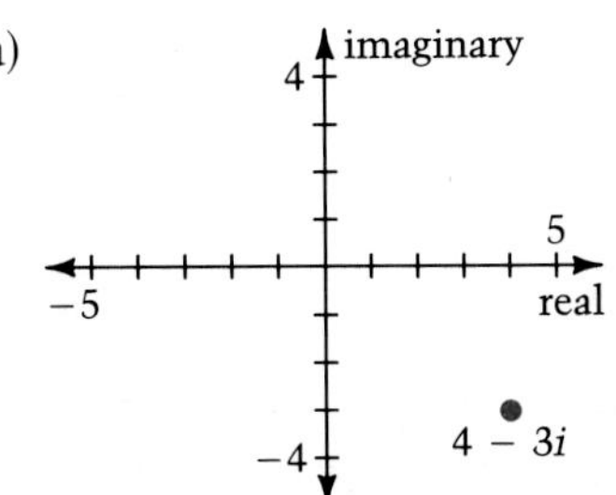


(b)

imaginary
4
−3 + 4i
5
−5
real
−4

(c)

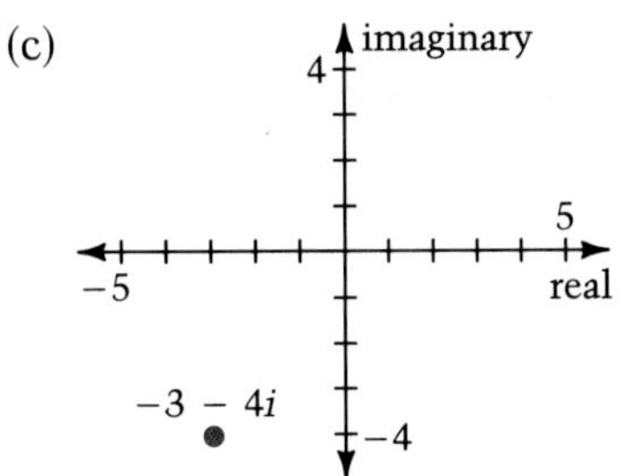


(d)

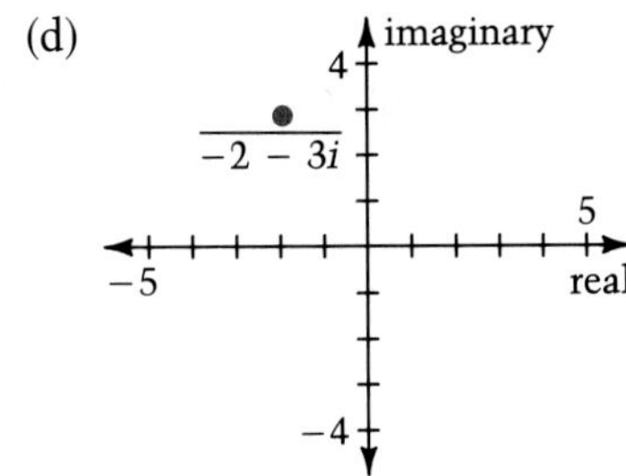


**Definition**

The **absolute value** of a complex number $a + bi$, denoted $|a + bi|$, is the distance between the origin and the point associated with $a + bi$. That is, $|a + bi| = \sqrt{a^2 + b^2}$.

**Example 5** Determine the absolute value of each of the following complex numbers.
(a) $2 - 3i$ (b) $-5i$ (c) $1 - i$

***Solution***

(a) $|2 - 3i| = \sqrt{2^2 + (-3)^2} = \sqrt{4 + 9} = \sqrt{13}$

(b) $|-5i| = \sqrt{0^2 + (-5)^2} = \sqrt{25} = 5$

(c) $|1 - i| = \sqrt{1^2 + (-1)^2} = \sqrt{1 + 1} = \sqrt{2}$

## Exercise Set 6.2

### Goal A

In exercises 1–14, write the reciprocal of each complex number in the form $a + bi$.

1. $2 + 3i$
2. $1 + 4i$
3. $5 - 2i$
4. $3 - 7i$
5. $-1 + i$
6. $-1 - i$
7. $-2 - 4i$
8. $6 - 3i$
9. $5i$
10. $-8i$
11. $-\frac{1}{7}i$
12. $\frac{2}{3}i$
13. $\sqrt{3} - 2i$
14. $-\sqrt{11} + 5i$

In exercises 15–34, simplify each quotient.

15. $\frac{6 + 2i}{2 - i}$
16. $\frac{8 + i}{2 - 3i}$
17. $\frac{2 + 3i}{1 + 4i}$
18. $\frac{1 + 4i}{3 + 2i}$
19. $\frac{5 - i}{2 + 5i}$
20. $\frac{3 - 4i}{6 + i}$
21. $\frac{3 - 10i}{1 - i}$
22. $\frac{6 + 5i}{1 + i}$
23. $\frac{4 - 2i}{1 + 2i}$
24. $\frac{3 + i}{1 - 3i}$
25. $\frac{34}{5 - 3i}$
26. $\frac{-53}{7 - 2i}$
27. $\frac{-2i}{5 + 4i}$
28. $\frac{10i}{3 + i}$
29. $\frac{3 - 7i}{-7i}$
30. $\frac{-9 + 8i}{3i}$
31. $\frac{1}{-6 + 2i}$
32. $\frac{1}{-2 - 4i}$
33. $\frac{6}{i}$
34. $\frac{11}{-i}$

### Goal B

In exercises 35–40, verify the following for the given values of the variables:
(a) $\overline{z + w} = \overline{z} + \overline{w}$ (b) $\overline{z - w} = \overline{z} - \overline{w}$
(c) $\overline{z \cdot w} = \overline{z} \cdot \overline{w}$ (d) $\overline{\left(\frac{z}{w}\right)} = \frac{\overline{z}}{\overline{w}}$

35. $z = 1 + 3i$ and $w = 2 - 4i$
36. $z = 5 + 2i$ and $w = 2 + 3i$
37. $z = 12 - 3i$ and $w = 5 + 9i$
38. $z = -11 + i$ and $w = 2 - 7i$
39. $z = -6 - 5i$ and $w = -1 - 4i$
40. $z = 9 - 9i$ and $w = -8 + 3i$

In exercises 41–46, verify the following for the given values of the variables: (a) $\overline{z^2} = (\overline{z})^2$; (b) $\overline{z^3} = (\overline{z})^3$.

41. $z = 2 - i$
42. $z = 3 + i$
43. $z = -2 + 6i$
44. $z = 3 + 4i$
45. $z = -5 - i$
46. $z = -2 - 8i$

### Goal C

In exercises 47–72, represent each complex number as a point in the complex plane.

47. $2 + 5i$
48. $3 + 2i$
49. $2 - 5i$
50. $3 - 2i$
51. $-2 + 5i$
52. $-3 + 2i$
53. $-2 - 5i$
54. $3 - i$
55. $1 + 4i$
56. $1 + 5i$
57. $-3 + i$
58. $-4 + i$
59. $4$
60. $5$
61. $-3$
62. $-6$
63. $6i$
64. $-2i$
65. $-i$
66. $i$
67. $\overline{5 - 4i}$
68. $\overline{2 - 6i}$
69. $\overline{4 + 2i}$
70. $\overline{5 + 5i}$
71. $\overline{-3 - 6i}$
72. $\overline{-4 - 3i}$

In exercises 73–94, determine the absolute value of each complex number.

73. $7 + 2i$
74. $5 + 3i$
75. $7 - 2i$
76. $5 - 3i$
77. $-8 + 6i$
78. $-3 + 4i$
79. $-8 - 6i$
80. $-3 - 4i$
81. $3 + 3i$
82. $4 + 2i$
83. $7 - i$
84. $-8 + i$
85. $2i$
86. $-7i$
87. $-i$
88. $i$

**89.** $\sqrt{3} + i\sqrt{13}$

**90.** $\sqrt{6} + i\sqrt{3}$

**91.** $3 - i\sqrt{7}$

**92.** $\sqrt{5} - 2i$

**93.** $\sqrt{14} + i$

**94.** $2 + i\sqrt{10}$

## Superset

In exercises 95–110, simplify each of the following expressions.

**95.** $\dfrac{1 + i}{(1 - 2i)^2}$

**96.** $\dfrac{1 - i}{(1 + 2i)^2}$

**97.** $\dfrac{5 - i}{(1 + i)(2 - 3i)}$

**98.** $\dfrac{-5 - i}{(1 - i)(2 + 3i)}$

**99.** $\dfrac{-1 + 8i}{(2 - i)(3 + 2i)}$

**100.** $\dfrac{14 + 2i}{(2 + i)(1 + 3i)}$

**101.** $\dfrac{4 - 3i}{(3 + 2i)(-1 + i)}$

**102.** $\dfrac{5 - i}{(3 - 2i)(1 + i)}$

**103.** $\dfrac{1}{2 + i} + \dfrac{-3}{1 - 3i}$

**104.** $\dfrac{-4}{1 + 3i} + \dfrac{2}{2 - i}$

**105.** $\dfrac{2 - 3i}{1 + i} - \dfrac{6 + 2i}{2 - i}$

**106.** $\dfrac{5 + 2i}{3 - i} - \dfrac{2 - 4i}{1 + i}$

**107.** $\left|\dfrac{1 - 2i}{3 + i}\right|$

**108.** $\left|\dfrac{1 + 2i}{3 - i}\right|$

**109.** $\dfrac{|1 - 2i|}{|3 + i|}$

**110.** $\dfrac{|1 + 2i|}{|3 - i|}$

In exercises 111–118, find all complex numbers $z$ that satisfy the given equation.

**111.** $(3 + 2i)z + i = 1 + 4i$

**112.** $(2 - 4i)z + 3i = -2 + 5i$

**113.** $(3 - 2i) + 5z = 3iz + (4 - 3i)$

**114.** $2iz - (8 - 5i) = z + (-10 + 7i)$

**115.** $2z + 5\overline{z} = 15 - 9i$

**116.** $7z - 3\overline{z} = -4 + 20i$

**117.** $z^2 = 8i$

**118.** $z^2 = -18i$

**119.** Let $z = a + bi$ and $w = c + di$. Show that $\overline{z - w} = \overline{z} - \overline{w}$.

**120.** Let $z = a + bi$ and $w = c + di$, where $w \neq 0$. Show that $\overline{\left(\dfrac{z}{w}\right)} = \dfrac{\overline{z}}{\overline{w}}$.

**121.** Let $z = a + bi$. Show that $\overline{\overline{z}} = z$.

**122.** Let $z = a + bi$. Show that $z \cdot \overline{z}$ is a real number.

**123.** Let $z = a + bi$. Show that $\frac{1}{2}(z + \overline{z}) = a$.

**124.** Let $z = a + bi$. Show that $\frac{1}{2}i(\overline{z} - z) = b$.

**125.** Let $z = a + bi$. Show that $|\overline{z}| = |z|$.

**126.** Let $z = a + bi$. Show that $|z|^2 = z \cdot \overline{z}$.

In exercises 127–130, determine whether each statement is true or false. If false, illustrate this with an example.

**127.** If $z$ and $w$ represent any complex numbers, then $|z - w| = |z| - |w|$.

**128.** If $z$ and $w$ represent any complex numbers, then $|z + w| = |z| + |w|$.

**129.** If $z$ and $w$ represent any complex numbers, then $|zw| = |z| \cdot |w|$.

**130.** If $z$ and $w$ represent any complex numbers, then $|z - w| = |w - z|$.

In exercises 131–138, represent $z$ and $\overline{z}$ as points in the complex plane.

**131.** $z = 3 + 5i$

**132.** $z = 4 + 2i$

**133.** $z = -3 + 5i$

**134.** $z = -4 + 2i$

**135.** $z = 2 - 6i$

**136.** $z = 1 - 3i$

**137.** $z = -2 - 6i$

**138.** $z = -1 - 3i$

**139.** Let $z = 2 + 5i$ and $w = 4 - 7i$. Represent $z$, $w$, and $z + w$ as points in the complex plane.

**140.** Repeat exercise 139 with the complex numbers $z = -2 + 7i$ and $w = 8 - 4i$.

**141.** Let $z = 2 + i$ and $w = 3 - 2i$. Represent $z$, $w$, and $zw$ as points in the complex plane.

**142.** Repeat exercise 141 with the complex numbers $z = -2 + 4i$ and $w = 1 + i$.

## 6.3 Zeros of Polynomial Functions

**Goal A** *to solve quadratic equations over the set of complex numbers*

We began this chapter by pointing out the need to extend the set of real numbers to the complex numbers so that any quadratic equation could be solved. The following examples show how to solve quadratic equations over the set of complex numbers.

**Example 1** Solve each of the equations over the set of complex numbers.

(a) $x^2 = -9$ (b) $x^2 - 4x + 13 = 0$ (c) $x^2 = 3x - 4$

***Solution*** (a) $x^2 = -9$

$x = \pm i\sqrt{9}$ ■ $-9$ has two square roots, $i\sqrt{9}$ and $-i\sqrt{9}$.

$x = \pm 3i$ ■ There are two solutions, $3i$ and $-3i$.

(b) $x^2 - 4x + 13 = 0$

$x = \dfrac{4 \pm \sqrt{16 - 52}}{2}$ ■ We used the quadratic formula with $a = 1$, $b = -4$, and $c = 13$.

$x = \dfrac{4}{2} \pm \dfrac{\sqrt{-36}}{2}$

$x = \dfrac{4}{2} \pm \dfrac{6i}{2}$

$x = 2 \pm 3i$ ■ There are two solutions, $2 + 3i$ and $2 - 3i$.

(c) $x^2 = 3x - 4$ ■ Write the given equation in the form $ax^2 + bx + c = 0$.

$x^2 - 3x + 4 = 0$

$x = \dfrac{3 \pm \sqrt{9 - 16}}{2}$ ■ We used the quadratic formula with $a = 1$, $b = -3$, and $c = 4$.

$x = \dfrac{3}{2} \pm \dfrac{\sqrt{-7}}{2}$

$x = \dfrac{3}{2} \pm \dfrac{i\sqrt{7}}{2}$ ■ Note the two solutions are $\frac{3}{2} + \frac{i\sqrt{7}}{2}$ and $\frac{3}{2} - \frac{i\sqrt{7}}{2}$.

Notice that in the preceding examples whenever $a + bi$ was a solution, so was $a - bi$. For any quadratic equation with real coefficients, if $a + bi$ is a solution, so is $a - bi$; that is, the complex roots of a quadratic equation with real coefficients come in conjugate pairs.

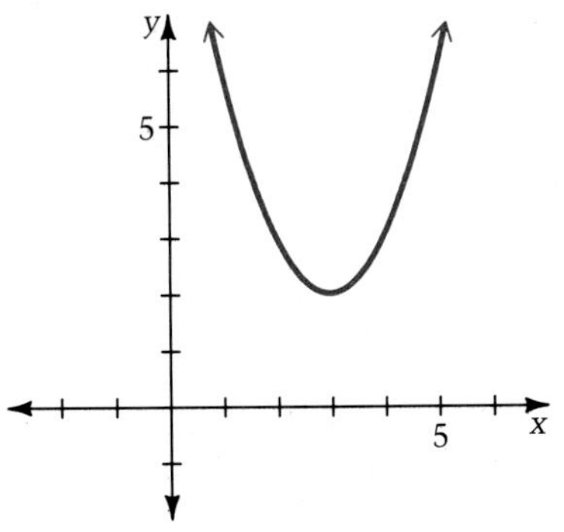


**Figure 6.8** $y = x^2 - 6x + 11$

**Goal B** *to determine the complex zeros of a quadratic function*

One of the reasons for defining the complex numbers is to provide a set of numbers over which any quadratic equation can be solved. Consider the graph of the function

$$y = x^2 - 6x + 11$$

shown in Figure 6.8. (Note that the function is graphed, as always, in the real plane.) The function $y = x^2 - 6x + 11$ has no real zeros. Since the $x$-intercepts of the graph of a function correspond to the real zeros of the function, the parabola at the left never intersects the $x$-axis.

To determine the zeros of the function

$$y = x^2 - 6x + 11,$$

we solve the equation $x^2 - 6x + 11 = 0$ for $x$.

$$x = \frac{-(-6) \pm \sqrt{36 - 44}}{2} = \frac{6 \pm \sqrt{-8}}{2} = \frac{6 \pm 2i\sqrt{2}}{2} = 3 \pm i\sqrt{2}$$

The two zeros of the function are imaginary numbers. Thus, we refer to them as **complex zeros.**

---

**Example 2** Determine all the zeros of each function.

(a) $f(x) = x^2 + 3x + 6$ (b) $f(x) = (x - 2)(2x^2 - x + 4)$

***Solution*** (a) $x^2 + 3x + 6 = 0$ ■ Set $f(x)$ equal to zero and solve.

$$x = -\frac{3}{2} \pm \frac{i\sqrt{15}}{2}$$

There are two complex zeros, $-\frac{3}{2} + \frac{i\sqrt{15}}{2}$ and $-\frac{3}{2} - \frac{i\sqrt{15}}{2}$.

(b) $(x - 2)(2x^2 - x + 4) = 0$ ■ Set $f(x)$ equal to zero and solve.

| $x - 2 = 0$ | $2x^2 - x + 4 = 0$ |
|---|---|
| $x = 2$ | $x = \dfrac{1 \pm \sqrt{1 - 32}}{2(2)}$ |
| | $x = \dfrac{1 \pm i\sqrt{31}}{4}$ |

There are three zeros: one real zero, 2; and two complex zeros, $\frac{1 \pm i\sqrt{31}}{4}$.

**Goal C** *to solve higher degree polynomial equations*

Recall that an $n$th degree polynomial equation of the form

$$a_nx^n + a_{n-1}x^{n-1} + \cdots + a_1x + a_0 = 0$$

can have at most $n$ real number solutions. If we solve such an equation over the set of complex numbers, however, we shall find that we must modify this statement. Consider the equation

$$x^3 - 8 = 0.$$

After factoring, we have

$$(x - 2)(x^2 + 2x + 4) = 0.$$

If we solve this over the set of real numbers, we obtain only one solution, $x = 2$. (The equation $x^2 + 2x + 4 = 0$ has no real solutions.) Now let us solve $x^3 - 8 = 0$ over the set of complex numbers.

$$(x - 2)(x^2 + 2x + 4) = 0$$

| $x - 2 = 0$ | $x^2 + 2x + 4 = 0$ |
|---|---|
| $x = 2$ | $x = \dfrac{-2 \pm \sqrt{4 - 16}}{2}$ |
| | $x = -1 \pm i\sqrt{3}$ |

Thus, we have found three solutions. It is no coincidence that we obtain three solutions when we solve a third degree polynomial equation over the set of complex numbers. We formalize this in the following rule.

**Rule**

Let

$$a_nx^n + a_{n-1}x^{n-1} + \cdots + a_1x + a_0 = 0$$

be a polynomial equation of degree $n > 0$ with real coefficients. The equation has exactly $n$ complex number solutions, provided each solution is counted according to its multiplicity.

Although we require that the polynomial equation in the above rule must have real coefficients, it can be shown that this rule is also true for polynomials with complex coefficients.

To understand the last sentence of the rule, let us look at the equation

$$(x - 7)^3(x + 5)^2(x^2 - x + 1) = 0.$$

Note that 7 is a solution of multiplicity three and $-5$ is a solution of multiplicity two. Thus, when counting solutions we should count 7 three times and $-5$ twice. Two more solutions of the equation are given by the factor $x^2 - x + 1$, namely $\frac{1}{2} \pm i\frac{\sqrt{3}}{2}$. Thus, the total number of solutions is seven. If you multiply the factors on the left side of the given equation, you will find that it is, as the rule predicts, a seventh degree polynomial.

**Example 3** Verify the previous rule for the equation $x(x - 1)^2(x^2 + 2x + 2) = 0$.

***Solution***

$$x(x - 1)^2(x^2 + 2x + 2) = 0$$

$$x = 0 \qquad \begin{aligned} (x - 1)^2 &= 0 \\ x &= 1 \end{aligned} \qquad \begin{aligned} x^2 + 2x + 2 &= 0 \\ x &= \frac{-2 \pm \sqrt{4 - 8}}{2} \\ x &= \frac{-2 \pm 2i}{2} \\ x &= -1 \pm i \end{aligned}$$

There are three roots of multiplicity one, 0, $-1 + i$, and $-1 - i$, and one root of multiplicity two, 1. Thus, there is a total of five roots. When the factors are multiplied, they yield $x^5 - x^3 - 2x^2 + 2x$, a fifth degree polynomial. The rule is therefore verified.

Recall that if you know that a number $c$ is a root of a polynomial equation $p(x) = 0$, then you also know that $x - c$ is a factor of $p(x)$. Dividing $p(x)$ by $x - c$ might then produce a quotient polynomial that is more easily factored than the original. This method is demonstrated in the following example.

**Example 4** The polynomial equation $x^4 - 4x^3 + 4x^2 + 3x - 6 = 0$ has 2 and $-1$ as solutions. Find all the solutions of the equation.

***Solution*** Since the equation is a fourth degree polynomial, we must find two roots (or one other root of multiplicity 2). Since 2 and $-1$ are solutions, $x - 2$ and $x + 1$ are factors of $x^4 - 4x^3 + 4x^2 + 3x - 6$. Thus, the product $(x - 2)(x + 1)$ or $x^2 - x - 2$ is also a factor. To determine the quotient polynomial we divide.

$$
\begin{array}{r}
x^2 - 3x + 3 \\
x^2 - x - 2\overline{)\,x^4 - 4x^3 + 4x^2 + 3x - 6} \\
\underline{-(x^4 - x^3 - 2x^2)} \\
-3x^3 + 6x^2 + 3x \\
\underline{-(-3x^3 + 3x^2 + 6x)} \\
3x^2 - 3x - 6 \\
\underline{-(3x^2 - 3x - 6)} \\
0
\end{array}
$$

Thus, $x^4 - 4x^3 + 4x^2 + 3x - 6 = (x^2 - x - 2)(x^2 - 3x + 3)$.

$$x^4 - 4x^3 + 4x^2 + 3x - 6 = 0$$
$$(x^2 - x - 2)(x^2 - 3x + 3) = 0$$

| $x^2 - x - 2 = 0$ | $x^2 - 3x + 3 = 0$ |
|---|---|
| $x = 2$ or $x = -1$ | $x = \dfrac{3 \pm \sqrt{9 - 12}}{2}$ |
| | $x = \dfrac{3}{2} \pm \dfrac{i\sqrt{3}}{2}$ |

■ Each factor is set equal to zero.

■ There are four solutions: 2, −1, $\frac{3}{2} + \frac{i\sqrt{3}}{2}$, and $\frac{3}{2} - \frac{i\sqrt{3}}{2}$.

For any polynomial equation with real coefficients, if a complex number $a + bi$ is a solution, then its conjugate $a - bi$ is also a solution. Thus, we say that the complex zeros of real polynomial equations come in conjugate pairs. The following example makes use of this fact.

**Example 5** Determine a third degree polynomial equation with real coefficients that has 3 and $2 - i$ as two of its solutions.

***Solution*** Since the polynomial must have real coefficients, the complex roots must come in conjugate pairs. Since we know that $2 - i$ is a root, then $2 + i$ must also be a root. To build a polynomial, we multiply the three known factors:

$$
\begin{aligned}
(x - 3)[x - (2 - i)][x - (2 + i)] &= (x - 3)(x^2 - (2 + i)x - (2 - i)x + 5) \\
&= (x - 3)(x^2 - 4x + 5) \\
&= x^3 - 7x^2 + 17x - 15
\end{aligned}
$$

A third degree polynomial equation with real coefficients and 3 and $2 - i$ as two of its solutions is $x^3 - 7x^2 + 17x - 15 = 0$.

We now consider a streamlined technique, called **synthetic division,** which is useful in dividing a polynomial by a first degree polynomial of the form $x - c$, for any real number $c$. Below on the left, we show a problem that employs the technique used in Example 4. On the right, we show the same process, but with three simplifications: (1) the variables have been omitted; (2) terms usually "brought down" for convenience of computation have been omitted; and (3) the first number in each of the products has been deleted since it invariably combines with the number above it to produce 0. We note that omitting the variables is legitimate since it is the coefficients alone that are needed in completing the division process. Of course, we must agree to write the coefficients in order of descending powers of the variable.

$$\begin{array}{rl}
 & 2x^2 + 5x - 1 \longleftarrow \textbf{quotient} \\
x - 3 \,\big)\!\!\!\! & \overline{\;2x^3 - x^2 - 16x + 7\;} \\
 & \underline{-(2x^3 - 6x^2)} \\
 & \quad 5x^2 - 16x \\
 & \quad \underline{-(5x^2 - 15x)} \\
 & \qquad -x + 7 \\
 & \qquad \underline{-(-x + 3)} \\
 & \qquad\qquad 4 \longleftarrow \textbf{remainder}
\end{array}
\qquad
\begin{array}{r|rrrr}
\textbf{quotient} \longrightarrow & 2 & 5 & -1 & \\
\hline
1 - 3\,) & 2 & -1 & -16 & 7 \\
 & & \underline{6} & & \\
 & & 5 & & \\
 & & & \underline{15} & \\
 & & & -1 & \\
 & & & & \underline{-3} \\
\textbf{remainder} \longrightarrow & & & & 4
\end{array}$$

We can further simplify the form on the right by moving all the numbers upward so as to occupy four lines. Since the divisor will always be of the form $x - c$, we need only write $c$ (in this case 3). We show this further simplification below on the left. Notice that the bottom row contains all but the first coefficient of the quotient (in this case 2), and ends with the remainder. By placing this first coefficient in the bottom row, that row will contain all the coefficients of the quotient, followed by the remainder. Thus, the top row is unnecessary. This final simplification is shown on the right below.

$$\begin{array}{rrrrr}
 & 2 & 5 & -1 & \\
\hline
3\,) & 2 & -1 & -16 & 7 \\
 & & 6 & 15 & -3 \\
\hline
 & & 5 & -1 & 4
\end{array}
\qquad\qquad
\begin{array}{r|rrrr}
3 & 2 & -1 & -16 & 7 \\
 & & 6 & 15 & -3 \\
\hline
 & 2 & 5 & -1 & 4
\end{array}$$

quotient with first coefficient missing — quotient — remainder

In the following example, we describe the steps involved in using synthetic division directly. Note that this technique relies only on the value of $c$ and the coefficients of the polynomial to be divided by $x - c$.

**Example 6** Use synthetic division to divide $2x^3 - x^2 - 16x + 7$ by $x - 3$.

***Solution*** Begin by writing the top row: $\underline{c|}$, from the divisor $x - c$ (in this case $c = 3$), followed by the coefficients of the dividend.

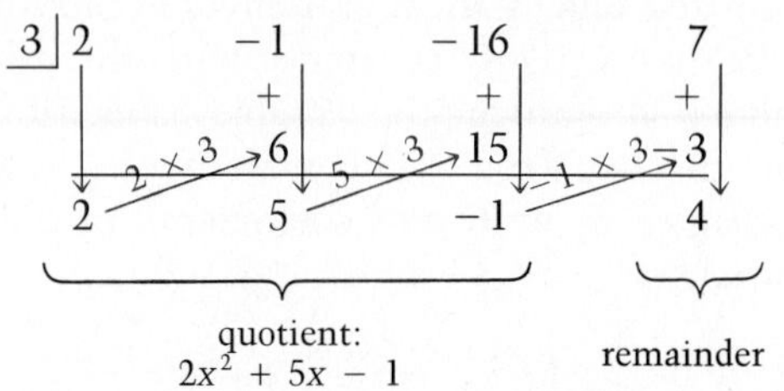


■ Bring down the first coefficient, 2. Then successively multiply by 3 and add to the next coefficient.

**Example 7** Use synthetic division to divide $x^4 - 3x^3 + 2x - 1$ by $x + 2$.

***Solution*** Since $x + 2 = x - (-2)$, in this case $c = -2$. For a missing power of $x$, record the coefficient 0.

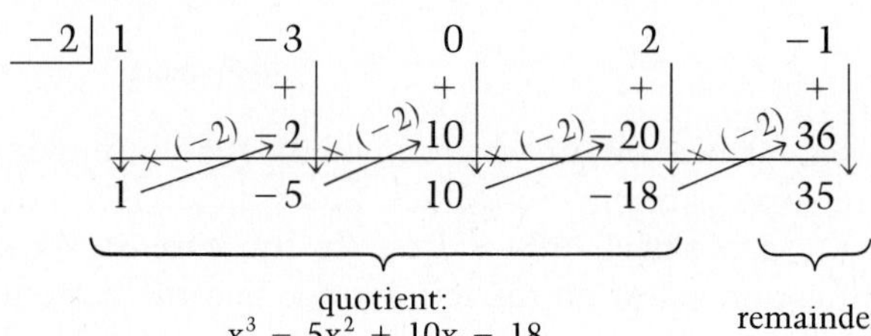


■ Since there is no $x^2$-term, its coefficient must be recorded as 0.

■ The degree of the quotient is one less than that of the divisor.

We leave as a Superset exercise the proof of the following.

**The Remainder Theorem**

If a polynomial $f(x)$ is divided by $x - c$, then the remainder is equal to $f(c)$.

By the Remainder Theorem, we can use synthetic division to determine function values. For example, if $f(x) = 2x^3 - x^2 - 16x + 7$, then $f(x) = 4$ (by Example 6), and if $g(x) = x^4 - 3x^3 + 2x - 1$, then $g(-2) = 35$ (by Example 7).

## Exercise Set 6.3

### Goal A

In exercises 1–12, solve each equation over the set of complex numbers.

**1.** $x^2 = -16$ **2.** $x^2 = -49$

**3.** $x^2 - 6x + 25 = 0$ **4.** $x^2 - 4x + 29 = 0$

**5.** $x^2 - 4x + 5 = 0$ **6.** $x^2 - 2x + 5 = 0$

**7.** $x^2 + 3x + 4 = 0$ **8.** $x^2 + x + 1 = 0$

**9.** $4x^2 - 8x + 5 = 0$ **10.** $4x^2 - 4x + 5 = 0$

**11.** $4x^2 + 8x + 11 = 0$ **12.** $3x^2 + 4x + 2 = 0$

### Goal B

In exercises 13–22, find all the zeros of each function.

**13.** $f(x) = 9x^2 + 64$ **14.** $f(x) = 4x^2 + 25$

**15.** $f(x) = 3x^2 - 4x + 2$ **16.** $f(x) = 3x^2 + x + 2$

**17.** $f(x) = -x^2 + 3x + 9$

**18.** $f(x) = -3x^2 - 3x - 1$

**19.** $f(x) = (x - 3)(x^2 - 6x + 11)$

**20.** $f(x) = (x + 8)(x^2 - 2x + 7)$

**21.** $f(x) = (3x - 1)(4x^2 + 5)$

**22.** $f(x) = (x - 11)(9x^2 + 10)$

### Goal C

In exercises 23–28, a polynomial equation and one or more of its solutions are given. Find all the solutions.

**23.** $x^3 - 4x^2 + 5x - 6 = 0;\ 3$

**24.** $x^3 + 8x^2 + 20x + 25 = 0;\ -5$

**25.** $x^4 + 5x^3 + 11x^2 + 13x + 6 = 0;\ -2, -1$

**26.** $x^4 - x^2 + 4x - 4 = 0;\ 1, -2$

**27.** $x^4 - 6x^3 + 13x^2 - 24x + 36 = 0;\ 2i$

**28.** $x^4 - 4x^3 + 16x^2 - 24x + 20 = 0;\ 1 + 3i$

**29.** Find a second degree polynomial equation with real coefficients that has $4 + 3i$ as one of its solutions.

**30.** Find a second degree polynomial equation with real coefficients that has $3 - 2i$ as one solution.

**31.** Find a third degree polynomial equation with real coefficients that has 4 and $-5i$ as solutions.

**32.** Find a third degree polynomial equation with real coefficients that has $2 - i\sqrt{5}$ and 4 as solutions.

In exercises 33–36, use synthetic division to verify the indicated function values.

**33.** $f(x) = 2x^3 - 3x^2 + 6x - 7;\ f(0) = -7, f(1) = -2$

**34.** $g(x) = x^4 - 3x^3 + 4x^2 - x - 1;\ g(-2) = 57, g(2) = 5$

**35.** $h(x) = x^5 - 3x^3 + x^2 - 4x + 10;\ h(-2) = 14, h(-1) = 17$

**36.** $G(x) = 1 - x + 8x^3 - 10x^5;\ G(-1) = 4, G(2) = -257$

### Superset

The Rational Root Theorem says the following: Let $a_nx^n + a_{n-1}x^{n-1} + \cdots + a_1x + a_0 = 0$ be a polynomial equation with integral coefficients. If this equation has any rational solutions, then they must be of the form $\frac{c}{d}$, where $c$ is a factor of $a_0$ and $d$ is a factor of $a_n$. In exercises 37–40, solve each equation over the set of complex numbers. (Hint: Look for rational solutions.)

**37.** $x^4 + x^3 + 2x^2 + 4x - 8 = 0$

**38.** $x^4 - 5x^3 + 8x^2 - 10x + 12 = 0$

**39.** $2x^4 + x^3 - 2x^2 - 4x - 3 = 0$

**40.** $3x^4 - x^3 + 3x - 1 = 0$

In exercises 41–46, solve each equation over the set of complex numbers.

**41.** $x^4 + 1 = 0$ **42.** $x^4 + 16 = 0$

**43.** $x^4 - 25 = 0$ **44.** $x^4 - 49 = 0$

**45.** $x^4 + 5x^2 - 20 = 0$ **46.** $x^4 + 2x^2 - 8 = 0$

**47.** The Division Algorithm says the following: When a polynomial $f(x)$ is divided by a nonzero polynomial $d(x)$, then there exist unique polynomials $q(x)$, the quotient, and $r(x)$, the remainder, such that $f(x) = d(x) \cdot q(x) + r(x)$, where $r(x) = 0$ or else the degree of $r(x)$ is one less than that of the divisor.

Use the Division Algorithm to prove the Remainder Theorem.

## 6.4 Trigonometric Form of Complex Numbers

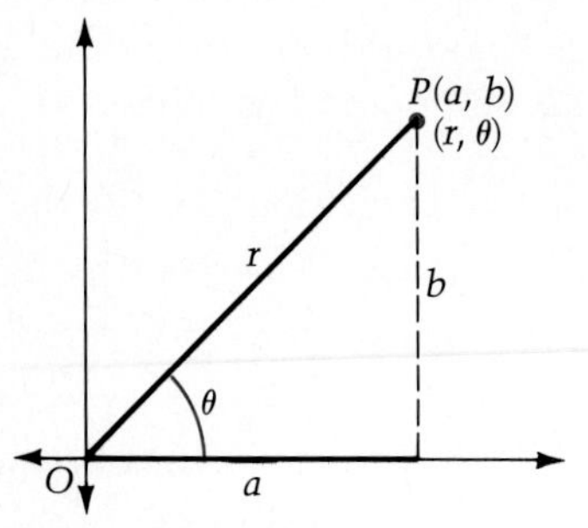

Figure 6.9

**Goal A** *to express a complex number in trigonometric form*

In the last chapter we described how each point $P(a, b)$ can be represented by polar coordinates $(r, \theta)$, where $r$ is the distance between $P$ and the origin, and $\theta$ is the angle that ray $OP$ makes with the positive $x$-axis. Any complex number $a + bi$ can be written in *polar* or *trigonometric form*, as illustrated below.

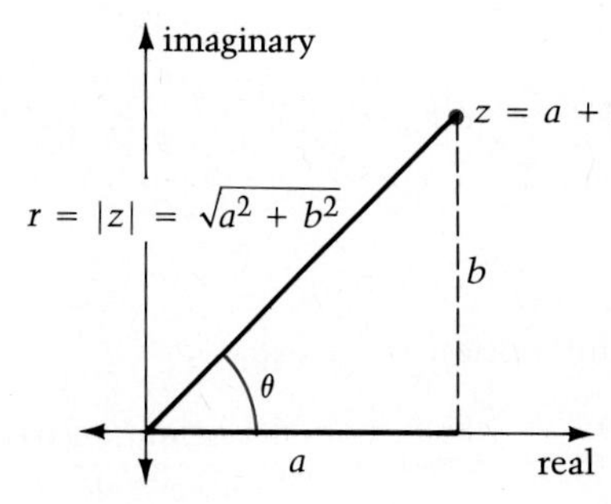

Figure 6.10

- $a = r \cos \theta$ since $\cos \theta = \dfrac{a}{r}$.
  $b = r \sin \theta$ since $\sin \theta = \dfrac{b}{r}$.
- Therefore, $z = (r \cos \theta) + (r \sin \theta)i$.

**Definition**
The complex number $z = a + bi$ has **trigonometric form**

$$z = r(\cos \theta + i \sin \theta)$$

where $r = |z| = \sqrt{a^2 + b^2}$, and $\tan \theta = \frac{b}{a}$. The number $r$ is called the **modulus** of $z$, and $\theta$ is called the **argument** of $z$.

**Example 1** Express each of the following complex numbers in trigonometric form. If rounding is necessary, express $\theta$ to the nearest multiple of $10'$.

(a) $2 + 2i$ (b) $-7 - 3i$

***Solution*** (a)

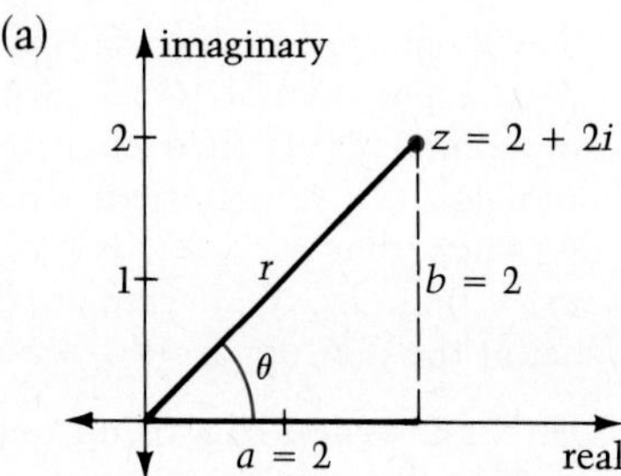


- Find $r$ and $\theta$, then apply the above definition.
- $r = |z| = \sqrt{2^2 + 2^2} = \sqrt{8} = 2\sqrt{2}$.
- Since $\tan \theta = \dfrac{2}{2} = 1$, and $\theta$ is a first quadrant angle, $\theta = \dfrac{\pi}{4}$.
- Thus, $z = 2\sqrt{2}\left(\cos \dfrac{\pi}{4} + i \sin \dfrac{\pi}{4}\right)$.

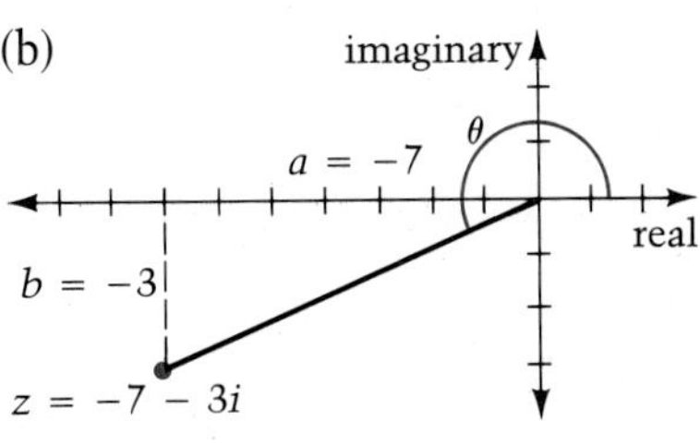


- $r = |z| = \sqrt{(-7)^2 + (-3)^2} = \sqrt{58}$.
- Since $\tan\theta = \dfrac{-3}{-7} \approx 0.4286$, $\theta$ is the third quadrant angle having 23°10′ (see Table 4) as its reference angle. Thus, $\theta \approx 180° + 23°10' = 203°10'$.

$$z = \sqrt{58}\,(\cos 203°10' + i \sin 203°10')$$

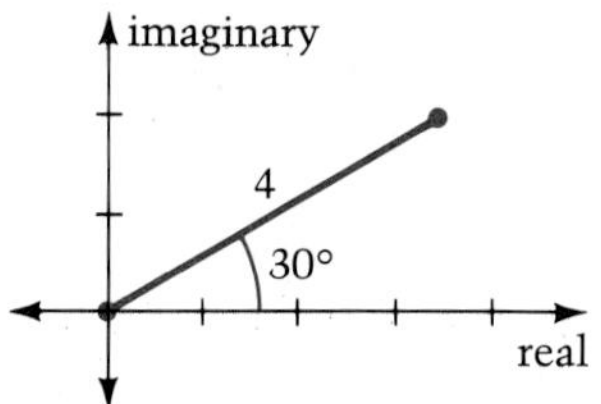

**Figure 6.11**

It is important to remember that the trigonometric form of a complex number is not unique. For example, all expressions of the form

$$z = 4[\cos(30° + k \cdot 360°) + i \sin(30° + k \cdot 360°)]$$

for every integer $k$, represent the same complex number, $2\sqrt{3} + 2i$.

**Goal B** *to determine the product and quotient of two complex numbers*

Suppose $z_1 = r_1(\cos\theta_1 + i \sin\theta_1)$ and $z_2 = r_2(\cos\theta_2 + i \sin\theta_2)$ are two complex numbers. Their product can be expressed as

$$\begin{aligned} z_1 z_2 &= r_1(\cos\theta_1 + i\sin\theta_1)\cdot r_2(\cos\theta_2 + i\sin\theta_2) \\ &= r_1r_2(\cos\theta_1\cos\theta_2 + i\cos\theta_1\sin\theta_2 \\ &\qquad + i\sin\theta_1\cos\theta_2 + i^2\sin\theta_1\sin\theta_2) \\ &= r_1r_2[\cos\theta_1\cos\theta_2 - \sin\theta_1\sin\theta_2 \\ &\qquad + i(\sin\theta_1\cos\theta_2 + \cos\theta_1\sin\theta_2)] \\ &= r_1r_2[\cos(\theta_1+\theta_2) + i\sin(\theta_1+\theta_2)] \end{aligned}$$

**Rule**

If $z_1 = r_1(\cos\theta_1 + i \sin\theta_1)$ and $z_2 = r_2(\cos\theta_2 + i \sin\theta_2)$ are two complex numbers, their product is given by

$$z_1z_2 = r_1r_2[\cos(\theta_1 + \theta_2) + i\sin(\theta_1 + \theta_2)].$$

The preceding rule tells us that the product of two complex numbers has modulus equal to the product of the two moduli, $r_1$ and $r_2$, and has argument equal to the sum of the two arguments $\theta_1$ and $\theta_2$.

**Example 2** Find $z_1z_2$ by using trigonometric forms. Express the answer in the form $a + bi$.
(a) $z_1 = 2(\cos 18° + i \sin 18°)$, $z_2 = 5(\cos 162° + i \sin 162°)$
(b) $z_1 = (-3 + 3i\sqrt{3})$, $z_2 = 1 - i\sqrt{3}$

***Solution*** (a) $z_1z_2 = 2 \cdot 5[\cos(18° + 162°) + i \sin(18° + 162°)]$
$= 10[\cos(180°) + i \sin(180°)] = 10[-1 + i(0)] = -10$

(b) First express each complex number in trigonometric form.

$$z_1 = 6\left(\cos \frac{2\pi}{3} + i \sin \frac{2\pi}{3}\right), \qquad z_2 = 2\left(\cos \frac{5\pi}{3} + i \sin \frac{5\pi}{3}\right)$$

$$z_1z_2 = 6 \cdot 2\left[\cos\left(\frac{2\pi}{3} + \frac{5\pi}{3}\right) + i \sin\left(\frac{2\pi}{3} + \frac{5\pi}{3}\right)\right] = 12\left[\cos\left(\frac{7\pi}{3}\right) + i \sin\left(\frac{7\pi}{3}\right)\right]$$
$$= 12\left(\cos \frac{\pi}{3} + i \sin \frac{\pi}{3}\right) = 12\left(\frac{1}{2} + i\frac{\sqrt{3}}{2}\right) = 6 + 6i\sqrt{3}$$

By using techniques of section 2, we can check the answer to part (b) of the above example.

$$\begin{aligned}(-3 + 3i\sqrt{3})(1 - i\sqrt{3}) &= (-3)(1) + (-3)(-i\sqrt{3}) \\ &\quad + (3i\sqrt{3})(1) + (3i\sqrt{3})(-i\sqrt{3}) \\ &= -3 + 3i\sqrt{3} + 3i\sqrt{3} - 9i^2 \\ &= (-3 + 9) + i(3\sqrt{3} + 3\sqrt{3}) = 6 + 6i\sqrt{3}\end{aligned}$$

In the exercises following this section, we suggest a way of proving the following rule for determining the quotient of two complex numbers.

**Rule**

If $z_1 = r_1(\cos \theta_1 + i \sin \theta_1)$ and $z_2 = r_2(\cos \theta_2 + i \sin \theta_2)$, then,

$$\frac{z_1}{z_2} = \frac{r_1}{r_2}[\cos(\theta_1 - \theta_2) + i \sin(\theta_1 - \theta_2)], \quad \text{for } z_2 \neq 0.$$

**Example 3** Find $\frac{z_1}{z_2}$ where $z_1 = 3 - i\sqrt{3}$ and $z_2 = 4 + 4i$ by dividing trigonometric forms. Express the answer in trigonometric form.

***Solution***

$$z_1 = 2\sqrt{3}\left(\cos \frac{11\pi}{6} + i \sin \frac{11\pi}{6}\right), \qquad z_2 = 4\sqrt{2}\left(\cos \frac{\pi}{4} + i \sin \frac{\pi}{4}\right)$$

■ Begin by expressing each complex number in trigonometric form.

$$\frac{z_1}{z_2} = \frac{2\sqrt{3}}{4\sqrt{2}}\left[\cos\left(\frac{11\pi}{6} - \frac{\pi}{4}\right) + i \sin\left(\frac{11\pi}{6} - \frac{\pi}{4}\right)\right]$$
$$= \frac{\sqrt{3}}{2\sqrt{2}} \cdot \frac{\sqrt{2}}{\sqrt{2}}\left[\cos\left(\frac{19\pi}{12}\right) + i \sin\left(\frac{19\pi}{12}\right)\right] = \frac{\sqrt{6}}{4}\left(\cos \frac{19\pi}{12} + i \sin \frac{19\pi}{12}\right)$$

At this point you should verify that the answer in Example 3 is the same complex number you would find if you applied the techniques of section 2 to the quotient $\left(\dfrac{3 - i\sqrt{3}}{4 + 4i}\right)$. You can determine $\cos \dfrac{19\pi}{12}$ and $\sin \dfrac{19\pi}{12}$ exactly by using the appropriate sum and difference identities.

**Goal C** *to determine powers and roots of a complex number*

By successively multiplying a complex number $z = r(\cos \theta + i \sin \theta)$ by itself, we observe a simple pattern:

$$\begin{aligned} z^2 &= r(\cos \theta + i \sin \theta) \cdot r(\cos \theta + i \sin \theta) \\ &= r^2[\cos(\theta + \theta) + i \sin(\theta + \theta)] \\ &= r^2(\cos 2\theta + i \sin 2\theta) \end{aligned}$$

$$\begin{aligned} z^3 = z^2 \cdot z &= r^2(\cos 2\theta + i \sin 2\theta) \cdot r(\cos\theta + i \sin \theta) \\ &= r^3(\cos 3\theta + i \sin 3\theta) \end{aligned}$$

This pattern holds for any positive integral power of $z$:

**De Moivre's Theorem**

For any positive integer $n$,

$$[r(\cos \theta + i \sin \theta)]^n = r^n(\cos n\theta + i \sin n\theta).$$

**Example 4** Use De Moivre's Theorem to express in the form $a + bi$: (a) $(-1 + i\sqrt{3})^5$ (b) $(2 + 2i)^6$

***Solution*** (a) $-1 + i\sqrt{3} = 2\left(\cos \dfrac{2\pi}{3} + i \sin \dfrac{2\pi}{3}\right)$ ■ $-1 + i\sqrt{3}$ is expressed in trigonometric form.

$$\begin{aligned} (-1 + i\sqrt{3})^5 &= \left[2\left(\cos \frac{2\pi}{3} + i \sin \frac{2\pi}{3}\right)\right]^5 = 2^5\left[\cos\left(5 \cdot \frac{2\pi}{3}\right) + i \sin\left(5 \cdot \frac{2\pi}{3}\right)\right] \\ &= 32\left(\cos \frac{10\pi}{3} + i \sin \frac{10\pi}{3}\right) = 32\left[-\frac{1}{2} + i\left(-\frac{\sqrt{3}}{2}\right)\right] = -16 - 16i\sqrt{3} \end{aligned}$$

(b) $2 + 2i = 2\sqrt{2}\left(\cos \dfrac{\pi}{4} + i \sin \dfrac{\pi}{4}\right)$ ■ $2 + 2i$ is expressed in trigonometric form.

$$\begin{aligned} (2 + 2i)^6 &= \left[2\sqrt{2}\left(\cos \frac{\pi}{4} + i \sin \frac{\pi}{4}\right)\right]^6 = (2\sqrt{2})^6\left[\cos\left(6 \cdot \frac{\pi}{4}\right) + i \sin\left(6 \cdot \frac{\pi}{4}\right)\right] \\ &= 512\left(\cos \frac{3\pi}{2} + i \sin \frac{3\pi}{2}\right) = 512[0 + i(-1)] = -512i \end{aligned}$$

In Example 4, we found that $(2 + 2i)^6 = -512i$; that is, $2 + 2i$ is a sixth root of $-512i$. In general, if $w$ and $z$ are complex numbers such that $w^n = z$, then $w$ is called an ***n*th root** of $z$.

We now consider how to determine such roots. For example, to find a cube root of $z = 1 - i\sqrt{3}$, we need $w = r(\cos\theta + i\sin\theta)$ such that

$$(1) \qquad [r(\cos\theta + i\sin\theta)]^3 = 1 - i\sqrt{3}.$$

By De Moivre's Theorem, $[r(\cos\theta + i\sin\theta)]^3 = r^3(\cos 3\theta + i\sin 3\theta)$. Also, we can rewrite $1 - i\sqrt{3}$ in general trigonometric form:

$$1 - i\sqrt{3} = 2[\cos(300° + k\cdot 360°) + i\sin(300° + k\cdot 360°)].$$

Thus, equation (1) can be rewritten as follows:

$$r^3(\cos 3\theta + i\sin 3\theta) = 2[\cos(300° + k\cdot 360°) + i\sin(300° + k\cdot 360°)].$$

We now determine $r$ and $\theta$ so that we can write $w$ in trigonometric form. It seems reasonable to set $r^3 = 2$ and $3\theta = 300° + k\cdot 360°$. Thus,

$$r = \sqrt[3]{2} \quad\text{and}\quad \theta = 100° + k\cdot 120°.$$

If

$$k = 0, \qquad w_0 = \sqrt[3]{2}(\cos 100° + i\sin 100°),$$
$$k = 1, \qquad w_1 = \sqrt[3]{2}(\cos 220° + i\sin 220°),$$
$$k = 2, \qquad w_2 = \sqrt[3]{2}(\cos 340° + i\sin 340°).$$

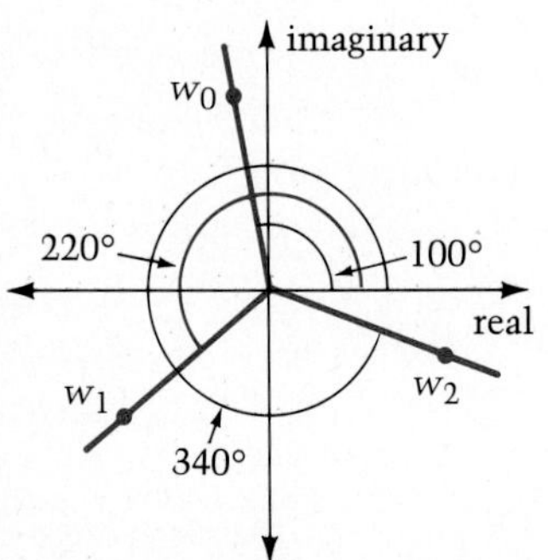


**Figure 6.12** The three cube roots of $1 - i\sqrt{3}$. Note that they are equally spaced (in 120° increments) and lie at the same distance ($\sqrt[3]{2}$) from the origin.

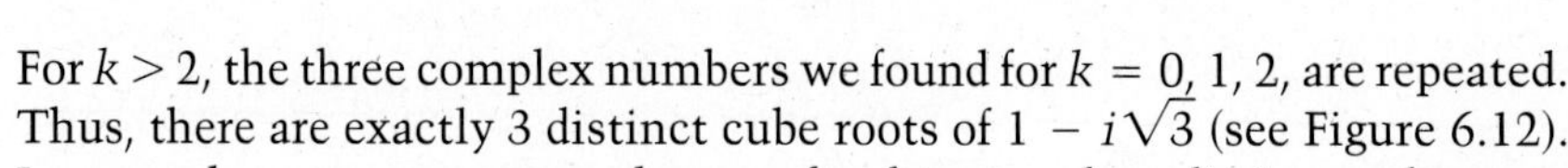

For $k > 2$, the three complex numbers we found for $k = 0, 1, 2$, are repeated. Thus, there are exactly 3 distinct cube roots of $1 - i\sqrt{3}$ (see Figure 6.12). In general, any nonzero complex number has exactly $n$ distinct $n$th roots.

**The *n*th Root Theorem**

The $n$th roots of the nonzero complex number $r(\cos\theta + i\sin\theta)$ are given by

$$\sqrt[n]{r}\left[\cos\left(\frac{\theta + k\cdot 360°}{n}\right) + i\sin\left(\frac{\theta + k\cdot 360°}{n}\right)\right]$$

where $k = 0, 1, 2, \ldots, n - 1$. We can replace 360° with $2\pi$ in the above statement to accommodate radian measure.

---

**Example 5** Find and plot the 4 fourth roots of $16(\cos 120° + i\sin 120°)$.

***Solution*** By the $n$th Root Theorem, there are exactly 4 fourth roots of this complex number:

$$w_0 = \sqrt[4]{16}\left(\cos\frac{120° + 0\cdot 360°}{4} + i\sin\frac{120° + 0\cdot 360°}{4}\right) = 2(\cos 30° + i\sin 30°)$$

$$w_1 = \sqrt[4]{16}\left(\cos \frac{120° + 1 \cdot 360°}{4} + i \sin \frac{120° + 1 \cdot 360°}{4}\right) = 2(\cos 120° + i \sin 120°)$$

$$w_2 = \sqrt[4]{16}\left(\cos \frac{120° + 2 \cdot 360°}{4} + i \sin \frac{120° + 2 \cdot 360°}{4}\right) = 2(\cos 210° + i \sin 210°)$$

$$w_3 = \sqrt[4]{16}\left(\cos \frac{120° + 3 \cdot 360°}{4} + i \sin \frac{120° + 3 \cdot 360°}{4}\right) = 2(\cos 300° + i \sin 300°)$$

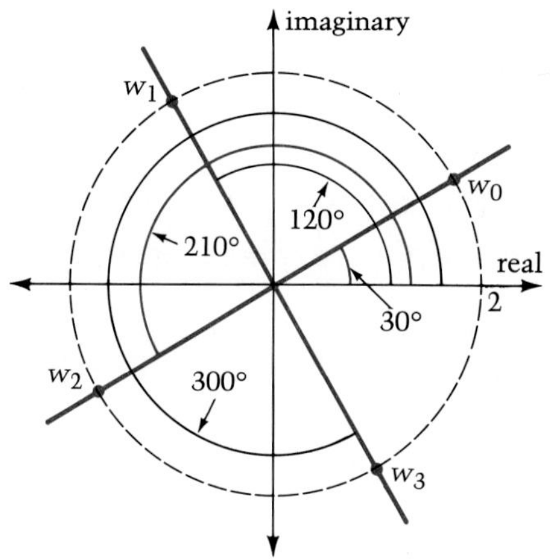


■ Note that the four fourth roots of $16(\cos 120° + i \sin 120°)$ are equally spaced $\left(\text{at } \frac{360°}{4} = 90° \text{ intervals}\right)$ on a circle centered at the origin, having radius 2.

If $z = 1$, then we refer to the $n$ $n$th roots of $z$ as the ***n*th roots of unity.**

**Example 6** Determine the sixth roots of unity.

***Solution*** First write $z = 1$ in trigonometric form. Since $z = 1 + 0i$, then $r = \sqrt{1^2 + 0^2} = 1$, and $\theta = 0°$. Thus $z = 1(\cos 0° + i \sin 0°)$. Each root has modulus $\sqrt[6]{1} = 1$, and argument determined by $\dfrac{0° + k \cdot 360°}{6}$ for $k = 0, 1, 2, 3, 4,$ and 5.

$$w_0 = 1(\cos 0° + i \sin 0°) = 1 \qquad w_1 = 1(\cos 60° + i \sin 60°) = \frac{1}{2} + \frac{\sqrt{3}}{2}i$$

$$w_2 = 1(\cos 120° + i \sin 120°) = -\frac{1}{2} + \frac{\sqrt{3}}{2}i \qquad w_3 = 1(\cos 180° + i \sin 180°) = -1$$

$$w_4 = 1(\cos 240° + i \sin 240°) = -\frac{1}{2} - \frac{\sqrt{3}}{2}i$$

$$w_5 = 1(\cos 300° + i \sin 300°) = \frac{1}{2} - \frac{\sqrt{3}}{2}i$$

## Exercise Set 6.4

### Goal A

In exercises 1–20, express each of the following complex numbers in trigonometric form.

**1.** $3i$ **2.** $-2i$ **3.** $-6$

**4.** $8$ **5.** $-1 + i\sqrt{3}$ **6.** $-\sqrt{3} + i$

**7.** $-2 - 2i$ **8.** $2 - 2i$ **9.** $-1 - i$

**10.** $1 + i$ **11.** $-3 + i\sqrt{3}$ **12.** $-3 - i\sqrt{3}$

**13.** $2\sqrt{3} - 2i$ **14.** $-2 + 2i\sqrt{3}$

**15.** $-3\sqrt{2} - 3i\sqrt{2}$ **16.** $3\sqrt{2} - 3i\sqrt{2}$

**17.** $5 + 12i$ **18.** $-8 + 15i$

**19.** $-24 - 7i$ **20.** $-7 - 4i$

### Goal B

In exercises 21–28, determine the product by multiplying the trigonometric forms. Express the answer in the form $a + bi$.

**21.** $(-2 + 2i)(-2 - 2i)$

**22.** $(-\sqrt{3} + i)(\sqrt{3} - i)$

**23.** $(3 + i\sqrt{3})(1 - i\sqrt{3})$

**24.** $(1 - i\sqrt{3})(-3 - i\sqrt{3})$

**25.** $(\sqrt{3} + i\sqrt{3})(-2 + 2i)$

**26.** $(3\sqrt{3} + 3i)(3 + 3i\sqrt{3})$

**27.** $(3 - 4i)(3i)$

**28.** $(5 + 12i)(-2i)$

In exercises 29–36, find $\frac{z_1}{z_2}$ by using the rule for dividing trigonometric forms.

**29.** $z_1 = 6(\cos 300° + i \sin 300°)$
$z_2 = 3(\cos 135° + i \sin 135°)$

**30.** $z_1 = 8\left(\cos \frac{7\pi}{6} + i \sin \frac{7\pi}{6}\right)$
$z_2 = 4\left(\cos \frac{2\pi}{3} + i \sin \frac{2\pi}{3}\right)$

**31.** $z_1 = 1 + 3i,\ z_2 = 1 - i$

**32.** $z_1 = 1 - i\sqrt{3},\ z_2 = 1 + i$

**33.** $z_1 = \sqrt{3} - 3i,\ z_2 = \sqrt{3} + i$

**34.** $z_1 = -3 - i\sqrt{3},\ z_2 = 1 - i$

**35.** $z_1 = 2 + 2i,\ z_2 = 3 - i\sqrt{3}$

**36.** $z_1 = -2 - 2i,\ z_2 = -3 + 3i$

### Goal C

In exercises 37–42, use De Moivre's Theorem to express in the form $a + bi$.

**37.** $(-2 + 2i)^4$ **38.** $(\sqrt{3} - i)^5$

**39.** $(\sqrt{3} + i)^3$ **40.** $(-1 - i)^6$

**41.** $(2\sqrt{3} - 2i)^{12}$ **42.** $(-3 - i\sqrt{3})^{10}$

In exercises 43–44, find the three cube roots of $z$.

**43.** $z = 3i$ **44.** $z = -2i$

In exercises 45–46, find the four fourth roots of $z$.

**45.** $z = -2 + 2i$ **46.** $z = -\sqrt{3} + i$

In exercises 47–48, find the five fifth roots of $z$.

**47.** $z = 1 + i$ **48.** $z = 1 + i\sqrt{3}$

**49.** Find (a) the fifth roots of unity, (b) the fourth roots of unity, (c) the eighth roots of unity.

**50.** Find (a) the eighth roots of $-1$, (b) the fifth roots of 32, (c) the fourth roots of 16.

### Superset

In exercises 51–56, solve the equation over the set of complex numbers.

**51.** $x^2 + 2x + 3 = 0$ **52.** $x^2 + 4x + 5 = 0$

**53.** $x^3 - 2i = 0$ **54.** $x^3 + 1 = 0$

**55.** $x^5 + 32 = 0$ **56.** $x^6 - 64 = 0$

**57.** Prove that if $z = r(\cos\theta + i\sin\theta)$, $\overline{z} = r(\cos(-\theta) + i\sin(-\theta))$.

**58.** Prove that if $z = r(\cos\theta + i\sin\theta)$, $\frac{1}{z} = \frac{1}{r}(\cos\theta - i\sin\theta)$ for $z \neq 0 + 0i$. Use this result to prove the rule for dividing two complex numbers.

# Chapter Review & Test

## Chapter Review

### 6.1 Introduction to Complex Numbers (pp. 206–215)

The number $i$ has the property that $i^2 = -1$. For this reason, $i$ can be considered a square root of $-1$. Thus, we write $i = \sqrt{-1}$. (p. 207)

If $p$ is a positive real number, then $\sqrt{-p} = i\sqrt{p}$. Note that $-p$ has two square roots, $i\sqrt{p}$ and $-i\sqrt{p}$. We refer to $i\sqrt{p}$ as the *principal square root of* $-p$. (p. 208)

If $a$ and $b$ are real numbers, then any number of the form $a + bi$ is called a *complex number.* The number $a$ is called the *real part* of the complex number, and $b$ is called the *imaginary part.* (p. 209)

If $a + bi$ and $c + di$ are complex numbers, then

$$(a + bi) + (c + di) = (a + c) + (b + d)i,$$
$$(a + bi) - (c + di) = (a - c) + (b - d)i, \quad \text{and}$$
$$(a + bi)(c + di) = (ac - bd) + (ad + bc)i. \text{ (p. 212)}$$

### 6.2 Properties of Complex Numbers (p. 216–223)

In general, for real numbers $a$ and $b$, the *complex conjugate* of $a + bi$ is $a - bi$. Since $(a + bi)(a - bi) = a^2 + b^2$, the product of a complex number and its conjugate is a real number. (p. 216)

Let $z$ be a complex number $a + bi$. Then $\overline{z}$ represents the complex conjugate $a - bi$. (p. 218)

If $\overline{z}$ and $\overline{w}$ are complex conjugates of the complex numbers $z$ and $w$, respectively, then:

**1.** $\overline{z + w} = \overline{z} + \overline{w}$ **2.** $\overline{z - w} = \overline{z} - \overline{w}$ **3.** $\overline{z \cdot w} = \overline{z} \cdot \overline{w}$

**4.** $\overline{\left(\frac{z}{w}\right)} = \frac{\overline{z}}{\overline{w}}$, for $w \neq 0$ **5.** $\overline{z^n} = (\overline{z})^n$, for every positive integer $n$

**6.** $\overline{z} = z$, if $z$ is a real number

### 6.3 Zeros of Polynomial Functions (pp. 224–229)

Let $a_nx^n + a_{n-1}x^{n-1} + \cdots + a_1x^1 + a_0 = 0$ be a polynomial equation of degree $n > 0$ with real coefficients. The equation has exactly $n$ complex number solutions, provided each solution is counted according to its multiplicity. (p. 226)

The complex zeros of real polynomial equations come in conjugate pairs. (p. 228)

### 6.4 Trigonometric Form of Complex Numbers (pp. 230–236)

The complex number $z = a + bi$ has trigonometric form

$$z = r(\cos\theta + i\sin\theta),$$

where $r = \sqrt{a^2 + b^2}$ and $\tan\theta = \frac{b}{a}$. The number $r$ is called the *modulus* of $z$ and $\theta$ is called the *argument* of $z$. (p. 230)

If $z_1 = r_1(\cos\theta_1 + i\sin\theta_1)$ and $z_2 = r_2(\cos\theta_2 + i\sin\theta_2)$, then

$$z_1 z_2 = r_1 r_2[\cos(\theta_1 + \theta_2) + i\sin(\theta_1 + \theta_2)],$$

$$\frac{z_1}{z_2} = \frac{r_1}{r_2}[\cos(\theta_1 - \theta_2) + i\sin(\theta_1 - \theta_2) \quad \text{for } z_2 \neq 0.$$

*De Moivre's Theorem:* For any positive integer $n$,

$$[r(\cos\theta + i\sin\theta)]^n = r^n(\cos n\theta + i\sin n\theta). \qquad \text{(p. 233)}$$

*The nth Root Theorem:* The $n$ $n$th roots of the nonzero complex number

$$r(\cos\theta + i\sin\theta)$$

are given by

$$\sqrt[n]{r}\left[\cos\left(\frac{\theta + k\cdot 360^\circ}{n}\right) + i\sin\left(\frac{\theta + k\cdot 360^\circ}{n}\right)\right]$$

where $k = 0, 1, 2, \ldots, n - 1$. We can replace $360^\circ$ with $2\pi$ in the above statement to accommodate radian measure. (p. 234)

If $z = 1$, then the $n$ $n$th roots of $z$ are referred to as the *nth roots of unity.* (p. 235)

## Chapter Test

**6.1A** Use the number $i$ to rewrite each of the following.

**1.** $\sqrt{-11}$ **2.** $\sqrt{-49}$

**3.** $\sqrt{-48}$ **4.** $-\sqrt{-28}$

Determine the real part and the imaginary part of each complex number.

**5.** $3 - 9i$ **6.** $2 + \frac{1}{6}i$

**7.** $-3i$ **8.** $\sqrt{19}$

**6.1B** Simplify each expression.

**9.** $(-i)^{27}$

**10.** $\sqrt{-3} \cdot \sqrt{-21} \cdot \sqrt{-7}$

**11.** $(9 - 5i) + (-3 - 4i)$

**12.** $(6 - i)(7 + 3i)$

**13.** $(\sqrt{8} + i)(\sqrt{8} - i)$

**14.** $(4i)^2(2 - 2i)$

**6.2A** Write the reciprocal of each complex number in the form $a + bi$.

**15.** $-8 + i$

**16.** $4 - i\sqrt{3}$

Simplify each quotient.

**17.** $\dfrac{9 + i}{2 - 3i}$

**18.** $\dfrac{10 - 5i}{-4i}$

**6.2B** Verify the following for the given values of the variables:

(a) $\overline{z + w} = \overline{z} + \overline{w}$

(b) $\overline{z - w} = \overline{z} - \overline{w}$

(c) $\overline{z \cdot w} = \overline{z} \cdot \overline{w}$

(d) $\overline{\left(\dfrac{z}{w}\right)} = \dfrac{\overline{z}}{\overline{w}}$

**19.** $z = 3 + i$ and $w = 4 - 3i$

**20.** $z = -1 - i$ and $w = 10 + 8i$

Verify the following for the given values of the variables:

(a) $\overline{z^2} = (\overline{z})^2$

(b) $\overline{z^3} = (\overline{z})^3$

**21.** $z = 6 + i$

**22.** $z = 1 - 3i$

**6.2C** Represent each complex number as a point in the complex plane.

**23.** $5 - 12i$

**24.** $-9$

**25.** $-7i$

**26.** $-6 + 3i$

Determine the absolute value of each complex number.

**27.** $4 + 5i$

**28.** $7 - 8i$

**29.** $-9i$

**30.** $-3 - i\sqrt{4}$

**6.3A** Solve each equation over the set of complex numbers.

**31.** $x^2 + 29 = 10x$

**32.** $-6x - 4 = 9x^2$

**6.3B** Determine all the zeros of each of the following functions.

**33.** $f(x) = x^2 + x + 3$

**34.** $f(x) = 2x^2 - 5x + 4$

**35.** $f(x) = (3x^2 + 7x - 6)(2x^2 - x + 4)$

**36.** $f(x) = (x - 5)^2(-2x^2 + 4x - 3)$

**6.3C** Verify the rule in Goal C for each polynomial equation.

**37.** $(x - 4)(x + 3)(x^2 - 5x + 11) = 0$

**38.** $x^5(2x - 1)(x + 3) = 0$

**39.** $x(x - 2)^3(x^2 + 4x + 16) = 0$

**40.** $x^4(x - 1)(x^3 + 27)^2 = 0$

In each of the following, a polynomial equation and one of its solutions are given. Find all the solutions of each polynomial equation.

**41.** $x^3 - 4x^2 + 6x - 4 = 0$; 2

**42.** $x^4 - 6x^3 + 13x - 24x + 36 = 0$; $2i$

**6.4A** Express each of the following complex numbers in trigonometric form.

**43.** $3 - i\sqrt{3}$

**44.** $-4 - 4i$

**6.4B** Find $z_1z_2$ and $\frac{z_1}{z_2}$ for each of the following complex numbers by using trigonometric forms. Express the answer in the form $a + bi$.

**45.** $z_1 = 6\sqrt{3} + 6i, z_2 = 1 - i\sqrt{3}$

**46.** $z_1 = -2 + 5i, z_2 = 7 - 2i$

**6.4C** Use De Moivre's Theorem to express in the form $a + bi$.

**47.** $\left(-\frac{\sqrt{3}}{2} + \frac{1}{2}i\right)^5$

**48.** $\left(-\frac{1}{2} + \frac{\sqrt{3}}{2}i\right)^4$

## Superset

**49.** Find real numbers $x$ and $y$ that satisfy each equation.

(a) $6 + 2xi = -3y - 10i$

(b) $(3x - y) + 2i = 9 + (x + 2y)i$

**50.** Simplify each expression.

(a) $\frac{1 + 5i}{(1 - 4i)^2}$

(b) $\frac{4 + i}{1 - i} - \frac{1 - 3i}{2 + i}$

**51.** Find all complex numbers $z$ that satisfy the equation $(2 - 7i)z + 4i = 5 - 3i$.

**52.** Let $z = 3 + i$ and $w = -2 - 5i$. Represent $z$, $w$, and $z + w$ as points in the complex plane.

**53.** Solve $x^4 + 81 = 0$ over the set of complex numbers.

# 7 Exponential and Logarithmic Functions

## 7.1 Exponential Functions

**Goal A** *to evaluate and simplify expressions containing exponents*

Our work with the functions in this chapter will rely upon an understanding of the rules of exponents. These rules are listed below. Even though we have defined $b^x$ for rational exponents only, the expression $b^x$ makes sense even when $x$ is not a rational number.

**Rules of Real Number Exponents**

**Rule 1** $b^0 = 1$ **Rule 2** $b^1 = b$

**Rule 3** $b^x b^y = b^{x+y}$ **Rule 4** $(b^x)^t = b^{xt}$

**Rule 5** $\dfrac{b^x}{b^y} = b^{x-y}$ **Rule 6** $b^{-x} = \dfrac{1}{b^x}$

where the base $b$ is positive, and the exponents $x$, $y$, and $t$ represent any real numbers.

**Example 1** Simplify the following expressions.

(a) $2(2^x)^3$ (b) $\dfrac{3^{x+1}}{3^2}$ (c) $4^{x+5} \cdot 8^{1-x}$

***Solution*** (a) $2(2^x)^3 = 2(2^{3x}) = 2^{3x+1}$

■ Rules 4 and 3 of Real Number Exponents are used. Note that $(2^x)^3$ is not the same as $2^{x+3}$.

(b) $\dfrac{3^{x+1}}{3^2} = 3^{(x+1)-2} = 3^{x-1}$ ■ Rule 5 of Real Number Exponents is used.

(c) $4^{x+5} \cdot 8^{1-x} = (2^2)^{x+5}(2^3)^{1-x}$ ■ Begin by rewriting each factor with the same base.

$= 2^{2x+10}2^{3-3x}$

$= 2^{-x+13}$

**Goal B** *to sketch the graph of an exponential function*

We now consider functions that are described by equations containing exponential expressions. These functions are used to solve problems involving population growth, compound interest, and radioactive decay.

**Definition**

An **exponential function** is a function of the form

$$f(x) = b^x,$$

where the base $b$ is any positive real number except 1.

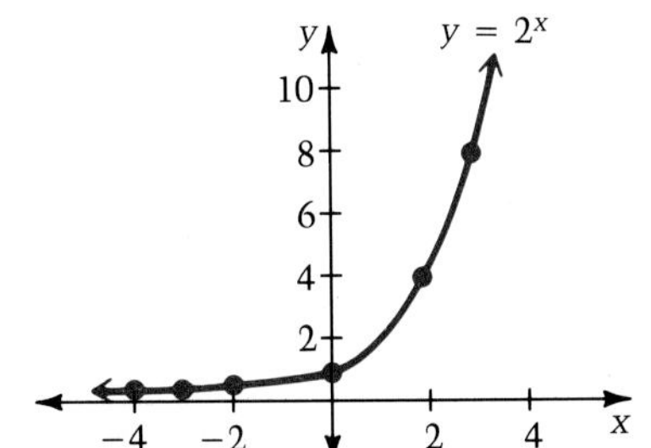


Figure 7.1

To sketch the graph of the exponential function $f(x) = 2^x$, we begin with a table of values.

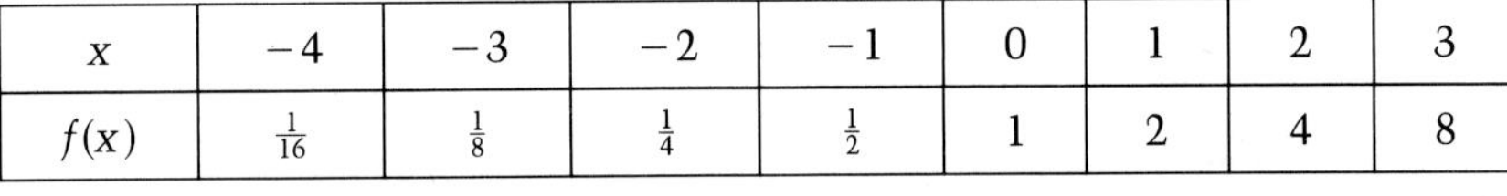

| $x$ | $-4$ | $-3$ | $-2$ | $-1$ | 0 | 1 | 2 | 3 |
|---|---|---|---|---|---|---|---|---|
| $f(x)$ | $\frac{1}{16}$ | $\frac{1}{8}$ | $\frac{1}{4}$ | $\frac{1}{2}$ | 1 | 2 | 4 | 8 |

We plot these points and draw a smooth curve as shown in Figure 7.1. Notice that as $x \to -\infty$, the corresponding values of $f(x)$ remain positive but get very close to zero. Thus, the $x$-axis is a horizontal asymptote.

Next we sketch the graph of $f(x) = (\frac{1}{2})^x$. By Rule 6 of Exponents we know that

$$\left(\frac{1}{2}\right)^x = \frac{1}{2^x} = 2^{-x}$$

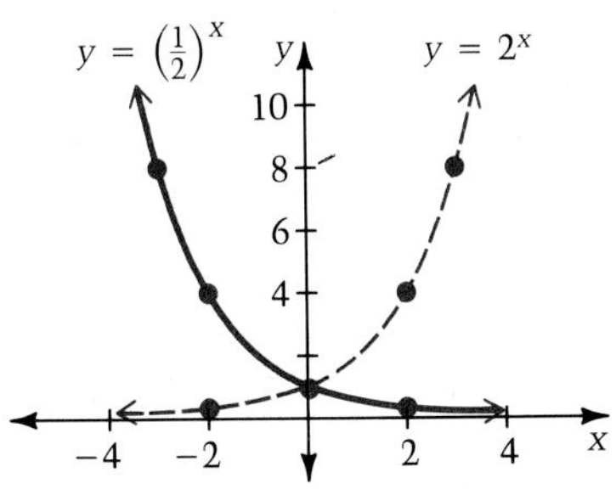


Figure 7.2

How does the graph of $y = 2^{-x}$ compare with the graph of $y = 2^x$? Recall that in general, replacing $x$ with $-x$ in an equation reflects a graph across the $y$-axis. Thus, $f(x) = (\frac{1}{2})^x$ (or $y = 2^{-x}$) is the reflection of $y = 2^x$ across the $y$-axis. We generalize these graphing results in Figure 7.3 for any positive base $b$ not equal to 1.

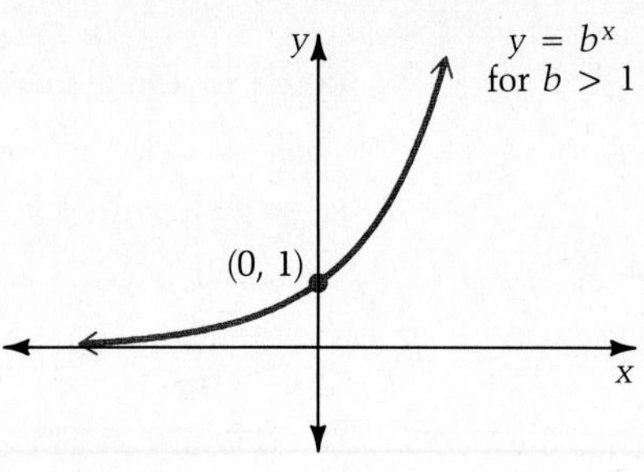

(a) Domain: $(-\infty, +\infty)$, Range: $(0, +\infty)$

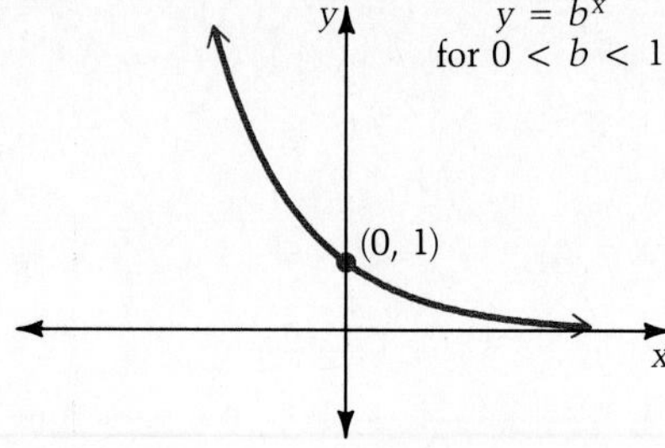

(b) Domain: $(-\infty, +\infty)$, Range: $(0, +\infty)$

**Figure 7.3**

In either case, the point (0, 1) is on the graph of $y = b^x$.

**Example 2** Sketch the graphs of each set of equations on a single set of coordinate axes.

(a) $y = 2^x$, $y = 3^x$, and $y = 10^x$ (b) $y = \left(\frac{1}{2}\right)^x$ and $y = \left(\frac{1}{3}\right)^x$

***Solution*** (a)

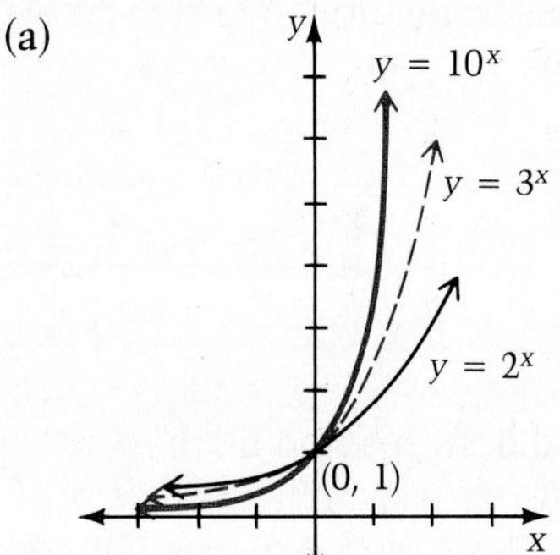


(b)

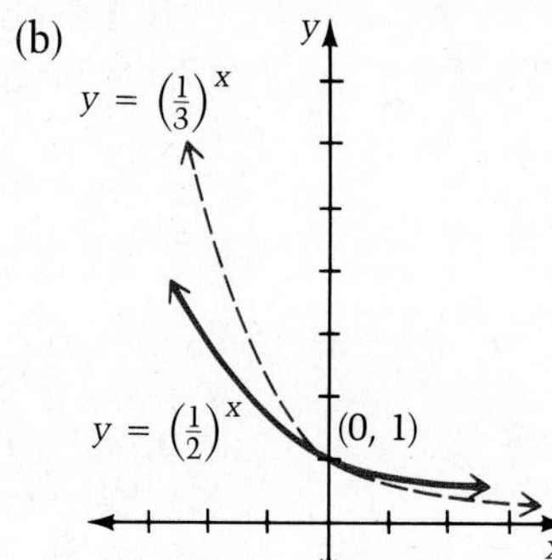


These graphs illustrate four important features of the function $f(x) = b^x$, where $b > 0$ and $b \neq 1$.

- If $x > 0$, then as the base $b$ increases, $b^x$ also increases.
- If $x < 0$, then as the base $b$ increases, $b^x$ decreases.
- For any base $b$, $b^0 = 1$.
- The $x$-axis is a horizontal asymptote.

**Example 3** Sketch the graph of $y = 2^x - 4$.

***Solution***

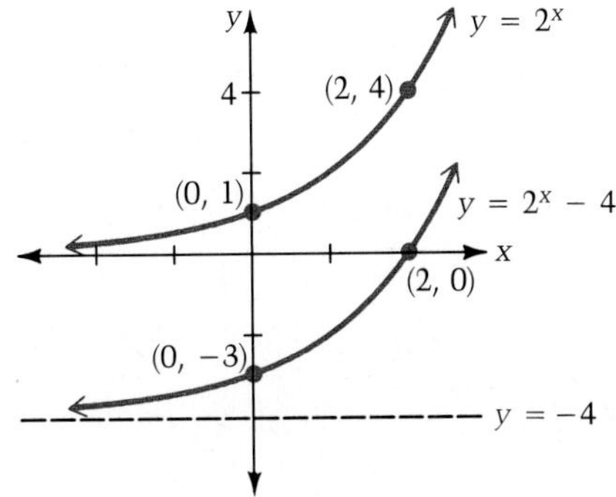


- Recall that the graph of $y = f(x) - 4$ is a translation of $y = f(x)$ four units down. The graph of $y = 2^x$ is lowered four units to obtain $y = 2^x - 4$.
- The line $y = -4$ is a horizontal asymptote.

**Goal C** *to solve exponential equations*

Before we use the Rules of Real Number Exponents to simplify and solve exponential equations, we must write the exponential expressions with the same base. Once this has been accomplished, we can use the following fact to write a simpler equivalent equation.

**Fact**

If $b > 0$, and $b \neq 1$, then $b^a = b^c$ is equivalent to $a = c$.

**Example 4** Solve the following equation for $x$: $2^{6-x} = \frac{1}{4}(8^x)$.

***Solution***

$2^{6-x} = \frac{1}{4}(8^x)$

- Begin by writing each term with the same base.

$2^{6-x} = \frac{1}{2^2}(2^3)^x$

$2^{6-x} = 2^{-2} \cdot 2^{3x}$

- Rules 6 and 4 of Real Number Exponents are used.

$2^{6-x} = 2^{-2+3x}$

- Since the bases are equal, by the fact above, the exponents are equal.

$6 - x = -2 + 3x$

$x = 2$

## Exercise Set 7.1 ☰ Calculator Exercises in the Appendix

### Goal A

In exercises 1–14, simplify each of the expressions.

1. $(2^x)^{-3}$
2. $(3^x)^{-2}$
3. $5^{x-7} \cdot 5^{x+3}$
4. $7^{4-x} \cdot 7^{x-9}$
5. $2^{x+2} \cdot 4^{-x}$
6. $3^{x-2} \cdot 9^x$
7. $\frac{3^{x+2}}{3^{4-x}}$
8. $\frac{2^{1-x}}{2^{5+x}}$
9. $4^{x+1} \cdot 8^{2-x}$
10. $9^{2-3x} \cdot 27^{x-4}$
11. $\frac{27^{2x-1}}{9^{x+2}}$
12. $\frac{8^{x-1}}{4^{3+2x}}$
13. $2^x + 4^x$
14. $9^x - 3^x$

### Goal B

In exercises 15–26, sketch the graph of each of the following equations.

15. $y = 4^x$
16. $y = 5^x$
17. $y = 3^{-x}$
18. $y = 5^{-x}$
19. $y = 3^x - 1$
20. $y = 2^x + 5$
21. $y = 3^{x-1}$
22. $y = 2^{x+5}$
23. $y = \left(\frac{1}{2}\right)^x - 1$
24. $y = 1 - \left(\frac{1}{3}\right)^x$
25. $y = -\left(\frac{1}{3}\right)^{x-2}$
26. $y = -\left(\frac{1}{2}\right)^{x+1}$

### Goal C

In exercises 27–38, solve each of the following equations for $x$.

27. $2^{x+1} = 16$
28. $3^{7-x} = 81$
29. $8^x = 4$
30. $9^x = 27$
31. $3^x = 9^{x+4}$
32. $2^x = 8^{5-x}$
33. $3^x = 9(3^{5-x})$
34. $2^x = 2^{3x}(4^2)$
35. $2^x = -4$
36. $3^{x-2} = 0$
37. $4^x = \frac{1}{2}(8^{x+1})$
38. $9^x = \frac{1}{27}(3^{1-x})$

### Superset

In exercises 39–44, solve each equation for $x$.

39. $2^{x^2} = 4$
40. $2^{x^3} = \frac{1}{2}$
41. $2^x(2^x - 1) = 0$
42. $3^x(9 - 3^x) = 0$
43. $2^{2x} - 5 \cdot 2^x + 4 = 0$
44. $3^{2x} - 12 \cdot 3^x + 27 = 0$
45. Carefully graph $y = 2^x$ and use your graph to estimate the value of:

    (a) $2^{\sqrt{2}}$ (b) $2^{\pi}$ (c) $2^{-\sqrt{3}}$
46. Carefully graph $y = 10^x$ and use your graph to estimate the value of $x$ when:

    (a) $y = 50$ (b) $y = 75$ (c) $y = 5$
47. Suppose some quantity $Q$ is related to another quantity $k$ by the relationship $Q = 3^k$. What happens to $Q$ if the value of $k$ is doubled? tripled? increased tenfold?
48. Does the equation $3^x = x$ have any solutions? (Hint: graph $y = 3^x$ and $y = x$ on the same set of axes.)

In exercises 49–54, sketch the graph of each of the following equations.

49. $y = 2^{|x|}$
50. $y = |2^x|$
51. $y = 2^{x^2}$
52. $y = 2^{1/x}$
53. $y = -2^{x^2}$
54. $y = 2^{-|x|}$

In exercises 55–58, determine whether each of the following functions is symmetric with respect to the $y$-axis or the origin.

55. $y = 2^x$
56. $y = 2^{x^2}$
57. $y = 2^x + 2^{-x}$
58. $y = 2^x - 2^{-x}$

## 7.2 Logarithmic Functions

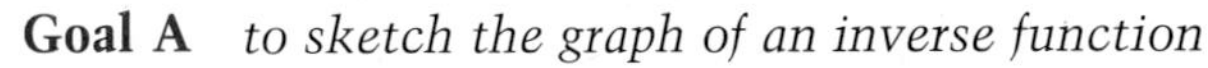

**Goal A** *to sketch the graph of an inverse function*

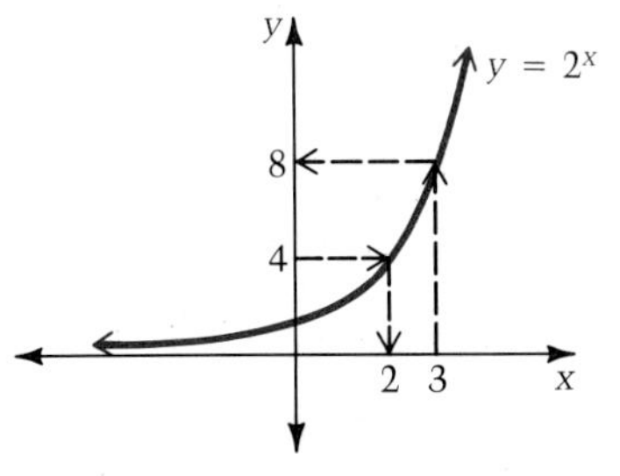


**Figure 7.4** "Given $x = 3$, find $y$." "Given $y = 4$, find $x$."

We now consider two questions that may be asked about a function $y = f(x)$. The first is straightforward: given a value of $x$, what is the value of $y$? To answer this, we simply evaluate the function. For example,

$$\text{if } y = 2^x \text{ and } x = 3, \text{ then } y = 8.$$

The second question reverses the first: given a value of $y$, what is the value of $x$? Sometimes we can answer this question easily. For instance,

$$\text{if } y = 2^x \text{ and } y = 4, \text{ then } x = 2.$$

Sometimes, however, the second question does not have a unique answer. For example,

$$\text{if } y = x^2 \text{ and } y = 16, \text{ then } x = 4 \text{ or } x = -4.$$

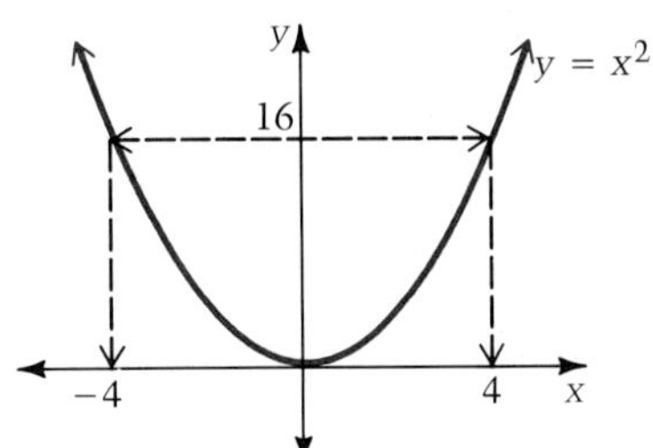


**Figure 7.5** "Given $y = 16$, find $x$." The answer is not unique.

If, for a given function $f$, the answer to "given $y$, find $x$" is always unique, then $f$ is said to be *one-to-one.* That is, a function is **one-to-one** if each $y$-value in the range of $f$ corresponds to exactly one $x$-value in the domain.

If a function $f$ is one-to-one, then there exists an *inverse function,* denoted $f^{-1}$, associated with $f$. The inverse function $f^{-1}$ reverses the process $f$; that is, if $f$ processes 3 and produces 10, then $f^{-1}$ processes 10 and produces 3.

**Definition**

If a function $f$ is one-to-one, then there exists a unique function $f^{-1}$, called the **inverse function of *f*,** such that

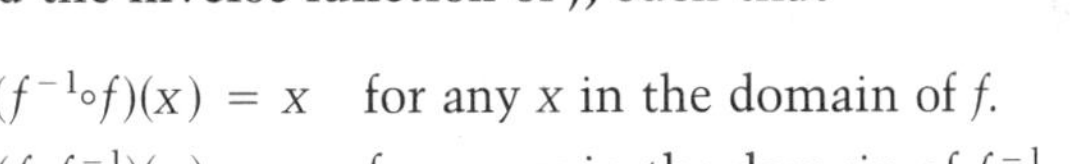

$$(f^{-1} \circ f)(x) = x \quad \text{for any } x \text{ in the domain of } f.$$

$$(f \circ f^{-1})(x) = x \quad \text{for any } x \text{ in the domain of } f^{-1}.$$

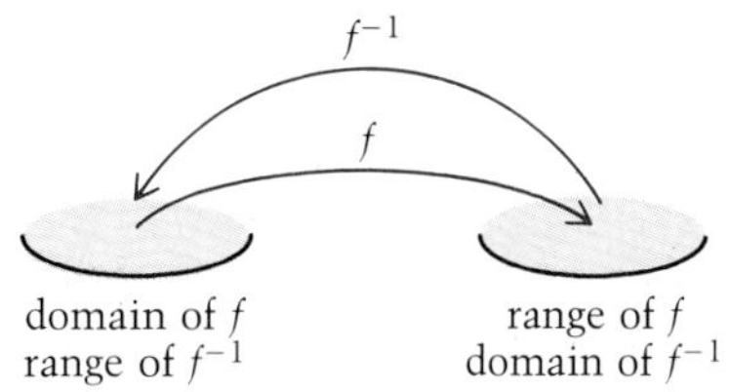


**Figure 7.6**

Thus, the function $f^{-1}$ interchanges the roles of the domain and the range of $f$: the domain of $f$ becomes the range of $f^{-1}$, and the range of $f$ becomes the domain of $f^{-1}$.

**Example 1** The function $f(x) = 3x - 2$ is one-to-one. Find its inverse function.

***Solution*** To find the equation describing the inverse function, first interchange $x$ and $y$. Then, since we usually write $x$ as the independent variable and $y$ as the dependent variable, we solve for $y$ in terms of $x$.

$y = 3x - 2$ ■ Interchange $x$ and $y$ to obtain the inverse.

$x = 3y - 2$ ■ Solve for $y$ in terms of $x$.

$$y = f^{-1}(x) = \frac{x + 2}{3}$$

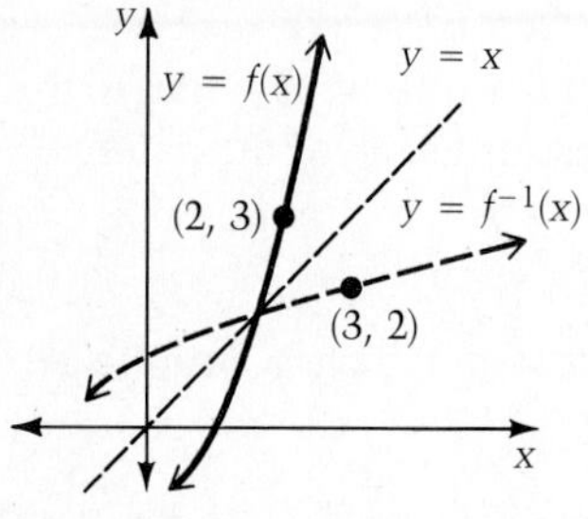

Figure 7.7

The graphs of a function $f$ and its inverse $f^{-1}$ are displayed in Figure 7.7. Note that the graph of $y = f^{-1}(x)$ is the reflection of $y = f(x)$ across the line $y = x$. Thus, to graph $f^{-1}$, we interchange the $x$- and $y$-coordinates of each ordered pair of $f$ and graph the resulting ordered pairs.

**Fact**

If a function $f$ has an inverse function $f^{-1}$, then we reflect the graph of $y = f(x)$ across the line $y = x$ to obtain the graph of $y = f^{-1}(x)$.

**Example 2** The function $f(x) = x^3$, with restricted domain $-1 \le x \le 2$, is one-to-one. Find the equation describing the inverse function and sketch the inverse.

***Solution***

$$y = x^3$$
$$x = y^3$$
$$y = f(x)^{-1} = x^{1/3}$$

■ To find the equation describing the inverse function, interchange $x$ and $y$ and then solve for $y$.

■ Sketch the function $y = f(x)$. Then reflect it across the line $y = x$ to obtain the graph of the inverse function.

■ Domain of $f$: $[-1, 2]$
Range of $f$: $[-1, 8]$

■ Domain of $f^{-1}$: $[-1, 8]$
Range of $f^{-1}$: $[-1, 2]$

**Goal B** *to rewrite an exponential equation as a logarithmic equation and vice versa*

To find the inverse of $y = b^x$, we use the same techniques we used in Goal A. The statement "given $y$, find $x$" for the function $y = b^x$ is the same as the statement "given $x$, find $y$" for the inverse function $x = b^y$ (interchange $x$ and $y$). The graphs of the inverse functions $x = b^y$ in Figure 7.8—one for the case where $b > 1$, and one for the case where $0 < b < 1$ —are found by reflecting $y = b^x$ across the line $y = x$.

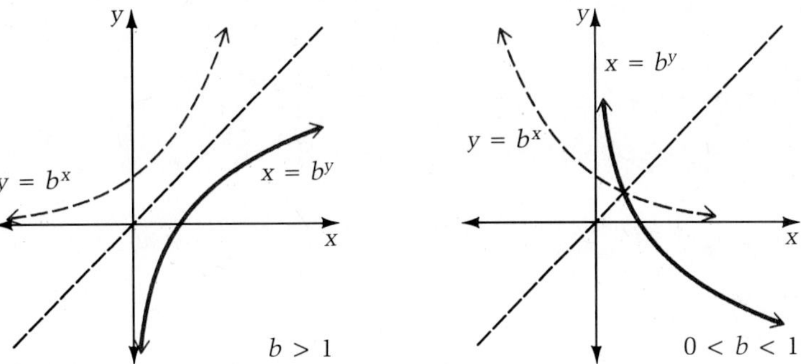


**Figure 7.8** $x = b^y$: domain is $(0, +\infty)$, range is $(-\infty, +\infty)$.

In order to describe these inverse functions in the usual form $y = f^{-1}(x)$, it is necessary to write the equation $x = b^y$ with $y$ in terms of $x$. We can describe $y$ in words:

$y$ is the power to which you raise $b$ in order to obtain $x$.

We define *logarithm* as an abbreviation for this description of $y$.

**Definition**
For $b > 0$ and $b \neq 1$, the function $f(x) = b^x$ has an inverse function, denoted $f^{-1}(x) = \log_b x$. This notation is an abbreviation for the **logarithm of x to the base *b*.** The domain of the logarithmic function is $(0, +\infty)$ and the range is $(-\infty, +\infty)$.

Notice that we have taken $y$ in the equation $x = b^y$ and defined it as a logarithm. Therefore, a logarithm is an exponent, and every exponential statement can be rewritten as an equivalent statement about logarithms. This equivalence is summarized below.

**Log/Exp Principle**
For $b > 0$ and $b \neq 1$, $b^A = C$ is equivalent to $\log_b C = A$.

**Example 3** Rewrite the following as logarithmic equations: (a) $x^3 = 64$ (b) $9^x = 3$.

***Solution*** (a)

$$x^3 = 64$$
$$\log_x 64 = 3$$

■ Use the Log/Exp Principle.

(b)

$$9^x = 3$$
$$\log_9 3 = x$$

■ Use the Log/Exp Principle.

**Example 4** Solve each of the following equations for $x$.

(a) $\log_3 x = 2$ (b) $\log_8 x = \frac{1}{3}$ (c) $\log_x 64 = 3$ (d) $\log_9 3 = x$

***Solution***

(a)
$$\log_3 x = 2$$
$$3^2 = x$$
$$x = 9$$

(b)
$$\log_8 x = \frac{1}{3}$$
$$8^{1/3} = x$$
$$x = 2$$

(c)
$$\log_x 64 = 3$$
$$x^3 = 64$$
$$x = 64^{1/3}$$
$$x = 4$$

(d)
$$\log_9 3 = x$$
$$9^x = 3$$
$$x = \frac{1}{2}$$

You should keep in mind that a logarithm is just an exponent. When we say $\log_3 9 = 2$, we are saying that 2 is the exponent to which we raise 3 in order to get 9. That is, $3^2 = 9$.

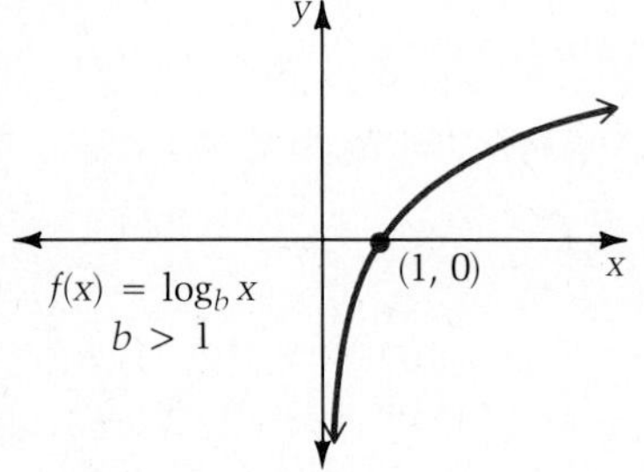


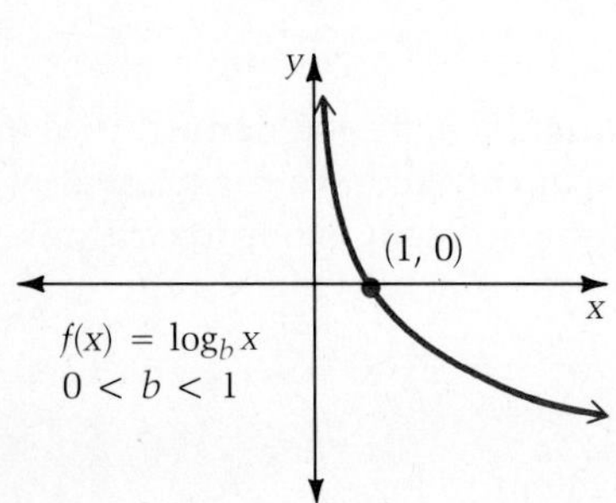

Figure 7.9

**Goal C** *to graph logarithmic functions*

The two general graphs of the function $f(x) = \log_b x$ are shown in Figure 7.9. Looking at these graphs, we recall the following important information about the logarithmic function:

- the domain is $(0, +\infty)$ and the range is $(-\infty, +\infty)$;
- $\log_b 1 = 0$ and $\log_b b = 1$;
- the function is one-to-one.

**Fact**

Since the logarithmic function is one-to-one, no two $x$-values correspond to the same $y$-value. Thus,

$$\text{if } \log_b A = \log_b C, \text{ then } A = C.$$

This fact will be useful when we solve logarithmic equations.

**Example 5** Sketch the graph of (a) $y = \log_2 x$ (b) $y = \log_2(x - 1)$.

***Solution*** (a)

| $x$ | $\frac{1}{4}$ | $\frac{1}{2}$ | 1 | 2 | 4 |
|---|---|---|---|---|---|
| $y$ | $-2$ | $-1$ | 0 | 1 | 2 |

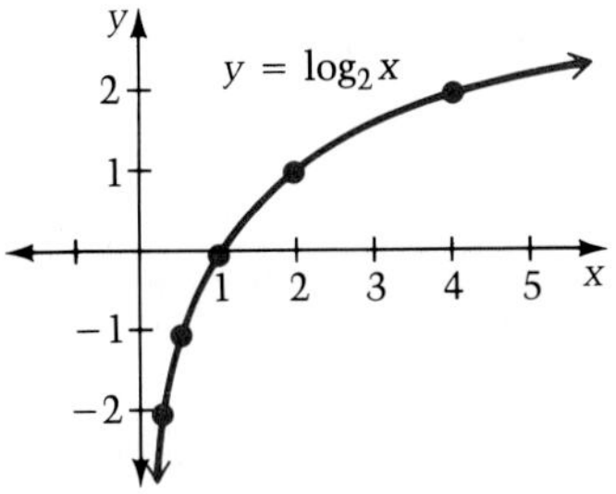

- To compute a table of values, we use the Log/Exp Principle to write $y = \log_2 x$ as $x = 2^y$.
- Plot the points and sketch the curve. The base for the logarithm in this example is $b = 2$. Note that this sketch is similar to the graph given on the previous page for the general case $b > 1$.

$y = \log_2 x$

(b)

$y = \log_2(x - 1)$

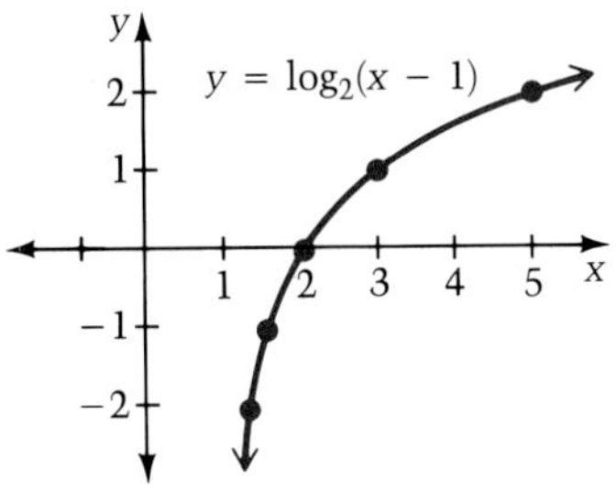

- Recall that the graph of $y = f(x - 1)$ is a translation of $y = f(x)$ one unit to the right. Thus, the graph of $y = \log_2 x$ is translated one unit to the right to obtain $y = \log_2(x - 1)$.

**Goal D** *to simplify and evaluate logarithmic expressions*

We use the following rules to simplify and evaluate logarithmic expressions. The proofs of these rules will be postponed until you have seen how they work.

**Rules of Logarithms**

**Rule 1** $\log_b 1 = 0$ **Rule 2** $\log_b b = 1$

**Rule 3** $\log_b(xy) = \log_b x + \log_b y$ **Rule 4** $\log_b(x^t) = t \log_b x$

**Rule 5** $\log_b\left(\dfrac{x}{y}\right) = \log_b x - \log_b y$

**Rule 6** $\log_b\left(\dfrac{1}{x}\right) = -\log_b x$

where $x$ and $y$ are positive real numbers, $t$ is any real number, and base $b$ is positive, but not equal to 1.

**Example 6** Given that $\log_2 x = 3.1$ and $\log_2 y = 1.6$, evaluate $\log_2 \frac{8x}{y^3}$.

***Solution***

$$\begin{aligned}
\log_2 \frac{8x}{y^3} &= \log_2 8x - \log_2 y^3 && \blacksquare \text{ Rule 5 of Logarithms is used.} \\
&= \log_2 8 + \log_2 x - \log_2 y^3 && \blacksquare \text{ Rule 3 of Logarithms is used.} \\
&= \log_2 8 + \log_2 x - 3 \log_2 y && \blacksquare \text{ Rule 4 of Logarithms is used.} \\
&= 3 + 3.1 - 3(1.6) && \blacksquare\ \log_2 8 = 3. \\
&= 1.3
\end{aligned}$$

As the next example illustrates, you *must* check solutions of logarithmic equations. Since logarithms are defined for positive numbers only, you must discard any solution that requires taking the logarithm of zero or a negative number.

**Example 7** Solve the following equation for $x$: $2 \log_2 x = \log_2(3x + 4)$.

***Solution***

$$2 \log_2 x = \log_2(3x + 4)$$ ■ Use Rule 4 of Logarithms.

$$\log_2 x^2 = \log_2(3x + 4)$$ ■ Use the fact that if $\log_b A = \log_b C$, then $A = C$.

$$x^2 = 3x + 4$$

$$x^2 - 3x - 4 = 0$$ ■ Use the Property of Zero Products.

$$(x - 4)(x + 1) = 0$$

| $x - 4 = 0$ | $x + 1 = 0$ |
|---|---|
| $x = 4$ | $x = -1$ |

■ Now, we *must* check solutions.

**Must Check:**

**$x = 4$**

$$2 \log_2 4 = \log_2(3 \cdot 4 + 4)$$

$$\log_2 4^2 = \log_2 16$$ ■ True.

**$x = -1$**

$$2 \log_2(-1) = \log_2(3(-1) + 4)$$ ■ $\log_2(-1)$ is undefined.

The only solution is $x = 4$.

The first fact below is a restatement of the definition of logarithm. The second fact is just a special case of Rule 4 of Logarithms. These two facts provide another way of saying that the logarithmic and exponential functions are inverse functions of one another.

**Fact**

**1** $b^{\log_b x} = x$ **2** $\log_b(b^x) = x$

We conclude by proving Rules 1, 3, and 4 of Logarithms. The proofs of Rules 2, 5, and 6 are left as exercises.

*Proof of Rule 1:* $\log_b 1 = 0$

Rule 1 of Real Number Exponents states that $b^0 = 1$

Then, by the Log/Exp Principle $\log_b 1 = 0$ ■

*Proof of Rule 3:* $\log_b(xy) = \log_b x + \log_b y$

Begin by letting $r = \log_b x$ and $s = \log_b y$

By the Log/Exp Principle $x = b^r$ and $y = b^s$

By Rule 3 of Real Number Exponents $xy = b^{r+s}$

By the Log/Exp Principle $\log_b(xy) = r + s$

Replacing $r$ with $\log_b x$ and $s$ with $\log_b y$, we obtain $\log_b(xy) = \log_b x + \log_b y$ ■

*Proof of Rule 4:* $\log_b(x^t) = t \cdot \log_b x$

Begin by letting $r = \log_b x$

By the Log/Exp Principle $x = b^r$

By Rule 4 of Real Number Exponents $x^t = b^{rt}$

By the Log/Exp Principle $\log_b(x^t) = rt$

Replacing $r$ with $\log_b x$, we obtain $\log_b(x^t) = t \cdot \log_b x$ ■

## Exercise Set 7.2

### Goal A

In exercises 1–8, the given function $y = f(x)$ is one-to-one. Find the equation describing the inverse function and sketch the inverse.

**1.** $f(x) = 6 - 2x, \quad 1 \le x \le 3$

**2.** $f(x) = 3x - 8, \quad 0 \le x \le 6$

**3.** $f(x) = x^2, \quad 0 < x \le 3$

**4.** $f(x) = x^3, \quad -3 \le x < 3$

**5.** $f(x) = \sqrt{4 - x}, \quad x \le 4$

**6.** $f(x) = 5 - x^2, \quad 0 < x < 2$

**7.** $f(x) = \frac{1}{3}(x - 1)^3, \quad 0 \le x < 3$

**8.** $f(x) = \frac{1}{4}x^4, \quad 0 < x < 3$

For exercises 9–16, identify the domain and range of the inverse function $y = f^{-1}(x)$ for each of the functions in Exercises 1–8.

## Goal B

In exercises 17–22, write each of the following as a logarithmic equation.

**17.** $x^2 = 25$ **18.** $x^4 = 16$ **19.** $16^x = 4$

**20.** $27^x = 3$ **21.** $5^3 = x$ **22.** $10^5 = x$

In exercises 23–32, use the Log/Exp Principle to solve each of the following equations for $x$.

**23.** $\log_2 x = 3$ **24.** $\log_4 x = 3$

**25.** $\log_8 x = \frac{1}{2}$ **26.** $\log_9 x = \frac{1}{2}$

**27.** $\log_x 4 = 2$ **28.** $\log_x 8 = 3$

**29.** $\log_8 4 = x$ **30.** $\log_9 27 = x$

**31.** $\log_x 3 = -2$ **32.** $\log_x 16 = -4$

## Goal C

In exercises 33–42, sketch the graph of each of the following equations.

**33.** $y = \log_3 x$ **34.** $y = \log_4 x$

**35.** $y = \log_{1/2} x$ **36.** $y = \log_{1/3} x$

**37.** $y = \log_2(x + 1)$ **38.** $y = \log_3(x - 2)$

**39.** $y = \log_3(-x)$ **40.** $y = \log_2(-x)$

**41.** $y = -\log_2(x - 2)$ **42.** $y = -\log_3(x + 1)$

## Goal D

In exercises 43–48, use the values $\log_2 x = 3.1$, $\log_2 y = 1.6$, and $\log_2 z = -2.7$ to evaluate each of the following expressions.

**43.** $\log_2(xyz)$ **44.** $\log_2\left(\frac{4}{xyz}\right)$

**45.** $\log_2\left(\frac{16x}{yz^3}\right)$ **46.** $\log_2\left(\frac{xy^2}{4z}\right)$

**47.** $\log_2\left(\frac{x}{y}\right) - \frac{\log_2 x}{\log_2 y}$

**48.** $\log_2 xy - (\log_2 x \cdot \log_2 y)$

In exercises 49–58, solve each of the following equations for $x$.

**49.** $4^{\log_8 x} = 2$ **50.** $9^{\log_3 x} = 3$

**51.** $\log_3(9^x) = -4$ **52.** $\log_4(2^x) = -3$

**53.** $\log_2(\log_3 x) = 2$ **54.** $\log_3(\log_2 x) = 2$

**55.** $\log_2(x^3) = \log_2(8x)$

**56.** $\log_2(x^2 - 6) = \log_2(-x)$

**57.** $\log_2(x^2) = 2 \cdot \log_2 x$

**58.** $\log_2 5x = \log_2 5 + \log_2 x$

## Superset

In exercises 59–64, sketch the graph of each of the following equations.

**59.** $y = \log_2(x^2)$

**60.** $y = 1 - \frac{1}{3}\log_2(x^3)$

**61.** $y = \log_2(8^x)$ **62.** $y = \log_8(2^x)$

**63.** $y = 4^{\log_2 x}$ **64.** $y = \log_2 |x - 4|$

In exercises 65–70, solve each of the following equations for $x$.

**65.** $\log_4(x + 7) = \log_2(x + 1)$

**66.** $\log_3(3x) = \log_3 27 \cdot \log_3 x$

**67.** $(3 - \log_2 x)\log_2 x = 0$

**68.** $\log_x 1 = 0$

**69.** $4 \log_2 x = \log_2(5x^2 - 4)$

**70.** $\log_2[\log_2(\log_2 x)] = 0$

**71.** Prove Rule 2 of Logarithms.

**72.** Prove Rule 5 of Logarithms.

**73.** Prove Rule 6 of Logarithms.

## 7.3 Common Logarithms

**Goal A** *to use a table of values to estimate common logarithms*

| $x$ | $\log x$ |
|---|---|
| $0.001 = 10^{-3}$ | $-3$ |
| $0.01 = 10^{-2}$ | $-2$ |
| $0.1 = 10^{-1}$ | $-1$ |
| $1 = 10^0$ | 0 |
| $10 = 10^1$ | 1 |

Frequently 10 is used as a base for logarithms because we use a base 10 system of enumeration. Base 10 logarithms are called **common logarithms.** To simplify the notation we shall write $\log_{10} x$ as $\log x$.

If $y = \log x$, then $y$ is the exponent to which you must raise 10 in order to get $x$. In this case, $x$ is called the **common antilogarithm** of $y$; that is, $x$ is the number you get when you raise 10 to the power $y$. The table at the left illustrates how we compute $\log x$ for certain special values of $x$. The graph of $y = \log x$ is shown in Figure 7.10.

A table of logarithmic values can be a valuable aid in performing complicated computations. Table 1 in the Appendix lists values for $\log x$. A small segment of the table is reproduced below.

| x | 0 | 1 | 2 | 3 | 4 | 5 | 6 | 7 | 8 | 9 |
|---|---|---|---|---|---|---|---|---|---|---|
| **1.0** | .0000 | .0043 | .0086 | .0128 | .0170 | .0212 | .0253 | .0294 | .0334 | .0374 |
| **1.1** | .0414 | .0453 | .0492 | .0531 | .0569 | .0607 | .0645 | .0682 | .0719 | .0755 |
| **1.2** | .0792 | .0828 | .0864 | .0899 | .0934 | .0969 | .1004 | .1038 | .1072 | .1106 |
| **1.3** | .1139 | .1173 | .1206 | .1239 | .1271 | .1303 | .1335 | .1367 | .1399 | .1430 |
| **1.4** | .1461 | .1492 | .1523 | .1553 | .1584 | .1614 | .1644 | .1673 | .1703 | .1732 |
| **1.5** | .1761 | .1790 | .1818 | .1847 | .1875 | .1903 | .1931 | .1959 | .1987 | .2014 |
| **1.6** | .2041 | .2068 | .2095 | .2122 | .2148 | .2175 | .2201 | .2227 | .2253 | .2279 |

**Figure 7.10** The Common Logarithmic Function.

The values of $x$ lie between 1 and 10 and are accurate to two decimal places. The values of $\log x$ lie between 0 and 1 and are accurate to four decimal places. Remember that $\log 1 = 0$ and $\log 10 = 1$. The first two digits of $x$ are given in the leftmost column, and the third digit is one of the column headings. The values of $\log x$ are the entries in the body of the table. To determine $\log 1.54$, we read down the leftmost column to 1.5, and then read across to the column labeled 4. The entry at this position in the table is 0.1875. Since the values in the table are approximations, $\log 1.54$ is approximately 0.1875, and we write $\log 1.54 \approx 0.1875$.

**Example 1** Use Table 1 in the Appendix to find (a) $\log 2.95$ (b) $\log 3.13$ (c) $\log 3.04$.

***Solution*** (a) $\log 2.95 \approx 0.4698$ (b) $\log 3.13 \approx 0.4955$ (c) $\log 3.04 \approx 0.4829$

We can use Table 1 to find the common logarithm of a number, even when the number is not between 1 and 10. To do so, we first write $x$ in **scientific notation.** That is, we write

$$x = \begin{bmatrix}\text{a number}\\ \text{between 1 and 10}\end{bmatrix} \times \begin{bmatrix}\text{a power}\\ \text{of ten}\end{bmatrix} = A \times 10^t.$$

We then use the Rules of Logarithms to write

$$\log x = \log(A \times 10^t) = \log A + \log 10^t = \log A + t.$$

For example, if $x = 178.0$, then in scientific notation $x = 1.78 \times 10^2$, and

$$\log x = \log(1.78 \times 10^2) = \log 1.78 + 2 \approx 2.2504.$$

**Example 2** Use the fact that $\log 3.14 \approx 0.4969$ to compute: (a) $\log 31.4$ (b) $\log 0.00314$.

***Solution***

(a)
$$\begin{aligned}\log 31.4 &= \log(3.14 \times 10^1)\\ &= \log 3.14 + \log 10\\ &\approx 0.4969 + 1\\ &= 1.4969\end{aligned}$$

(b)
$$\begin{aligned}\log 0.00314 &= \log(3.14 \times 10^{-3})\\ &= \log 3.14 + \log 10^{-3}\\ &\approx 0.4969 - 3\\ &= -2.5031\end{aligned}$$

**Goal B** *to use a table to estimate common antilogarithms*

To find a common antilogarithm, we follow our procedure for finding common logarithms in reverse. Suppose we are given that $\log x = 0.4843$. To find $x$, we must find the entry 0.4843 in Table 1. This entry appears in the row labeled 3.0 and the column labeled 5. Thus, if $\log x = 0.4843$, then $x \approx 3.05$.

**Example 3** Use Table 1 to solve the following for $x$: (a) $\log x = 0.9175$ (b) $\log x = 0.0792$.

***Solution*** (a) if $\log x = 0.9175$, then $x \approx 8.27$ (b) if $\log x = 0.0792$, then $x \approx 1.20$

To estimate the antilogarithm of a number that is not between 0 and 1, we first write the number as a sum of the form:

$$y = (\text{an integer}) + (\text{a number between 0 and 1}).$$

For any number there is only one way to form this sum. For example,

$$2.4969 = 2 + 0.4969 \qquad -1.4969 = -2 + 0.5031$$
$$5.4969 = 5 + 0.4969 \qquad -3.4969 = -4 + 0.5031$$

We can then use Table 1 to find the antilogarithm of the decimal part of the sum. (The integer becomes the power of 10 in the answer.) For example, suppose we must solve $\log x = 2.4969$ for $x$. First we write

$$\log x = 2 + 0.4969.$$

We know that $2 = \log 10^2$ and using Table 1, we determine that $0.4969 = \log 3.14$. Therefore,

$$\log x \approx \log 10^2 + \log 3.14.$$

Then by Rule 3 of Logarithms, we have

$$\log x \approx \log(10^2 \times 3.14).$$

Thus, $x \approx 3.14 \times 10^2 = 314$.

**Example 4** Find the antilogarithm of 1.1430. That is, solve $\log x = 1.1430$ for $x$.

***Solution***

$\log x = 1.1430$

$= 1 + 0.1430$

$\approx \log 10^1 + \log 1.39$ ■ From Table 1 we found that $\log 1.39 = 0.1430$.

$= \log(10^1 \times 1.39)$ ■ Rule 3 of Logarithms is used.

$x \approx 1.39 \times 10^1 = 13.9$ ■ $\log A = \log C$ implies $A = C$.

**Example 5** Find the antilogarithm of $-2.7033$. That is, solve $\log x = -2.7033$ for $x$.

***Solution***

$$\log x = -2.7033 = -3 + 0.2967 \approx \log 10^{-3} + \log 1.98 = \log(10^{-3} \times 1.98)$$

Thus, $x \approx 1.98 \times 10^{-3}$ or 0.00198.

## Exercise Set 7.3

### Goal A

In exercises 1–12, use Table 1 in the Appendix to find the following.

**1.** log 4.83 **2.** log 2.57 **3.** log 8.81

**4.** log 9.00 **5.** log 1.87 **6.** log 5.99

**7.** log 3.26 **8.** log 6.71 **9.** log 2.33

**10.** log 4.00 **11.** log 1.60 **12.** log 7.04

In exercises 13–24, use scientific notation to rewrite each of the following numbers.

**13.** 47.24 **14.** 0.1008 **15.** 0.3502

**16.** 32.15 **17.** 0.015 **18.** 348.9

**19.** 0.0908 **20.** 0.0029 **21.** 3294.1

**22.** 6804.5 **23.** 0.0003 **24.** 0.00001

In exercises 25–36, use Table 1 to compute each of the following.

**25.** log 32.9 **26.** log 0.142 **27.** log 0.68

**28.** log 75.1 **29.** log 0.057 **30.** log 920

**31.** log 349 **32.** log 0.010

**33.** log 5160 **34.** log 0.0082

**35.** log 14,600 **36.** log 0.0045

### Goal B

In exercises 37–54, use Table 1 to solve each of the following equations for $x$.

**37.** $\log x = 0.5490$ **38.** $\log x = 0.6702$

**39.** $\log x = 0.9304$ **40.** $\log x = 0.4771$

**41.** $\log x = 0.7396$ **42.** $\log x = 0.2122$

**43.** $\log x = 2.7520$ **44.** $\log x = 7.9800$

**45.** $\log x = 2.4800$ **46.** $\log x = 1.3054$

**47.** $\log x = 4.6637$ **48.** $\log x = 3.0414$

**49.** $\log x = -0.7520$ **50.** $\log x = -1.3054$

**51.** $\log x = -2.4609$ **52.** $\log x = -2.5391$

**53.** $\log x = -0.9957$ **54.** $\log x = -3.0044$

### Superset

In exercises 55–64, determine whether each statement is true or false. If false, illustrate this with an example.

**55.** $\log x + \log y = \log(x + y)$

**56.** $\log x - \log y = \log(x - y)$

**57.** $(\log x)(\log y) = \log(xy)$

**58.** $(\log x)^2 = 2 \log x$

**59.** $\log x^2 = 2 \log x$

**60.** $x + \log x^{-1} = x - \log x$

**61.** $\log 100x = 100 \log x$

**62.** $\log 10^x = x$

**63.** $\log 10^x = \log x^{10}$

**64.** $\log(10 + x) = 1 + \log x$

In exercises 65–68, use the fact that

$$\log_b x = \frac{\log x}{\log b}$$

to estimate the given expression.

**65.** $\log_3 10$ **66.** $\log_5 15$

**67.** $\log_2 5$ **68.** $\log_2 10$

**69.** Suppose some quantity $Q$ is related to another quantity $x$ by the relationship $Q = \log_2 x$. What happens to the value of $Q$ if $x$ is doubled? tripled? quadrupled?

**70.** Suppose some quantity $Q$ is related to another quantity $x$ by the relationship $Q = \log_b x$. What happens to the value of $Q$ if $x$ is doubled? tripled? quadrupled?

| $x$ | $f(x)$ |
|---|---|
| −1 | 1.9 |
| 1 | 3.4 |
| 3 | 4.8 |

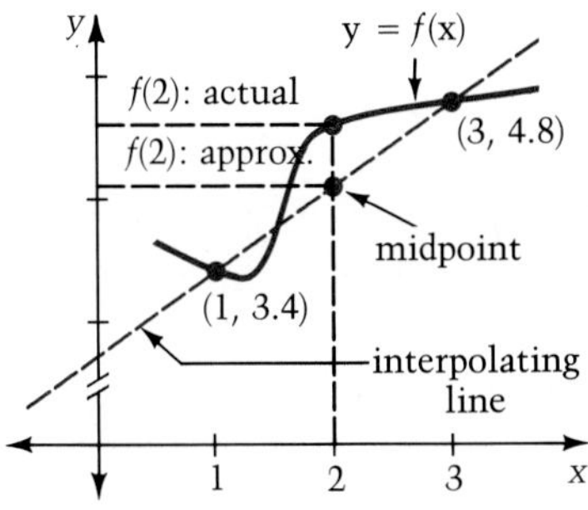


**Figure 7.11**

## 7.4 Linear Interpolation

**Goal A** *to use linear interpolation to approximate function values*

Frequently a table of values (such as Table 1 in the Appendix) does not include the specific value we need. For example, suppose we are given the table on the left and we want to determine the value of $f(2)$. Although the table provides no function value for $x = 2$, we can use the information we have to estimate $f(2)$. Having no other information about the function $f$, we can reasonably assume that, since 2 is midway between 1 and 3, the value $f(2)$ is midway between $f(1)$ and $f(3)$.

As shown in Figure 7.11, the geometric meaning of our assumption is that the point $(2, f(2))$ is near the midpoint of the line segment that joins the points $(1, f(1))$ and $(3, f(3))$. As you can see, the midpoint of the line segment and the point $(2, f(2))$ do not coincide, but the $y$-coordinate of the midpoint is a good approximation of $f(2)$.

The use of a line to approximate function values is called **linear interpolation.** For the example above, the equation of the line is found by using the points (1, 3.4) and (3, 4.8). We find the approximation by substituting $x = 2$ into the equation of this line.

**Example 1** Use linear interpolation and the function values given in the table at the right to approximate $f(6.2)$ to the nearest tenth.

| $x$ | 1 | 3 | 5 | 7 |
|---|---|---|---|---|
| $f(x)$ | 3.4 | 8.7 | 15.2 | 23.5 |

***Solution*** Since 6.2 is an $x$-value lying between 5 and 7 in the table, we use the points (5, 15.2) and (7, 23.5) to determine the equation of the interpolating line.

$$\text{slope} = \frac{23.5 - 15.2}{7 - 5} = \frac{8.3}{2} = 4.15$$

■ The slope of the interpolating line is computed. Now use the point-slope form to write an equation of the interpolating line.

$$y - 15.2 = 4.15(x - 5)$$

$$y - 15.2 = 4.15(6.2 - 5)$$

■ $x = 6.2$ is substituted into the equation.

$$y = 4.15(1.2) + 15.2$$

$$y = 20.18$$

The approximate value of $f(6.2)$ to the nearest tenth is 20.2.

This method also works when we are given a value of $y$ and are looking for $x$. We now consider such an example.

**Example 2** Given that $f(x) = 4.3$, use linear interpolation and the function values given in the table in Example 1 to approximate $x$ to the nearest tenth.

***Solution*** Since 4.3 is a $y$-value lying between 3.4 and 8.7 in the table, we use the points (1, 3.4) and (3, 8.7) to determine the equation of the interpolating line.

$$\text{slope} = \frac{8.7 - 3.4}{3 - 1} = \frac{5.3}{2} = 2.65$$

$$y - 3.4 = 2.65(x - 1)$$

■ The point-slope form is used.

$$4.3 - 3.4 = 2.65(x - 1)$$

■ $y = 4.3$ is substituted into the equation.

$$x \approx 1.34$$

The approximate value of $x$ to the nearest tenth is 1.3.

**Goal B** *to use linear interpolation to estimate common logarithms and antilogarithms*

When we apply linear interpolation to logarithms, we shall use the values in Table 1 in the Appendix. For a given problem, it is helpful to make a small table of those values that bracket the value of interest.

**Example 3** Find log 85.37.

***Solution*** $\log 85.37 = \log(8.537 \times 10^1) = \log 8.537 + \log 10 = \log 8.537 + 1$

Interpolation

| $x$ | 8.53 | 8.537 | 8.54 |
|---|---|---|---|
| $\log x$ | 0.9309 | ? | 0.9315 |

■ Consult Table 1, and record the bracketing values in a small table.

$$\text{slope} = \frac{0.9315 - 0.9309}{8.54 - 8.53} = 0.06$$

$$y - 0.9309 = 0.06(x - 8.53)$$

■ The point-slope form is used.

$$y = 0.06(0.007) + 0.9309$$

$$y = 0.93132$$

■ Thus, 0.9313 is the approximation for log 8.537.

$$\log 85.37 \approx 0.9313 + 1 = 1.9313$$

In the following example, the interpolation is done off to the side of the problem, much as auxiliary computations are often done.

**Example 4** Solve $\log A = -1.4132$ for $A$.

***Solution*** Remember that we must first write $-1.4132$ as the sum of an integer and a number between 0 and 1.

$$-1.4132 = -2 + 0.5868$$

| $x$ | 3.86 | ? | 3.87 |
|---|---|---|---|
| $\log x$ | 0.5866 | 0.5868 | 0.5877 |

Interpolation

$$\text{slope} = \frac{0.5877 - 0.5866}{3.87 - 3.86} = 0.11$$

$$y - 0.5866 = 0.11(x - 3.86)$$

$$0.5868 - 0.5866 = 0.11(x - 3.86)$$

$$0.0002 = 0.11(x - 3.86)$$

$$x \approx 3.862$$

$$\log A \approx \log 10^{-2} + \log 3.862 = \log(10^{-2} \times 3.862)$$

Thus, $A \approx 3.862 \times 10^{-2}$ or 0.03862.

## Exercise Set 7.4 ⊟ Calculator Exercises in the Appendix

### Goal A

In exercises 1–8, use linear interpolation and the function values given in the table to approximate $A$ to the nearest tenth.

1.

| $x$ | 1 | 3 | 7 | 13 |
|---|---|---|---|---|
| $f(x)$ | 2.8 | 14.3 | 38.5 | 64.2 |

(a) $A = f(5.1)$ (b) $f(A) = 14.3$
(c) $A = f(2.8)$ (d) $f(A) = 53.1$

2.

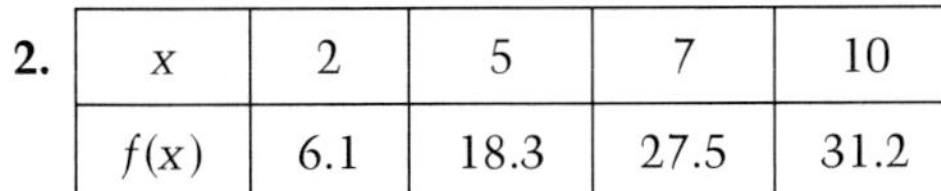

| $x$ | 2 | 5 | 7 | 10 |
|---|---|---|---|---|
| $f(x)$ | 6.1 | 18.3 | 27.5 | 31.2 |

(a) $A = f(2.7)$ (b) $f(A) = 19$
(c) $A = f(9.9)$ (d) $f(A) = 28.3$

3.

| $x$ | 2.1 | 3.8 | 5.6 | 8.1 |
|---|---|---|---|---|
| $f(x)$ | $-7.1$ | $-0.4$ | 6.9 | 15.2 |

(a) $A = f(3)$ (b) $f(A) = 8$
(c) $A = f(4.1)$ (d) $f(A) = -0.1$

4.

| $x$ | $-7.3$ | $-2.4$ | 5.9 | 12.1 |
|---|---|---|---|---|
| $f(x)$ | $-17.0$ | 3.8 | 9.6 | 16.3 |

(a) $A = f(-5.1)$ (b) $f(A) = 0$
(c) $A = f(1.7)$ (d) $f(A) = 14.6$

5.

| $x$ | $-2$ | 1 | 5 | 10 |
|---|---|---|---|---|
| $f(x)$ | 8.6 | 1.4 | $-4.3$ | $-7.8$ |

(a) $A = f(8)$ (b) $f(A) = 0$
(c) $A = f(0.2)$ (d) $f(A) = 5$

6.

| $x$ | $-15.3$ | $4.7$ | $11.5$ | $21.6$ |
|---|---|---|---|---|
| $f(x)$ | $5.2$ | $1.8$ | $-3.6$ | $-10.4$ |

(a) $A = f(-0.6)$ (b) $f(A) = -2.7$
(c) $A = f(20)$ (d) $f(A) = 4.4$

7.

| $x$ | $-4.1$ | $-1.9$ | $-0.1$ | $1.3$ |
|---|---|---|---|---|
| $f(x)$ | $7.3$ | $0.8$ | $-3.4$ | $-4.9$ |

(a) $A = f(-2.7)$ (b) $f(A) = 6.5$
(c) $A = f(0)$ (d) $f(A) = -0.6$

8.

| $x$ | $-9.8$ | $-3.2$ | $-0.2$ | $4.5$ |
|---|---|---|---|---|
| $f(x)$ | $11.7$ | $2.1$ | $-5.2$ | $-7.6$ |

(a) $A = f(-1.5)$ (b) $f(A) = 3.9$
(c) $A = f(1.3)$ (d) $f(A) = -1.9$

## Goal B

In exercises 9–38, use Table 1 in the Appendix and linear interpolation to solve each of the following equations for $A$.

**9.** $\log 7.513 = A$ **10.** $\log 1.395 = A$

**11.** $\log 4.739 = A$ **12.** $\log 6.952 = A$

**13.** $\log 0.3045 = A$ **14.** $\log 52.36 = A$

**15.** $\log 27.98 = A$ **16.** $\log 0.4861 = A$

**17.** $\log 478.2 = A$ **18.** $\log 0.02948 = A$

**19.** $\log 0.05173 = A$ **20.** $\log 332.2 = A$

**21.** $\log 0.051042 = A$ **22.** $\log 0.60158 = A$

**23.** $\log 0.91375 = A$ **24.** $\log 0.01112 = A$

**25.** $\log A = 0.9501$ **26.** $\log A = 0.8778$

**27.** $\log A = 0.3292$ **28.** $\log A = 0.6131$

**29.** $\log A = 3.9844$ **30.** $\log A = 4.3299$

**31.** $\log A = -0.6573$ **32.** $\log A = -0.7313$

**33.** $\log A = -0.2397$ **34.** $\log A = -0.4658$

**35.** $\log A = -1.5797$ **36.** $\log A = -9.2412$

**37.** $\log A = -3.2846$ **38.** $\log A = -5.2108$

## Superset

In exercises 39–52, use logarithms to solve each of the following equations for $x$ to the nearest hundredth.

**39.** $x^2 = 2$ **40.** $x^5 = 2$

**41.** $8^x = 2$ **42.** $3^x = 8$

**43.** $12^x = 5$ **44.** $15^x = 9$

**45.** $x^3 = 18$ **46.** $x^2 = 5$

**47.** $2^x = \frac{1}{3}$ **48.** $x = (2.7)^{3.1}$

**49.** $x = \left(1 + \frac{1}{10}\right)^{10}$ **50.** $3^x = \frac{1}{2}$

**51.** $x = \frac{(32.8)(17.1)}{53.2}$

**52.** $x = \frac{(14.3)(0.28)}{(31.5)(0.036)}$

**53.** The product of the first $n$ positive integers is denoted by $n!$ (read "$n$ factorial"):

$$n! = 1 \cdot 2 \cdot 3 \cdot 4 \cdots (n-2)(n-1)n.$$

For example, $7! = 1 \cdot 2 \cdot 3 \cdot 4 \cdot 5 \cdot 6 \cdot 7 = 5040$. Use the fact that

$$\log n! = \log 1 + \log 2 + \log 3 + \cdots + \log n$$

and the logarithm table to estimate the number of digits in the number:

(a) $5!$ (b) $10!$ (c) $20!$

**54.** Complete the following table, where

$$f(x) = x - \frac{x^2}{2} + \frac{x^3}{3}$$

| $x$ | 0.05 | 0.10 | 0.15 | 0.20 | 0.25 |
|---|---|---|---|---|---|
| $\log(1 + x)$ | | | | | |
| $f(x)$ | | | | | |
| $0.4343 \cdot f(x)$ | | | | | |

What do you notice?

| $t$ | $N(t)$ |
|---|---|
| 0 | 1 |
| 1 | 2 |
| 2 | 4 |
| 3 | 8 |
| 4 | 16 |
| 5 | 32 |

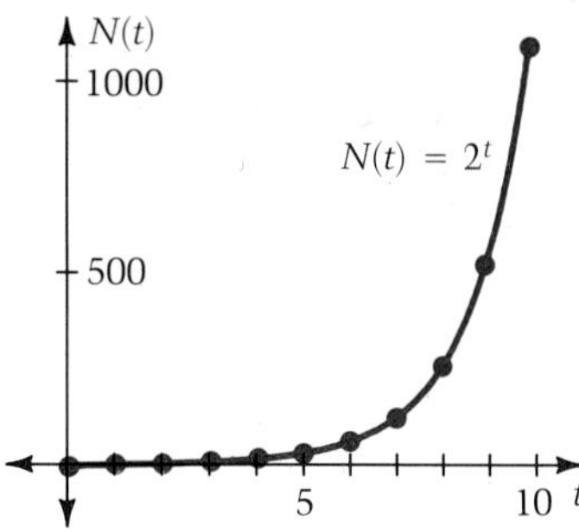


Figure 7.12

## 7.5 Applications and the Natural Logarithm

**Goal A** *to use common logarithms to solve applied problems*

Exponential and logarithmic functions are used to solve a variety of applied problems. Interest on savings and loans, population growth, radioactive decay, and earthquake intensity are some of those subject areas employing exponential or logarithmic models.

Consider the following hypothetical example of population growth. Suppose we are studying the growth of bacteria, and we begin with a single cell that splits after an hour of growth to form two cells. After another hour each of the two cells splits to form two more cells, making a total of four cells. This doubling process continues, and at the end of the third hour there is a total of eight cells. The data for the first 5 hours are shown in the table at the left ($t$ is the number of hours elapsed, and $N(t)$ is the number of cells at time $t$). A graph for $0 \le t \le 10$ is shown in Figure 7.12.

As both the graph and the table suggest, the function that models this situation is $N(t) = 2^t$, and its domain is the set of positive integers. We show a smooth curve to indicate the exponential nature of the process. Note that in this situation, the initial population was 1 cell.

If there are $B_0$ cells initially, and if the population doubles every hour, then after $t$ hours, the size of the population, $N(t)$, is given by the equation

$$N(t) = B_0 2^t.$$

**Example 1** The bacteria *E. coli* is found in many organisms, including humans. Suppose that, under certain conditions, the number of these bacteria present in an experimental colony is given by the equation $N(t) = B_0 2^t$. How many *E. coli* are present 6 hours after the start of the experiment, if there were 1,500,000 initially?

***Solution***

$$N(t) = B_0 2^t$$

$$N(6) = 1{,}500{,}000 \cdot 2^6$$ ■ Values are substituted for $B_0$ and $t$.

$$N(6) = 96{,}000{,}000$$

**Example 2** How long will it take the colony in Example 1 to grow to 20,000,000 bacteria?

***Solution***

$$N(t) = B_0 2^t$$

$$20{,}000{,}000 = 1{,}500{,}000 \cdot 2^t$$ ■ Values are substituted for $N(t)$ and $B_0$.

$$13.3 \approx 2^t$$

$$\log 13.3 \approx \log 2^t$$

■ We take the common log of both sides and then use Rule 4 of Logarithms to solve for $t$.

$$\log 13.3 \approx t \cdot \log 2$$

$$\frac{\log 13.3}{\log 2} \approx t$$

■ $\log 13.3 \approx 1.1239$ and $\log 2 \approx 0.3010$.

$$t \approx \frac{1.1239}{0.3010} \approx 3.7$$

There will be 20,000,000 bacteria in the colony after approximately 3.7 hours.

Example 2 demonstrates a technique for solving for exponents: first take the logarithm of both sides of the equation; then use Rule 4 of Logarithms to write the exponent as a coefficient in the equation.

Exponential and logarithmic models are used to solve problems concerning investments and loans. If $P$ dollars are invested at an interest rate $i$ (expressed as a decimal), and the interest is compounded $n$ times per year, then the amount, $A$, of money in the account after $t$ years is

$$A = P\left(1 + \frac{i}{n}\right)^{nt}.$$

**Example 3** Suppose \$2000 is deposited in an account that advertises a 9.2% interest rate compounded quarterly. What will the value of the account be in 25 years?

***Solution***

$$A = P\left(1 + \frac{i}{n}\right)^{nt}$$

$$A = 2000\left(1 + \frac{0.092}{4}\right)^{4(25)}$$

$$A = 2000(1.023)^{100}$$

■ We shall use logarithms to perform this complicated computation. If you have a calculator with a $y^x$ key, you can find $(1.023)^{100}$ directly.

$$\log A = \log[2000(1.023)^{100}]$$

$$\log A = \log 2000 + 100 \log 1.023$$

$$\log A \approx 3.3010 + 100(0.0099)$$

$$\log A \approx 4.291$$

$$A \approx 1.95 \times 10^4$$

■ We took the antilogarithm of both sides to find $A$.

The value of the account will be approximately \$19,500 in 25 years.

**Goal B** *to use natural logarithms to solve applied problems*

Let us perform an experiment to determine the effect of increasing the number of times interest is compounded each year. To make things simple, suppose that \$1.00 is invested for one year at an interest rate of 100% (i.e., $P = 1$, $t = 1$, and $i = 1$). In the table below, we show the value of the investment after one year as we vary the number of times the interest is compounded. (It is customary to use a 360-day year when compounding interest.)

| $n$ | **Interest is compounded:** | $A = \left(1 + \frac{1}{n}\right)^n$ |
|---|---|---|
| 1 | annually | 2 |
| 2 | semiannually | 2.25 |
| 4 | quarterly | ≈ 2.441406 |
| 12 | monthly | ≈ 2.613035 |
| 52 | weekly | ≈ 2.692597 |
| 360 | daily | ≈ 2.714516 |
| 8,640 | hourly | ≈ 2.718117 |
| 518,400 | each minute | ≈ 2.718262 |
| 31,104,000 | each second | ≈ 2.718282 |

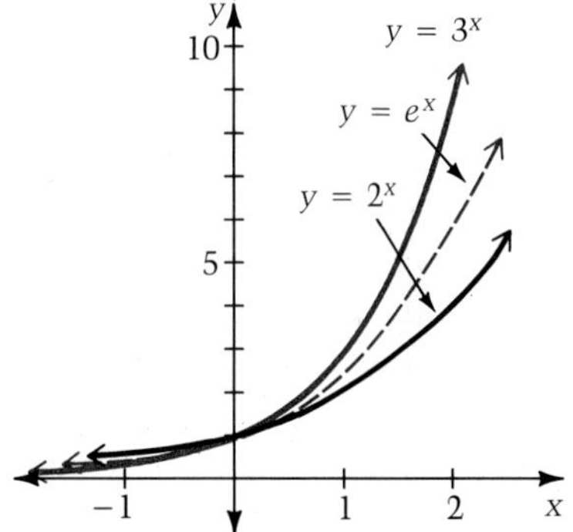

Figure 7.13

Notice that as $n$ increases, the value of the investment approaches some constant value. Like $\pi$, this constant value is an irrational number. It occurs so frequently that we give it a name. We call it ***e:***

$$e \approx 2.718282.$$

The function $y = e^x$ is called the **natural exponential function,** and the function $y = \log_e x$, usually written $y = \ln x$, is called the **natural logarithmic function.** Values for $e^x$, $e^{-x}$, and $\ln x$ are given in Tables 2 and 3 in the Appendix. As you might expect, since $e$ is between 2 and 3, the graph of $y = e^x$ lies between the graphs of $y = 2^x$ and $y = 3^x$.

When interest is advertised as being "compounded continuously", the model used to determine the value of an account after $t$ years is

$$A = Pe^{it},$$

where $P$ is the amount initially invested, and $i$ is the annual interest rate.

**Example 4** An amount of \$12,000 is placed in an account advertising an 8% annual interest rate, compounded continuously.

(a) How much is the account worth after 5 years have passed?
(b) How long is it before the account is worth twice the initial investment?

***Solution***

(a) $A = Pe^{it}$ ■ Substitute values for $P$, $i$, and $t$.

$A = 12{,}000 \cdot e^{(0.08)(5)}$

$A \approx 12{,}000 \cdot 1.49182$ ■ $e^{0.4} \approx 1.49182$ (see Table 2).

$A \approx 17{,}901$

After 5 years, the value of the account is approximately \$17,901.

(b) $24{,}000 = 12{,}000 \cdot e^{0.08t}$

$2 = e^{0.08t}$

$\ln 2 = \ln e^{0.08t}$

$\ln 2 = 0.08t \ln e$ ■ Rule 4 of Logarithms is used.

$\ln 2 = (0.08t)(1)$ ■ $\ln e = \log_e e = 1$ by Rule 2 of Logarithms.

$\frac{0.6931}{0.08} \approx t$ ■ $\ln 2 \approx 0.6931$ (see Table 3).

$t \approx 8.66$

The amount in the account will double its value in roughly $8\frac{2}{3}$ years.

The previous example describes an **exponential growth** model because the value of the account is increasing. Of equal interest is the **exponential decay** model. The most common exponential decay problems involve radioactive substances.

Carbon-14 is a radioactive substance present in all living matter. While an organism is alive, the amount of Carbon-14 it contains remains constant. Once the organism dies, however, the number of Carbon-14 atoms begins to decrease. This process is called **radioactive decay.** By comparing the level of radiation in a fossil with the level present in a similar living sample, we can estimate how long ago the fossil was a living organism. Essential to such an investigation is the formula

$$y = y_0 e^{kt},$$

where $y_0$ is the amount of radioactive substance present initially, $y$ is the amount present after $t$ years, and $k$ is the *decay constant.*

If $T$ is the half-life in years of the radioactive substance under study (i.e., the time it takes for a given sample to reduce to half its size), then

$$k = \frac{-1}{T} \ln 2 \approx \frac{-0.6931}{T}.$$

**Example 5** Carbon-14 has a half-life of 5750 years. Suppose that 100 grams of Carbon-14 were present in an organism when it lived 3000 years ago. How much Carbon-14 would remain now?

***Solution***

$k \approx \dfrac{-0.6931}{5750}$ ■ Begin by determining the decay constant.

$k \approx -0.00012$

$y \approx y_0 e^{(-0.00012)t}$ ■ The value of $k$ is substituted into the radioactive decay model.

$y \approx 100e^{(-0.00012)3000}$ ■ Values of $y_0$ and $t$ are substituted.

$y \approx 100e^{-0.36}$

$y \approx 100(0.69768)$ ■ $e^{-0.36} \approx 0.69768$ (see Table 2).

$y \approx 69.768$

There would be approximately 69.8 grams of Carbon-14 present today.

## Exercise Set 7.5 ⊟ Calculator Exercises in the Appendix

### Goal A

1. (Refer to Example 1) How many bacteria are present 7 hours after the start of the experiment, if there were 1350 initially?
2. (Refer to Example 1) How many bacteria are present 4 hours after the start of the experiment, if there were 58,000 initially?
3. (Refer to Example 1) How long will it take a colony of 5000 bacteria to triple in number?
4. (Refer to Example 1) How long ago were there less than 1000 bacteria in the colony, if there are 7500 present now?
5. If \$1000 is placed in an account that earns 8% compounded quarterly, what is the value of the account 10 years later?
6. If \$500 is placed in an account that earns 6% compounded semiannually, what is the value of the account 12 years later?
7. How much money must be placed in an account that earns 7% compounded semiannually, so that there will be \$2500 in the account 5 years from now?
8. How much money must be placed in an account that earns 6% compounded quarterly, so that there will be \$1000 in the account 8 years from now?

**9.** If you put 1¢ in your bank account today, 2¢ tomorrow, 4¢ the day after tomorrow, 8¢ the next day, and so on, in how many days will you have to deposit at least $100?

**10.** (Refer to exercise 9) Suppose instead, you deposit 1¢, 3¢, 9¢, 27¢, and so on. In how many days will you have to deposit at least $100?

## Goal B

**11.** If $1000 is placed in an account that earns 8% compounded continuously, what is the value of the account 10 years later?

**12.** If $500 is placed in an account that earns 6% compounded continuously, what is the value of the account 12 years later?

**13.** Suppose $1762 is placed in an account that earns 7% compounded continuously. How long will it be before the account is worth $2500?

**14.** Suppose $619 is placed in an account that earns 6% compounded continuously. How long will it be before the account is worth $1000?

**15.** Suppose there were 150 gm of Carbon-14 in an organism when it lived 10,000 years ago. How much Carbon-14 remains today?

**16.** Suppose there are 20 gm of Carbon-14 remaining in an organism known to have lived 2500 years ago. How many grams of Carbon-14 were in the organism when it lived?

**17.** Suppose we know that there were 180 gm of Carbon-14 in an organism when it lived and only 25 gm remain today. How long ago did the organism live?

**18.** Suppose there are 200 gm of Carbon-14 in an organism that dies today. In how many years will there be less than 75 gm left in the organism?

**19.** If inflation persists at 6% annually compounded continuously, how much will an item that costs $10 today cost in 5 years?

**20.** The population of the United States tends to grow at a rate proportional to the size of the population. Thus, we may represent the population size by an exponential model:

$$A(t) = 180e^{0.013t}$$

where $t$ is the number of years since 1960 and $A(t)$ is the population in millions. What will the population of the United States be in the year 2000?

## Superset

**21.** How much money must be placed in an account that earns 6% compounded continuously in order that $1000 may be withdrawn at the end of each year forever?

**22.** (Refer to exercise 21) How much must be placed in the account if the interest is compounded quarterly?

**23.** The **effective annual rate** of interest in an account that holds $A(t)$ dollars after $t$ years is defined by

$$\frac{[A(t) - A(0)]}{A(0)} \times 100\%.$$

What is the effective annual rate of interest on money invested at 6% compounded quarterly?

**24.** (Refer to exercise 23) What is the effective annual rate of interest on money invested at 6% compounded continuously?

**25.** At what rate, compounded continuously, must a sum of money be invested so that the actual earnings are 8% per year?

**26.** (Refer to exercise 20) In what year will the population of the United States exceed the current world population of 4 billion?

# Chapter Review & Test

## Chapter Review

### 7.1 Exponential Functions (pp. 242–246)

Suppose $x$, $y$, and $t$ are any real numbers and $b$ is positive. Then the *Rules of Real Number Exponents* are as follows (p. 242):

$$b^0 = 1 \quad b^1 = b \quad b^x b^y = b^{x+y} \quad (b^x)^t = b^{xt} \quad \frac{b^x}{b^y} = b^{x-y} \quad b^{-x} = \frac{1}{b^x}$$

A function of the form $y = b^x$, where $b$ is any positive constant except 1, is called an *exponential function*; $b$ is called the *base*. (p. 243) Four important features of the exponential function $f(x) = b^x$ are (p. 244):

- If $x > 0$, then as $b$ increases, $b^x$ increases.
- If $x < 0$, then as $b$ increases, $b^x$ decreases.
- For any base $b$, $b^0 = 1$.
- The $x$-axis is a horizontal asymptote.

### 7.2 Logarithmic Functions (pp. 247–254)

A function $f$ is *one-to-one* if each $y$-value in the range of $f$ corresponds to exactly one $x$-value in the domain. If a function $f$ is one-to-one, then there exists a unique *inverse function* $f^{-1}$ such that $(f^{-1} \circ f)(x) = x$ for any $x$ in the domain of $f$, and $(f \circ f^{-1})(x) = x$ for any $x$ in the domain of $f^{-1}$. The domain of $f$ becomes the range of $f^{-1}$, and the range of $f$ becomes the domain of $f^{-1}$. (p. 247) The graph of $y = f^{-1}(x)$ is the reflection of $y = f(x)$ across the line $y = x$. (p. 248)

The exponential function $f(x) = b^x$ has an inverse called the *logarithmic function with base b* and denoted $f(x) = \log_b x$. Thus, a logarithm is an exponent. The domain of the logarithmic function is $(0, +\infty)$, and the range is $(-\infty, +\infty)$. (p. 249) Since the logarithmic function is one-to-one, the equation $\log_b A = \log_b C$ implies $A = C$. (p. 250)

*Log/Exp Principle:* For $b > 0$ and $b \neq 1$, $b^A = C$ is equivalent to $\log_b C = A$. (p. 249)

Suppose $x$ and $y$ are positive real numbers, $t$ is any real number, and base $b$ is positive, but not equal to 1. The *Rules of Logarithms* are as follows (p. 251):

$$\log_b 1 = 0 \qquad \log_b b = 1 \qquad \log_b(xy) = \log_b x + \log_b y$$

$$\log_b(x^t) = t \log_b x \qquad \log_b\left(\frac{x}{y}\right) = \log_b x - \log_b y \qquad \log_b\left(\frac{1}{x}\right) = -\log_b x$$

Also remember that: $b^{\log_b x} = x$ and $\log_b b^x = x$. (p. 253)

### 7.3 Common Logarithms (pp. 255–258)

Base 10 logarithms are called *common logarithms;* we write $\log_{10} x$ as $\log x$. (p. 255) (Table 1 in the Appendix lists values of $\log x$.)

If $y = \log x$, then $x$ is called the *common antilogarithm* of $y$; that is, $x$ is the number you get when you raise 10 to the power $y$: $x = 10^y$. (p. 255)

### 7.4 Linear Interpolation (pp. 259–262)

The use of a line to approximate function values is called *linear interpolation.* (p. 259) We can use linear interpolation to estimate common logarithms and antilogarithms. (p. 260)

### 7.5 Applications and the Natural Logarithm (pp. 263–268)

Exponential and logarithmic functions serve as models for problems involving population growth, compound interest, and radioactive decay. (p. 263)

The function $y = e^x$, where $e \approx 2.718282$, is called the *natural exponential function;* the function $y = \log_e x$, usually written $y = \ln x$, is called the *natural logarithmic function.* (p. 265) (Values of $e^x$, $e^{-x}$, and $\ln x$ are given in Tables 2 and 3 in the Appendix.)

## Chapter Test

**7.1A** Simplify each of the following expressions.

**1.** $4^{x-1} \cdot 8^{2-3x}$

**2.** $\dfrac{9^{x+2}}{27^{2x}}$

**7.1B** Sketch the graph of each of the following equations.

**3.** $y = 3^{x-2}$

**4.** $y = \left(\dfrac{1}{2}\right)^x - 4$

**7.1C** Solve each of the following equations for $x$.

**5.** $3^{x+1} = 0$

**6.** $4^{-2x} = 8^{5x-2}$

**7.2A** The given function $y = f(x)$ is one-to-one. Find the equation describing the inverse function, sketch the inverse, and identify its domain and range.

**7.** $f(x) = 7 - 2x, \quad 1 \leq x \leq 4$

**8.** $f(x) = \sqrt{9 - x}, \quad x \leq 9$

**7.2B** Write each of the following as a logarithmic equation.

**9.** $x^3 = 125$ **10.** $4^x = 64$ **11.** $7^4 = x$

Use the Log/Exp Principle to solve each of the following equations for $x$.

**12.** $\log_3 27 = x$ **13.** $\log_x 16 = -2$

**7.2C** Sketch the graph of each of the following equations.

**14.** $y = \log_2(x - 3)$ **15.** $y = \log_{1/4} x$

**7.2D** Use the values $\log_2 x = 4.3$, $\log_2 y = 2.8$, and $\log_2 z = -3.1$ to evaluate each expression.

**16.** $\log_2\left(\frac{8x}{yz}\right)$ **17.** $\log_2(xy^2z^3)$

Solve each of the following equations for $x$.

**18.** $\log_4 2^x = -5$ **19.** $\log_3(\log_2 x) = 0$

**20.** $\log_2 x^3 = \log_2 9x$ **21.** $\log_2 8x = \log_2 8 + \log_2 x$

**7.3A** Use Table 1 in the Appendix to find each of the following.

**22.** $\log 7.14$ **23.** $\log 3.96$

Use scientific notation to rewrite each of the following numbers.

**24.** 473.89 **25.** 0.00451

Use Table 1 to compute each of the following.

**26.** $\log 27.4$ **27.** $\log 0.947$

**7.3B** Use Table 1 to solve each of the following equations for $x$.

**28.** $\log x = 0.6096$ **29.** $\log x = 1.4609$ **30.** $\log x = -1.5391$

**7.4A** Use linear interpolation and the function values given in the table below to approximate $A$ to the nearest tenth.

| $x$ | $-3$ | 2 | 8 | 15 |
|---|---|---|---|---|
| $f(x)$ | 4.2 | 1.7 | $-2.4$ | $-6.3$ |

**31.** $f(3) = A$ **32.** $f(A) = -3$

**7.4B** Use Table 1 and linear interpolation to solve each of the following equations for $A$.

**33.** $\log 0.2635 = A$

**34.** $\log 4728 = A$

**35.** $\log 539.6 = A$

**36.** $\log A = 1.9452$

**37.** $\log A = -2.3921$

**38.** $\log A = -3.3929$

**7.5A** Solve each of the following.

**39.** The number of bacteria present in an experimental colony is given by the equation $N(t) = B_0 2^t$, where $t$ is measured in hours. If there are 1,600,000 bacteria cells present initially, how long will it take the colony to grow to 2,400,000 bacteria?

**40.** If \$500 is placed in an account that earns 8% compounded quarterly, what is the value of the account 6 years later?

**7.5B** Solve each of the following.

**41.** If \$500 is placed in an account that earns 8% compounded continuously, what is the value of the account 6 years later?

**42.** Suppose there were 80 gm of Carbon-14 in an organism when it lived 300,000 years ago. How much Carbon-14 remains today? (Use the formula $y = y_0 e^{kt}$ where $k = -\frac{1}{T}\ln 2$.)

**Superset**

**43.** Solve $3^{x^5} = -1$ for $x$.

**44.** Solve $\log_2(x^2 - 5) = \log_2(4x)$ for $x$.

**45.** Sketch the graph of $y = 9^{\log_3 x}$.

**46.** Suppose some quantity $Q$ is related to another quantity $x$ by the relationship $Q = \log_3 x$. What happens to the value of $Q$ if $x$ is doubled? tripled? quadrupled?

**47.** Use logarithms to solve each of the following equations for $x$ to the nearest hundredth.

(a) $x^2 = 3$ (b) $2^x = 6$

**48.** What is the effective annual rate of interest on money invested at 10% compounded continuously?

## Cumulative Test: *Chapters 6 and 7*

In exercises 1–2, simplify each expression.

**1.** $(5 - 3i)(-2 + 4i)$ **2.** $(i - 7)(2 - i)i$

In exercises 3–4, simplify each quotient.

**3.** $\dfrac{6 + 5i}{-3i}$ **4.** $\dfrac{5i}{3 - 2i}$

In exercises 5–6, represent the following complex numbers as a point in the complex plane.

**5.** $5 - 3i$ **6.** $8 + i$

In exercises 7–8, determine the absolute value of each complex number.

**7.** $6i$ **8.** $-5 + 9i$

In exercises 9–10, solve the following equations over the set of complex numbers.

**9.** $2x^2 - 6x + 5 = 0$ **10.** $5x^2 = -2x - 3$

In exercises 11–12, a polynomial equation and one of its solutions is given. Find all the solutions to the polynomial equation.

**11.** $x^3 + x^2 - 16x - 16 = 0, \ -1$

**12.** $x^4 + 4x^3 - 8x^2 + 16x - 48 = 0, \ 2i$

In exercises 13–14, express the following complex numbers in trigonometric form.

**13.** $2 - 2\sqrt{3}i$ **14.** $-\sqrt{2} + \sqrt{2}i$

In exercises 15–16, using trigonometric forms determine $z_1z_2$ and $\dfrac{z_1}{z_2}$. Express in the form $a + bi$.

**15.** $z_1 = 2 - 2i\sqrt{3}, z_2 = -1 + i$ **16.** $z_1 = -\sqrt{2} + i\sqrt{2}, z_2 = \sqrt{3} + i$

In exercises 17–18, use DeMoivre's Theorem to express in the form $a + bi$.

**17.** $z = (1 + i)^4$ **18.** $z = (\sqrt{3} + i)^5$

In exercises 19–20, simplify each of the following expressions.

**19.** $2^{5x-2} \cdot 8^{-x-1}$ **20.** $\dfrac{5^{2x+1}}{25^{-x/2}}$

In exercises 21–22, sketch the graph of each equation.

**21.** $y = 2^{x-2}$ **22.** $y = \left(\dfrac{1}{3}\right)^x + 3$

In exercises 23–24, solve each equation for $x$.

**23.** $4 = 8(2^{x-3})$

**24.** $25^x = \dfrac{5^{x+1}}{125}$

In exercises 25–26, the given function $y = f(x)$ is one-to-one. Find the equation describing the inverse function, sketch the inverse, and identify its domain and range.

**25.** $f(x) = 2x - 5;\ 0 \le x \le 3$

**26.** $f(x) = \sqrt{x - 1},\ 1 \le x \le 5$

In exercises 27–28, rewrite the following as logarithmic equations.

**27.** $16^x = 2$

**28.** $x^{1/3} = 5$

In exercises 29–30, sketch the graph of each equation.

**29.** $y = \log_{1/4} x$

**30.** $y = \log_{1/4} 2x$

In exercises 31–32, solve the following equations for $x$.

**31.** $\log_2(x^2 - 6) = \log_2 x$

**32.** $2 \log_2 x = \log_2(6x - 8)$

In exercises 33–34, use Table 1 to solve each equation for $x$.

**33.** $\log x = -1.0128$

**34.** $\log x = 3.6911$

In exercises 35–36, use Table 1 and linear interpolation to solve each equation for $A$.

**35.** $\log 0.8214 = A$

**36.** $\log A = -2.8090$

In exercises 37–38, use the formula $A = P(1 + \frac{i}{n})^{nt}$ to solve the following compound interest problems.

**37.** Suppose \$3000 is deposited into an account that earns 9% interest compounded semi-annually. What will the value of the account be in 10 years?

**38.** What would the value of the account be in 10 years if the interest is compounded monthly? daily? (Use 360 days per year.)

### Superset

**39.** Solve the following logarithmic equations.

(a) $\log_2(\log_2 4^x) = 1$ (b) $\log_3 x - \log_9 2x = 0$

**40.** Use the fact that $\log_b x = \dfrac{\log x}{\log b}$ to estimate $\log_2 12$.

**41.** Simplify each expression. Write your answer in the form $a + bi$, $a$, or $bi$.

(a) $\dfrac{5 + i}{3 - 2i} - \dfrac{7 + 2i}{3i}$ (b) $\dfrac{|1 - i|}{|-2 - i|}$

**42.** Solve $5x^3 - 10x^2 + 25x - 50 = 0$ over the set of complex numbers.

# Appendix

## Calculator Exercises

### Exercise Set 1.2

In exercises 1–6, determine $f(x)$ for the $x$-values 1, 1.2, 1.4, 1.6, 1.8, and 2.

**1.** $f(x) = 6x - 5$

**2.** $f(x) = 0.02x - 3.76$

**3.** $f(x) = x^2 - 5x + 3.66$

**4.** $f(x) = 3.5x^2 - 0.05x + 1$

**5.** $f(x) = \sqrt{\dfrac{x - 0.08}{2}}$ (round to nearest thousandth)

**6.** $f(x) = \sqrt{\dfrac{16 - 5.2x}{0.05}}$ (round to nearest thousandth)

### Exercise Set 1.4

In exercises 1–8, determine the slope-intercept form ($y = mx + b$) of the line that is described. Round the values of $m$ and $b$ to the nearest hundredth after all computations are performed.

**1.** line passes through (3.41, 8.22) and (−1.73, 4.56)

**2.** line passes through (−7.80, −5.49) and (5, 2.11)

3. line passes through (–1, 5.37) and (2.12, 1.48)

4. line passes through (0, 0.82) and (1.03, −0.08)

5. line passes through (5.71, 3.08) and is parallel to the graph of $y = -0.89x + 1.7$

6. line passes through (6.88, −0.21) and is parallel to the graph of $4.80x - 7.68y + 8 = 0$

7. line passes through (0.53, −0.72) and is perpendicular to the graph of $1.20x + 6.30y - 8.95 = 0$

8. line passes through (−1.78, −6.81) and is perpendicular to the graph of $6.30x - 2.8y - 5 = 0$

In exercises 9–16, (a) carefully graph the given linear function over the interval $0 \le x \le 5$; (b) use your graph to estimate $f(2.5)$; (c) use your calculator to determine $f(2.5)$ exactly.

9. $f(x) = 2x - 7$
10. $f(x) = 5 - 3x$
11. $f(x) = -2.2x + 7.4$
12. $f(x) = 9.6x - 13.8$
13. $f(x) = 0.2x + 5.1$
14. $f(x) = 0.03x - 1.2$
15. $f(x) = \pi - \sqrt{2}x$
16. $f(x) = \sqrt{2}x - \pi$

## Exercise Set 1.5

In exercises 1–8, (a) carefully graph the given quadratic function over the interval $0 \le x \le 5$; (b) use your graph to estimate $f(2.5)$; (c) use your calculator to determine $f(2.5)$ exactly.

1. $f(x) = 4 - 0.2x^2$
2. $f(x) = 3 + 0.3x^2$
3. $f(x) = x^2 - 3x + 4$
4. $f(x) = x^2 - x - 3$
5. $f(x) = 0.4x^2 - 1.1x$
6. $f(x) = 10.3x - 2.1x^2$
7. $f(x) = 0.3x^2 - x - 0.43$
8. $f(x) = 0.2x^2 - 0.08x - 3.92$

The height in feet $h(t)$ of an object shot upwards (from ground level) with a velocity of $v_0$ ft/sec can be described as a function of time ($t$): $h(t) = v_0t - 16t^2$. In exercises 9–12, determine the time at which the object reaches its maximum height, and the maximum height achieved. (After computation, round answers to the nearest tenth.)

9. $v_0 = 80.5$ ft/sec
10. $v_0 = 22.6$ ft/sec
11. $v_0 = 50.2$ ft/sec
12. $v_0 = 197.4$ ft/sec

## Exercise Set 2.1

In exercises 1–10, $\triangle ABC$ is a right triangle, $c$ is the length of the hypotenuse and $a$ and $b$ are the lengths of the legs. Given the lengths of two sides, use the Pythagorean Theorem to find the length of the third side.

**1.** $a = 42.72 \quad b = 31.68$
**2.** $a = 34.35 \quad b = 63.45$
**3.** $a = 293.4 \quad c = 568.3$
**4.** $b = 349.6 \quad c = 638.2$
**5.** $a = 0.7184 \quad b = 0.6319$
**6.** $a = 1.231 \quad b = 1.893$
**7.** $b = 2462 \quad c = 4091$
**8.** $a = 5000 \quad c = 5663$
**9.** $a = 12.427 \quad c = 22.645$
**10.** $b = 79.351 \quad c = 83.381$

## Exercise Set 2.2

In exercises 1–10, use the given information about the right triangle $ABC$ from exercises 1–10 in Exercise Set 2.1 to find the values of the six trigonometric ratios of $\angle A$.

In exercises 11–14, given an equilateral triangle $ABC$ with side of length $s$ and altitude of length $h$, find the indicated part.

**11.** If $s = 6.348$, find $h$.
**12.** If $h = 15.259$, find $s$.
**13.** If $h = 25.9808$, find the perimeter of $\triangle ABC$.
**14.** If $h = 9.4873$, find the area of $\triangle ABC$.

In exercises 15–20, $\alpha$ and $\beta$ are complementary. From the given information, find the indicated trigonometric ratios.

**15.** $\sin \alpha = 0.3846$, find $\cos \alpha$ and $\cos \beta$.
**16.** $\cos \alpha = 0.4286$, find $\sin \alpha$ and $\tan \beta$.
**17.** $\csc \alpha = 1.1333$, find $\tan \alpha$ and $\cos \beta$.
**18.** $\sec \alpha = 1.3810$, find $\tan \beta$ and $\tan \alpha$.
**19.** $\cos \alpha = 0.9459$, find $\tan \alpha$ and $\sin \beta$.
**20.** $\sin \alpha = 0.2800$, find $\cos \alpha$ and $\tan \alpha$.

## Exercise Set 2.3

In exercises 1–6, the given point is on the terminal side of angle $\alpha$ in standard position. Determine the values of the six trigonometric functions of $\alpha$.

**1.** $(-5\sqrt{7}, 3\sqrt{3})$
**2.** $(-4\sqrt{5}, -7\sqrt{5})$

3. (25.34, −72.67)

4. (2.872, −6.449)

5. (−13.856, −5.657)

6. (−11.180, 6.471)

## Exercise Set 2.4

In exercises 1–4, two approximate trigonometric function values of an angle are given. Find the other four trigonometric function values rounded to four decimal places.

1. $\sin 32° = 0.5299$, $\cos 32° = 0.8480$
2. $\tan 66° = 2.2460$, $\sec 66° = 2.4586$
3. $\tan 55° = 1.4281$, $\sin 55° = 0.8192$
4. $\tan 23° = 0.4245$, $\csc 23° = 2.5593$

In exercises 5–8, use the Reciprocal and Pythagorean Identities to determine the values of the other five trigonometric functions of $\alpha$.

5. $\sin \alpha = 0.5124$

6. $\cos \alpha = 0.3051$

7. $\cos \alpha = 0.8236$

8. $\sin \alpha = 0.7645$

A calculator will compute the trigonometric function values of angles measured either in degrees or radians. You can select the units for the angular measure with a degree/radian key (DRG on some calculators, DEG on others). Be sure the calculator is in the degree mode for the next exercise set.

In exercises 9–14, use a calculator to determine the following.

9. $\sin 31.2°$

10. $\cos 43.8°$

11. $\tan 21.66°$

12. $\tan 49.33°$

13. $\cos 52.25°$

14. $\sin 80.46°$

To use a calculator to evaluate the cotangent, secant and cosecant functions, it is necessary to use the Reciprocal Identities. For example:

$$\csc 36° = \frac{1}{\sin 36°}.$$

To take a reciprocal of a number $x$, use the reciprocal key, $1/x$. (Be sure that the calculator is in the degree mode.)

In exercises 15–20, use a calculator to determine the following.

15. $\csc 56.3°$

16. $\csc 10.5°$

17. $\cot 48.2°$

18. $\sec 15.7°$

19. $\sec 45.5°$

20. $\cot 78.9°$

To find an angle $\alpha$ to two decimal places such that $\sin \alpha = 0.5402$, we use the inverse key, INV.

ENTER .5402 PRESS INV PRESS sin
DISPLAY 32.69725475

Thus, $\alpha = 32.70°$. (If you did not get this answer, check to be sure you have your calculator in the degree mode.)

The inverse key followed by the sine key "undoes" what the sine key "does". The INV key, when followed by the sin, cos, or tan key calculates the smallest angle in Quadrant I having the positive trigonometric function value entered in the display (negative trigonometric function values entered in the display produce an angle in Quadrant IV).

In exercises 21–26, use a calculator to determine the acute angle $\alpha$ in degrees to two decimal places.

**21.** $\sin \alpha = 0.4695$ **22.** $\cos \alpha = 0.8480$

**23.** $\cos \alpha = 0.7698$ **24.** $\sin \alpha = 0.4669$

**25.** $\cot \alpha = 1.072$ **26.** $\sec \alpha = 3.078$

In exercises 27–36, evaluate each of the following to four decimal places.

**27.** $\cos 408°$ **28.** $\tan 597°$ **29.** $\tan 212°$ **30.** $\sin 320°$

**31.** $\sin(-1134°)$ **32.** $\cos(-666°)$ **33.** $\cos 196.2°$ **34.** $\sin 196.2°$

**35.** $\csc(-287°)$ **36.** $\sec(-136°)$

In exercises 37–44, find $\alpha$ in degrees to two decimal places.

**37.** $\cos \alpha = -0.7145$, $\alpha$ in Quadrant II

**38.** $\sin \alpha = -0.7593$, $\alpha$ in Quadrant III

**39.** $\tan \alpha = -10.988$, $\alpha$ in Quadrant IV

**40.** $\tan \alpha = -4.915$, $\alpha$ in Quadrant II

**41.** $\sin \alpha = -0.6211$, $\alpha$ in Quadrant IV

**42.** $\cos \alpha = 0.9898$, $\alpha$ in Quadrant IV

**43.** $\cot \alpha = 3.420$, $\alpha$ in Quadrant III

**44.** $\csc \alpha = -1.070$, $\alpha$ in Quadrant III

## Exercise Set 3.1

If $P$ is the terminal point of a standard arc of length $s$, then

$$P \text{ is in Quadrant I when } 0 < s < \frac{\pi}{2},$$

$$P \text{ is in Quadrant II when } \frac{\pi}{2} < s < \pi,$$

$$P \text{ is in Quadrant III when } \pi < s < \frac{3\pi}{2},$$

$$P \text{ is in Quadrant IV when } \frac{3\pi}{2} < s < 2\pi,$$

$$P \text{ is in Quadrant I when } 2\pi < s < \frac{5\pi}{2}, \text{ etc.}$$

In exercises 1–10, given that the real number is represented as a standard arc, determine the quadrant in which the standard arc terminates.

**1.** 25 **2.** 32 **3.** $7\sqrt{11}$ **4.** $6\sqrt{3}$ **5.** $-416$

**6.** $-128$ **7.** $3\sqrt{5} + 21$ **8.** $2\sqrt{7} - 34$ **9.** $\sqrt{\pi} + 24$ **10.** $\pi^2 + 10$

In exercises 11–14, the point $(x, y)$ is on the unit circle. You are given one of its coordinates and the quadrant in which the point lies. Determine an approximate value of the other coordinate to four decimal places.

**11.** $x = 0.3672$, Quadrant IV **12.** $y = 0.7218$, Quadrant II

**13.** $y = -0.4516$, Quadrant III **14.** $x = -0.6711$, Quadrant II

In exercises 15–22, convert the measure of the angle from degrees to radians or vice versa. Approximate all answers to three decimal places.

**15.** 35.8° **16.** 169.2° **17.** 248°36′ **18.** −312°48′

**19.** 12.5664 **20.** 15.7080 **21.** −5.4264 **22.** −18.1681

## Exercise Set 3.2

In exercises 1–14, determine the function values to four decimal places. (Be sure the calculator is in the radian mode.)

**1.** sin 1.0417 **2.** cos 0.4189 **3.** cos 2.0944 **4.** sin 5.2360

**5.** sin −7.3304 **6.** cos −8.6394 **7.** tan 26.1799 **8.** tan −13.1947

**9.** cos 10.5000 **10.** sin 5.0000 **11.** sec 8.500 **12.** csc 3.7000

**13.** cot 0.1768 **14.** cot 27.3200

## Exercise Set 3.3

In exercises 1–6, evaluate the given function for the $x$-values 1, $-3$, 21, 7.1416, $-15.7080$, and 235.

**1.** $f(x) = \sin(x + \pi)$

**2.** $f(x) = \cos(x - \pi)$

**3.** $f(x) = -\sin\left(x - \frac{\pi}{4}\right)$

**4.** $f(x) = -\cos\left(x + \frac{\pi}{4}\right)$

**5.** $f(x) = \tan\left(x - \frac{\pi}{2}\right)$

**6.** $f(x) = \cot\left(x - \frac{\pi}{6}\right)$

## Exercise Set 3.4

In exercises 1–4, evaluate the given function for the $x$-values 1, $-2.5$, 4.512, 24.67, $-31.56$, and 49.223.

**1.** $f(x) = 2\sin\left(\frac{1}{2}x + \frac{\pi}{4}\right)$

**2.** $f(x) = 2\cos(2x + \pi)$

**3.** $f(x) = \sqrt{3}\cos\left(-\frac{1}{2}x - \frac{\pi}{2}\right)$

**4.** $f(x) = \frac{3}{2}\sin\left(-x + \frac{\pi}{2}\right)$

## Exercise Set 3.5

In exercises 1–8, evaluate the expression. Express answers to four decimal places, with angles expressed in radians.

**1.** $\sin^{-1}\left(\frac{\sqrt{5}}{12}\right)$

**2.** $\arcsin\left(\frac{\sqrt{7}}{3}\right)$

**3.** $\arctan(-\sqrt{11})$

**4.** $\tan^{-1}\left(\frac{\sqrt{17}}{\sqrt{3}}\right)$

**5.** $\sin(\arctan(3.1562))$

**6.** $\cos(\arcsin(-0.7246))$

**7.** $\tan^{-1}(\tan(36.7^\circ))$

**8.** $\tan(\tan^{-1}(-24))$

## Exercise Set 4.2

We saw that the trigonometric values for angles of 15° and 75° can be calculated exactly using the special angles and the Sum and Difference Identities. In exercises 1–6, compare the trigonometric function values found on the calculator, with the value found by using the special angles and identities.

**1.** $\sin 15^\circ$ **2.** $\cos 15^\circ$ **3.** $\cos 75^\circ$ **4.** $\sin 75^\circ$ **5.** $\tan 15^\circ$ **6.** $\tan 75^\circ$

In exercises 7–12, suppose $\sin \alpha = -0.3472$, $\alpha$ a third quadrant angle, and $\cos \beta = -0.6561$, $\beta$ a second quadrant angle. Compute the given expression in two ways: first determine $\alpha$ and $\beta$, then evaluate the trigonometric function value. Second, use the appropriate sum or difference identity.

**7.** $\sin(\alpha + \beta)$ **8.** $\cos(\alpha + \beta)$ **9.** $\cos(\alpha - \beta)$

**10.** $\sin(\alpha - \beta)$ **11.** $\tan(\alpha + \beta)$ **12.** $\tan(\alpha - \beta)$

## Exercise Set 4.3

In exercises 1–4, suppose that $\theta$ is acute and $\sin \theta = 0.3846$. Compute the given trigonometric function value.

**1.** $\sin 2\theta$ **2.** $\cos 2\theta$ **3.** $\cot 2\theta$ **4.** $\tan 2\theta$

In exercises 5–8, suppose that $\theta$ is a second quadrant angle and $\cos \theta = -0.6154$. Compute the given trigonometric function value.

**5.** $\cos \frac{\theta}{2}$ **6.** $\sin \frac{\theta}{2}$ **7.** $\tan \frac{\theta}{2}$ **8.** $\cot \frac{\theta}{2}$

## Exercise Set 4.5

In exercises 1–10, solve the equation for $x$ in the interval $[0°, 360°)$.

**1.** $12 \sin^2 x - \sin x - 1 = 0$ **2.** $15 \cos^2 x + \cos x - 6 = 0$

**3.** $\sin x + 3 \cos x = 0$ **4.** $4 \cos^2 x + 2 \sin x = 3$

**5.** $2 \cos 2x - 2 \sin^2 x = 1$ **6.** $\tan^2 2x + 2 \tan 2x = 1$

**7.** $\sec^2 x + 2 \tan x - 2 = 0$ **8.** $3 \tan^2 x + 2 \sec x = 5$

**9.** $\cos^2 x - 0.97 \cos x + 0.18 = 0$ **10.** $\sin^2 x + 0.24 \sin x - 0.1081 = 0$

## Exercise Set 5.1

In exercises 1–10, solve the right triangle $ABC$ with right angle at $C$.

**1.** $a = 23.47$, $A = 32.8°$ **2.** $a = 1.831$, $B = 58.2°$

**3.** $b = 1293.0$, $A = 47.73°$ **4.** $b = 2.828$, $A = 12.62°$

**5.** $c = 19.800$, $A = 28.19°$ **6.** $c = 505.07$, $B = 41.38°$

**7.** $a = 2.3261$, $b = 2.4495$ **8.** $a = 84.853$, $c = 305.123$

**9.** $b = 0.9847$, $a = 0.2231$ **10.** $b = 398.24$, $c = 1763.66$

**11.** A 52 ft long guy wire runs from level ground to the top of an antenna. The guy wire makes an angle of 68°36′ with the antenna. How far out from the base of the antenna is the guy wire anchored?

12. A railroad track passes below a long, level highway bridge and at a right angle to it. At the same instant that a train traveling 60 mph passes under the bridge, a car traveling 45 mph on the bridge is directly above the train. If the highway bridge is 100 feet above the track, find the distance between the train and car 10 seconds later. (Note that 60 mph = 88 ft/s and 45 mph = 66 ft/s.)

## Exercise Set 5.2

In exercises 1–10, solve triangle $ABC$.

1. $A = 101.7°$, $B = 19.3°$, $c = 118.5$
2. $A = 42.8°$, $C = 51.1°$, $b = 92.75$
3. $B = 75.6°$, $C = 33.2°$, $a = 4384.0$
4. $A = 48.6°$, $B = 94.1°$, $c = 282.84$
5. $a = 14.95$, $b = 17.43$, $A = 58.6°$
6. $b = 0.4731$, $a = 0.4347$, $B = 78.4°$
7. $a = 6241$, $c = 8233$, $C = 31.7°$
8. $a = 3.3569$, $c = 2.7973$, $A = 62.5°$
9. $b = 756.2$, $c = 541.6$, $B = 112.3°$
10. $b = 464.5$, $c = 637.3$, $C = 73.8°$
11. A wind storm caused a 64 ft tall telephone pole to lean due east at an angle of 63°48′ to the ground. A 106 ft long guy wire is then attached to the top of the pole and anchored in the ground due west of the foot of the pole. How far from the foot of the pole is the guy wire anchored?
12. Two lighthouses are located 49 mi apart on the coast of Massachusetts, one near Salem and the other in Provincetown on Cape Cod. The Provincetown lighthouse is on a bearing of S46°E from the Salem lighthouse. A fishing boat in distress radios an SOS signal. The bearing of the signal from the Salem station is S55°12′E and from Provincetown N47°36′E. How far is the ship from Provincetown?

## Exercise Set 5.3

In exercises 1–10, solve the triangle $ABC$.

1. $a = 5808$, $b = 7920$, $C = 109.3°$
2. $a = 39.12$, $b = 36.37$, $C = 80.7°$

3. $b = 9.591$, $c = 6.856$, $A = 51.5°$

4. $b = 207.36$, $c = 214.48$, $A = 98.4°$

5. $a = 0.9674$, $c = 0.8834$, $B = 61.6°$

6. $a = 8347.0$, $c = 6723.0$, $B = 68.71°$

7. $a = 2.2365$, $b = 3.4641$, $c = 3.1623$

8. $a = 46.90$, $b = 38.73$, $c = 33.17$

9. $a = 0.2500$, $b = 0.1887$, $c = 0.3448$

10. $a = 3218.8$, $b = 3555.3$, $c = 3806.7$

11. Determine the perimeter of a regular octagon inscribed in a circle of radius 24.6 in.

12. Determine the perimeter of a regular 12-gon inscribed in a circle of radius 35 cm.

In exercises 13–16, find the area of triangle $ABC$.

13. $A = 41.8°$, $b = 35.49$, $c = 42.78$

14. $B = 69.6°$, $a = 479.6$, $c = 404.9$

15. $C = 105.1°$, $a = 3.784$, $b = 3.465$

16. $a = 8.485$, $b = 5.568$, $c = 7.616$

17. Find the area of the octagon in exercise 11.

18. Two sides and the included angle of a parallelogram have measures 17.88, 21.91 and 54.4°, respectively. Find the area of the parallelogram.

## Exercise Set 5.4

In exercises 1–4, points $A$ and $B$ are given. Describe the vector from $A$ to $B$ as an ordered pair and determine its magnitude and direction.

1. $A(2.37, -1.89)$, $B(-2.75, 3.45)$

2. $A(-2.64, -3.31)$, $B(3.73, 1.80)$

3. $A(-5.57, -2.48)$, $B(1.96, -3.73)$

4. $A(7.57, 3.09)$, $B(-3.78, -1.07)$

In exercises 5–8, vector **a** has magnitude $\|\mathbf{a}\|$ and direction $\theta$, vector **b** has magnitude $\|\mathbf{b}\|$ and direction $\varphi$. Express vector $\mathbf{c} = \mathbf{a} + \mathbf{b}$ as an ordered pair.

5. $\|\mathbf{a}\| = 9.89$, $\theta = 87°$
$\|\mathbf{b}\| = 2.79$, $\varphi = 73°$

6. $\|\mathbf{a}\| = 65.4$, $\theta = 42°$
$\|\mathbf{b}\| = 78.8$, $\varphi = 64°$

7. $\|\mathbf{a}\| = 979.8$, $\theta = 205.6°$
$\|\mathbf{b}\| = 346.4$, $\varphi = 81.4°$

8. $\|\mathbf{a}\| = 286.2$, $\theta = 101.2°$
$\|\mathbf{b}\| = 303.0$, $\varphi = 293.6°$

## Exercise Set 5.5

In exercises 1–4, the magnitude $\|\mathbf{a}\|$ and direction $\theta$ of the vector **a** are given. Find $\mathbf{a}_x$ and $\mathbf{a}_y$.

**1.** $\|\mathbf{a}\| = 9.89,\ \theta = 87°$

**2.** $\|\mathbf{a}\| = 65.4,\ \theta = 42°$

**3.** $\|\mathbf{a}\| = 979.8,\ \theta = 205.6°$

**4.** $\|\mathbf{a}\| = 286.2,\ \theta = 101.2$

**5.** Along a stretch of the Mississippi River, the water flows directly south with a current of 5 mph. A person who can row a boat at 4 mph in still water attempts to row due west across the river. What is the actual direction and speed of the boat?

**6.** Two forces of 26 newtons and 40 newtons act on a point in the plane. If the angle between the forces is 73.6°, find the magnitude of the resultant force and its angle with the smaller of the two given forces.

**7.** An object weighing 100 lb is suspended between two trees by two ropes attached at the top of the object. The rope off to the left makes an angle of 58° with the vertical and the rope off to the right an angle of 27° with the vertical. Find the tension (magnitude of the force vector) along each rope.

## Exercise Set 5.6

In exercises 1–4, an object is in simple harmonic motion described by the given equation, where $x$ is measured in cm and $t$ in seconds. Determine where the object is and whether it is moving up or down at $t = 0$, $t = 1.35$, $t = 2.57$, and $t = 12.39$. (Note: Take displacement below the equilibrium point as positive and above as negative.)

**1.** $x = 0.56 \sin 4.2t$

**2.** $x = 9.32 \cos(0.37t)$

**3.** $x = 0.24 \cos\left(\frac{t}{0.25} + 1.05\right)$

**4.** $x = 2.36 \sin(1.25t - 0.79)$

In exercises 5–8, the point is given in polar form. Determine the rectangular coordinates of the point.

**5.** $(-5.38, 231.7°)$

**6.** $(4.32, 325.9°)$

**7.** $(2.19, 2.59)$

**8.** $(-3.07, 0.93)$

In exercises 9–12, the point is written with rectangular coordinates. Determine the polar coordinates of the given point for $\theta$ in the interval $[0, \pi)$.

**9.** $(-3.83, 5.25)$

**10.** $(4.14, -3.86)$

**11.** $(-5.20, -3.12)$

**12.** $(7.46, 2.39)$

In exercises 13–16, for each of the polar equations determine the values of $r$ for the following values of $\theta$: 0, 1, 1.5, 2.25, 3.14, 4, 4.71, 5.12, 6.28.

**13.** $r = 3 + \sin\theta$ **14.** $r = 1 - 2\cos\theta$

**15.** $r = 1.4\theta$ **16.** $r = 1.5\sin 5\theta$

## Exercise Set 7.1

**1.** Use the $y^x$-key on your calculator to determine $2^x$ for the $x$-values 3.10, 3.11, 3.12, 3.13, 3.14, 3.15. Which of these six values most closely approximates $2^\pi$?

**2.** Which value is greater, $3^\pi$ or $\pi^3$?

In exercises 3–12, use a calculator to estimate each of the following to the nearest hundredth.

**3.** $2^{\sqrt{2}}$ **4.** $2^{\sqrt{3}}$ **5.** $3^{\sqrt{2}+\sqrt{3}}$ **6.** $3^{1-\sqrt{2}}$

**7.** $\pi^{2-\sqrt{2}}$ **8.** $\pi^{3+\sqrt{3}}$ **9.** $(2.72)^{-0.25}$ **10.** $(2.72)^{-3.14}$

**11.** $\sqrt{2}^{\sqrt{2}}$ **12.** $\pi^\pi$

In exercises 13–16, an equation and three values are given. Determine which of the three values best approximates a solution of the equation.

**13.** $3^x = 1 + x$; $-0.41$, $-0.17$, $1.15$

**14.** $2^{x-1} = 1 - x$; $-1.15$, $0.35$, $1.5$

**15.** $2^{x-2} = \sqrt{x - 1}$; 1.0, 1.5, 2.2

**16.** $2^{x+1} = x^2 - x + 1$; $-0.8$, $-0.4$, $0.4$

## Exercise Set 7.4

In exercises 1–12, solve the given equation by taking the logarithm of each side and then solving for $x$. Round answers to the nearest thousandth. Check answers by using the $y^x$ key on your calculator.

**1.** $6^x = 3$ **2.** $8^x = 50$

**3.** $(2.1)^x = 10$ **4.** $2^x = 100$

**5.** $3^x = 4^{x-1}$ **6.** $(3.5)^x = (2.2)^{x+1}$

**7.** $x^5 = 7x, x > 0$ **8.** $x^8 = 3x, x \neq 0$

**9.** $(x - 1)^5 = 0.0015, x > 1$ **10.** $(x + 1)^8 = 37.92, x > -1$

**11.** $15x^6 = x, x > 0$ **12.** $0.01x^7 = x, x > 0$

## Exercise Set 7.5

**1.** Suppose \$18,500 is placed in an account that advertises a 9.13% annual interest rate compounded continuously.
   (a) How much is the account worth after $4\frac{1}{2}$ years?
   (b) How long will it take for the account to be worth three times the initial investment?

**2.** If 86.2 gm of Carbon-14 were present in an organism when it was alive, and 38.7 gm are present in the remains today, how long ago did the organism live?

**3.** A new water treatment installation continuously reduces the percentage of pollutant in the water by $\frac{1}{50}$ per day ($k = -0.02$). How many days does it take to cut the pollutant in the water to less than 1% of what it was originally?

**4.** A certain species of bird is in danger of extinction. It is estimated that only 900 of these creatures are still alive. Estimates five years ago placed the population at 1200. Experts claim that once the population drops below 200, this species' situation will be irreversible. In how many years will this happen? (Assume that the population changes at a rate proportional to its size.)

# Tables

## Table 1 Common Logarithms

| x | 0 | 1 | 2 | 3 | 4 | 5 | 6 | 7 | 8 | 9 |
|---|---|---|---|---|---|---|---|---|---|---|
| **1.0** | .0000 | .0043 | .0086 | .0128 | .0170 | .0212 | .0253 | .0294 | .0334 | .0374 |
| **1.1** | .0414 | .0453 | .0492 | .0531 | .0569 | .0607 | .0645 | .0682 | .0719 | .0755 |
| **1.2** | .0792 | .0828 | .0864 | .0899 | .0934 | .0969 | .1004 | .1038 | .1072 | .1106 |
| **1.3** | .1139 | .1173 | .1206 | .1239 | .1271 | .1303 | .1335 | .1367 | .1399 | .1430 |
| **1.4** | .1461 | .1492 | .1523 | .1553 | .1584 | .1614 | .1644 | .1673 | .1703 | .1732 |
| **1.5** | .1761 | .1790 | .1818 | .1847 | .1875 | .1903 | .1931 | .1959 | .1987 | .2014 |
| **1.6** | .2041 | .2068 | .2095 | .2122 | .2148 | .2175 | .2201 | .2227 | .2253 | .2279 |
| **1.7** | .2304 | .2330 | .2355 | .2380 | .2405 | .2430 | .2455 | .2480 | .2504 | .2529 |
| **1.8** | .2553 | .2577 | .2601 | .2625 | .2648 | .2672 | .2695 | .2718 | .2742 | .2765 |
| **1.9** | .2788 | .2810 | .2833 | .2856 | .2878 | .2900 | .2923 | .2945 | .2967 | .2989 |
| **2.0** | .3010 | .3032 | .3054 | .3075 | .3096 | .3118 | .3139 | .3160 | .3181 | .3201 |
| **2.1** | .3222 | .3243 | .3263 | .3284 | .3304 | .3324 | .3345 | .3365 | .3385 | .3404 |
| **2.2** | .3424 | .3444 | .3464 | .3483 | .3502 | .3522 | .3541 | .3560 | .3579 | .3598 |
| **2.3** | .3617 | .3636 | .3655 | .3674 | .3692 | .3711 | .3729 | .3747 | .3766 | .3784 |
| **2.4** | .3802 | .3820 | .3838 | .3856 | .3874 | .3892 | .3909 | .3927 | .3945 | .3962 |
| **2.5** | .3979 | .3997 | .4014 | .4031 | .4048 | .4065 | .4082 | .4099 | .4116 | .4133 |
| **2.6** | .4150 | .4166 | .4183 | .4200 | .4216 | .4232 | .4249 | .4265 | .4281 | .4298 |
| **2.7** | .4314 | .4330 | .4346 | .4362 | .4378 | .4393 | .4409 | .4425 | .4440 | .4456 |
| **2.8** | .4472 | .4487 | .4502 | .4518 | .4533 | .4548 | .4564 | .4579 | .4594 | .4609 |
| **2.9** | .4624 | .4639 | .4654 | .4669 | .4683 | .4698 | .4713 | .4728 | .4742 | .4757 |
| **3.0** | .4771 | .4786 | .4800 | .4814 | .4829 | .4843 | .4857 | .4871 | .4886 | .4900 |
| **3.1** | .4914 | .4928 | .4942 | .4955 | .4969 | .4983 | .4997 | .5011 | .5024 | .5038 |
| **3.2** | .5051 | .5065 | .5079 | .5092 | .5105 | .5119 | .5132 | .5145 | .5159 | .5172 |
| **3.3** | .5185 | .5198 | .5211 | .5224 | .5237 | .5250 | .5263 | .5276 | .5289 | .5302 |
| **3.4** | .5315 | .5328 | .5340 | .5353 | .5366 | .5378 | .5391 | .5403 | .5416 | .5428 |
| **3.5** | .5441 | .5453 | .5465 | .5478 | .5490 | .5502 | .5514 | .5527 | .5539 | .5551 |
| **3.6** | .5563 | .5575 | .5587 | .5599 | .5611 | .5623 | .5635 | .5647 | .5658 | .5670 |
| **3.7** | .5682 | .5694 | .5705 | .5717 | .5729 | .5740 | .5752 | .5763 | .5775 | .5786 |
| **3.8** | .5798 | .5809 | .5821 | .5832 | .5843 | .5855 | .5866 | .5877 | .5888 | .5899 |
| **3.9** | .5911 | .5922 | .5933 | .5944 | .5955 | .5966 | .5977 | .5988 | .5999 | .6010 |
| **4.0** | .6021 | .6031 | .6042 | .6053 | .6064 | .6075 | .6085 | .6096 | .6107 | .6117 |
| **4.1** | .6128 | .6138 | .6149 | .6160 | .6170 | .6180 | .6191 | .6201 | .6212 | .6222 |
| **4.2** | .6232 | .6243 | .6253 | .6263 | .6274 | .6284 | .6294 | .6304 | .6314 | .6325 |
| **4.3** | .6335 | .6345 | .6355 | .6365 | .6375 | .6385 | .6395 | .6405 | .6415 | .6425 |
| **4.4** | .6435 | .6444 | .6454 | .6464 | .6474 | .6484 | .6493 | .6503 | .6513 | .6522 |
| **4.5** | .6532 | .6542 | .6551 | .6561 | .6571 | .6580 | .6590 | .6599 | .6609 | .6618 |
| **4.6** | .6628 | .6637 | .6646 | .6656 | .6665 | .6675 | .6684 | .6693 | .6702 | .6712 |
| **4.7** | .6721 | .6730 | .6739 | .6749 | .6758 | .6767 | .6776 | .6785 | .6794 | .6803 |
| **4.8** | .6812 | .6821 | .6830 | .6839 | .6848 | .6857 | .6866 | .6875 | .6884 | .6893 |
| **4.9** | .6902 | .6911 | .6920 | .6928 | .6937 | .6946 | .6955 | .6964 | .6972 | .6981 |
| **5.0** | .6990 | .6998 | .7007 | .7016 | .7024 | .7033 | .7042 | .7050 | .7059 | .7067 |
| **5.1** | .7076 | .7084 | .7093 | .7101 | .7110 | .7118 | .7126 | .7135 | .7143 | .7152 |
| **5.2** | .7160 | .7168 | .7177 | .7185 | .7193 | .7202 | .7210 | .7218 | .7226 | .7235 |
| **5.3** | .7243 | .7251 | .7259 | .7267 | .7275 | .7284 | .7292 | .7300 | .7308 | .7316 |
| **5.4** | .7324 | .7332 | .7340 | .7348 | .7356 | .7364 | .7372 | .7380 | .7388 | .7396 |

## Table 1 Common Logarithms (*continued*)

| *x* | 0 | 1 | 2 | 3 | 4 | 5 | 6 | 7 | 8 | 9 |
|---|---|---|---|---|---|---|---|---|---|---|
| **5.5** | .7404 | .7412 | .7419 | .7427 | .7435 | .7443 | .7451 | .7459 | .7466 | .7474 |
| **5.6** | .7482 | .7490 | .7497 | .7505 | .7513 | .7520 | .7528 | .7536 | .7543 | .7551 |
| **5.7** | .7559 | .7566 | .7574 | .7582 | .7589 | .7597 | .7604 | .7612 | .7619 | .7627 |
| **5.8** | .7634 | .7642 | .7649 | .7657 | .7664 | .7672 | .7679 | .7686 | .7694 | .7701 |
| **5.9** | .7709 | .7716 | .7723 | .7731 | .7738 | .7745 | .7752 | .7760 | .7767 | .7774 |
| **6.0** | .7782 | .7789 | .7796 | .7803 | .7810 | .7818 | .7825 | .7832 | .7839 | .7846 |
| **6.1** | .7853 | .7860 | .7868 | .7875 | .7882 | .7889 | .7896 | .7903 | .7910 | .7917 |
| **6.2** | .7924 | .7931 | .7938 | .7945 | .7952 | .7959 | .7966 | .7973 | .7980 | .7987 |
| **6.3** | .7993 | .8000 | .8007 | .8014 | .8021 | .8028 | .8035 | .8041 | .8048 | .8055 |
| **6.4** | .8062 | .8069 | .8075 | .8082 | .8089 | .8096 | .8102 | .8109 | .8116 | .8122 |
| **6.5** | .8129 | .8136 | .8142 | .8149 | .8156 | .8162 | .8169 | .8176 | .8182 | .8189 |
| **6.6** | .8195 | .8202 | .8209 | .8215 | .8222 | .8228 | .8235 | .8241 | .8248 | .8254 |
| **6.7** | .8261 | .8267 | .8274 | .8280 | .8287 | .8293 | .8299 | .8306 | .8312 | .8319 |
| **6.8** | .8325 | .8331 | .8338 | .8344 | .8351 | .8357 | .8363 | .8370 | .8376 | .8382 |
| **6.9** | .8388 | .8395 | .8401 | .8407 | .8414 | .8420 | .8426 | .8432 | .8439 | .8445 |
| **7.0** | .8451 | .8457 | .8463 | .8470 | .8476 | .8482 | .8488 | .8494 | .8500 | .8506 |
| **7.1** | .8513 | .8519 | .8525 | .8531 | .8537 | .8543 | .8549 | .8555 | .8561 | .8567 |
| **7.2** | .8573 | .8579 | .8585 | .8591 | .8597 | .8603 | .8609 | .8615 | .8621 | .8627 |
| **7.3** | .8633 | .8639 | .8645 | .8651 | .8657 | .8663 | .8669 | .8675 | .8681 | .8686 |
| **7.4** | .8692 | .8698 | .8704 | .8710 | .8716 | .8722 | .8727 | .8733 | .8739 | .8745 |
| **7.5** | .8751 | .8756 | .8762 | .8768 | .8774 | .8779 | .8785 | .8791 | .8797 | .8802 |
| **7.6** | .8808 | .8814 | .8820 | .8825 | .8831 | .8837 | .8842 | .8848 | .8854 | .8859 |
| **7.7** | .8865 | .8871 | .8876 | .8882 | .8887 | .8893 | .8899 | .8904 | .8910 | .8915 |
| **7.8** | .8921 | .8927 | .8932 | .8938 | .8943 | .8949 | .8954 | .8960 | .8965 | .8971 |
| **7.9** | .8976 | .8982 | .8987 | .8993 | .8998 | .9004 | .9009 | .9015 | .9020 | .9025 |
| **8.0** | .9031 | .9036 | .9042 | .9047 | .9053 | .9058 | .9063 | .9069 | .9074 | .9079 |
| **8.1** | .9085 | .9090 | .9096 | .9101 | .9106 | .9112 | .9117 | .9122 | .9128 | .9133 |
| **8.2** | .9138 | .9143 | .9149 | .9154 | .9159 | .9165 | .9170 | .9175 | .9180 | .9186 |
| **8.3** | .9191 | .9196 | .9201 | .9206 | .9212 | .9217 | .9222 | .9227 | .9232 | .9238 |
| **8.4** | .9243 | .9248 | .9253 | .9258 | .9263 | .9269 | .9274 | .9279 | .9284 | .9289 |
| **8.5** | .9294 | .9299 | .9304 | .9309 | .9315 | .9320 | .9325 | .9330 | .9335 | .9340 |
| **8.6** | .9345 | .9350 | .9355 | .9360 | .9365 | .9370 | .9375 | .9380 | .9385 | .9390 |
| **8.7** | .9395 | .9400 | .9405 | .9410 | .9415 | .9420 | .9425 | .9430 | .9435 | .9440 |
| **8.8** | .9445 | .9450 | .9455 | .9460 | .9465 | .9469 | .9474 | .9470 | .9484 | .9489 |
| **8.9** | .9494 | .9499 | .9504 | .9509 | .9513 | .9518 | .9523 | .9528 | .9533 | .9538 |
| **9.0** | .9542 | .9547 | .9552 | .9557 | .9562 | .9566 | .9571 | .9576 | .9581 | .9586 |
| **9.1** | .9590 | .9595 | .9600 | .9605 | .9609 | .9614 | .9619 | .9624 | .9628 | .9633 |
| **9.2** | .9638 | .9643 | .9647 | .9652 | .9657 | .9661 | .9666 | .9671 | .9675 | .9680 |
| **9.3** | .9685 | .9689 | .9694 | .9699 | .9703 | .9708 | .9713 | .9717 | .9722 | .9727 |
| **9.4** | .9731 | .9736 | .9741 | .9745 | .9750 | .9754 | .9759 | .9763 | .9768 | .9773 |
| **9.5** | .9777 | .9782 | .9786 | .9791 | .9795 | .9800 | .9805 | .9809 | .9814 | .9818 |
| **9.6** | .9823 | .9827 | .9832 | .9836 | .9841 | .9845 | .9850 | .9854 | .9859 | .9863 |
| **9.7** | .9868 | .9872 | .9877 | .9881 | .9886 | .9890 | .9894 | .9899 | .9903 | .9908 |
| **9.8** | .9912 | .9917 | .9921 | .9926 | .9930 | .9934 | .9939 | .9943 | .9948 | .9952 |
| **9.9** | .9956 | .9961 | .9965 | .9969 | .9974 | .9978 | .9983 | .9987 | .9991 | .9996 |

## Table 2 Values of $e^x$ and $e^{-x}$

| $x$ | $e^x$ | $e^{-x}$ | $x$ | $e^x$ | $e^{-x}$ | $x$ | $e^x$ | $e^{-x}$ |
|---|---|---|---|---|---|---|---|---|
| .00 | 1.00000 | 1.00000 | .40 | 1.49182 | .67032 | .80 | 2.22554 | .44032 |
| .01 | 1.01005 | .99005 | .41 | 1.50682 | .66365 | .85 | 2.33965 | .42741 |
| .02 | 1.02020 | .98020 | .42 | 1.52196 | .65705 | .90 | 2.45960 | .40657 |
| .03 | 1.03045 | .97045 | .43 | 1.53726 | .65051 | .95 | 2.58571 | .38674 |
| .04 | 1.04081 | .96079 | .44 | 1.55271 | .64404 | 1.00 | 2.71828 | .36788 |
| .05 | 1.05127 | .95123 | .45 | 1.56831 | .63763 | 1.10 | 3.00416 | .33287 |
| .06 | 1.06184 | .94176 | .46 | 1.58407 | .63128 | 1.20 | 3.32011 | .30119 |
| .07 | 1.07251 | .93239 | .47 | 1.59999 | .62500 | 1.30 | 3.66929 | .27253 |
| .08 | 1.08329 | .92312 | .48 | 1.61607 | .61878 | 1.40 | 4.05519 | .24659 |
| .09 | 1.09417 | .91393 | .49 | 1.63232 | .61263 | 1.50 | 4.48168 | .22313 |
| .10 | 1.10517 | .90484 | .50 | 1.64872 | .60653 | 1.60 | 4.95302 | .20189 |
| .11 | 1.11628 | .89583 | .51 | 1.66529 | .60050 | 1.70 | 5.47394 | .18268 |
| .12 | 1.12750 | .88692 | .52 | 1.68203 | .59452 | 1.80 | 6.04964 | .16529 |
| .13 | 1.13883 | .87810 | .53 | 1.69893 | .58860 | 1.90 | 6.68589 | .14956 |
| .14 | 1.15027 | .86936 | .54 | 1.71601 | .58275 | 2.00 | 7.38905 | .13533 |
| .15 | 1.16183 | .86071 | .55 | 1.73325 | .57695 | 2.10 | 8.16616 | .12245 |
| .16 | 1.17351 | .85214 | .56 | 1.75067 | .57121 | 2.20 | 9.02500 | .11080 |
| .17 | 1.18530 | .84366 | .57 | 1.76827 | .56553 | 2.30 | 9.97417 | .10025 |
| .18 | 1.19722 | .83527 | .58 | 1.78604 | .55990 | 2.40 | 11.02316 | .09071 |
| .19 | 1.20925 | .82696 | .59 | 1.80399 | .55433 | 2.50 | 12.18248 | .08208 |
| .20 | 1.22140 | .81873 | .60 | 1.82212 | .54881 | 3.00 | 20.08551 | .04978 |
| .21 | 1.23368 | .81058 | .61 | 1.84043 | .54335 | 3.50 | 33.11545 | .03020 |
| .22 | 1.24608 | .80252 | .62 | 1.85893 | .53794 | 4.00 | 54.59815 | .01832 |
| .23 | 1.25860 | .79453 | .63 | 1.87761 | .53259 | 4.50 | 90.01713 | .01111 |
| .24 | 1.27125 | .78663 | .64 | 1.89648 | .52729 | 5.00 | 148.41316 | .00674 |
| .25 | 1.28403 | .77880 | .65 | 1.91554 | .52205 | 5.50 | 224.69193 | .00409 |
| .26 | 1.29693 | .77105 | .66 | 1.93479 | .51685 | 6.00 | 403.42879 | .00248 |
| .27 | 1.30996 | .76338 | .67 | 1.95424 | .51171 | 6.50 | 665.14163 | .00150 |
| .28 | 1.32313 | .75578 | .68 | 1.97388 | .50662 | 7.00 | 1096.63316 | .00091 |
| .29 | 1.33643 | .74826 | .69 | 1.99372 | .50158 | 7.50 | 1808.04241 | .00055 |
| .30 | 1.34986 | .74082 | .70 | 2.01375 | .49659 | 8.00 | 2980.95799 | .00034 |
| .31 | 1.36343 | .73345 | .71 | 2.03399 | .49164 | 8.50 | 4914.76884 | .00020 |
| .32 | 1.37713 | .72615 | .72 | 2.05443 | .48675 | 9.00 | 8130.08392 | .00012 |
| .33 | 1.39097 | .71892 | .73 | 2.07508 | .48191 | 9.50 | 13359.72683 | .00007 |
| .34 | 1.40495 | .71177 | .74 | 2.09594 | .47711 | 10.00 | 22026.46579 | .00005 |
| .35 | 1.41907 | .70469 | .75 | 2.11700 | .47237 | | | |
| .36 | 1.43333 | .69768 | .76 | 2.13828 | .46767 | | | |
| .37 | 1.44773 | .69073 | .77 | 2.15977 | .46301 | | | |
| .38 | 1.46228 | .68386 | .78 | 2.18147 | .45841 | | | |
| .39 | 1.47698 | .67706 | .79 | 2.20340 | .45384 | | | |

## Table 3 Natural Logarithms

| $x$ | $\ln x$ | $x$ | $\ln x$ | $x$ | $\ln x$ |
|---|---|---|---|---|---|
| | | 4.5 | 1.5041 | 9.0 | 2.1972 |
| 0.1 | −2.3026 | 4.6 | 1.5261 | 9.1 | 2.2083 |
| 0.2 | −1.6094 | 4.7 | 1.5476 | 9.2 | 2.2192 |
| 0.3 | −1.2040 | 4.8 | 1.5686 | 9.3 | 2.2300 |
| 0.4 | −0.9163 | 4.9 | 1.5892 | 9.4 | 2.2407 |
| 0.5 | −0.6931 | 5.0 | 1.6094 | 9.5 | 2.2513 |
| 0.6 | −0.5108 | 5.1 | 1.6292 | 9.6 | 2.2618 |
| 0.7 | −0.3567 | 5.2 | 1.6487 | 9.7 | 2.2721 |
| 0.8 | −0.2231 | 5.3 | 1.6677 | 9.8 | 2.2824 |
| 0.9 | −0.1054 | 5.4 | 1.6864 | 9.9 | 2.2925 |
| 1.0 | 0.0000 | 5.5 | 1.7047 | 10 | 2.3026 |
| 1.1 | 0.0953 | 5.6 | 1.7228 | 11 | 2.3979 |
| 1.2 | 0.1823 | 5.7 | 1.7405 | 12 | 2.4849 |
| 1.3 | 0.2624 | 5.8 | 1.7579 | 13 | 2.5649 |
| 1.4 | 0.3365 | 5.9 | 1.7750 | 14 | 2.6391 |
| 1.5 | 0.4055 | 6.0 | 1.7918 | 15 | 2.7081 |
| 1.6 | 0.4700 | 6.1 | 1.8083 | 16 | 2.7726 |
| 1.7 | 0.5306 | 6.2 | 1.8245 | 17 | 2.8332 |
| 1.8 | 0.5878 | 6.3 | 1.8405 | 18 | 2.8904 |
| 1.9 | 0.6419 | 6.4 | 1.8563 | 19 | 2.9444 |
| 2.0 | 0.6931 | 6.5 | 1.8718 | 20 | 2.9957 |
| 2.1 | 0.7419 | 6.6 | 1.8871 | 25 | 3.2189 |
| 2.2 | 0.7885 | 6.7 | 1.9021 | 30 | 3.4012 |
| 2.3 | 0.8329 | 6.8 | 1.9169 | 35 | 3.5553 |
| 2.4 | 0.8755 | 6.9 | 1.9315 | 40 | 3.6889 |
| 2.5 | 0.9163 | 7.0 | 1.9459 | 45 | 3.8067 |
| 2.6 | 0.9555 | 7.1 | 1.9601 | 50 | 3.9120 |
| 2.7 | 0.9933 | 7.2 | 1.9741 | 55 | 4.0073 |
| 2.8 | 1.0296 | 7.3 | 1.9879 | 60 | 4.0943 |
| 2.9 | 1.0647 | 7.4 | 2.0015 | 65 | 4.1744 |
| 3.0 | 1.0986 | 7.5 | 2.0149 | 70 | 4.2485 |
| 3.1 | 1.1314 | 7.6 | 2.0281 | 75 | 4.3175 |
| 3.2 | 1.1632 | 7.7 | 2.0412 | 80 | 4.3820 |
| 3.3 | 1.1939 | 7.8 | 2.0541 | 85 | 4.4427 |
| 3.4 | 1.2238 | 7.9 | 2.0669 | 90 | 4.4998 |
| 3.5 | 1.2528 | 8.0 | 2.0794 | 100 | 4.6052 |
| 3.6 | 1.2809 | 8.1 | 2.0919 | 110 | 4.7005 |
| 3.7 | 1.3083 | 8.2 | 2.1041 | 120 | 4.7875 |
| 3.8 | 1.3350 | 8.3 | 2.1163 | 130 | 4.8676 |
| 3.9 | 1.3610 | 8.4 | 2.1282 | 140 | 4.9416 |
| 4.0 | 1.3863 | 8.5 | 2.1401 | 150 | 5.0106 |
| 4.1 | 1.4110 | 8.6 | 2.1518 | 160 | 5.0752 |
| 4.2 | 1.4351 | 8.7 | 2.1633 | 170 | 5.1358 |
| 4.3 | 1.4586 | 8.8 | 2.1748 | 180 | 5.1930 |
| 4.4 | 1.4816 | 8.9 | 2.1861 | 190 | 5.2470 |

## Table 4 Values of Trigonometric Functions

| α (degrees) | α (radians) | sin α | cos α | tan α | cot α | sec α | csc α | | |
|---|---|---|---|---|---|---|---|---|---|
| **0°00′** | .0000 | .0000 | 1.0000 | .0000 | — | 1.000 | — | 1.5708 | **90°00′** |
| 10 | .0029 | .0029 | 1.0000 | .0029 | 343.8 | 1.000 | 343.8 | 1.5679 | 50 |
| 20 | .0058 | .0058 | 1.0000 | .0058 | 171.9 | 1.000 | 171.9 | 1.5650 | 40 |
| 30 | .0087 | .0087 | 1.0000 | .0087 | 114.6 | 1.000 | 114.6 | 1.5621 | 30 |
| 40 | .0116 | .0116 | .9999 | .0116 | 85.94 | 1.000 | 85.95 | 1.5592 | 20 |
| 50 | .0145 | .0145 | .9999 | .0145 | 68.75 | 1.000 | 68.76 | 1.5563 | 10 |
| **1°00′** | .0175 | .0175 | .9998 | .0175 | 57.29 | 1.000 | 57.30 | 1.5533 | **89°00′** |
| 10 | .0204 | .0204 | .9998 | .0204 | 49.10 | 1.000 | 49.11 | 1.5504 | 50 |
| 20 | .0233 | .0233 | .9997 | .0233 | 42.96 | 1.000 | 42.98 | 1.5475 | 40 |
| 30 | .0262 | .0262 | .9997 | .0262 | 38.19 | 1.000 | 38.20 | 1.5446 | 30 |
| 40 | .0291 | .0291 | .9996 | .0291 | 34.37 | 1.000 | 34.38 | 1.5417 | 20 |
| 50 | .0320 | .0320 | .9995 | .0320 | 31.24 | 1.001 | 31.26 | 1.5388 | 10 |
| **2°00′** | .0349 | .0349 | .9994 | .0349 | 28.64 | 1.001 | 28.65 | 1.5359 | **88°00′** |
| 10 | .0378 | .0378 | .9993 | .0378 | 26.43 | 1.001 | 26.45 | 1.5330 | 50 |
| 20 | .0407 | .0407 | .9992 | .0407 | 24.54 | 1.001 | 24.56 | 1.5301 | 40 |
| 30 | .0436 | .0436 | .9990 | .0437 | 22.90 | 1.001 | 22.93 | 1.5272 | 30 |
| 40 | .0465 | .0465 | .9989 | .0466 | 21.47 | 1.001 | 21.49 | 1.5243 | 20 |
| 50 | .0495 | .0494 | .9988 | .0495 | 20.21 | 1.001 | 20.23 | 1.5213 | 10 |
| **3°00′** | .0524 | .0523 | .9986 | .0524 | 19.08 | 1.001 | 19.11 | 1.5184 | **87°00′** |
| 10 | .0553 | .0552 | .9985 | .0553 | 18.07 | 1.002 | 18.10 | 1.5155 | 50 |
| 20 | .0582 | .0581 | .9983 | .0582 | 17.17 | 1.002 | 17.20 | 1.5126 | 40 |
| 30 | .0611 | .0610 | .9981 | .0612 | 16.35 | 1.002 | 16.38 | 1.5097 | 30 |
| 40 | .0640 | .0640 | .9980 | .0641 | 15.60 | 1.002 | 15.64 | 1.5068 | 20 |
| 50 | .0669 | .0669 | .9978 | .0670 | 14.92 | 1.002 | 14.96 | 1.5039 | 10 |
| **4°00′** | .0698 | .0698 | .9976 | .0699 | 14.30 | 1.002 | 14.34 | 1.5010 | **86°00′** |
| 10 | .0727 | .0727 | .9974 | .0729 | 13.73 | 1.003 | 13.76 | 1.4981 | 50 |
| 20 | .0756 | .0756 | .9971 | .0758 | 13.20 | 1.003 | 13.23 | 1.4952 | 40 |
| 30 | .0785 | .0785 | .9969 | .0787 | 12.71 | 1.003 | 12.75 | 1.4923 | 30 |
| 40 | .0814 | .0814 | .9967 | .0816 | 12.25 | 1.003 | 12.29 | 1.4893 | 20 |
| 50 | .0844 | .0843 | .9964 | .0846 | 11.83 | 1.004 | 11.87 | 1.4864 | 10 |
| **5°00′** | .0873 | .0872 | .9962 | .0875 | 11.43 | 1.004 | 11.47 | 1.4835 | **85°00′** |
| 10 | .0902 | .0901 | .9959 | .0904 | 11.06 | 1.004 | 11.10 | 1.4806 | 50 |
| 20 | .0931 | .0929 | .9957 | .0934 | 10.71 | 1.004 | 10.76 | 1.4777 | 40 |
| 30 | .0960 | .0958 | .9954 | .0963 | 10.39 | 1.005 | 10.43 | 1.4748 | 30 |
| 40 | .0989 | .0987 | .9951 | .0992 | 10.08 | 1.005 | 10.13 | 1.4719 | 20 |
| 50 | .1018 | .1016 | .9948 | .1022 | 9.788 | 1.005 | 9.839 | 1.4690 | 10 |
| **6°00′** | .1047 | .1045 | .9945 | .1051 | 9.514 | 1.006 | 9.567 | 1.4661 | **84°00′** |
| 10 | .1076 | .1074 | .9942 | .1080 | 9.255 | 1.006 | 9.309 | 1.4632 | 50 |
| 20 | .1105 | .1103 | .9939 | .1110 | 9.010 | 1.006 | 9.065 | 1.4603 | 40 |
| 30 | .1134 | .1132 | .9936 | .1139 | 8.777 | 1.006 | 8.834 | 1.4573 | 30 |
| 40 | .1164 | .1161 | .9932 | .1169 | 8.556 | 1.007 | 8.614 | 1.4544 | 20 |
| 50 | .1193 | .1190 | .9929 | .1198 | 8.345 | 1.007 | 8.405 | 1.4515 | 10 |
| | | cos α | sin α | cot α | tan α | csc α | sec α | α (radians) | α (degrees) |

## Table 4 Values of Trigonometric Functions (*continued*)

| $\alpha$ (degrees) | $\alpha$ (radians) | $\sin \alpha$ | $\cos \alpha$ | $\tan \alpha$ | $\cot \alpha$ | $\sec \alpha$ | $\csc \alpha$ | | |
|---|---|---|---|---|---|---|---|---|---|
| **7°00′** | .1222 | .1219 | .9925 | .1228 | 8.144 | 1.008 | 8.206 | 1.4486 | **83°00′** |
| 10 | .1251 | .1248 | .9922 | .1257 | 7.953 | 1.008 | 8.016 | 1.4457 | 50 |
| 20 | .1280 | .1276 | .9918 | .1287 | 7.770 | 1.008 | 7.834 | 1.4428 | 40 |
| 30 | .1309 | .1305 | .9914 | .1317 | 7.596 | 1.009 | 7.661 | 1.4399 | 30 |
| 40 | .1338 | .1334 | .9911 | .1346 | 7.429 | 1.009 | 7.496 | 1.4370 | 20 |
| 50 | .1376 | .1363 | .9907 | .1376 | 7.269 | 1.009 | 7.337 | 1.4341 | 10 |
| **8°00′** | .1396 | .1392 | .9903 | .1405 | 7.115 | 1.010 | 7.185 | 1.4312 | **82°00′** |
| 10 | .1425 | .1421 | .9899 | .1435 | 6.968 | 1.010 | 7.040 | 1.4283 | 50 |
| 20 | .1454 | .1449 | .9894 | .1465 | 6.827 | 1.011 | 6.900 | 1.4254 | 40 |
| 30 | .1484 | .1478 | .9890 | .1495 | 6.691 | 1.011 | 6.765 | 1.4224 | 30 |
| 40 | .1513 | .1507 | .9886 | .1524 | 6.561 | 1.012 | 6.636 | 1.4195 | 20 |
| 50 | .1542 | .1536 | .9881 | .1554 | 6.435 | 1.012 | 6.512 | 1.4166 | 10 |
| **9°00′** | .1571 | .1564 | .9877 | .1584 | 6.314 | 1.012 | 6.392 | 1.4137 | **81°00′** |
| 10 | .1600 | .1593 | .9872 | .1614 | 6.197 | 1.013 | 6.277 | 1.4108 | 50 |
| 20 | .1629 | .1622 | .9868 | .1644 | 6.084 | 1.013 | 6.166 | 1.4079 | 40 |
| 30 | .1658 | .1650 | .9863 | .1673 | 5.976 | 1.014 | 6.059 | 1.4050 | 30 |
| 40 | .1687 | .1679 | .9858 | .1703 | 5.871 | 1.014 | 5.955 | 1.4021 | 20 |
| 50 | .1716 | .1708 | .9853 | .1733 | 5.769 | 1.015 | 5.855 | 1.3992 | 10 |
| **10°00′** | .1745 | .1736 | .9848 | .1763 | 5.671 | 1.015 | 5.759 | 1.3963 | **80°00′** |
| 10 | .1774 | .1765 | .9843 | .1793 | 5.576 | 1.016 | 5.665 | 1.3934 | 50 |
| 20 | .1804 | .1794 | .9838 | .1823 | 5.485 | 1.016 | 5.575 | 1.3904 | 40 |
| 30 | .1833 | .1822 | .9833 | .1853 | 5.396 | 1.017 | 5.487 | 1.3875 | 30 |
| 40 | .1862 | .1851 | .9827 | .1883 | 5.309 | 1.018 | 5.403 | 1.3846 | 20 |
| 50 | .1891 | .1880 | .9822 | .1914 | 5.226 | 1.018 | 5.320 | 1.3817 | 10 |
| **11°00′** | .1920 | .1908 | .9816 | .1944 | 5.145 | 1.019 | 5.241 | 1.3788 | **79°00′** |
| 10 | .1949 | .1937 | .9811 | .1974 | 5.066 | 1.019 | 5.164 | 1.3759 | 50 |
| 20 | .1978 | .1965 | .9805 | .2004 | 4.989 | 1.020 | 5.089 | 1.3730 | 40 |
| 30 | .2007 | .1994 | .9799 | .2035 | 4.915 | 1.020 | 5.016 | 1.3701 | 30 |
| 40 | .2036 | .2022 | .9793 | .2065 | 4.843 | 1.021 | 4.945 | 1.3672 | 20 |
| 50 | .2065 | .2051 | .9787 | .2095 | 4.773 | 1.022 | 4.876 | 1.3643 | 10 |
| **12°00′** | .2094 | .2079 | .9781 | .2126 | 4.705 | 1.022 | 4.810 | 1.3614 | **78°00′** |
| 10 | .2123 | .2108 | .9775 | .2156 | 4.638 | 1.023 | 4.745 | 1.3584 | 50 |
| 20 | .2153 | .2136 | .9769 | .2186 | 4.574 | 1.024 | 4.682 | 1.3555 | 40 |
| 30 | .2182 | .2164 | .9763 | .2217 | 4.511 | 1.024 | 4.620 | 1.3526 | 30 |
| 40 | .2211 | .2193 | .9757 | .2247 | 4.449 | 1.025 | 4.560 | 1.3497 | 20 |
| 50 | .2240 | .2221 | .9750 | .2278 | 4.390 | 1.026 | 4.502 | 1.3468 | 10 |
| **13°00′** | .2269 | .2250 | .9744 | .2309 | 4.331 | 1.026 | 4.445 | 1.3439 | **77°00′** |
| 10 | .2298 | .2278 | .9737 | .2339 | 4.275 | 1.027 | 4.390 | 1.3410 | 50 |
| 20 | .2327 | .2306 | .9730 | .2370 | 4.219 | 1.028 | 4.336 | 1.3381 | 40 |
| 30 | .2356 | .2334 | .9724 | .2401 | 4.165 | 1.028 | 4.284 | 1.3352 | 30 |
| 40 | .2385 | .2363 | .9717 | .2432 | 4.113 | 1.029 | 4.232 | 1.3323 | 20 |
| 50 | .2414 | .2391 | .9710 | .2462 | 4.061 | 1.030 | 4.182 | 1.3294 | 10 |
| | | $\cos \alpha$ | $\sin \alpha$ | $\cot \alpha$ | $\tan \alpha$ | $\csc \alpha$ | $\sec \alpha$ | $\alpha$ (radians) | $\alpha$ (degrees) |

Table 4 Values of Trigonometric Functions (*continued*)

| $\alpha$ (degrees) | $\alpha$ (radians) | sin $\alpha$ | cos $\alpha$ | tan $\alpha$ | cot $\alpha$ | sec $\alpha$ | csc $\alpha$ | | |
|---|---|---|---|---|---|---|---|---|---|
| **14°00′** | .2443 | .2419 | .9703 | .2493 | 4.011 | 1.031 | 4.134 | 1.3265 | **76°00′** |
| 10 | .2473 | .2447 | .9696 | .2524 | 3.962 | 1.031 | 4.086 | 1.3235 | 50 |
| 20 | .2502 | .2476 | .9689 | .2555 | 3.914 | 1.032 | 4.039 | 1.3206 | 40 |
| 30 | .2531 | .2504 | .9681 | .2586 | 3.867 | 1.033 | 3.994 | 1.3177 | 30 |
| 40 | .2560 | .2532 | .9674 | .2617 | 3.821 | 1.034 | 3.950 | 1.3148 | 20 |
| 50 | .2589 | .2560 | .9667 | .2648 | 3.776 | 1.034 | 3.906 | 1.3119 | 10 |
| **15°00′** | .2618 | .2588 | .9659 | .2679 | 3.732 | 1.035 | 3.864 | 1.3090 | **75°00′** |
| 10 | .2647 | .2616 | .9652 | .2711 | 3.689 | 1.036 | 3.822 | 1.3061 | 50 |
| 20 | .2676 | .2644 | .9644 | .2742 | 3.647 | 1.037 | 3.782 | 1.3032 | 40 |
| 30 | .2705 | .2672 | .9636 | .2773 | 3.606 | 1.038 | 3.742 | 1.3003 | 30 |
| 40 | .2734 | .2700 | .9628 | .2805 | 3.566 | 1.039 | 3.703 | 1.2974 | 20 |
| 50 | .2763 | .2728 | .9621 | .2836 | 3.526 | 1.039 | 3.665 | 1.2945 | 10 |
| **16°00′** | .2793 | .2756 | .9613 | .2867 | 3.487 | 1.040 | 3.628 | 1.2915 | **74°00′** |
| 10 | .2822 | .2784 | .9605 | .2899 | 3.450 | 1.041 | 3.592 | 1.2886 | 50 |
| 20 | .2851 | .2812 | .9596 | .2931 | 3.412 | 1.042 | 3.556 | 1.2857 | 40 |
| 30 | .2880 | .2840 | .9588 | .2962 | 3.376 | 1.043 | 3.521 | 1.2828 | 30 |
| 40 | .2909 | .2868 | .9580 | .2994 | 3.340 | 1.044 | 3.487 | 1.2799 | 20 |
| 50 | .2938 | .2896 | .9572 | .3026 | 3.305 | 1.045 | 3.453 | 1.2770 | 10 |
| **17°00′** | .2967 | .2924 | .9563 | .3057 | 3.271 | 1.046 | 3.420 | 1.2741 | **73°00′** |
| 10 | .2996 | .2952 | .9555 | .3089 | 3.237 | 1.047 | 3.388 | 1.2712 | 50 |
| 20 | .3025 | .2979 | .9546 | .3121 | 3.204 | 1.048 | 3.356 | 1.2683 | 40 |
| 30 | .3054 | .3007 | .9537 | .3153 | 3.172 | 1.049 | 3.326 | 1.2654 | 30 |
| 40 | .3083 | .3035 | .9528 | .3185 | 3.140 | 1.049 | 3.295 | 1.2625 | 20 |
| 50 | .3113 | .3062 | .9520 | .3217 | 3.108 | 1.050 | 3.265 | 1.2595 | 10 |
| **18°00′** | .3142 | .3090 | .9511 | .3249 | 3.078 | 1.051 | 3.236 | 1.2566 | **72°00′** |
| 10 | .3171 | .3118 | .9502 | .3281 | 3.047 | 1.052 | 3.207 | 1.2537 | 50 |
| 20 | .3200 | .3145 | .9492 | .3314 | 3.018 | 1.053 | 3.179 | 1.2508 | 40 |
| 30 | .3229 | .3173 | .9483 | .3346 | 2.989 | 1.054 | 3.152 | 1.2479 | 30 |
| 40 | .3258 | .3201 | .9474 | .3378 | 2.960 | 1.056 | 3.124 | 1.2450 | 20 |
| 50 | .3287 | .3228 | .9465 | .3411 | 2.932 | 1.057 | 3.098 | 1.2421 | 10 |
| **19°00′** | .3316 | .3256 | .9455 | .3443 | 2.904 | 1.058 | 3.072 | 1.2392 | **71°00′** |
| 10 | .3345 | .3283 | .9446 | .3476 | 2.877 | 1.059 | 3.046 | 1.2363 | 50 |
| 20 | .3374 | .3311 | .9436 | .3508 | 2.850 | 1.060 | 3.021 | 1.2334 | 40 |
| 30 | .3403 | .3338 | .9426 | .3541 | 2.824 | 1.061 | 2.996 | 1.2305 | 30 |
| 40 | .3432 | .3365 | .9417 | .3574 | 2.798 | 1.062 | 2.971 | 1.2275 | 20 |
| 50 | .3462 | .3393 | .9407 | .3607 | 2.773 | 1.063 | 2.947 | 1.2246 | 10 |
| **20°00′** | .3491 | .3420 | .9397 | .3640 | 2.747 | 1.064 | 2.924 | 1.2217 | **70°00′** |
| 10 | .3520 | .3448 | .9387 | .3673 | 2.723 | 1.065 | 2.901 | 1.2188 | 50 |
| 20 | .3549 | .3475 | .9377 | .3706 | 2.699 | 1.066 | 2.878 | 1.2159 | 40 |
| 30 | .3578 | .3502 | .9367 | .3739 | 2.675 | 1.068 | 2.855 | 1.2130 | 30 |
| 40 | .3607 | .3529 | .9356 | .3772 | 2.651 | 1.069 | 2.833 | 1.2101 | 20 |
| 50 | .3636 | .3557 | .9346 | .3805 | 2.628 | 1.070 | 2.812 | 1.2072 | 10 |
| | | cos $\alpha$ | sin $\alpha$ | cot $\alpha$ | tan $\alpha$ | csc $\alpha$ | sec $\alpha$ | $\alpha$ (radians) | $\alpha$ (degrees) |

## Table 4 Values of Trigonometric Functions (*continued*)

| α (degrees) | α (radians) | sin α | cos α | tan α | cot α | sec α | csc α | | |
|---|---|---|---|---|---|---|---|---|---|
| **21°00′** | .3665 | .3584 | .9336 | .3839 | 2.605 | 1.071 | 2.790 | 1.2043 | **69°00′** |
| 10 | .3694 | .3611 | .9325 | .3872 | 2.583 | 1.072 | 2.769 | 1.2014 | 50 |
| 20 | .3723 | .3638 | .9315 | .3906 | 2.560 | 1.074 | 2.749 | 1.1985 | 40 |
| 30 | .3752 | .3665 | .9304 | .3939 | 2.539 | 1.075 | 2.729 | 1.1956 | 30 |
| 40 | .3782 | .3692 | .9293 | .3973 | 2.517 | 1.076 | 2.709 | 1.1926 | 20 |
| 50 | .3811 | .3719 | .9283 | .4006 | 2.496 | 1.077 | 2.689 | 1.1897 | 10 |
| **22°00′** | .3840 | .3746 | .9272 | .4040 | 2.475 | 1.079 | 2.669 | 1.1868 | **68°00′** |
| 10 | .3869 | .3773 | .9261 | .4074 | 2.455 | 1.080 | 2.650 | 1.1839 | 50 |
| 20 | .3898 | .3800 | .9250 | .4108 | 2.434 | 1.081 | 2.632 | 1.1810 | 40 |
| 30 | .3927 | .3827 | .9239 | .4142 | 2.414 | 1.082 | 2.613 | 1.1781 | 30 |
| 40 | .3956 | .3854 | .9228 | .4176 | 2.394 | 1.084 | 2.595 | 1.1752 | 20 |
| 50 | .3985 | .3881 | .9216 | .4210 | 2.375 | 1.085 | 2.577 | 1.1723 | 10 |
| **23°00′** | .4014 | .3907 | .9205 | .4245 | 2.356 | 1.086 | 2.559 | 1.1694 | **67°00′** |
| 10 | .4043 | .3934 | .9194 | .4279 | 2.337 | 1.088 | 2.542 | 1.1665 | 50 |
| 20 | .4072 | .3961 | .9182 | .4314 | 2.318 | 1.089 | 2.525 | 1.1636 | 40 |
| 30 | .4102 | .3987 | .9171 | .4348 | 2.300 | 1.090 | 2.508 | 1.1606 | 30 |
| 40 | .4131 | .4014 | .9159 | .4383 | 2.282 | 1.092 | 2.491 | 1.1577 | 20 |
| 50 | .4160 | .4041 | .9147 | .4417 | 2.264 | 1.093 | 2.475 | 1.1548 | 10 |
| **24°00′** | .4189 | .4067 | .9135 | .4452 | 2.246 | 1.095 | 2.459 | 1.1519 | **66°00′** |
| 10 | .4218 | .4094 | .9124 | .4487 | 2.229 | 1.096 | 2.443 | 1.1490 | 50 |
| 20 | .4247 | .4120 | .9112 | .4522 | 2.211 | 1.097 | 2.427 | 1.1461 | 40 |
| 30 | .4276 | .4147 | .9100 | .4557 | 2.194 | 1.099 | 2.411 | 1.1432 | 30 |
| 40 | .4305 | .4173 | .9088 | .4592 | 2.177 | 1.100 | 2.396 | 1.1403 | 20 |
| 50 | .4334 | .4200 | .9075 | .4628 | 2.161 | 1.102 | 2.381 | 1.1374 | 10 |
| **25°00′** | .4363 | .4226 | .9063 | .4663 | 2.145 | 1.103 | 2.366 | 1.1345 | **65°00′** |
| 10 | .4392 | .4253 | .9051 | .4699 | 2.128 | 1.105 | 2.352 | 1.1316 | 50 |
| 20 | .4422 | .4279 | .9038 | .4734 | 2.112 | 1.106 | 2.337 | 1.1286 | 40 |
| 30 | .4451 | .4305 | .9026 | .4770 | 2.097 | 1.108 | 2.323 | 1.1257 | 30 |
| 40 | .4480 | .4331 | .9013 | .4806 | 2.081 | 1.109 | 2.309 | 1.1228 | 20 |
| 50 | .4509 | .4358 | .9001 | .4841 | 2.066 | 1.111 | 2.295 | 1.1199 | 10 |
| **26°00′** | .4538 | .4384 | .8988 | .4877 | 2.050 | 1.113 | 2.281 | 1.1170 | **64°00′** |
| 10 | .4567 | .4410 | .8975 | .4913 | 2.035 | 1.114 | 2.268 | 1.1141 | 50 |
| 20 | .4596 | .4436 | .8962 | .4950 | 2.020 | 1.116 | 2.254 | 1.1112 | 40 |
| 30 | .4625 | .4462 | .8949 | .4986 | 2.006 | 1.117 | 2.241 | 1.1083 | 30 |
| 40 | .4654 | .4488 | .8936 | .5022 | 1.991 | 1.119 | 2.228 | 1.1054 | 20 |
| 50 | .4683 | .4514 | .8923 | .5059 | 1.977 | 1.121 | 2.215 | 1.1025 | 10 |
| **27°00′** | .4712 | .4540 | .8910 | .5095 | 1.963 | 1.122 | 2.203 | 1.0996 | **63°00′** |
| 10 | .4741 | .4566 | .8897 | .5132 | 1.949 | 1.124 | 2.190 | 1.0966 | 50 |
| 20 | .4771 | .4592 | .8884 | .5169 | 1.935 | 1.126 | 2.178 | 1.0937 | 40 |
| 30 | .4800 | .4617 | .8870 | .5206 | 1.921 | 1.127 | 2.166 | 1.0908 | 30 |
| 40 | .4829 | .4643 | .8857 | .5243 | 1.907 | 1.129 | 2.154 | 1.0879 | 20 |
| 50 | .4858 | .4669 | .8843 | .5280 | 1.894 | 1.131 | 2.142 | 1.0850 | 10 |
| | | cos α | sin α | cot α | tan α | csc α | sec α | α (radians) | α (degrees) |

## Table 4 Values of Trigonometric Functions (*continued*)

| $\alpha$ (degrees) | $\alpha$ (radians) | sin $\alpha$ | cos $\alpha$ | tan $\alpha$ | cot $\alpha$ | sec $\alpha$ | csc $\alpha$ | | |
|---|---|---|---|---|---|---|---|---|---|
| **28°00′** | .4887 | .4695 | .8829 | .5317 | 1.881 | 1.133 | 2.130 | 1.0821 | **62°00′** |
| 10 | .4916 | .4720 | .8816 | .5354 | 1.868 | 1.134 | 2.118 | 1.0792 | 50 |
| 20 | .4945 | .4746 | .8802 | .5392 | 1.855 | 1.136 | 2.107 | 1.0763 | 40 |
| 30 | .4974 | .4772 | .8788 | .5430 | 1.842 | 1.138 | 2.096 | 1.0734 | 30 |
| 40 | .5003 | .4797 | .8774 | .5467 | 1.829 | 1.140 | 2.085 | 1.0705 | 20 |
| 50 | .5032 | .4823 | .8760 | .5505 | 1.816 | 1.142 | 2.074 | 1.0676 | 10 |
| **29°00′** | .5061 | .4848 | .8746 | .5543 | 1.804 | 1.143 | 2.063 | 1.0647 | **61°00′** |
| 10 | .5091 | .4874 | .8732 | .5581 | 1.792 | 1.145 | 2.052 | 1.0617 | 50 |
| 20 | .5120 | .4899 | .8718 | .5619 | 1.780 | 1.147 | 2.041 | 1.0588 | 40 |
| 30 | .5149 | .4924 | .8704 | .5658 | 1.767 | 1.149 | 2.031 | 1.0559 | 30 |
| 40 | .5178 | .4950 | .8689 | .5696 | 1.756 | 1.151 | 2.020 | 1.0530 | 20 |
| 50 | .5207 | .4975 | .8675 | .5735 | 1.744 | 1.153 | 2.010 | 1.0501 | 10 |
| **30°00′** | .5236 | .5000 | .8660 | .5774 | 1.732 | 1.155 | 2.000 | 1.0472 | **60°00′** |
| 10 | .5265 | .5025 | .8646 | .5812 | 1.720 | 1.157 | 1.990 | 1.0443 | 50 |
| 20 | .5294 | .5050 | .8631 | .5851 | 1.709 | 1.159 | 1.980 | 1.0414 | 40 |
| 30 | .5323 | .5075 | .8616 | .5890 | 1.698 | 1.161 | 1.970 | 1.0385 | 30 |
| 40 | .5352 | .5100 | .8601 | .5930 | 1.686 | 1.163 | 1.961 | 1.0356 | 20 |
| 50 | .5381 | .5125 | .8587 | .5969 | 1.675 | 1.165 | 1.951 | 1.0327 | 10 |
| **31°00′** | .5411 | .5150 | .8572 | .6009 | 1.664 | 1.167 | 1.942 | 1.0297 | **59°00′** |
| 10 | .5440 | .5175 | .8557 | .6048 | 1.653 | 1.169 | 1.932 | 1.0268 | 50 |
| 20 | .5469 | .5200 | .8542 | .6088 | 1.643 | 1.171 | 1.923 | 1.0239 | 40 |
| 30 | .5498 | .5225 | .8526 | .6128 | 1.632 | 1.173 | 1.914 | 1.0210 | 30 |
| 40 | .5527 | .5250 | .8511 | .6168 | 1.621 | 1.175 | 1.905 | 1.0181 | 20 |
| 50 | .5556 | .5275 | .8496 | .6208 | 1.611 | 1.177 | 1.896 | 1.0152 | 10 |
| **32°00′** | .5585 | .5299 | .8480 | .6249 | 1.600 | 1.179 | 1.887 | 1.0123 | **58°00′** |
| 10 | .5614 | .5324 | .8465 | .6289 | 1.590 | 1.181 | 1.878 | 1.0094 | 50 |
| 20 | .5643 | .5348 | .8450 | .6330 | 1.580 | 1.184 | 1.870 | 1.0065 | 40 |
| 30 | .5672 | .5373 | .8434 | .6371 | 1.570 | 1.186 | 1.861 | 1.0036 | 30 |
| 40 | .5701 | .5398 | .8418 | .6412 | 1.560 | 1.188 | 1.853 | 1.0007 | 20 |
| 50 | .5730 | .5422 | .8403 | .6453 | 1.550 | 1.190 | 1.844 | .9977 | 10 |
| **33°00′** | .5760 | .5446 | .8387 | .6494 | 1.540 | 1.192 | 1.836 | .9948 | **57°00′** |
| 10 | .5789 | .5471 | .8371 | .6536 | 1.530 | 1.195 | 1.828 | .9919 | 50 |
| 20 | .5818 | .5495 | .8355 | .6577 | 1.520 | 1.197 | 1.820 | .9890 | 40 |
| 30 | .5847 | .5519 | .8339 | .6619 | 1.511 | 1.199 | 1.812 | .9861 | 30 |
| 40 | .5876 | .5544 | .8323 | .6661 | 1.501 | 1.202 | 1.804 | .9832 | 20 |
| 50 | .5905 | .5568 | .8307 | .6703 | 1.492 | 1.204 | 1.796 | .9803 | 10 |
| **34°00′** | .5934 | .5592 | .8290 | .6745 | 1.483 | 1.206 | 1.788 | .9774 | **56°00′** |
| 10 | .5963 | .5616 | .8274 | .6787 | 1.473 | 1.209 | 1.781 | .9745 | 50 |
| 20 | .5992 | .5640 | .8258 | .6830 | 1.464 | 1.211 | 1.773 | .9716 | 40 |
| 30 | .6021 | .5664 | .8241 | .6873 | 1.455 | 1.213 | 1.766 | .9687 | 30 |
| 40 | .6050 | .5688 | .8225 | .6916 | 1.446 | 1.216 | 1.758 | .9657 | 20 |
| 50 | .6080 | .5712 | .8208 | .6959 | 1.437 | 1.218 | 1.751 | .9628 | 10 |
| | | cos $\alpha$ | sin $\alpha$ | cot $\alpha$ | tan $\alpha$ | csc $\alpha$ | sec $\alpha$ | $\alpha$ (radians) | $\alpha$ (degrees) |

**Table 4 Values of Trigonometric Functions (*continued*)**

| $\alpha$ (degrees) | $\alpha$ (radians) | sin $\alpha$ | cos $\alpha$ | tan $\alpha$ | cot $\alpha$ | sec $\alpha$ | csc $\alpha$ | | |
|---|---|---|---|---|---|---|---|---|---|
| **35°00′** | .6109 | .5736 | .8192 | .7002 | 1.428 | 1.221 | 1.743 | .9599 | **55°00′** |
| 10 | .6138 | .5760 | .8175 | .7046 | 1.419 | 1.223 | 1.736 | .9570 | 50 |
| 20 | .6167 | .5783 | .8158 | .7089 | 1.411 | 1.226 | 1.729 | .9541 | 40 |
| 30 | .6196 | .5807 | .8141 | .7133 | 1.402 | 1.228 | 1.722 | .9512 | 30 |
| 40 | .6225 | .5831 | .8124 | .7177 | 1.393 | 1.231 | 1.715 | .9483 | 20 |
| 50 | .6254 | .5854 | .8107 | .7221 | 1.385 | 1.233 | 1.708 | .9454 | 10 |
| **36°00′** | .6283 | .5878 | .8090 | .7265 | 1.376 | 1.236 | 1.701 | .9425 | **54°00′** |
| 10 | .6312 | .5901 | .8073 | .7310 | 1.368 | 1.239 | 1.695 | .9396 | 50 |
| 20 | .6341 | .5925 | .8056 | .7355 | 1.360 | 1.241 | 1.688 | .9367 | 40 |
| 30 | .6370 | .5948 | .8039 | .7400 | 1.351 | 1.244 | 1.681 | .9338 | 30 |
| 40 | .6400 | .5972 | .8021 | .7445 | 1.343 | 1.247 | 1.675 | .9308 | 20 |
| 50 | .6429 | .5995 | .8004 | .7490 | 1.335 | 1.249 | 1.668 | .9279 | 10 |
| **37°00′** | .6458 | .6018 | .7986 | .7536 | 1.327 | 1.252 | 1.662 | .9250 | **53°00′** |
| 10 | .6487 | .6041 | .7969 | .7581 | 1.319 | 1.255 | 1.655 | .9221 | 50 |
| 20 | .6516 | .6065 | .7951 | .7627 | 1.311 | 1.258 | 1.649 | .9192 | 40 |
| 30 | .6545 | .6088 | .7934 | .7673 | 1.303 | 1.260 | 1.643 | .9163 | 30 |
| 40 | .6574 | .6111 | .7916 | .7720 | 1.295 | 1.263 | 1.636 | .9134 | 20 |
| 50 | .6603 | .6134 | .7898 | .7766 | 1.288 | 1.266 | 1.630 | .9105 | 10 |
| **38°00′** | .6632 | .6157 | .7880 | .7813 | 1.280 | 1.269 | 1.624 | .9076 | **52°00′** |
| 10 | .6661 | .6180 | .7862 | .7860 | 1.272 | 1.272 | 1.618 | .9047 | 50 |
| 20 | .6690 | .6202 | .7844 | .7907 | 1.265 | 1.275 | 1.612 | .9018 | 40 |
| 30 | .6720 | .6225 | .7826 | .7954 | 1.257 | 1.278 | 1.606 | .8988 | 30 |
| 40 | .6749 | .6248 | .7808 | .8002 | 1.250 | 1.281 | 1.601 | .8959 | 20 |
| 50 | .6778 | .6271 | .7790 | .8050 | 1.242 | 1.284 | 1.595 | .8930 | 10 |
| **39°00′** | .6807 | .6293 | .7771 | .8098 | 1.235 | 1.287 | 1.589 | .8901 | **51°00′** |
| 10 | .6836 | .6316 | .7753 | .8146 | 1.228 | 1.290 | 1.583 | .8872 | 50 |
| 20 | .6865 | .6338 | .7735 | .8195 | 1.220 | 1.293 | 1.578 | .8843 | 40 |
| 30 | .6894 | .6361 | .7716 | .8243 | 1.213 | 1.296 | 1.572 | .8814 | 30 |
| 40 | .6923 | .6383 | .7698 | .8292 | 1.206 | 1.299 | 1.567 | .8785 | 20 |
| 50 | .6952 | .6406 | .7679 | .8342 | 1.199 | 1.302 | 1.561 | .8756 | 10 |
| **40°00′** | .6981 | .6428 | .7660 | .8391 | 1.192 | 1.305 | 1.556 | .8727 | **50°00′** |
| 10 | .7010 | .6450 | .7642 | .8441 | 1.185 | 1.309 | 1.550 | .8698 | 50 |
| 20 | .7039 | .6472 | .7623 | .8491 | 1.178 | 1.312 | 1.545 | .8668 | 40 |
| 30 | .7069 | .6494 | .7604 | .8541 | 1.171 | 1.315 | 1.540 | .8639 | 30 |
| 40 | .7098 | .6517 | .7585 | .8591 | 1.164 | 1.318 | 1.535 | .8610 | 20 |
| 50 | .7127 | .6539 | .7566 | .8642 | 1.157 | 1.322 | 1.529 | .8581 | 10 |
| **41°00′** | .7156 | .6561 | .7547 | .8693 | 1.150 | 1.325 | 1.524 | .8552 | **49°00′** |
| 10 | .7185 | .6583 | .7528 | .8744 | 1.144 | 1.328 | 1.519 | .8523 | 50 |
| 20 | .7214 | .6604 | .7509 | .8796 | 1.137 | 1.332 | 1.514 | .8494 | 40 |
| 30 | .7243 | .6626 | .7490 | .8847 | 1.130 | 1.335 | 1.509 | .8465 | 30 |
| 40 | .7272 | .6648 | .7470 | .8899 | 1.124 | 1.339 | 1.504 | .8436 | 20 |
| 50 | .7301 | .6670 | .7451 | .8952 | 1.117 | 1.342 | 1.499 | .8407 | 10 |
| | | cos $\alpha$ | sin $\alpha$ | cot $\alpha$ | tan $\alpha$ | csc $\alpha$ | sec $\alpha$ | $\alpha$ (radians) | $\alpha$ (degrees) |

## Table 4 Values of Trigonometric Functions (*continued*)

| α (degrees) | α (radians) | sin α | cos α | tan α | cot α | sec α | csc α | | |
|---|---|---|---|---|---|---|---|---|---|
| **42°00′** | .7330 | .6691 | .7431 | .9004 | 1.111 | 1.346 | 1.494 | .8378 | **48°00′** |
| 10 | .7359 | .6713 | .7412 | .9057 | 1.104 | 1.349 | 1.490 | .8348 | 50 |
| 20 | .7389 | .6734 | .7392 | .9110 | 1.098 | 1.353 | 1.485 | .8319 | 40 |
| 30 | .7418 | .6756 | .7373 | .9163 | 1.091 | 1.356 | 1.480 | .8290 | 30 |
| 40 | .7447 | .6777 | .7353 | .9217 | 1.085 | 1.360 | 1.476 | .8261 | 20 |
| 50 | .7476 | .6799 | .7333 | .9271 | 1.079 | 1.364 | 1.471 | .8232 | 10 |
| **43°00′** | .7505 | .6820 | .7314 | .9325 | 1.072 | 1.367 | 1.466 | .8203 | **47°00′** |
| 10 | .7534 | .6841 | .7294 | .9380 | 1.066 | 1.371 | 1.462 | .8174 | 50 |
| 20 | .7563 | .6862 | .7274 | .9435 | 1.060 | 1.375 | 1.457 | .8145 | 40 |
| 30 | .7592 | .6884 | .7254 | .9490 | 1.054 | 1.379 | 1.453 | .8116 | 30 |
| 40 | .7621 | .6905 | .7234 | .9545 | 1.048 | 1.382 | 1.448 | .8087 | 20 |
| 50 | .7560 | .6926 | .7214 | .9601 | 1.042 | 1.386 | 1.444 | .8058 | 10 |
| **44°00′** | .7679 | .6947 | .7193 | .9657 | 1.036 | 1.390 | 1.440 | .8029 | **46°00′** |
| 10 | .7709 | .6967 | .7173 | .9713 | 1.030 | 1.394 | 1.435 | .7999 | 50 |
| 20 | .7738 | .6988 | .7153 | .9770 | 1.024 | 1.398 | 1.431 | .7970 | 40 |
| 30 | .7767 | .7009 | .7133 | .9827 | 1.018 | 1.402 | 1.427 | .7941 | 30 |
| 40 | .7796 | .7030 | .7112 | .9884 | 1.012 | 1.406 | 1.423 | .7912 | 20 |
| 50 | .7825 | .7050 | .7092 | .9942 | 1.006 | 1.410 | 1.418 | .7883 | 10 |
| **45°00′** | .7854 | .7071 | .7071 | 1.000 | 1.000 | 1.414 | 1.414 | .7854 | **45°00′** |
| | | cos α | sin α | cot α | tan α | csc α | sec α | α (radians) | α (degrees) |

# Answers to Exercises

## Chapter 1

### Exercise Set 1.1, page 4

**1.** $A(3, -4)$ **3.** $B(-3, 4)$ **5.** $C(0, 5)$ **11.** $A\left(\frac{1}{2}, \frac{\sqrt{3}}{2}\right)$ **13.** $B\left(\frac{\sqrt{2}}{2}, -\frac{\sqrt{2}}{2}\right)$ **15.** $C\left(-\frac{\sqrt{3}}{2}, -\frac{1}{2}\right)$

**7.** $D\left(-\frac{2}{3}, 7\right)$ **9.** $E\left(\frac{3}{4}, \frac{1}{3}\right)$ **17.** $D(0, \pi)$ **19.** $E\left(\frac{\pi}{2}, -1\right)$

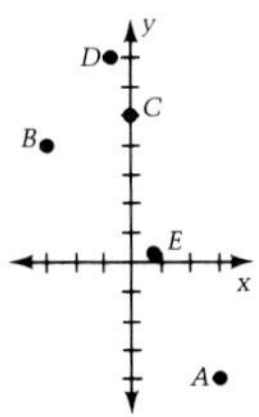


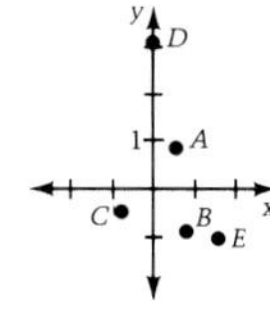


**21.** $2\sqrt{5}$ **23.** 8 **25.** 10 **27.** 15 **29.** $\sqrt{6}$ **31.** 1 **33.** 2 **35.** 1

**37.** yes, no **39.** no, no **43.** yes

### Exercise Set 1.2, page 11

**1.** function: {(1, A), (2, B), (3, C), (4, A)}; domain: {1, 2, 3, 4}; range: {A, B, C}

**3.** function: {(81, 10), (82, 20), (83, 20), (84, 35), (85, 30)}; domain: {81, 82, 83, 84, 85}; range: {10, 20, 30, 35}

**5.** 14, 8, $-7$, $8 - 3c^2$, $2 - 3h$ **7.** 20, 6, 6, $c^4 - 5c^2 + 6$, $-h + h^2$ **9.** 10, 10, 10, 10, 10

**11.** yes **13.** yes **15.** 3 **17.** 1 **19.** 23 **21.** 3 **23.** $-\dfrac{199}{25}$ **25.** $-\dfrac{2}{9}$ **31.** $f^{-1}(x) = \dfrac{5 - x}{3}$

**33.** $f^{-1}(x) = 2x - 4$ **35.** (a) 26, (b) 12, (c) $4t^2 - 8t + 5$ **37.** odd

**39.** neither **41.** even

**Exercise Set 1.3, page 19**

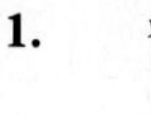

**1.**

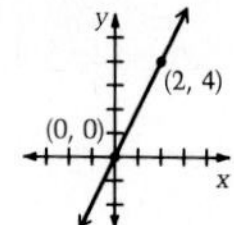


**3.**

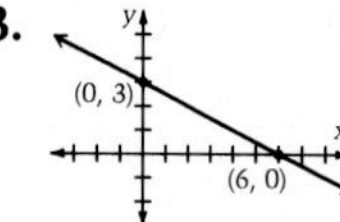


**5.**

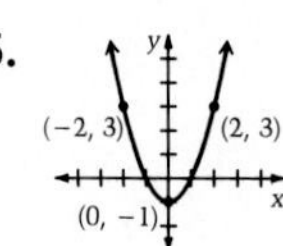


**7.**

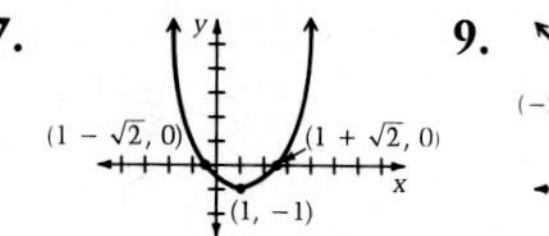


**9.**

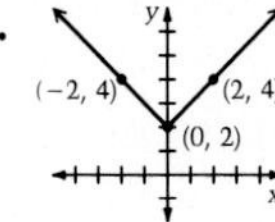


**11.**

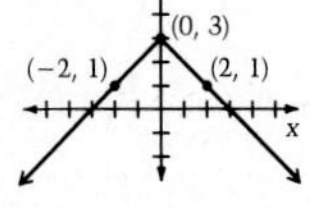


**13.** Domain: $[-3, +\infty)$

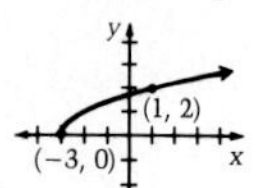


**15.** Domain: $(-\infty, 4]$

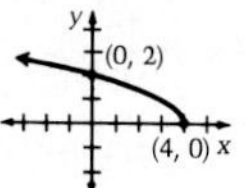


**17.** Domain: all real numbers except $-4$

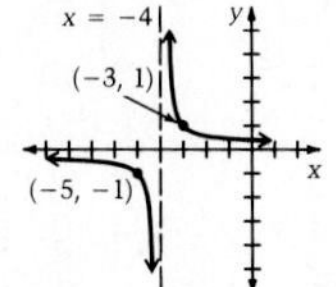


**19.** (a) (3, 2), (b) $(-3, -2)$, (c) $(-3, 2)$

**21.** (a) $(4, -3)$, (b) $(-4, 3)$, (c) $(-4, -3)$



**23.** (a) $(0, -5)$, (b) $(0, 5)$, (c) $(0, -5)$

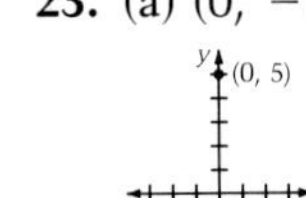


**25.** $(-2, 3)$ **27.** $(3, 4)$

**29.** $(5, 0)$

**31.**

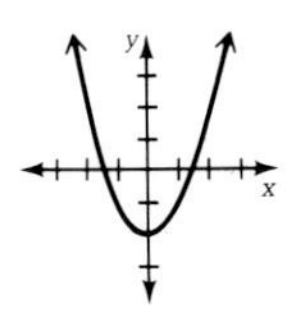

**33.**

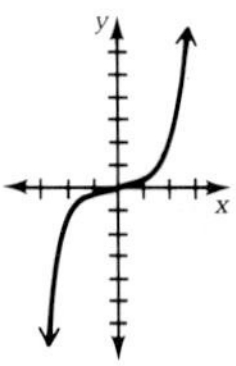

**35.** $f^{-1}(x) = \dfrac{8 - x}{2}$
domain: $[0, 6]$
range: $[1, 4]$

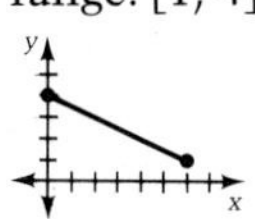

**37.** $f^{-1}(x) = \sqrt{x} - 1$
domain: $[1, 9)$
range: $[0, 2)$

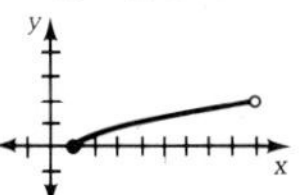

**39.** $f^{-1}(x) = 5 - x^2$
domain: $[0, +\infty)$
range: $(-\infty, 5]$

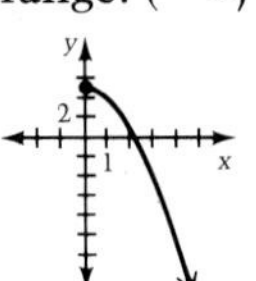


**41.**

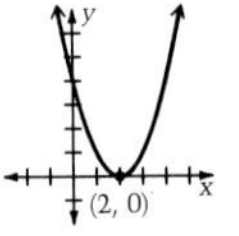


**43.**

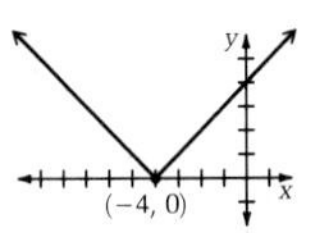


**45.**

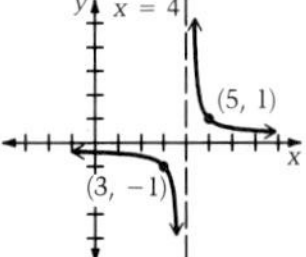


**47.**

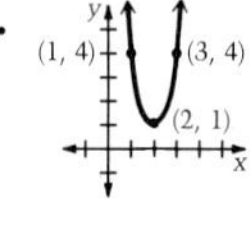

**49.** 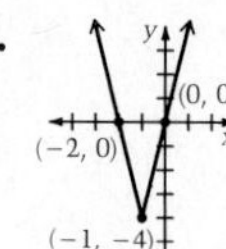


**51.** (5, 2) **53.** (4, 1) **55.** 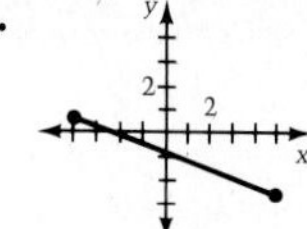

**57.** 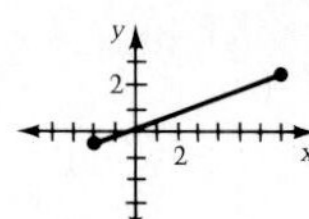

**59.** 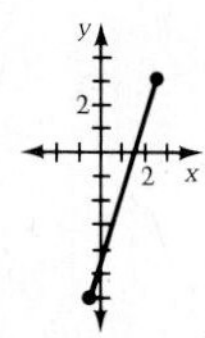

## Exercise Set 1.4, page 24

**1.** $m = 5$ 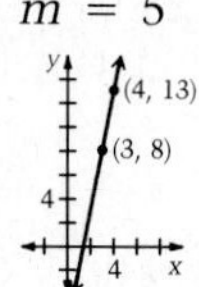


**3.** $m = -\frac{1}{4}$ 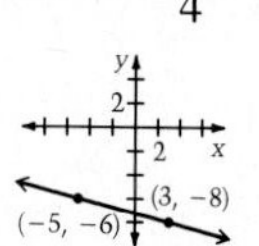


**5.** $m = 0$ 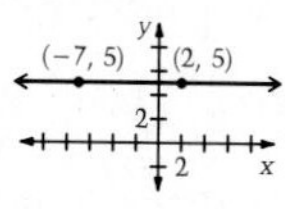


**7.** $m = 0$ 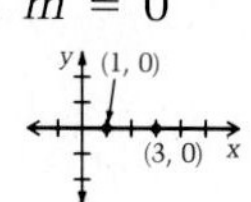


**9.** $m = \frac{1}{8}$ 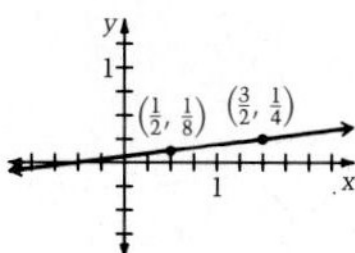


**11.** $m = 1.1$ 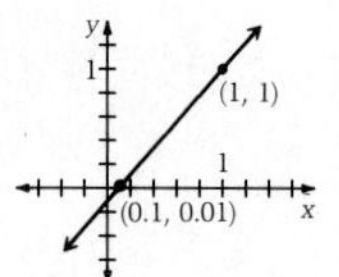


**13.** (a) $y - 8 = 5(x - 3)$, (b) $y = 5x - 7$, (c) $5x - y - 7 = 0$

**15.** (a) $y - (-6) = -\frac{1}{4}(x - (-5))$, (b) $y = -\frac{1}{4}x - \frac{29}{4}$, (c) $\frac{1}{4}x + y + \frac{29}{4} = 0$

**17.** (a) $y - 5 = 0(x - 2)$, (b) $y = 5$, (c) $y - 5 = 0$ **19.** (a) $y - 0 = 0(x - 3)$, (b) $y = 0$, (c) $y = 0$

**21.** (a) $y - \frac{1}{8} = \frac{1}{8}\left(x - \frac{1}{2}\right)$, (b) $y = \frac{1}{8}x + \frac{1}{16}$, (c) $\frac{1}{8}x - y + \frac{1}{16} = 0$

**23.** (a) $y - 1 = 1.1(x - 1)$, (b) $y = 1.1x - 0.1$, (c) $1.1x - y - 0.1 = 0$

**25.** (a) $y - 5 = -2(x - 3)$, (b) $y = -2x + 11$, (c) $2x + y - 11 = 0$

**27.** (a) $y - (-2) = 0(x - 2)$, (b) $y = -2$, (c) $y + 2 = 0$

**29.** (a) $y - 3 = \frac{1}{2}(x - 0)$, (b) $y = \frac{1}{2}x + 3$, (c) $\frac{1}{2}x - y + 3 = 0$

**31.** (a) $y - (0) = -2(x - 0)$, (b) $y = -2x$, (c) $2x + y = 0$

**33.** (a) $y - 0 = -2(x - 0)$, (b) $y = -2x$, (c) $2x + y = 0$

**35.** (a) $y - (-2) = -\frac{1}{6}(x - 5)$, (b) $y = -\frac{1}{6}x - \frac{7}{6}$, (c) $\frac{1}{6}x + y + \frac{7}{6} = 0$

**37.** The slope of the segment with endpoints (6, 3) and (5, 1) is 2. The slope of the segment with endpoints (1, 3) and (5, 1) is $-\frac{1}{2}$. Since the slopes are negative reciprocals, the segments are perpendicular and the triangle is a right triangle.

**39.** $y = \frac{7}{3} - \frac{x}{3}$, Domain: $[1, +\infty)$ **41.** $y = \frac{8}{7}x + \frac{100}{7}$

**Exercise Set 1.5, page 28**

**1.** 0, 5 **3.** $-10$, 10 **5.** $-1$, $-8$ **7.** $-2$, 5 **9.** $-4$, $-1$ **11.** 4 **13.** no real number solutions

**15.** $-1 \pm \frac{\sqrt{5}}{2}$

**17.** $y = 1(x - 3)^2 + 10$. The vertex is (3, 10), the axis of symmetry is $x = 3$, and the parabola opens upward.

**19.** $y = -\frac{1}{3}(x - (-1))^2 + \left(-\frac{5}{3}\right)$. The vertex is $(-1, -\frac{5}{3})$, the axis of symmetry is $x = -1$, and the parabola opens downward.

**21.** $y = \frac{1}{3}(x - 9)^2 + (-1)$. The vertex is $(9, -1)$, the axis of symmetry is $x = 9$, and the parabola opens upward.

**23.** $y = -\frac{2}{7}(x - (-6))^2 + 0$. The vertex is $(-6, 0)$, the axis of symmetry is $x = -6$, and the parabola opens downward.

**25.** $y = 1(x - 0)^2 + 4$. The vertex is (0, 4), the axis of symmetry is $x = 0$, and the parabola opens upward.

**27.** 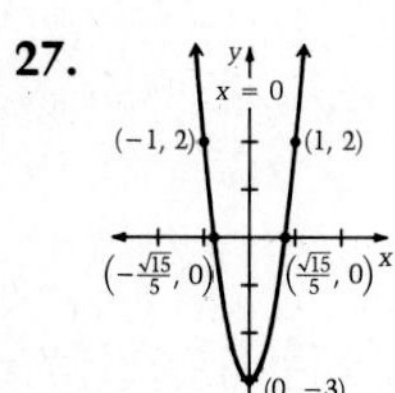


**29.** 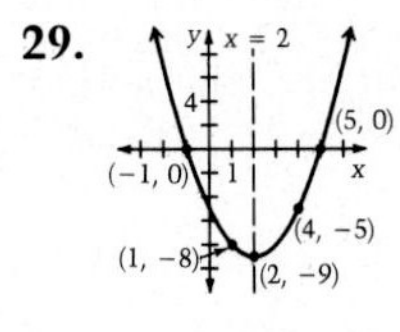


**31.** 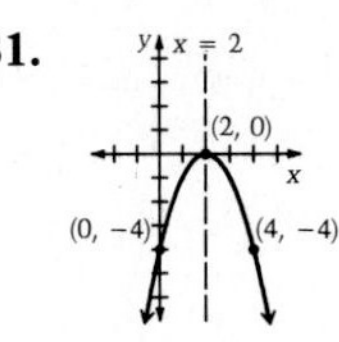


**33.** 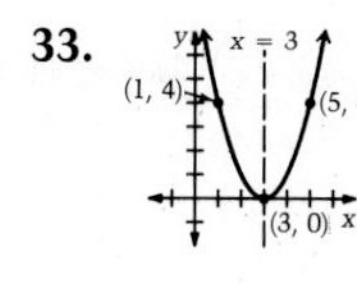


**35.** 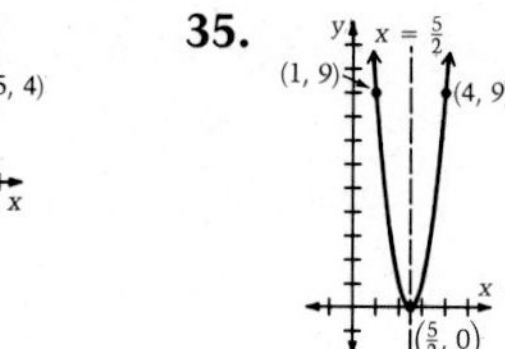


**37.** y x = 5 (5, −4) x −5 (8, −22) (2, −22) (10, −54) (0, −54)

**39.** 0, maximum **41.** 9, maximum **43.** $\pm 2\sqrt{6}$ **45.** $\frac{125}{9}$

**47.** (a) $x$, $12 - x$, (b) $P(x) = 12x - x^2$, (c) 6, 6

## Chapter Test, page 30

**1.** A(4, 1) **3.** $B\left(-\frac{\sqrt{3}}{2}, -\frac{1}{2}\right)$ **5.** $3\sqrt{10}$ **7.** $\frac{1}{2}$

**9.** function: {(1, 25), (2, 20), (3, 5), (4, 15), (5, 25), (6, 20), (7, 10), (8, 5)}; domain: {1, 2, 3, 4, 5, 6, 7, 8}; range: {5, 10, 15, 20, 25}

**11.** $15, 9, -h^4 - 5h^2 + 9, -b^2 - 3b + 13$ **13.** $-\frac{17}{3}, -5, \frac{h^2 - 15}{3}, \frac{b - 16}{3}$ **15.** no **17.** 1 **19.** $\frac{1}{32}$

**23.**

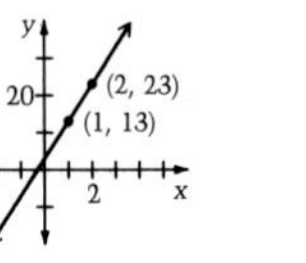


**25.**

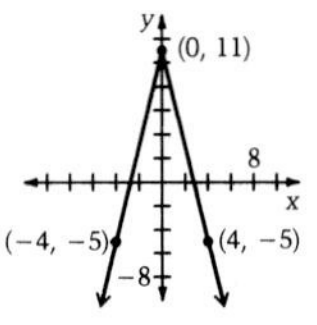


**27.** (a) (0, 3) (b) (0, −3) (c) (−3, 0) (d) (0, 3)

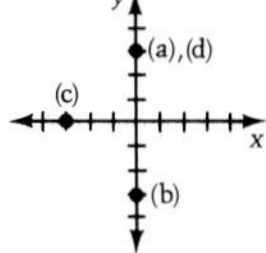


**29.** $f^{-1}(x) = \frac{x - 5}{2}$
domain: [3, 9]
range: [−1, 2]

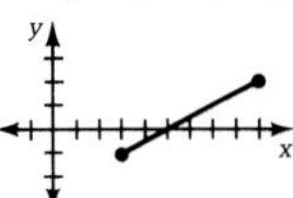


**31.**

**33.** $m = -5$

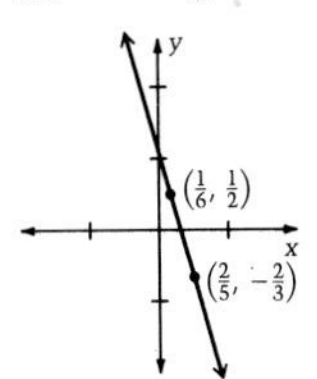


**35.** (a) $y - 4 = 3(x - (-2))$
(b) $y = 3x + 10$
(c) $3x - y + 10 = 0$

**37.** −4, −3

**39.** $y = 6(x - (-4))^2 + 9$. The vertex is $(-4, 9)$, the axis of symmetry is $x = -4$, and the parabola opens up.

**41.** 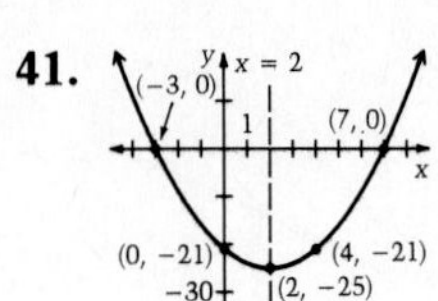


**43.** (a) no (b) yes **45.** $-22, \frac{11}{3}$

**Exercise Set 2.1, page 39**

**1.** 67° **3.** 30° **5.** 18 **7.** 10 **9.** 36 **11.** 12 **13.** $\sqrt{13}$ **15.** $\frac{3}{4}$ **17.** 5 **19.** $\frac{32}{5}$ **21.** 10

**23.** 24 **25.** $\frac{36\sqrt{5}}{5}$ **27.** $\frac{18\sqrt{5}}{5}$ **29.** 42 ft **31.** 17 mi **33.** $\frac{19\sqrt{2}}{2}$ in by $\frac{19\sqrt{2}}{2}$ in, $\frac{361}{2}$ in$^2$ **35.** 30°

**37.** $\frac{5\sqrt{3}}{2}$ **39.** $\frac{5}{2}$ **41.** $\frac{25\sqrt{3}}{8}$ **43.** 240 **45.** $12\sqrt{2}$

**Exercise Set 2.2, page 46**

**1.** $\sin\alpha = \frac{3}{5}$ $\csc\alpha = \frac{5}{3}$
$\cos\alpha = \frac{4}{5}$ $\sec\alpha = \frac{5}{4}$
$\tan\alpha = \frac{3}{4}$ $\cot\alpha = \frac{4}{3}$

**3.** $\sin\alpha = \frac{8}{17}$ $\csc\alpha = \frac{17}{8}$
$\cos\alpha = \frac{15}{17}$ $\sec\alpha = \frac{17}{15}$
$\tan\alpha = \frac{8}{15}$ $\cot\alpha = \frac{15}{8}$

**5.** $\sin\alpha = \frac{12}{37}$ $\csc\alpha = \frac{37}{12}$
$\cos\alpha = \frac{35}{37}$ $\sec\alpha = \frac{37}{35}$
$\tan\alpha = \frac{12}{35}$ $\cot\alpha = \frac{35}{12}$

**7.** $\sin\alpha = \frac{3}{5}$ $\csc\alpha = \frac{5}{3}$
$\cos\alpha = \frac{4}{5}$ $\sec\alpha = \frac{5}{4}$
$\tan\alpha = \frac{3}{4}$ $\cot\alpha = \frac{4}{3}$

**9.** $\sin\alpha = \frac{\sqrt{2}}{2}$ $\csc\alpha = \sqrt{2}$
$\cos\alpha = \frac{\sqrt{2}}{2}$ $\sec\alpha = \sqrt{2}$
$\tan\alpha = 1$ $\cot\alpha = 1$

**11.** $\sin\alpha = \frac{1}{2}$ $\csc\alpha = 2$
$\cos\alpha = \frac{\sqrt{3}}{2}$ $\sec\alpha = \frac{2\sqrt{3}}{3}$
$\tan\alpha = \frac{\sqrt{3}}{3}$ $\cot\alpha = \sqrt{3}$

**13.** $\sin\alpha = \frac{\sqrt{3}}{2}$ $\csc\alpha = \frac{2\sqrt{3}}{3}$
$\cos\alpha = \frac{1}{2}$ $\sec\alpha = 2$
$\tan\alpha = \sqrt{3}$ $\cot\alpha = \frac{\sqrt{3}}{3}$

**15.** $\sin\alpha = \frac{1}{2}$ $\csc\alpha = 2$
$\cos\alpha = \frac{\sqrt{3}}{2}$ $\sec\alpha = \frac{2\sqrt{3}}{3}$
$\tan\alpha = \frac{\sqrt{3}}{3}$ $\cot\alpha = \sqrt{3}$

**17.** $\tan\alpha = \frac{a}{b} = \cot\beta$

**19.** $\sin\beta = \frac{b}{c} = \cos\alpha$ **21.** $3\sqrt{3}$ **23.** $4\sqrt{3}$ **25.** 30 **27.** $14\sqrt{3}$ **29.** $24\sqrt{2}$ **31.** 15

**33.** 45° **35.** 60° **37.** $\frac{4}{5}, \frac{3}{4}$ **39.** $\frac{15}{8}, \frac{15}{17}$ **41.** $\frac{4}{5}, \frac{4}{5}$ **43.** $\frac{24}{25}, \frac{7}{24}$ **45.** $\frac{\sqrt{19}}{10}, \frac{9}{10}$

**49.** 6 **51.** $30\sqrt{3}$ **53.** $36\sqrt{3}$ ft, 36 ft

**Exercise Set 2.3, page 52**

**1.** 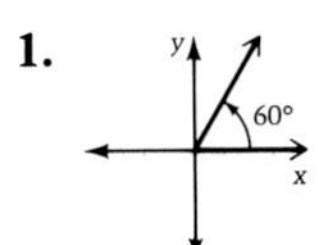


**3.** 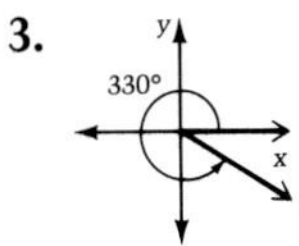


**5.** 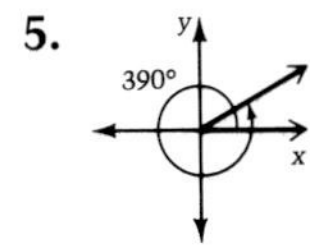


**7.** 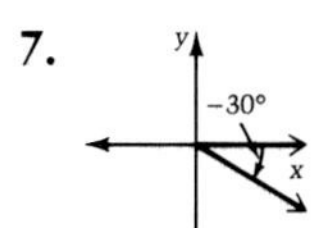


**9.** 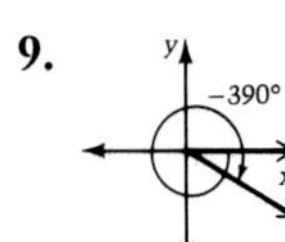

**11.** 

**13.** 45° **15.** 3° **17.** 322° **19.** 60° **21.** 20°, −340° **23.** 225°, −135°

**25.** 310°, −50° **27.** 89°, −271° **29.** 360°, −360° **31.** 135°, −225° **33.** $\frac{1}{2}°$, $-359\frac{1}{2}°$

**35.** $\sin\alpha = \frac{3}{5}$ $\csc\alpha = \frac{5}{3}$
$\cos\alpha = -\frac{4}{5}$ $\sec\alpha = -\frac{5}{4}$
$\tan\alpha = -\frac{3}{4}$ $\cot\alpha = -\frac{4}{3}$

**37.** $\sin\alpha = -\frac{5}{13}$ $\csc\alpha = -\frac{13}{5}$
$\cos\alpha = -\frac{12}{13}$ $\sec\alpha = -\frac{13}{12}$
$\tan\alpha = \frac{5}{12}$ $\cot\alpha = \frac{12}{5}$

**39.** $\sin\alpha = \frac{3}{5}$ $\csc\alpha = \frac{5}{3}$
$\cos\alpha = -\frac{4}{5}$ $\sec\alpha = -\frac{5}{4}$
$\tan\alpha = -\frac{3}{4}$ $\cot\alpha = -\frac{4}{3}$

**41.** $\sin\alpha = \frac{\sqrt{2}}{2}$ $\csc\alpha = \sqrt{2}$
$\cos\alpha = \frac{\sqrt{2}}{2}$ $\sec\alpha = \sqrt{2}$
$\tan\alpha = 1$ $\cot\alpha = 1$

**43.** $\sin\alpha = -\frac{\sqrt{2}}{2}$ $\csc\alpha = -\sqrt{2}$
$\cos\alpha = \frac{\sqrt{2}}{2}$ $\sec\alpha = \sqrt{2}$
$\tan\alpha = -1$ $\cot\alpha = -1$

**45.** $\sin\alpha = -\frac{2\sqrt{5}}{5}$ $\csc\alpha = -\frac{\sqrt{5}}{2}$
$\cos\alpha = \frac{\sqrt{5}}{5}$ $\sec\alpha = \sqrt{5}$
$\tan\alpha = -2$ $\cot\alpha = -\frac{1}{2}$

**47.** $\sin\alpha = -\frac{2}{3}$ $\csc\alpha = -\frac{3}{2}$
$\cos\alpha = \frac{\sqrt{5}}{3}$ $\sec\alpha = \frac{3\sqrt{5}}{5}$
$\tan\alpha = -\frac{2\sqrt{5}}{5}$ $\cot\alpha = -\frac{\sqrt{5}}{2}$

**49.** 1 **51.** undefined **53.** undefined **55.** 0 **57.** 0

**59.** undefined **61.** 180° **63.** 108° **65.** $\sin\varphi = \frac{5\sqrt{26}}{26}$ $\csc\varphi = \frac{\sqrt{26}}{5}$ $\cos\varphi = -\frac{\sqrt{26}}{26}$ $\sec\varphi = -\sqrt{26}$ $\cot\varphi = -\frac{1}{5}$ **67.** $\sin\varphi = \frac{3}{5}$ $\sec\varphi = -\frac{5}{4}$ $\cos\varphi = -\frac{4}{5}$ $\cot\varphi = -\frac{4}{3}$ $\tan\varphi = -\frac{4}{3}$

**69.** $\sin\varphi = \frac{3\sqrt{34}}{34}$ $\csc\varphi = \frac{\sqrt{34}}{3}$ $\cos\varphi = -\frac{5\sqrt{34}}{34}$ $\sec\varphi = -\frac{\sqrt{34}}{5}$ $\tan\varphi = -\frac{3}{5}$ **71.** $\sin\varphi = \frac{4}{5}$ $\csc\varphi = \frac{5}{4}$ $\tan\varphi = -\frac{4}{3}$ $\sec\varphi = -\frac{5}{3}$ $\cot\varphi = -\frac{3}{4}$ **73.** $\cos\varphi = -\frac{\sqrt{21}}{5}$ $\csc\varphi = -\frac{5}{2}$ $\tan\varphi = \frac{2\sqrt{21}}{21}$ $\sec\varphi = -\frac{5\sqrt{21}}{21}$ $\cot\varphi = \frac{\sqrt{21}}{2}$

**75.** $\sin\varphi = -\frac{4}{5}$ $\csc\varphi = -\frac{5}{4}$ $\tan\varphi = \frac{4}{3}$ $\sec\varphi = -\frac{5}{3}$ $\cot\varphi = \frac{3}{4}$ **77.** $\sin\varphi = -\frac{3}{5}$ $\sec\varphi = \frac{5}{4}$ $\cos\varphi = \frac{4}{5}$ $\cot\varphi = -\frac{4}{3}$ $\tan\varphi = -\frac{3}{4}$ **79.** $\sin\varphi = -\frac{\sqrt{51}}{10}$ $\csc\varphi = -\frac{10\sqrt{51}}{51}$ $\tan\varphi = -\frac{\sqrt{51}}{7}$ $\sec\varphi = \frac{10}{7}$ $\cot\varphi = -\frac{7\sqrt{51}}{51}$

**Exercise Set 2.4, page 59**

**1.** $\tan 32° = 0.62$ $\sec 32° = 1.18$ $\csc 32° = 1.89$ $\cot 32° = 1.60$ **3.** $\sin 66° = 0.91$ $\csc 66° = 1.09$ $\cos 66° = 0.41$ $\cot 66° = 0.44$ **5.** 36° **7.** 52° **9.** 45°

**11.** sin 63° **13.** tan 52° **15.** sec 87° **17.** cos 71° **19.** 0.85 **21.** 2.56 **23.** 2.38

**25.** $\cos\alpha = \frac{\sqrt{3}}{2}$ $\csc\alpha = 2$ $\tan\alpha = \frac{\sqrt{3}}{3}$ $\sec\alpha = \frac{2\sqrt{3}}{3}$ $\cot\alpha = \sqrt{3}$

**27.** $\sin\alpha = \frac{3\sqrt{13}}{13}$ $\csc\alpha = \frac{\sqrt{13}}{3}$ $\cos\alpha = \frac{2\sqrt{13}}{13}$ $\sec\alpha = \frac{\sqrt{13}}{2}$ $\cot\alpha = \frac{2}{3}$

**29.** $\sin\alpha = \frac{2\sqrt{5}}{5}$ $\csc\alpha = \frac{\sqrt{5}}{2}$ $\cos\alpha = \frac{\sqrt{5}}{5}$ $\sec\alpha = \sqrt{5}$ $\tan\alpha = 2$

**31.** $\sin\alpha = \frac{3}{5}$ $\csc\alpha = \frac{5}{3}$ $\tan\alpha = \frac{3}{4}$ $\sec\alpha = \frac{5}{4}$ $\cot\alpha = \frac{4}{3}$

**33.** 0.5175 **35.** 0.3973 **37.** 1.828 **39.** 0.8952 **41.** 0.6111

**43.** 1.402 **45.** 28° **47.** 43° **49.** 39°40′ **51.** 23° **53.** 0.6368 **55.** 0.6100 **57.** 0.3181

**59.** 0.1352 **61.** 25.88 **63.** 3°27′ **65.** 69°34′ **67.** 26°34′ **69.** yes **71.** yes **73.** yes **75.** yes

**77.** $\frac{\sqrt{3}}{2}, \frac{1}{2}$ **79.** $-\sqrt{3}, \frac{\sqrt{3}}{3}$ **81.** $\sin(60° + 30°) = 1 = \sin 60° \cos 30° + \sin 30° \cos 60°$

**83.** $\cos 90° = 0 = \cos 45° \cos 45° - \sin 45° \sin 45°$

**Exercise Set 2.5, page 64**

**1.** positive **3.** positive **5.** positive **7.** positive **9.** positive **11.** positive **13.** negative

**15.** 20° **17.** 40° **19.** 55° **21.** 70° **23.** 84° **25.** 5° **27.** 85°35′ **29.** $-\frac{1}{2}$ **31.** $-\sqrt{3}$ **33.** 1

**35.** $\frac{2\sqrt{3}}{3}$ **37.** $-\frac{2\sqrt{3}}{3}$ **39.** $\frac{\sqrt{3}}{3}$ **41.** $\frac{1}{2}$ **43.** $-0.4040$ **45.** 0.6691 **47.** $-1.192$ **49.** $-0.9994$

**51.** 5.759 **53.** $-0.8090$ **55.** $-1.494$ **57.** 0.9957 **59.** 0.1853 **61.** negative **63.** negative **65.** negative **67.** negative **69.** negative **71.** positive **73.** negative **75.** negative

## Chapter Test, page 66

**1.** $63°$ **3.** 20 **5.** 30 **7.** $\sin\alpha = \frac{8}{17}$ $\csc\alpha = \frac{17}{8}$ $\cos\alpha = \frac{15}{17}$ $\sec\alpha = \frac{17}{15}$ $\tan\alpha = \frac{8}{15}$ $\cot\alpha = \frac{15}{8}$ **9.** 48 **11.** $\frac{3\sqrt{10}}{20}, \frac{2\sqrt{10}}{3}$ **13.**

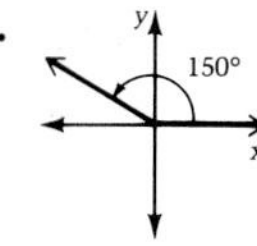


**15.**

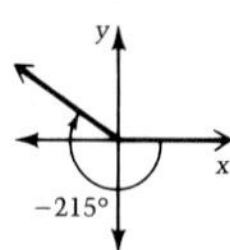


**17.** $75°$ **19.** $285°$ **21.** $\frac{12}{13}, -\frac{12}{5}$ **23.** 0 **25.** $13°$ **27.** 1.05

**29.** $\sin\alpha = \frac{3\sqrt{13}}{13}$ $\csc\alpha = \frac{\sqrt{13}}{3}$ $\cos\alpha = \frac{2\sqrt{13}}{13}$ $\sec\alpha = \frac{\sqrt{13}}{2}$ $\cot\alpha = \frac{2}{3}$ **31.** 0.7386 **33.** negative **35.** positive **37.** $35°$ **39.** $-1$

**41.** 0.9793 **43.** $8\sqrt{3}$ **45.** $\sin\varphi = -\frac{5}{13}$ $\csc\varphi = -\frac{13}{5}$

$\tan\varphi = \frac{5}{12}$ $\sec\varphi = -\frac{13}{12}$

$\cot\varphi = \frac{12}{5}$

**47.** $\sin(2\cdot 120°) = -\frac{\sqrt{3}}{2} = 2\sin 120° \cos 120°$ **49.** negative

## Chapter 3

### Exercise Set 3.1, page 76

**1.** yes **3.** no **5.**

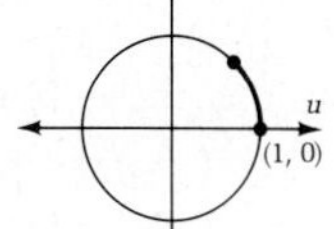


**7.**

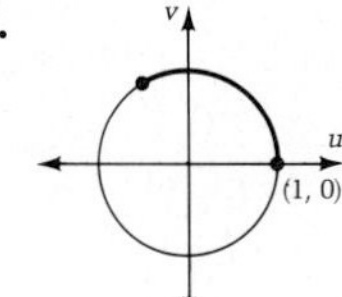


**9.**

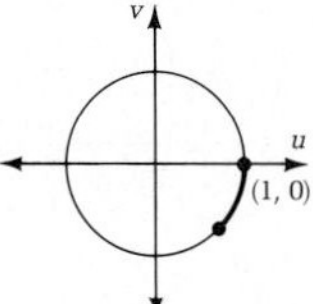


**11.**

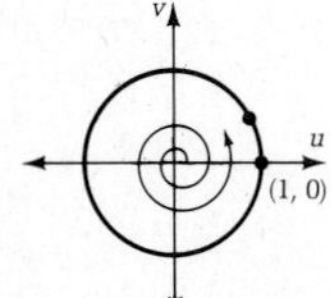


**13.** $\frac{\pi}{3}$ **15.** $\frac{7\pi}{4}$ **17.** $\frac{7\pi}{6}$ **19.** $-\frac{\pi}{6}$ **21.** $\frac{5\pi}{12}$ **23.** $-3\pi$ **25.** 270°

**27.** 150° **29.** −540° **31.** −750° **33.** $\frac{270°}{\pi}$ **35.** $2\pi$ in **37.** 8 m **39.** $-\frac{7\pi}{5}$

**41.** (a) Minute hand: $-24\pi$, Hour hand: $-2\pi$ (b) $-2\pi, -\frac{\pi}{6}$ (c) $-\pi, -\frac{\pi}{12}$ (d) $-\frac{\pi}{6}, -\frac{\pi}{72}$

**43.** $120\pi, \frac{3\pi}{5}$ **45.** 5280 radians **47.** (a) $\left(\frac{\sqrt{3}}{2}, \frac{1}{2}\right)$ (b) $\left(\frac{\sqrt{2}}{2}, \frac{\sqrt{2}}{2}\right)$ (c) $\left(\frac{1}{2}, \frac{\sqrt{3}}{2}\right)$

**Exercise Set 3.2, page 84**

**1.** undefined **3.** 0 **5.** $\frac{\sqrt{3}}{2}$ **7.** 2 **9.** 1 **11.** $-\frac{1}{2}$ **13.** $\frac{2\sqrt{3}}{3}$ **15.** −1 **17.** $-\frac{\sqrt{2}}{2}$

**19.** $-\sqrt{3}$ **21.** $\sqrt{2}$ **23.** 0 **25.** $\frac{\sqrt{3}}{3}$ **27.** $-\sqrt{3}$ **29.** $\frac{2\sqrt{3}}{3}$ **31.** −0.53 **33.** −1.65 **35.** 1.05

**37.** −0.36 **39.** $\sin\frac{13\pi}{6}\sec\frac{5\pi}{3} = 1 = \tan\frac{9\pi}{4}$ **41.** $\tan\frac{\pi}{3} = \sqrt{3} = \dfrac{1-\cos\frac{4\pi}{3}}{\sin\frac{\pi}{3}}$

**43.** $\cos\frac{5\pi}{6} = -\frac{\sqrt{3}}{2} = \cos\frac{\pi}{2}\cos\frac{\pi}{3} - \sin\frac{\pi}{2}\sin\frac{\pi}{3}$ **45.** $\frac{3\pi}{4}$ **47.** $\frac{5\pi}{6}$ **49.** $\frac{\pi}{4}$

**Exercise Set 3.3, page 95**

**1.**

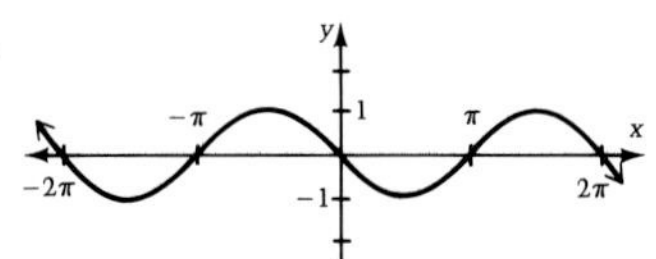

**3.**

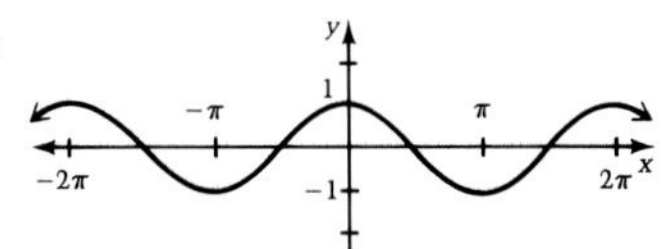

**5.**

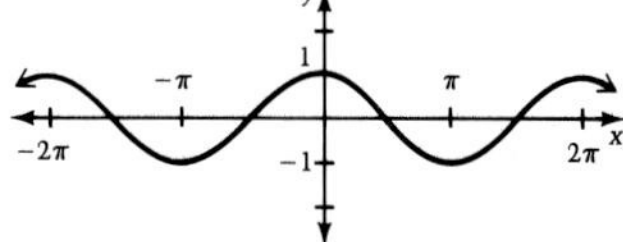

**7.**

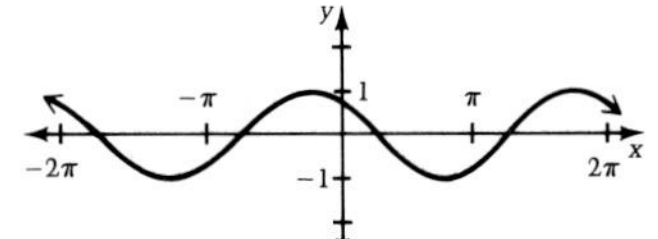

**9.**

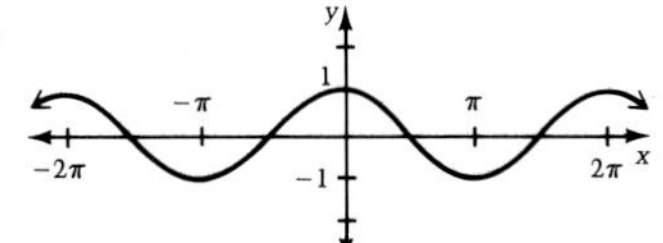

**11.**

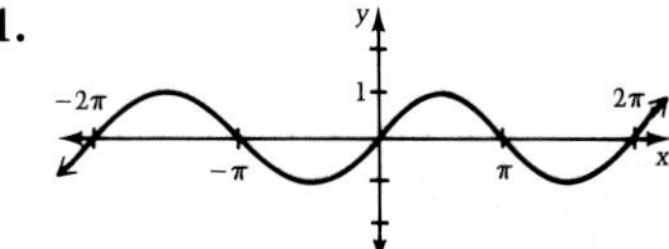

**13.** 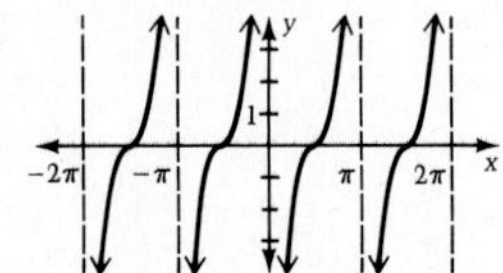


**15.** 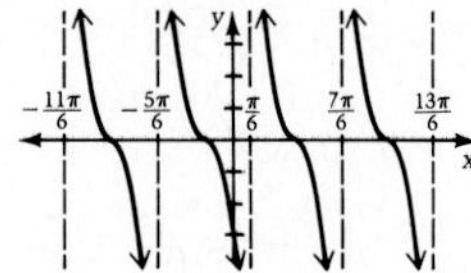


**17.** 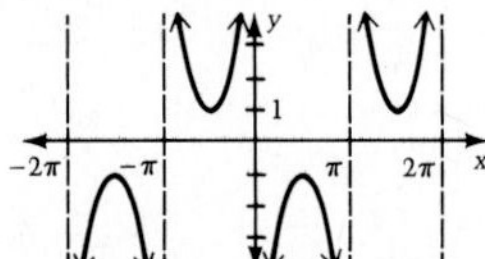


**19.** 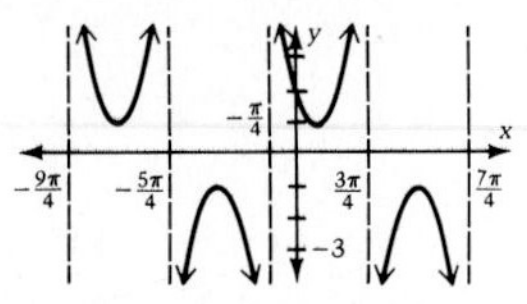


**21.** 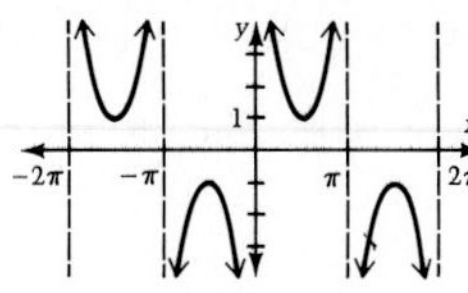


**23.** False 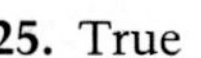 **25.** True **27.** True

**29.** False **31.** 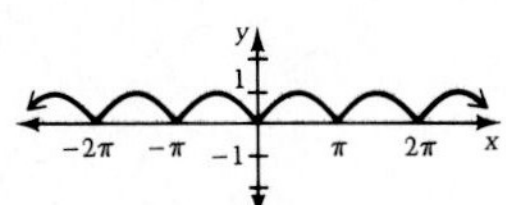


**33.** 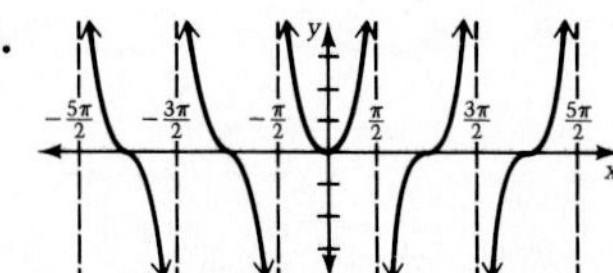


**35.** 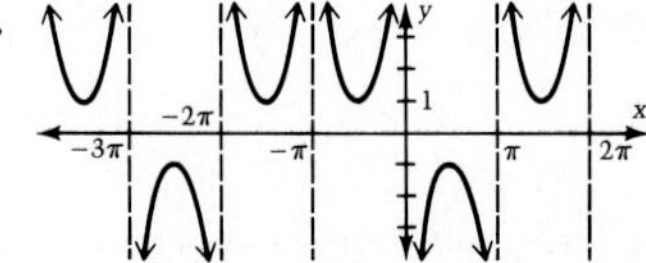


**37.** 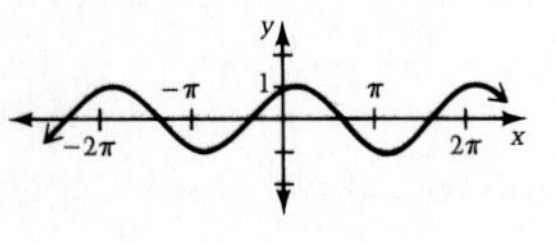


**39.** 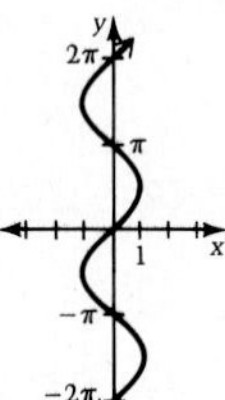


**41.** 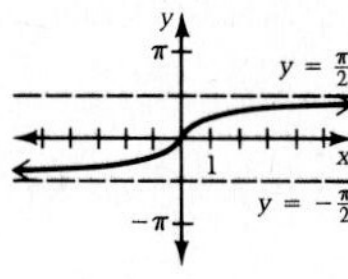


## Exercise Set 3.4, page 102

**1.** 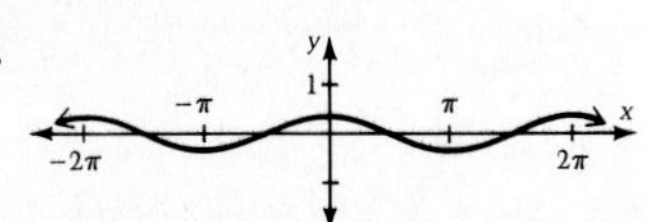


**3.** 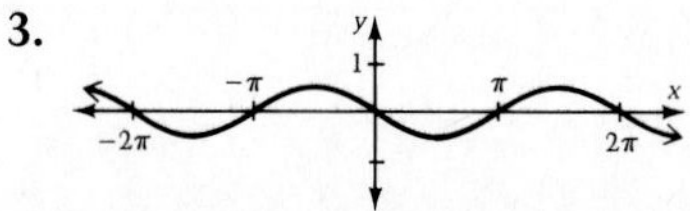


**5.** 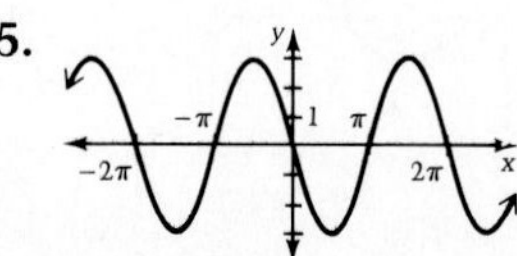

7. 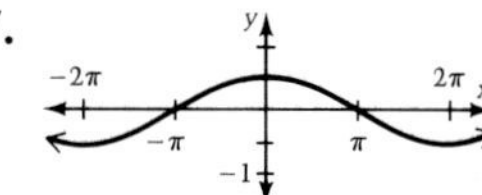

9. 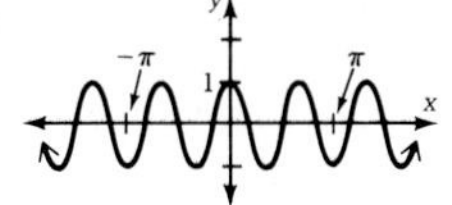

11. 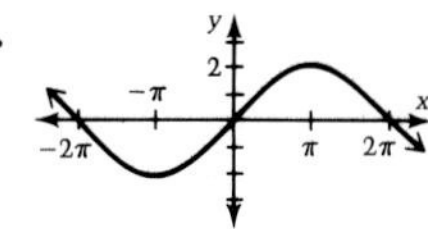

13. 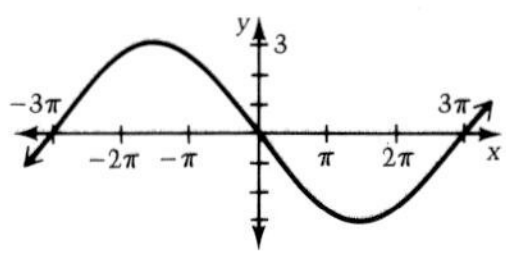

15. 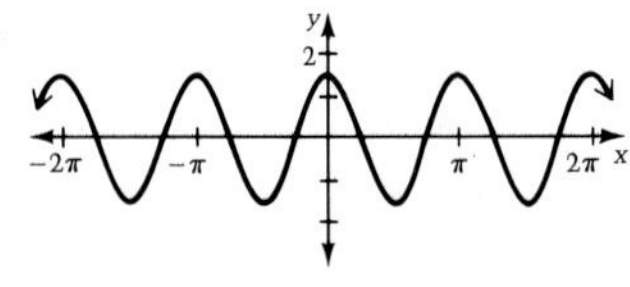

17. amplitude: 2
period: $2\pi$
phase shift: $-\frac{\pi}{2}$

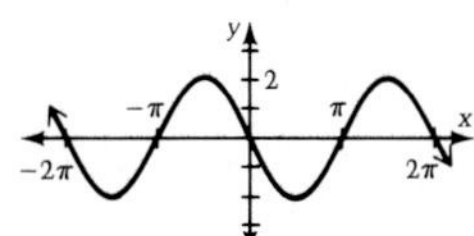

19. amplitude: 2
period: $4\pi$
phase shift: $\frac{\pi}{3}$

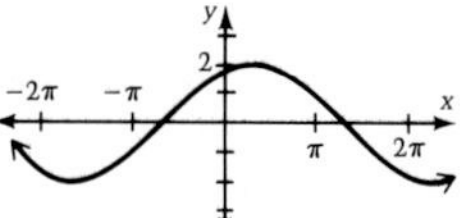

21. amplitude: 2
period: $4\pi$
phase shift $-\frac{\pi}{2}$

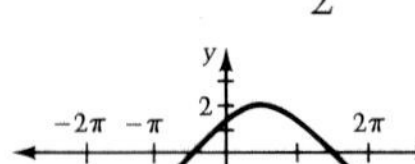

23. amplitude: $\sqrt{3}$
period: $4\pi$
phase shift: $-\pi$

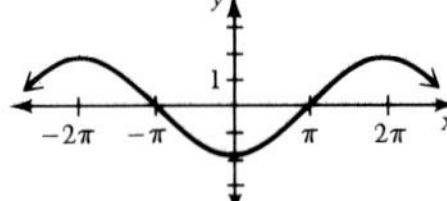

25. amplitude: 3
period: $\pi$
phase shift: $\frac{\pi}{2}$

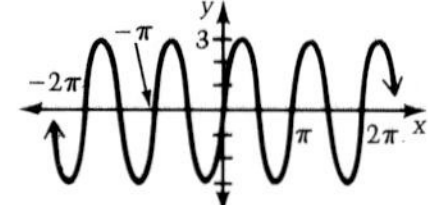

27. amplitude: 3
period: $4\pi$
phase shift: $-\pi$

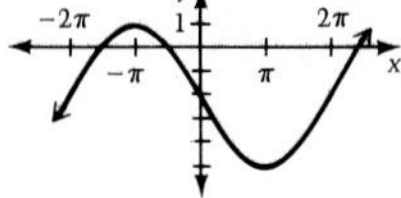

29. amplitude: 2
period: $2\pi$
phase shift: $\frac{\pi}{2}$

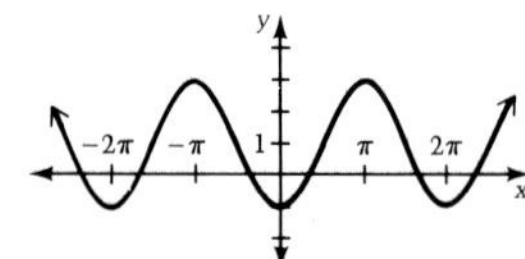

31. 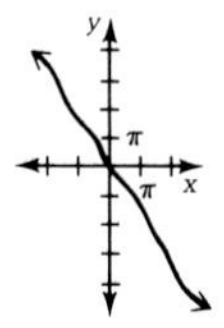

33. 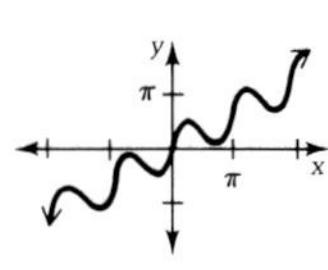

35. 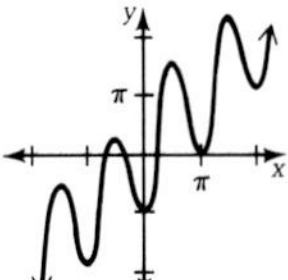

**37.** 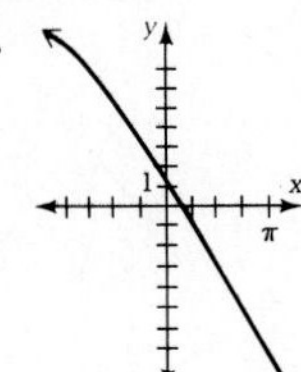


**39.** 8, −8

**41.** 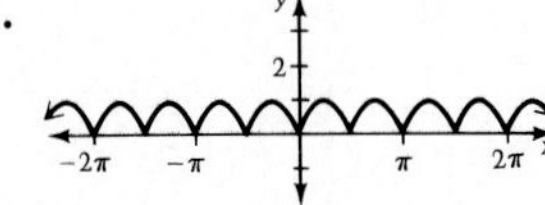


**43.** 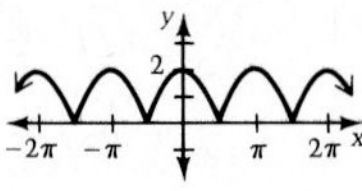


**45.** 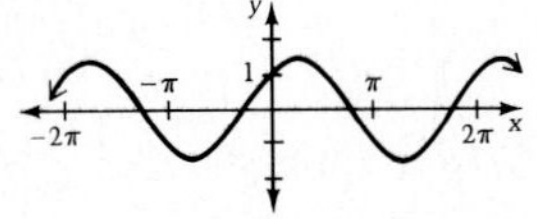


**47.** 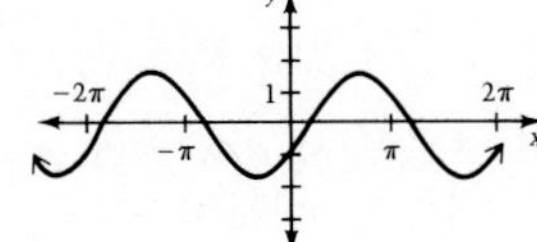


**49.** False **51.** True

**Exercise Set 3.5, page 110**

**1.** $\frac{\pi}{3}$ **3.** $\pi$ **5.** $-\frac{\pi}{6}$ **7.** $-\frac{\pi}{4}$ **9.** 0 **11.** $\frac{1}{2}$ **13.** $-\sqrt{3}$ **15.** 0 **17.** $-\frac{\sqrt{3}}{2}$ **19.** $-\frac{\sqrt{2}}{2}$

**21.** $-\frac{\pi}{4}$ **23.** $\frac{\pi}{6}$ **25.** $\frac{4}{5}$ **27.** $\frac{4}{5}$ **29.** $\frac{4}{5}$ **31.** $\frac{5}{13}$ **33.** $-\frac{2\sqrt{5}}{15}$ **35.** $\frac{z^2}{1 + z^2}$ **37.** $1 - z^2$

**39.** $-\frac{z\sqrt{1 - z^2}}{1 - z^2}$ **41.** $\frac{2\pi}{3}$ **43.** $-\frac{5}{13}$ **45.** $\frac{\sqrt{5}}{3}$ **47.** $\frac{\sqrt{5}}{2}$ **49.** $-\frac{\sqrt{165}}{11}$ **51.** 0.3229 **53.** 1.4312

**55.** 0.0814 **57.**

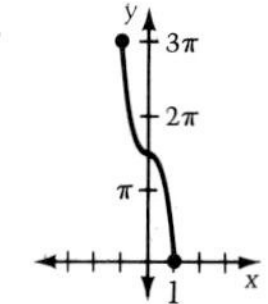


**59.**

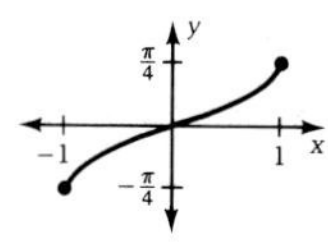


## Chapter Test, page 112

**1.** yes **3.**

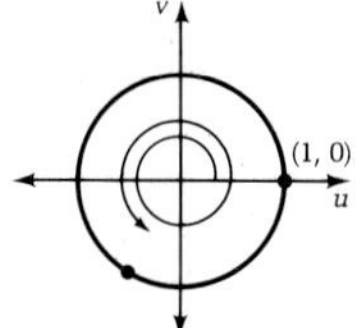


**5.**

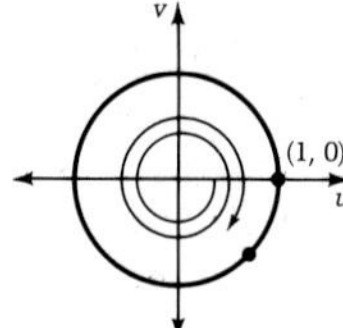


**7.** $\frac{4\pi}{3}$ **9.** 810° **11.** $\frac{3\pi}{2}$, 270° **13.** $\frac{9\pi}{4}$

**15.** $-\frac{1}{2}$ **17.** $\frac{\sqrt{3}}{3}$ **19.** 0.78 **21.**

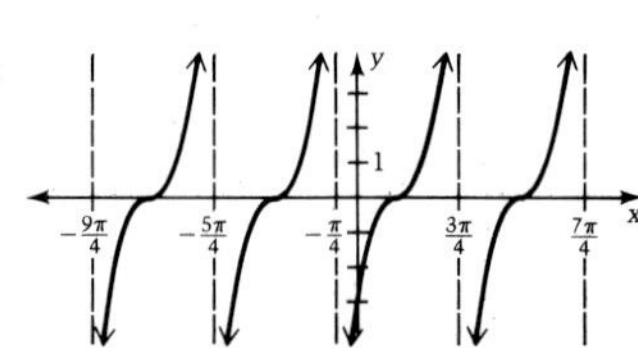


**23.** False **25.** False

**27.** amplitude: 3
period: $\pi$

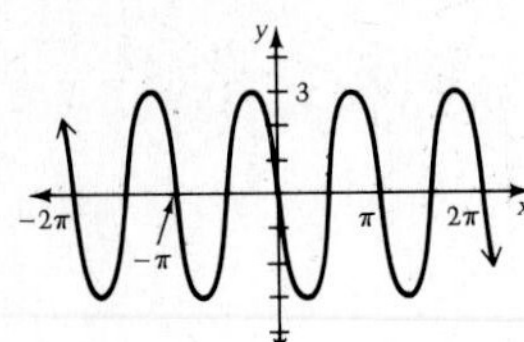

**29.** amplitude: 4
period: $\pi$
phase shift: $-\frac{\pi}{2}$

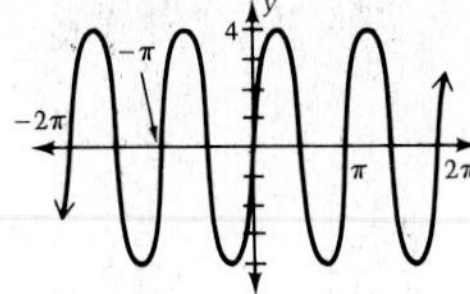

**31.**

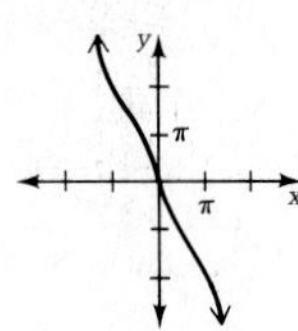

**33.** $\frac{\pi}{6}$ **35.** $\frac{4}{5}$

**37.** True **39.** False **41.** 6, $-6$ **43.** $\frac{1}{2}, -\frac{1}{2}$ **45.** $4\pi$

## Cumulative Test: Chapters 1–3, page 115

**1.** 10 **3.** 6, 11, $h^4 - 2h^2 + 3$, $h^2 - 10h + 27$ **5.** yes **7.** $y = 4x^2 - 20x + 21$
Domain: all real numbers

**9.** $f^{-1}(x) = \frac{x + 5}{2}$
Domain: $[-5, 1]$
Range: $[0, 3]$

**11.**

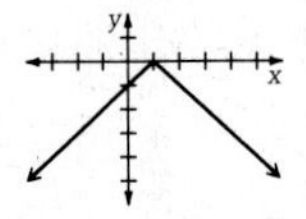

**13.**

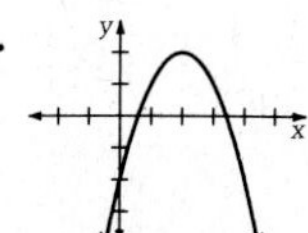

**15.** (a) $y - (-2) = -\frac{3}{2}(x - 0)$ (b) $y = -\frac{3}{2}x - 2$ (c) $\frac{3}{2}x + y + 2 = 0$

**17.**

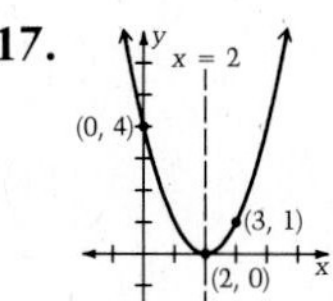


**19.** $c = 17$

**21.** $\frac{40\sqrt{3}}{3}$ **23.** $-\frac{5\sqrt{29}}{29}, -\frac{2\sqrt{29}}{29}, \frac{5}{2}$ **25.** $-\frac{\sqrt{2}}{2}$ **27.** $\sqrt{3}$ **29.** $-\frac{7\pi}{2}$ **31.** $480°$ **37.** $3h + 14$

**39.** (a) $\sin\theta = -\frac{40}{41}$ $\csc\theta = -\frac{41}{40}$ (b) $\cos\theta = -\frac{24}{25}$ $\csc\theta = -\frac{25}{7}$

$\tan\theta = -\frac{40}{9}$ $\sec\theta = \frac{41}{9}$ $\tan\theta = \frac{7}{24}$ $\sec\theta = -\frac{25}{24}$

$\cot\theta = -\frac{9}{40}$ $\cot\theta = \frac{24}{7}$

## Chapter 4

### Exercise Set 4.1, page 122

**9.** $\sin\alpha$ **11.** $\tan\beta$ **13.** $\sin\beta$ **15.** $\tan^2\varphi$ **17.** $\tan\alpha + 1$ **39.** $\frac{\sqrt{1+\cot^2\alpha}}{1+\cot^2\alpha}$

**41.** $-\frac{\sqrt{1-\sin^2\beta}}{1-\sin^2\beta}$ **43.** $\frac{\sec\gamma\sqrt{\sec^2\gamma - 1}}{\sec^2\gamma - 1}$

### Exercise Set 4.2, page 129

**1.** $\frac{\sqrt{3}}{2}$ **3.** $\frac{\sqrt{6}-\sqrt{2}}{4}$ **5.** $\frac{\sqrt{6}-\sqrt{2}}{4}$ **7.** $\frac{\sqrt{2}-\sqrt{6}}{4}$ **9.** $\frac{\sqrt{2}}{2}$ **11.** $-\frac{\sqrt{2}}{2}$ **13.** $\frac{\sqrt{6}-\sqrt{2}}{4}$

**15.** $\frac{\sqrt{6}+\sqrt{2}}{4}$ **17.** $\frac{\sqrt{6}-\sqrt{2}}{4}$ **19.** $\frac{-\sqrt{2}-\sqrt{6}}{4}$ **21.** $\frac{\sqrt{3}+3}{3-\sqrt{3}}$ **23.** $-2-\sqrt{3}$ **25.** $\frac{\sqrt{3}-3}{3+\sqrt{3}}$

**27.** (a) $-\frac{56}{65}$ (b) $\frac{33}{65}$ (c) $-\frac{56}{33}$ **29.** (a) $-\frac{9\sqrt{13}}{169}$ (b) $-\frac{46\sqrt{13}}{169}$ (c) $\frac{9}{46}$ **31.** (a) $\frac{16}{65}$ (b) $-\frac{63}{65}$ (c) $-\frac{16}{63}$

**33.** (a) $-\frac{3\sqrt{13}}{13}$ (b) $\frac{2\sqrt{13}}{13}$ (c) $-\frac{3}{2}$ **45.** $\frac{\sqrt{3}}{2}\cos\alpha - \frac{1}{2}\sin\alpha$ **47.** $\frac{\tan\alpha + 1}{1 - \tan\alpha}$ **49.** $\cos\varphi$

**51.** $\frac{\sqrt{2}}{2}(\cos\alpha - \sin\alpha)$ **53.** $\tan(\alpha - \beta) = \frac{\tan\alpha - \tan\beta}{1 + \tan\alpha\tan\beta}$ **55.** $\sec(\alpha + \beta) = \frac{\sec\alpha\sec\beta}{1 - \tan\alpha\tan\beta}$

**57.** $\sin 2\alpha = 2\sin\alpha\cos\alpha$

**Exercise Set 4.3, page 135**

**1.** $\frac{\sqrt{3}}{2}, -\frac{1}{2}, -\sqrt{3}$ **3.** $-\frac{\sqrt{3}}{2}, -\frac{1}{2}, \sqrt{3}$ **5.** $-1$, 0, not defined **7.** $\frac{\sqrt{3}}{2}, -\frac{1}{2}, -\sqrt{3}$ **9.** 1, 0, not defined

**11.** $-\frac{\sqrt{3}}{2}, \frac{1}{2}, -\sqrt{3}$ **13.** $\frac{\sqrt{3}}{2}, -\frac{1}{2}, -\sqrt{3}$ **15.** 1, 0, not defined **17.** 1, 0, defined

**19.** (a) $-\frac{24}{25}$ (b) $\frac{7}{25}$ (c) $-\frac{24}{7}$ **21.** (a) $-\frac{120}{169}$ (b) $-\frac{119}{169}$ (c) $\frac{120}{119}$ **23.** (a) $\frac{720}{1681}$ (b) $\frac{1519}{1681}$ (c) $\frac{720}{1519}$

**25.** (a) $\frac{4}{5}$ (b) $\frac{3}{5}$ (c) $\frac{4}{3}$ **27.** $\frac{1}{2}, \frac{\sqrt{3}}{2}, \frac{\sqrt{3}}{3}$ **29.** $\frac{\sqrt{2}}{2}, \frac{\sqrt{2}}{2}, 1$

**31.** $\frac{\sqrt{2+\sqrt{3}}}{2}, -\frac{\sqrt{2-\sqrt{3}}}{2}, -\frac{1}{2-\sqrt{3}}$ **33.** $-\frac{\sqrt{2+\sqrt{2}}}{2}, -\frac{\sqrt{2-\sqrt{2}}}{2}, \frac{\sqrt{2}}{2-\sqrt{2}}$

**35.** $-\frac{\sqrt{2-\sqrt{2}}}{2}, \frac{\sqrt{2+\sqrt{2}}}{2}, -\frac{\sqrt{2}}{2+\sqrt{2}}$ **37.** 0, 1, 0 **39.** $\frac{\sqrt{3}}{2}, -\frac{1}{2}, -\sqrt{3}$

**41.** $\frac{\sqrt{2-\sqrt{3}}}{2}, \frac{\sqrt{2+\sqrt{3}}}{2}, \frac{1}{2+\sqrt{3}}$ **43.** $\frac{\sqrt{2-\sqrt{3}}}{2}, -\frac{\sqrt{2+\sqrt{3}}}{2}, -\frac{1}{2+\sqrt{3}}$

**45.** $-\frac{\sqrt{2+\sqrt{3}}}{2}, \frac{\sqrt{2-\sqrt{3}}}{2}, -\frac{1}{2-\sqrt{3}}$ **47.** $\frac{3\sqrt{10}}{10}, \frac{\sqrt{10}}{10}, 3$ **49.** $\frac{2\sqrt{13}}{13}, -\frac{3\sqrt{13}}{13}, -\frac{2}{3}$

**51.** $\frac{\sqrt{82}}{82}, \frac{9\sqrt{82}}{82}, \frac{1}{9}$ **53.** $\sqrt{\frac{5 + 2\sqrt{5}}{10}}, -\sqrt{\frac{5 - 2\sqrt{5}}{10}}, \frac{1}{\sqrt{5} - 2}$ **59.** $\cos 3\alpha = 4\cos^3\alpha - 3\cos\alpha$

**61.** $\sin 4\alpha = 4\sin\alpha\cos\alpha - 8\sin^3\alpha\cos\alpha$ **63.** $2\cos^2 6\alpha - 1$ **65.** $2\sin 4\alpha\cos 4\alpha$

**67.** $\frac{3\tan 2\alpha - \tan^3 2\alpha}{1 - 3\tan^2 2\alpha}$ **69.** $\cot 12\alpha + \csc 12\alpha$ **71.** $\pm 2\sin 4\alpha\sqrt{1 - \sin^2 4\alpha}$ **73.** $\cos\alpha$ **75.** $\sin\alpha$

**77.** $\cos\alpha$ **79.** 1 **81.** $1 - \sin\alpha$ **83.** $\cos 2\alpha$

**Exercise Set 4.4, page 141**

**1.** $\cos 3\alpha$ **3.** $\sin 3\alpha$ **5.** $\tan 4\varphi$ **7.** $1 - \cos\alpha$ **21.** yes **23.** no **25.** yes **27.** $\frac{\sqrt{6}}{2}$ **29.** $\frac{\sqrt{2}}{2}$

**31.** $-\frac{\sqrt{2}}{2}$ **33.** $-\frac{\sqrt{6}}{2}$ **39.** $2\sin 2x\cos x$ **41.** $2\sin 4x\cos 2x$ **43.** $-2\sin 4x\sin 2x$

**45.** $\frac{1}{2}\sin 5x + \frac{1}{2}\sin(-x)$ **47.** $\frac{1}{2}\cos x - \frac{1}{2}\cos 5x$ **49.** $\frac{2\sin x\cos x + 1}{\cos^2 x - \sin^2 x}$

**Exercise Set 4.5, page 149**

**1.** $\frac{4\pi}{3} + 2n\pi, \frac{5\pi}{3} + 2n\pi$ **3.** $\frac{\pi}{6} + 2n\pi, \frac{5\pi}{6} + 2n\pi, 0 + n\pi$ **5.** $\frac{\pi}{6} + n\pi, \frac{5\pi}{6} + n\pi$

**7.** $\frac{\pi}{3} + 2n\pi, \frac{5\pi}{3} + 2n\pi$ **9.** $0 + 2n\pi, \frac{3\pi}{2} + 2n\pi$ **11.** $\frac{\pi}{6}, \frac{5\pi}{6}, \frac{3\pi}{2}$ **13.** $\frac{\pi}{6}, \frac{5\pi}{6}$ **15.** $0, \pi$

**17.** $\frac{\pi}{2}, \frac{7\pi}{6}, \frac{11\pi}{6}$ **19.** $\frac{\pi}{8}, \frac{5\pi}{8}, \frac{9\pi}{8}, \frac{13\pi}{8}$ **21.** $-60°, 60°$ **23.** $0, \frac{\pi}{4}, \frac{3\pi}{4}, \pi$ **25.** $0°, 60°, 180°$

**27.** 30°, 90°, 150°, 210°, 270°, 330°, 45°, 135°, 225°, 315° **29.** $\frac{\pi}{2}, \frac{7\pi}{6}, \frac{3\pi}{2}, \frac{11\pi}{6}$ **31.** $0, \frac{\pi}{6}, \frac{5\pi}{6}, \pi$

**33.** 0°, 180°, 48°40′, 131°20′ **35.** 270°, 19°30′, 160°30′ **37.** 210°, 330°, 41°50′, 138°10′

**39.** 10°30′, 169°30′, 223°, 317° **41.** no solution

**43.** 27°20′, 62°40′, 117°20′, 152°40′, 207°20′, 242°40′, 297°20′, 332°40′ **45.** $-\frac{\pi}{6}$ **47.** $\frac{\pi}{3}$ **49.** $\frac{24}{25}$

**51.** $-\frac{7}{25}$ **53.** $-\frac{33}{65}$ **55.** 61°16′ **57.** $\frac{1}{2}$ **59.** $\theta = \frac{\pi}{3}$ or $\theta = \frac{5\pi}{3}$, $r = \frac{1}{2}$ **61.** 11°19′, 191°19′

**Chapter Test, page 152**

**5.** $\frac{3 - \sqrt{3}}{3 + \sqrt{3}}$ **9.** $-1$, 0, not defined **11.** $-\sqrt{\frac{3 - \sqrt{5}}{6}}$ **13.** $-\frac{\sqrt{10}}{10}$ **15.** $\csc x$ **17.** $\sin x$

**19.** $\cos 2\varphi$ **25.** $-\frac{\sqrt{2}}{2}$ **27.** $\frac{1}{2}\sin 7x + \frac{1}{2}\sin x$ **29.** $\frac{2\pi}{3} + 2n\pi, \frac{4\pi}{3} + 2n\pi, 0 + 2n\pi$ **31.** $\frac{\pi}{2} + 2n\pi$

**33.** 90°, 199°30′, 340°30′ **35.** 0°, 98°10′, 261°10′ **37.** $\frac{7\pi}{12}$ **39.** 0 **41.** $\frac{\sqrt{1 + \cot^2 \alpha}}{1 + \cot^2 \alpha}$

**43.** (a) $-\sin x$ (b) $-\sin \beta$ **45.** $\frac{3 \tan 2x - \tan^3 2x}{1 - 3 \tan^2 2x}$

## Chapter 5

**Exercise Set 5.1, page 161**

**1.** $\angle A = 45°, a = 32\sqrt{2}, b = 32\sqrt{2}$ **3.** $a = 44\sqrt{3}, \angle B = 30°, c = 88$ **5.** $\angle A = 30°, \angle B = 60°, c = 30$

**7.** $\angle A = 60°, b = 8\sqrt{3}, c = 16\sqrt{3}$ **9.** $\angle A = 45°, \angle B = 45°, b = 25\sqrt{2}$

**11.** $\angle B = 54°, c = 120, b = 96$ **13.** $\angle A = 78°, b = 1.9, c = 9.2$

**15.** $a = 7.6, \angle B = 52°, b = 9.7$ **17.** $a = 0.132, \angle B = 61°40', c = 0.277$

**19.** $\angle A = 18°42', a = 45.69, b = 135.0$ **21.** 21′ **23.** 5800 ft **25.** 5.2 ft **27.** 57°

**29.** (a) 60° (b) 40° (c) 76° (d) 22° **31.** 2700′ **33.** 17°

**Exercise Set 5.2, page 167**

**1.** $b = 20$, $\angle C = 105°$, $c = 27$ **3.** $\angle A = 15°$, $a = 2.9$, $c = 8.0$ **5.** $b = 9.9$, $\angle C = 56°$, $c = 20$

**7.** $\angle A = 80°$, $a = 46$, $b = 28$ **9.** $\angle A = 41°30'$, $b = 1.043$, $c = 1.011$

**11.** yes, $\angle B = 90°$, $\angle C = 60°$, $c = 12\sqrt{3}$ **13.** no **15.** yes, $\angle B = 51°$, $\angle C = 9°$, $c = 9.8$

**17.** yes, $\angle B = 63°50'$, $\angle C = 63°40'$, $c = 36.4$ or $\angle B = 116°10'$, $\angle C = 11°20'$, $c = 7.97$ **19.** 431 yards

**21.** N 0°30′ E **23.** from first observer: 4.4 mi, from second observer: 6.2 mi **25.** $x = \dfrac{m\tan\theta\tan\varphi}{\tan\varphi - \tan\theta}$

**Exercise Set 5.3, page 172**

**1.** $\angle A = 69°$, $\angle B = 51°$, $c = 11$ **3.** $a = 31$, $\angle B = 34°$, $\angle C = 26°$ **5.** $\angle A = 72°$, $b = 95$, $\angle C = 54°$

**7.** $a = 27.8$, $\angle B = 13°54'$, $\angle C = 18°30'$ **9.** $\angle A = 37°$, $\angle B = 53°$, $\angle C = 90°$

**11.** $\angle A = 66°$, $\angle B = 62°$, $\angle C = 52°$ **13.** $\angle A = 29°$, $\angle B = 47°$, $\angle C = 104°$

**15.** $\angle A = 62°40'$, $\angle B = 36°20'$, $\angle C = 81°$ **17.** 170 m **19.** 3.9, 7.2 **21.** 47° **23.** 99.8

**25.** 52.1 **27.** 672.0 **29.** 7737 **31.** $\angle A = 65°30'$, $\angle B = 69°50'$, $\angle C = 44°40'$, $a = 15.5$ or $\angle A = 114°30'$, $\angle B = 38°10'$, $\angle C = 27°20'$, $a = 23.6$ **33.** 47.0

**Exercise Set 5.4, page 178**

**1.** $2\sqrt{34}$, 30.96° **3.** $2\sqrt{13}$, 326.31° **5.** 13, 247.38° **7.** 11, 0°

**9.**

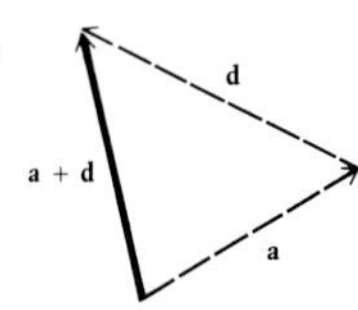


**11.**

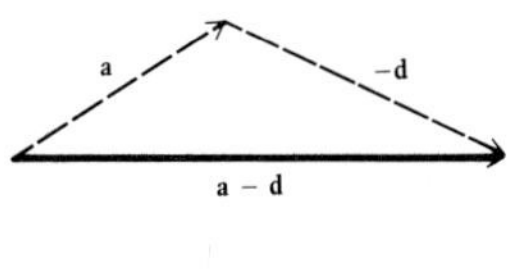


**13.**

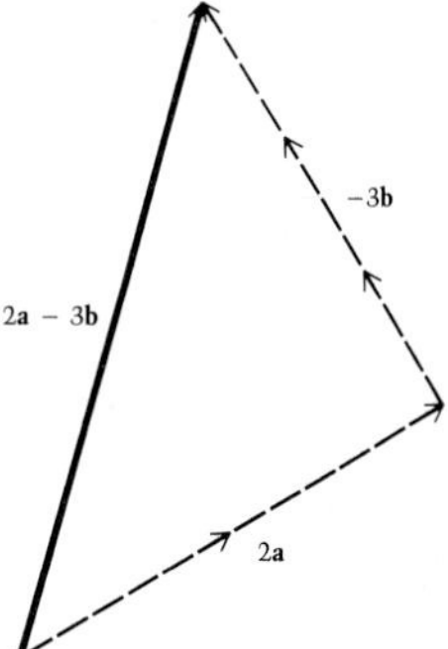


**15.**

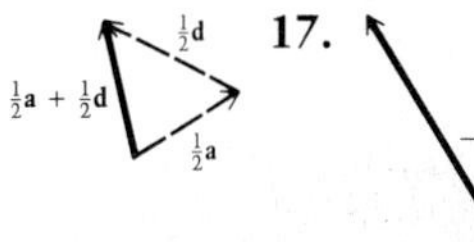


**17.**

**19.** 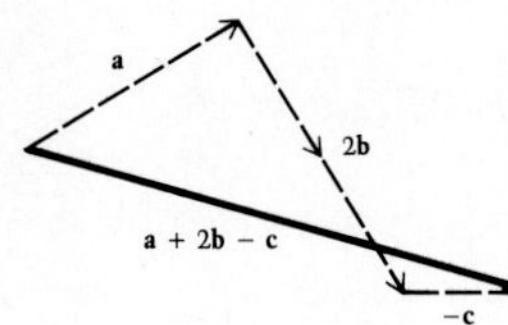


**21.** $\langle 5, 12\rangle$ **23.** $\langle 3, -4\rangle$ **25.** $\langle 1, -8\rangle$ **27.** $(2, 2)$ **29.** $\left(-2\frac{1}{2}, -1\frac{1}{2}\right)$

**31.** $\left(\frac{5}{2}, \frac{5}{2}\right)$ **33.** $\langle -20, 4\rangle$ **35.** $\langle 5, 19\rangle$ **37.** $\langle 13, 17\rangle$ **39.** $\langle -21, 2\rangle$

**41.** $10, \left\langle -\frac{3}{5}, \frac{4}{5}\right\rangle, \left\langle \frac{6}{5}, -\frac{8}{5}\right\rangle$ **43.** $4, \langle 0, -1\rangle, \langle 0, 2\rangle$ **45.** $2\sqrt{3}, \left\langle \frac{\sqrt{6}}{3}, \frac{\sqrt{3}}{3}\right\rangle, \left\langle -\frac{2\sqrt{6}}{3}, -\frac{2\sqrt{3}}{3}\right\rangle$

**47.** $\frac{10}{3}, \left\langle \frac{4}{5}, -\frac{3}{5}\right\rangle, \left\langle -\frac{8}{5}, \frac{6}{5}\right\rangle$ **49.** $-i + 8j, \sqrt{2}, \frac{1}{\sqrt{2}}i + \frac{1}{\sqrt{2}}j$ **51.** $3i - 16j, \sqrt{13}, \frac{3}{\sqrt{13}}i - \frac{2}{\sqrt{13}}j$

**53.** $-\frac{19}{2}i + \frac{13}{2}j, \frac{\sqrt{26}}{2}, \frac{1}{\sqrt{26}}i + \frac{5}{\sqrt{26}}j$ **55.** $\langle -6, 6\sqrt{3}\rangle$ **57.** $\langle -5\sqrt{2}, 5\sqrt{2}\rangle$ **59.** $\langle 2.7, -7.3\rangle$

**61.** $r = 1, s = -2$

**Exercise Set 5.5, page 185**

**1.** 569 mph, N 82°6′ E **3.** 427 mph, N 16°19′ W **5.** 391 N = 88 lbs. **7.** 937 N = 211 lbs.

**9.** −2, 98° **11.** 6, 18° **13.** 0, 90° **15.** −11, 112° **17.** perpendicular **19.** neither **21.** parallel

**23.** 4827 foot lbs. **25.** 54,067 foot lb.

**Exercise Set 5.6, page 190**

**1.** (a) 0, stationary
(b) $\frac{\sqrt{2}}{4}$, stationary
(c) $\frac{1}{2}$, stationary
(d) 0, down
(e) $-\frac{1}{2}$, stationary
(f) 0, down
(g) 0, up

**3.** (a) $-3$, stationary
(b) 3, stationary
(c) $-3$, stationary
(d) $-3$, stationary
(e) $-3$, stationary
(f) $-3$, stationary
(g) $-3$, stationary

**5.** (a) 0, stationary
(b) 0, down
(c) 0, down
(d) 0, down
(e) 0, down
(f) 0, down
(g) 0, down

**7.** $x = 6 \sin 2t$
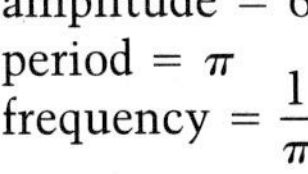
amplitude $= 6$
period $= \pi$
frequency $= \frac{1}{\pi}$

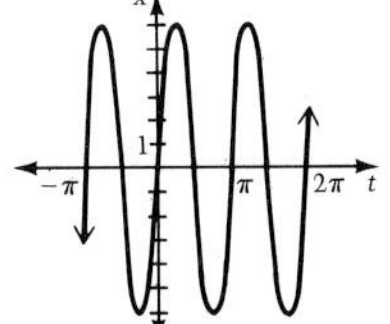


**9.** amplitude $= 3$
period $= 4\pi$
frequency $= \frac{1}{4\pi}$

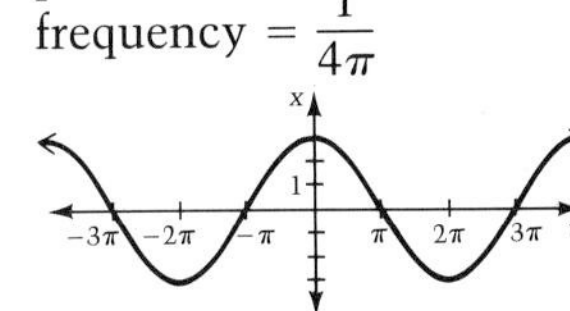


**11.** $x = \sin\left(t + \frac{\pi}{6}\right)$
amplitude $= 1$, period $= 2\pi$, frequency $= \frac{1}{2\pi}$

**13.** $x = \sqrt{2}\sin\left(t - \frac{\pi}{4}\right)$
amplitude $= \sqrt{2}$, period $= 2\pi$, frequency $= \frac{1}{2\pi}$

**15.** $x = \frac{13}{2}\sin\left(2t - 1.18\right)$
amplitude $= \frac{13}{2}$, period $= \pi$, frequency $= \frac{1}{\pi}$

**17.** $x = 6\cos\frac{2\pi}{3}t$, frequency $= \frac{1}{3}$

**19.** $x = 6\cos 1.25\pi t$, frequency $= 0.625$

**21.** $x = 0.60\cos 2t$, frequency $= \frac{1}{\pi}$

**23.** $x = 0.02\cos\sqrt{9.8}t$, frequency $= \frac{\sqrt{9.8}}{2\pi}$

**25.** $x = 3 \sin \frac{3}{2}t$ **27.** $x = \frac{3}{2} \sin \frac{2}{3}t$ **29.** multiplied by $\sqrt{2}$ **31.** $x = 0.01 \cos 400\pi t$ **33.** no

**Exercise Set 5.7, page 198**

**1.** $A(2, 120°)$ **3.** $C(3, 135°)$ **5.** $E(2, -120°)$
**7.** $G(-1, -270°)$ **9.** $I(-2, -420°)$ **11.** $K(3, \pi)$
**13.** $M\left(-2, \frac{5\pi}{6}\right)$ **15.** $Q\left(-4, -\frac{\pi}{2}\right)$

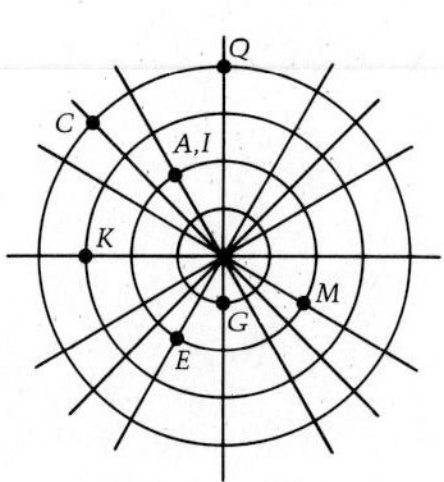


**17.** $(-1, \sqrt{3})$ **19.** $\left(-\frac{3\sqrt{2}}{2}, \frac{3\sqrt{2}}{2}\right)$ **21.** $(-1, -\sqrt{3})$ **23.** $(0, -1)$ **25.** $(-1, \sqrt{3})$

**27.** $(-3, 0)$ **29.** $(\sqrt{3}, -1)$ **31.** $(0, 4)$ **33.** $\left(4, \frac{\pi}{2}\right), \left(-4, \frac{3\pi}{2}\right)$ **35.** $\left(\sqrt{2}, \frac{5\pi}{4}\right), \left(-\sqrt{2}, \frac{\pi}{4}\right)$

**37.** $\left(2, \frac{11\pi}{6}\right), \left(-2, \frac{5\pi}{6}\right)$ **39.** $\left(2, \frac{3\pi}{4}\right), \left(-2, \frac{7\pi}{4}\right)$ **41.** $(5, 2.21), (-5, 5.36)$

**43.** $(2\sqrt{13}, 0.98), (-2\sqrt{13}, 4.12)$ **45.** $x^2 + y^2 = 9$ **47.** $y = -x$ **49.** $x - 2y = 5$

**51.** $y^2 + 2x = 1$ **53.** $x^2 - 3y^2 - 8y - 4 = 0$ **55.** $\theta = \frac{\pi}{2}$ **57.** $r(\cos\theta + \sin\theta) = 1$

**59.** $r - 2\cos\theta = 0$ **61.** $r^2\cos^2\theta = 6r\sin\theta + 9$ **63.** $r^2 \sin 2\theta = 1$ **65.** $r^2 = 16r\sin\theta$

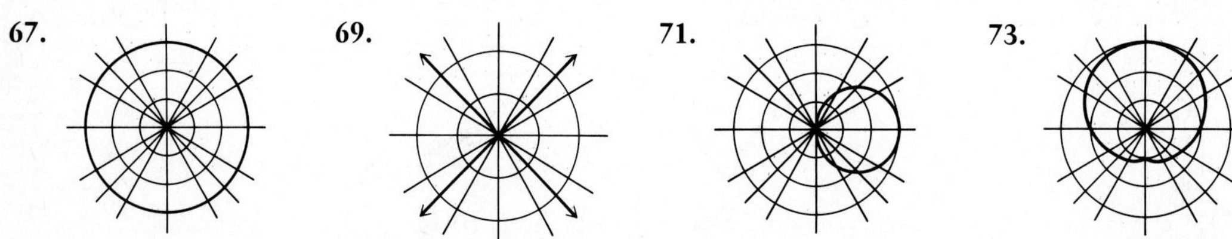

**75.** $2r \cos \theta - 3r \sin \theta = -4$ **77.** (a) 4.91 (b) 4.91 **79.**

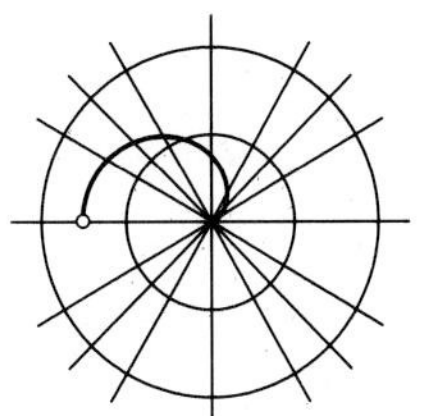

**81.**

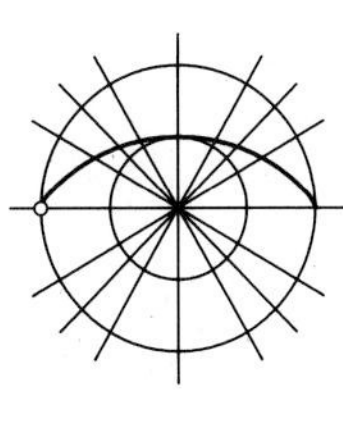

## Chapter Test, page 200

**1.** $\angle A = 30°$, $a = 4\sqrt{3}$, $c = 8\sqrt{3}$ **3.** $a = 64.6$, $\angle B = 56°18'$, $c = 116$

**5.** $\angle A = 37°40'$, $\angle B = 52°20'$, $c = 59.3$ **7.** 70° **9.** $a = 35$, $b = 53$, $\angle C = 56°$

**11.** yes, $\angle B = 117°06'$, $\angle C = 38°36'$, $b = 13.4$ or $\angle B = 14°18'$, $\angle C = 141°24'$, $b = 3.7$

**13.** 472 yd **15.** $\angle A = 23°40'$, $\angle B = 37°20'$, $c = 3.32$ **17.** $\angle A = 39°34'$, $\angle B = 54°41'$, $\angle C = 85°45'$

**19.** 49.1 **21.** (a) magnitude $= 3\sqrt{2}$, direction 135° (b) $\langle -5, 1 \rangle$ **23.** $\langle 5, 9 \rangle$ **25.** $\frac{5\sqrt{2}}{2}$, $\left\langle -4, 7\frac{1}{2} \right\rangle$

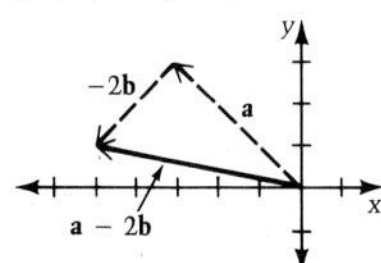


**27.** $\mathbf{v_x} = -7.3i$, $\mathbf{v_y} = 4.6j$ **29.** 494 mph, N 74°27′ E **31.** −4, 117°

**33.** amplitude = 4
period = $4\pi$
frequency = $\frac{1}{4\pi}$

**35.** $(-2\sqrt{2}, -2\sqrt{2})$

**37.** $x^2 + \left(y - \frac{3}{2}\right)^2 = \frac{9}{4}$

**39.**

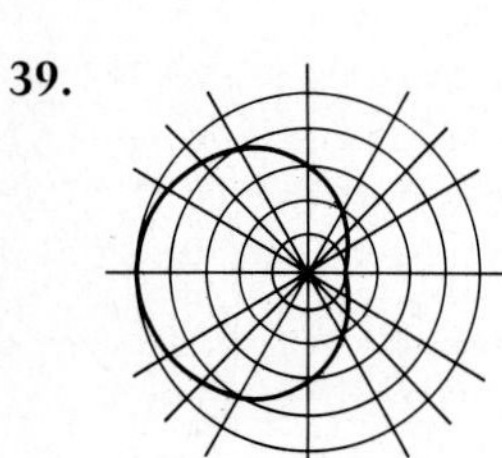

**41.** $2\sqrt{5}$ **43.** $-2\mathbf{a} - 3\mathbf{b}$ **45.** 54°44′

## Cumulative Test: Chapters 4 and 5, page 203

**1.** $\frac{\sqrt{6} - \sqrt{2}}{4}$ **3.** $\sqrt{3} - 2$ **5.** $\sin 2\alpha = 1$, $\cos 2\alpha = 0$, $\tan 2\alpha$ not defined **7.** $\sin 2\alpha = 1$, $\cos 2\alpha = 0$, $\tan 2\alpha$ not defined **9.** $\sin \frac{\alpha}{2} = \sqrt{\frac{15 + \sqrt{161}}{30}}$, $\cos \frac{\alpha}{2} = \sqrt{\frac{15 - \sqrt{161}}{30}}$

**11.** 1.58 **17.** $\frac{\pi}{2}$ **19.** $\frac{\pi}{18}, \frac{5\pi}{18}, \frac{13\pi}{18}, \frac{17\pi}{18}$ **21.** $\frac{12}{13}$ **23.** $\frac{\sqrt{3}}{2}$ **25.** 24 **27.** 26.9

**29.** 3705 $\text{yd}^2$ **31.** $\|\mathbf{c}\| = 5\sqrt{2}$, $\mathbf{u}_\mathbf{c} = \left\langle \frac{7\sqrt{2}}{10}, \frac{\sqrt{2}}{10} \right\rangle$, $-\frac{1}{2}\mathbf{c} = \left\langle -\frac{7}{2}, -\frac{1}{2} \right\rangle$ **33.** N 36°52′ E at 25 yd/sec

**35.** 0, 90° **37.** amplitude = 4, period = $4\pi$, frequency = $\frac{1}{4\pi}$, $x = 2\sqrt{3}$ **39.** $\left(-2, \frac{\pi}{6}\right)$

**41.** $\sin\theta = 0.8$, $\cos\theta = 0.6$ **43.** 51.4 inches

## Chapter 6

### Exercise Set 6.1, page 214

**1.** $i\sqrt{13}$ **3.** $i\sqrt{15}$ **5.** $2i$ **7.** $10i$ **9.** $2i\sqrt{6}$ **11.** $7i\sqrt{2}$ **13.** $-3i$ **15.** $-2i\sqrt{5}$ **17.** $-3i\sqrt{7}$

**19.** 2, 5 **21.** $-3$, 4 **23.** $-1$, $-2$ **25.** $\frac{1}{2}$, 6 **27.** 0, 6 **29.** $-\sqrt{3}$, 0 **31.** $\frac{1}{4}$, $-\sqrt{10}$ **33.** 1

**35.** $-i$ **37.** $-1$ **39.** $-i$ **41.** $-8i$ **43.** $-81$ **45.** $-2\sqrt{14}$ **47.** $4i$ **49.** $-6\sqrt{2}$ **51.** $8 + 9i$

**53.** $-2 + 8i$ **55.** $10 - 15i$ **57.** $-2 - 10i$ **59.** $-3 - 3i$ **61.** $5 - i$ **63.** $2 + 14i$

**65.** $-22 + 14i$ **67.** $2 + 15i$ **69.** $-12 + 4i$ **71.** $-24 + 32i$ **73.** $3 + 6i$ **75.** $13 + 11i$

**77.** $-24 - 31i$ **79.** $-6 - 10i$ **81.** 41 **83.** 2 **85.** $-5 + 12i$ **87.** $24 - 10i$ **89.** $-4 - 3i$

**91.** $-54 + 54i$ **93.** $28 - 4i$ **95.** $-2$, 3 **97.** $\frac{3}{4}$, $-2$ **99.** $-2$, 5 **101.** 20, 12 **103.** 0

**105.** $i$ **107.** True **109.** False **111.** True **113.** True

**115.** $1^4 = 1$, $i^4 = 1$, $(-1)^4 = 1$, $(-i)^4 = (-1)^4 i^4 = 1$

**117.** $1^6 = 1$, $\left(\frac{1}{2} + \frac{\sqrt{3}}{2}i\right)^6 = \left[\left(\frac{1}{2} + \frac{\sqrt{3}}{2}i\right)^2\left(\frac{1}{2} + \frac{\sqrt{3}}{2}i\right)\right]^2 = \left[\left(-\frac{1}{2} + \frac{\sqrt{3}}{2}i\right)\left(\frac{1}{2} + \frac{\sqrt{3}}{2}i\right)\right]^2 = (-1)^2 = 1$,
$\left(-\frac{1}{2} + \frac{\sqrt{3}}{2}i\right)^6 = \left[\left(-\frac{1}{2} + \frac{\sqrt{3}}{2}i\right)^2\left(-\frac{1}{2} + \frac{\sqrt{3}}{2}i\right)\right]^2 = \left[\left(-\frac{1}{2} - \frac{\sqrt{3}}{2}i\right)\left(-\frac{1}{2} + \frac{\sqrt{3}}{2}i\right)\right]^2 = (1)^2 = 1$, $(-1)^6 = 1$,
$\left(-\frac{1}{2} - \frac{\sqrt{3}}{2}i\right)^6 = \left[-\left(\frac{1}{2} + \frac{\sqrt{3}}{2}i\right)^6\right] = (-1)^6\left(\frac{1}{2} + \frac{\sqrt{3}}{2}i\right)^6 = (1)(1) = 1$,
$\left(\frac{1}{2} - \frac{\sqrt{3}}{2}i\right)^6 = \left[-\left(-\frac{1}{2} + \frac{\sqrt{3}}{2}i\right)\right]^6 = (-1)^6\left(-\frac{1}{2} + \frac{\sqrt{3}}{2}i\right)^6 = (1)(1) = 1$

**119.** True **121.** False, $(i)(i) = -1$ **123.** True

**Exercise Set 6.2, page 222**

**1.** $\frac{2}{13} - \frac{3}{13}i$ **3.** $\frac{5}{29} + \frac{2}{29}i$ **5.** $-\frac{1}{2} - \frac{1}{2}i$ **7.** $-\frac{1}{10} + \frac{1}{5}i$ **9.** $0 - \frac{1}{5}i$ **11.** $0 + 7i$

**13.** $\frac{\sqrt{3}}{7} + \frac{2}{7}i$ **15.** $2 + 2i$ **17.** $\frac{14}{17} - \frac{5}{17}i$ **19.** $\frac{5}{29} - \frac{27}{29}i$ **21.** $\frac{13}{2} - \frac{7}{2}i$ **23.** $-2i$

**25.** $5 + 3i$ **27.** $-\frac{8}{41} - \frac{10}{41}i$ **29.** $1 + \frac{3}{7}i$ **31.** $-\frac{3}{20} - \frac{1}{20}i$ **33.** $-6i$

**35.** (a) $\overline{z + w} = 3 + i = \bar{z} + \bar{w}$ (b) $\overline{z - w} = -1 - 7i = \bar{z} - \bar{w}$ (c) $\overline{z \cdot w} = 14 - 2i = \bar{z} \cdot \bar{w}$
(d) $\overline{\left(\frac{z}{w}\right)} = -\frac{1}{2} - \frac{1}{2}i = \frac{\bar{z}}{\bar{w}}$

**37.** (a) $\overline{z + w} = 17 - 6i = \bar{z} + \bar{w}$ (b) $\overline{z - w} = 7 + 12i = \bar{z} - \bar{w}$ (c) $\overline{z \cdot w} = 87 - 93i = \bar{z} \cdot \bar{w}$
(d) $\overline{\left(\frac{z}{w}\right)} = \frac{33}{106} + \frac{123}{106}i = \frac{\bar{z}}{\bar{w}}$

**39.** (a) $\overline{z + w} = -7 + 9i = \bar{z} + \bar{w}$ (b) $\overline{z - w} = -5 + i = \bar{z} - \bar{w}$ (c) $\overline{z \cdot w} = -14 - 29i = \bar{z} \cdot \bar{w}$
(d) $\overline{\left(\frac{z}{w}\right)} = \frac{26}{17} + \frac{19}{17}i = \frac{\bar{z}}{\bar{w}}$

**41.** (a) $\overline{z^2} = 3 + 4i = (\bar{z})^2$ (b) $\overline{z^3} = 2 + 11i = (\bar{z})^3$ **43.** (a) $\overline{z^2} = -32 + 24i = (\bar{z})^2$ (b) $\overline{z^3} = 208 + 144i = (\bar{z})^3$

**45.** (a) $\overline{z^2} = 24 - 10i = (\bar{z})^2$ (b) $\overline{z^3} = -110 + 74i = (\bar{z})^3$ **47.** $P(2, 5)$ **49.** $P(2, -5)$ **51.** $P(-2, 5)$

**53.** $P(-2, -5)$ **55.** $P(1, 4)$ **57.** $P(-3, 1)$ **59.** $P(4, 0)$ **61.** $P(-3, 0)$ **63.** $P(0, 6)$ **65.** $P(0, -1)$

**67.** $P(5, 4)$ **69.** $P(4, -2)$ **71.** $P(-3, 6)$ **73.** $\sqrt{53}$ **75.** $\sqrt{53}$ **77.** 10 **79.** 10 **81.** $3\sqrt{2}$

**83.** $5\sqrt{2}$ **85.** 2 **87.** 1 **89.** 4 **91.** 4 **93.** $\sqrt{15}$ **95.** $-\frac{7}{25} + \frac{1}{25}i$ **97.** 1 **99.** $i$

**101.** $-\frac{23}{26} + \frac{11}{26}i$ **103.** $\frac{1}{10} - \frac{11}{10}i$ **105.** $-\frac{5}{2} - \frac{9}{2}i$ **107.** $\frac{\sqrt{2}}{2}$ **109.** $\frac{\sqrt{2}}{2}$ **111.** $\frac{9}{13} + \frac{7}{13}i$

**113.** $\frac{4}{17} - \frac{1}{17}i$ **115.** $\frac{15}{7} + 3i$ **117.** $2 + 2i,\ -2 - 2i$

**119.** $\overline{z - w} = \overline{(a + bi) - (c + di)} = \overline{(a - c) + (b - d)i} = (a - c) - (b - d)i = (a - bi) - (c - di)$
$= \overline{(a + bi)} - \overline{(c + di)} = \bar{z} - \bar{w}$

**121.** $\bar{\bar{z}} = \overline{\overline{(a + bi)}} = \overline{a - bi} = a + bi = z$

**123.** $\frac{1}{2}(z + \bar{z}) = \frac{1}{2}[(a + bi) + \overline{(a + bi)}] = \frac{1}{2}(a + bi + a - bi) = a$

**125.** $|\bar{z}| = |\overline{a + bi}| = |a - bi| = \sqrt{a^2 + (-b)^2} = \sqrt{a^2 + b^2} = |a + bi| = |z|$

**127.** False, $|(-1) - 1| \neq |-1| - |1|$ **129.** True **131.** $z$: $P(3, 5)$, $\bar{z}$: $P(3, -5)$

**133.** $z$: $P(-3, 5)$, $\bar{z}$: $P(-3, -5)$ **135.** $z$: $P(2, -6)$, $\bar{z}$: $P(2, 6)$ **137.** $z$: $P(-2, -6)$, $\bar{z}$: $P(-2, 6)$

**139.** $z$: $P(2, 5)$, $w$: $P(4, -7)$, $z + w$: $P(6, -2)$ **141.** $z$: $P(2, 1)$, $w$: $P(3, -2)$, $zw$: $P(8, -1)$

**Exercise Set 6.3, page 231**

**1.** $\pm 4i$ **3.** $3 \pm 4i$ **5.** $2 \pm i$ **7.** $-\frac{3}{2} \pm \frac{\sqrt{7}}{2}i$ **9.** $1 \pm \frac{1}{2}i$ **11.** $-1 \pm \frac{\sqrt{7}}{2}i$ **13.** $\pm\frac{8}{3}i$

**15.** $\frac{2}{3} \pm \frac{\sqrt{2}}{3}i$ **17.** $\frac{3 \pm 3\sqrt{5}}{2}$ **19.** $3, 3 \pm i\sqrt{2}$ **21.** $\frac{1}{3}, \pm\frac{\sqrt{5}}{2}i$ **23.** $3, \frac{1}{2} \pm \frac{\sqrt{7}}{2}i$

**25.** $-2, -1, -1 \pm i\sqrt{2}$ **27.** $2i, -2i, 3$ **29.** $x^2 - 8x + 25 = 0$ **31.** $x^3 - 4x^2 + 25x - 100 = 0$

**37.** $1, -2, \pm 2i$ **39.** $-1, \frac{3}{2}, -\frac{1}{2} \pm \frac{\sqrt{3}}{2}i$ **41.** $\pm i^{3/2}, \pm i^{1/2}$ **43.** $\pm i\sqrt{5}, \pm\sqrt{5}$ **45.** $\pm\left(\frac{-5 \pm \sqrt{105}}{2}\right)^{1/2}$

**Exercise Set 6.4, page 238**

**1.** $3\left[\cos\frac{\pi}{2} + i\sin\frac{\pi}{2}\right]$ **3.** $6[\cos\pi + i\sin\pi]$ **5.** $2\left[\cos\frac{2\pi}{3} + i\sin\frac{2\pi}{3}\right]$ **7.** $2\sqrt{2}\left[\cos\frac{5\pi}{4} + i\sin\frac{5\pi}{4}\right]$

**9.** $\sqrt{2}\left[\cos\frac{5\pi}{4} + i\sin\frac{5\pi}{4}\right]$ **11.** $2\sqrt{3}\left[\cos\frac{5\pi}{6} + i\sin\frac{5\pi}{6}\right]$ **13.** $4\left[\cos\frac{11\pi}{6} + i\sin\frac{11\pi}{6}\right]$

**15.** $6\left(\cos\frac{5\pi}{4} + i\sin\frac{5\pi}{4}\right)$ **17.** $13(\cos 67°23' + i\sin 67°23')$ **19.** $25(\cos 196°16' + i\sin 196°16')$

**21.** $8$ **23.** $6 - 2i\sqrt{3}$ **25.** $-4\sqrt{3}$ **27.** $12 + 9i$ **29.** $2(\cos 165° + i\sin 165°)$

**31.** $\sqrt{5}(\cos(-243°26') + i\sin(-243°26'))$ **33.** $\sqrt{3}(\cos 270° + i\sin 270°)$
$\sqrt{3}\left(\cos\frac{3\pi}{2} + i\sin\frac{3\pi}{2}\right)$

**35.** $\frac{\sqrt{6}}{3}(\cos(-285°) + i\sin(-285°))$ **37.** $-64$ **39.** $8i$ **41.** $4^{12}$
$\frac{\sqrt{6}}{3}\left(\cos -\frac{57\pi}{36} + i\sin -\frac{57\pi}{36}\right)$

**43.** $\sqrt[3]{3}(\cos 30° + i \sin 30°)$
$\sqrt[3]{3}(\cos 150° + i \sin 150°)$
$\sqrt[3]{3}(\cos 270° + i \sin 270°)$

**45.** $\sqrt[8]{8}(\cos 33°45' + i \sin 33°45')$
$\sqrt[8]{8}(\cos 123°45' + i \sin 123°45')$
$\sqrt[8]{8}(\cos 213°45' + i \sin 213°45')$
$\sqrt[8]{8}(\cos 303°45' + i \sin 303°45')$

**47.** $\sqrt[10]{2}(\cos 9° + i \sin 9°)$
$\sqrt[10]{2}(\cos 81° + i \sin 81°)$
$\sqrt[10]{2}(\cos 153° + i \sin 153°)$
$\sqrt[10]{2}(\cos 225° + i \sin 225°)$
$\sqrt[10]{2}(\cos 297° + i \sin 297°)$

**49.** (a) $1(\cos 0° + i \sin 0°)$
$1(\cos 72° + i \sin 72°)$
$1(\cos 144° + i \sin 144°)$
$1(\cos 216° + i \sin 216°)$
$1(\cos 288° + i \sin 288°)$

(b) $1(\cos 0° + i \sin 0°) = 1$
$1(\cos 90° + i \sin 90°) = i$
$1(\cos 180° + i \sin 180°) = -1$
$1(\cos 270° + i \sin 270°) = -i$

(c) $1(\cos 0° + i \sin 0°) = 1$
$1(\cos 45° + i \sin 45°) = \frac{\sqrt{2}}{2} + \frac{\sqrt{2}}{2}i$
$1(\cos 90° + i \sin 90°) = i$
$1(\cos 135° + i \sin 135°) = -\frac{\sqrt{2}}{2} + \frac{\sqrt{2}}{2}i$
$1(\cos 180° + i \sin 180°) = -1$
$1(\cos 225° + i \sin 225°) = -\frac{\sqrt{2}}{2} - \frac{\sqrt{2}}{2}i$
$1(\cos 270° + i \sin 270°) = -i$
$1(\cos 315° + i \sin 315°) = \frac{\sqrt{2}}{2} - \frac{\sqrt{2}}{2}i$

**51.** $1 \pm \sqrt{2}i$

**53.** $\sqrt[3]{2}(\cos 30° + i \sin 30°)$
$\sqrt[3]{2}(\cos 150° + i \sin 150°)$
$\sqrt[3]{2}(\cos 270° + i \sin 270°)$

**55.** $2(\cos 36° + i \sin 36°)$
$2(\cos 108° + i \sin 108°)$
$2(\cos 180° + i \sin 180°)$
$2(\cos 252° + i \sin 252°)$
$2(\cos 324° + i \sin 324°)$

**Chapter Test, page 240**

**1.** $i\sqrt{11}$ **3.** $4i\sqrt{3}$ **5.** 3, −9 **7.** 0, −3 **9.** $i$ **11.** $6 - 9i$ **13.** 9 **15.** $-\frac{8}{65} - \frac{1}{65}i$

**17.** $\frac{15}{13} + \frac{29}{13}i$ **19.** (a) $\overline{z + w} = 7 + 2i = \bar{z} + \bar{w}$ (b) $\overline{z - w} = -1 - 4i = \bar{z} - \bar{w}$ (c) $\overline{z \cdot w} = 15 + 5i = \bar{z} \cdot \bar{w}$
(d) $\overline{\left(\frac{z}{w}\right)} = \frac{9}{25} - \frac{13}{25}i = \frac{\bar{z}}{\bar{w}}$

**21.** (a) $\overline{z^2} = 35 - 12i = (\bar{z})^2$ (b) $\overline{z^3} = 198 - 107i = (\bar{z})^3$ **23.** $P(5, -12)$ **25.** $P(0, -7)$ **27.** $\sqrt{41}$

**29.** 9 **31.** $5 \pm 2i$ **33.** $-\frac{1}{2} \pm \frac{\sqrt{11}}{2}i$ **35.** $\frac{2}{3}, -3, \frac{1}{4} \pm \frac{\sqrt{31}}{4}i$

**37.** Degree of polynomial: 4; Roots: 4, $-3, \frac{5}{2} + \frac{\sqrt{19}}{2}i$, and $\frac{5}{2} - \frac{\sqrt{19}}{2}i$

**39.** Degree of polynomial: 6; Roots: 0, 2 (multiplicity three), $-2 + 2i\sqrt{3}$, and $-2 - 2i\sqrt{3}$

**41.** 2, $1 + i$ **43.** $2\sqrt{3}\left(\cos \frac{11\pi}{6} + i \sin \frac{11\pi}{6}\right)$ **45.** $24i, 3\sqrt{3} - 3i$ **47.** $\frac{\sqrt{3}}{2} + \frac{1}{2}i$

**49.** (a) −5, −2 (b) $\frac{20}{7}, -\frac{3}{7}$ **51.** $\frac{59}{53} + \frac{21}{53}i$ **53.** $\pm 3i^{3/2}, \pm 3i^{1/2}$

## Chapter 7

### Exercise Set 7.1, page 248

**1.** $2^{-3x}$ **3.** $5^{2x-4}$ **5.** $2^{2-x}$ **7.** $3^{2x-2}$ **9.** $2^{8-x}$ **11.** $3^{4x-7}$ **13.** $2^x(1 + 2^x)$

**15.** 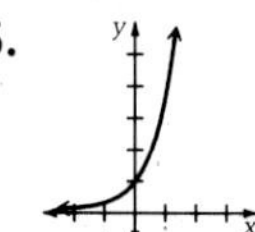
**17.** 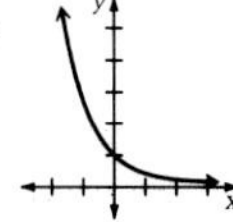
**19.** 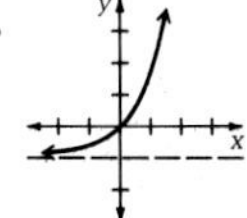
**21.** 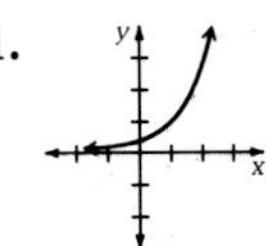
**23.** **25.** 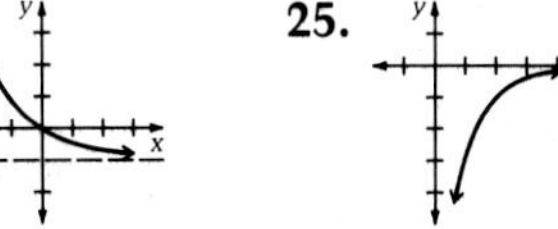

**27.** 3 **29.** $\frac{2}{3}$ **31.** $-8$ **33.** $\frac{7}{2}$ **35.** No solution **37.** $-2$ **39.** $\pm\sqrt{2}$ **41.** 0 **43.** 2, 0

**45.** (a) $\approx 2.7$ (b) $\approx 8.8$ (c) $\approx 0.3$ **47.** $Q$ is squared. $Q$ is cubed. $Q$ is raised to the tenth power.

**49.** 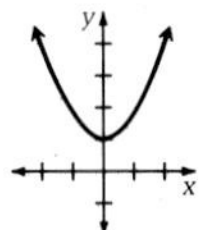
**51.** 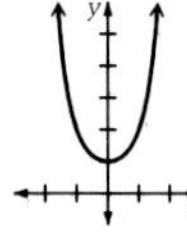
**53.** 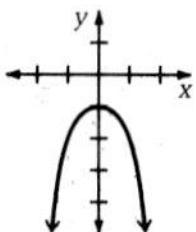
**55.** no, no **57.** yes, no

### Exercise Set 7.2, page 255

**1.** $f^{-1}(x) = 3 - \frac{x}{2}$ **3.** $f^{-1}(x) = \sqrt{x}$ **5.** $f^{-1}(x) = 4 - x^2$ **7.** $f^{-1}(x) = \sqrt[3]{3x} + 1$

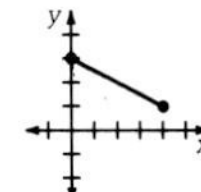
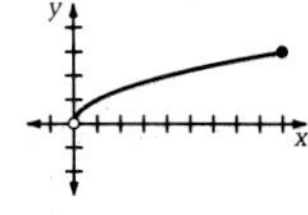
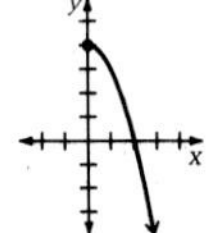
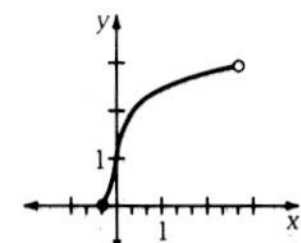

**9.** [0, 4], [1, 3] **11.** (0, 9], (0, 3] **13.** $[0, +\infty), (-\infty, 4]$ **15.** $\left[-\frac{1}{3}, \frac{8}{3}\right), [0, 3)$ **17.** $\log_x 25 = 2$

**19.** $\log_{16} 4 = x$ **21.** $\log_5 x = 3$ **23.** 8 **25.** $2\sqrt{2}$ **27.** 2 **29.** $\frac{2}{3}$ **31.** $\frac{\sqrt{3}}{3}$

**33.** 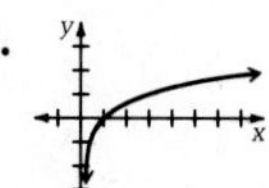 **35.** 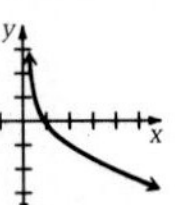 **37.** 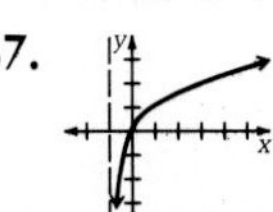 **39.** 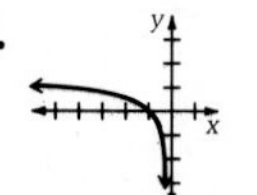 **41.** 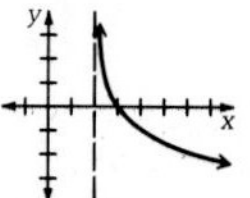 **43.** 2

**45.** 13.6 **47.** $-0.4375$ **49.** $2\sqrt{2}$ **51.** $-2$ **53.** 81 **55.** $2\sqrt{2}$ **57.** All real numbers $x > 0$

**59.** 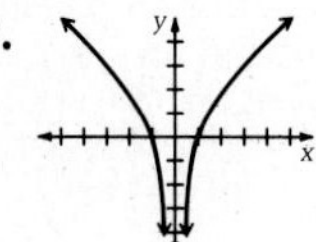 **61.** 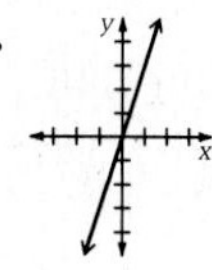 **63.** 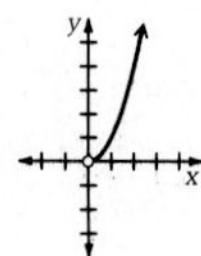 **65.** 2 **67.** 8, 1 **69.** 2, 1

**71.** By Rule 2 of Exponents $b^1 = b$. By Log/Exp Principle $\log_b b = 1$. **73.** By Rule 5 of Logarithms $\log_b\left(\frac{1}{x}\right) = \log_b 1 - \log_b x$. By Rule 1 of Logarithms $\log_b\left(\frac{1}{x}\right) = -\log_b x$.

**Exercise Set 7.3, page 260**

**1.** 0.6839 **3.** 0.9450 **5.** 0.2718 **7.** 0.5132 **9.** 0.3674 **11.** 0.2041 **13.** $4.724 \times 10^1$

**15.** $3.502 \times 10^{-1}$ **17.** $1.5 \times 10^{-2}$ **19.** $9.08 \times 10^{-2}$ **21.** $3.2941 \times 10^3$ **23.** $3.0 \times 10^{-4}$

**25.** 1.5172 **27.** $-0.1675$ **29.** $-1.2441$ **31.** 2.5428 **33.** 3.7126 **35.** 4.1644 **37.** 3.54

**39.** 8.52 **41.** 5.49 **43.** 565 **45.** 302 **47.** 46,100 **49.** 0.177 **51.** 0.00346 **53.** 0.101

**55.** False, $\log 1 + \log 1 \neq \log(1 + 1)$ **57.** False, $(\log 10)(\log 10) \neq \log[(10)(10)]$ **59.** True

**61.** False, $\log[100(1)] \neq 100 \log 1$ **63.** False, $\log(10^1) \neq \log(1^{10})$ **65.** 2.0960 **67.** 2.3223

**69.** $Q$ is increased by 1. $Q$ is increased by $\log_2 3$. $Q$ is increased by 2.

**Exercise Set 7.4, page 263**

**1.** (a) 27.0 (b) 3.0 (c) 13.2 (d) 10.4 **3.** (a) −3.6 (b) 5.9 (c) 0.8 (d) 3.9 **5.** (a) −6.4 (b) 2.0 (c) 3.3 (d) −0.5

**7.** (a) 3.2 (b) −3.8 (c) −3.5 (d) −1.3 **9.** 0.8758 **11.** 0.6757 **13.** −0.5164 **15.** 1.4469

**17.** 2.6796 **19.** −1.2863 **21.** −1.2921 **23.** −0.0392 **25.** 8.915 **27.** 2.134 **29.** 9,648

**31.** 0.2202 **33.** 0.5759 **35.** 0.02632 **37.** $5.193 \times 10^{-4}$ **39.** 1.41 **41.** 0.33 **43.** 0.65

**45.** 2.62 **47.** −1.59 **49.** 2.59 **51.** 10.54 **53.** (a) 3 (b) 7 (c) 19

**Exercise Set 7.5, page 269**

**1.** 172,800 **3.** 1.6 hours **5.** $2208 **7.** $1774 **9.** 14 **11.** $2226 **13.** 5 years **15.** 45.2 grams

**17.** 16,451 years ago **19.** $13.50 **21.** at least $16,172 **23.** 6.14% **25.** 7.7%

**Chapter Test, page 272**

**1.** $2^{4-7x}$ **3.** **5.** No solution **7.** $f^{-1}(x) = \dfrac{7 - x}{2}$ **9.** $\log_x 125 = 3$ **11.** $\log_7 x = 4$

Domain: $[-1, 5]$
Range: $[1, 4]$

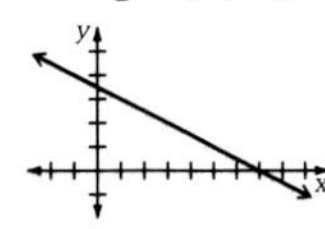

**13.** $\frac{1}{4}$ **15.** **17.** 0.6 **19.** 2 **21.** All real numbers $x > 0$ **23.** 0.5977 **25.** $4.51 \times 10^{-3}$

**27.** $-0.0237$ **29.** 28.9 **31.** 1.0 **33.** $-0.5792$ **35.** 2.7321 **37.** $4.054 \times 10^{-3}$ **39.** 0.6 hours

**41.** \$808 **43.** No solution **45.** **47.** (a) 1.73 (b) 2.59

## Cumulative Test: Chapters 6 and 7, page 275

**1.** $2 + 26i$ **3.** $-\frac{5}{3} + 2i$ **5.** $P(5, -3)$ **7.** 6 **9.** $\frac{3}{2} \pm \frac{1}{2}i$ **11.** $-1, -4, 4$

**13.** $4\left(\cos \frac{5\pi}{3} + i \sin \frac{5\pi}{3}\right)$ **15.** $4\sqrt{2}\left(\cos \frac{5\pi}{12} + i \sin \frac{5\pi}{12}\right), 2\sqrt{2}\left(\cos \frac{\pi}{12} - i \sin \frac{\pi}{12}\right)$ **17.** $-4$

**19.** $2^{2x-5}$ **21.**

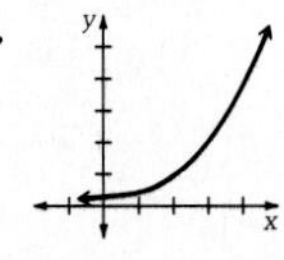

**23.** 2 **25.** $f^{-1}(x) = \frac{x+5}{2}$, Domain: $[-5, 1]$, Range: $[0, 3]$

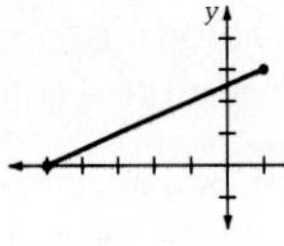

**27.** $\log_{16} 2 = x$

**29.**

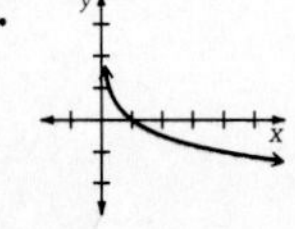

**31.** 3 **33.** 0.971 **35.** $-0.0855$ **37.** \$7235

**39.** (a) 1 (b) 2 **41.** (a) $\frac{1}{3} + \frac{10}{3}i$ (b) $\frac{\sqrt{10}}{5}$

## Calculator Exercises

### Exercise Set 1.2, page A1

**1.** 1, 2.2, 3.4, 4.6, 5.8, 7 **3.** −0.34, −0.9, −1.38, −1.78, −2.1, −2.34

**5.** 0.678, 0.748, 0.812, 0.872, 0.927, 0.980

### Exercise Set 1.4, page A1

**1.** $y = 0.71x + 5.79$ **3.** $y = -1.25x + 4.12$ **5.** $y = -0.89x + 8.16$ **7.** $y = 5.25x - 3.50$

**9.** (b) −2 (c) −2 **11.** (b) 1.9 (c) 1.9 **13.** (b) 5.6 (c) 5.6 **15.** (b) −0.4 (c) ≈ −0.39

### Exercise Set 1.5, page A2

**1.** (b) 2.8 (c) 2.75 **3.** (b) 2.8 (c) 2.75 **5.** (b) −0.3 (c) −0.25 **7.** (b) −1.1 (c) −1.055

**9.** 2.5 sec, 101.3 ft **11.** 1.6 sec, 39.4 ft

### Exercise Set 2.1, page A3

**1.** $c = 53.19$ **3.** $b = 486.7$ **5.** $c = 0.9568$ **7.** $a = 3267$ **9.** $b = 18.931$

### Exercise Set 2.2, page A3

**1.** $\sin A = 0.8032$ $\csc A = 1.2451$
$\cos A = 0.5956$ $\sec A = 1.6790$
$\tan A = 1.3485$ $\cot A = 0.7416$

**3.** $\sin A = 0.5163$ $\csc A = 1.9370$
$\cos A = 0.8564$ $\sec A = 1.1677$
$\tan A = 0.6029$ $\cot A = 1.6588$

**5.** $\sin A = 0.7508$ $\csc A = 1.3318$
$\cos A = 0.6604$ $\sec A = 1.5142$
$\tan A = 1.1369$ $\cot A = 0.8796$

**7.** $\sin A = 0.7986$ $\csc A = 1.2522$
$\cos A = 0.6018$ $\sec A = 1.6617$
$\tan A = 1.3270$ $\cot A = 0.7536$

**9.** $\sin A = 0.5488$ $\csc A = 1.8222$
$\cos A = 0.8360$ $\sec A = 1.1962$
$\tan A = 0.6565$ $\cot A = 1.5233$

**11.** $h = 5.498$ **13.** 90 **15.** $\cos\alpha = 0.9231$, $\cos\beta = 0.3846$

**17.** $\tan\alpha = 1.8755$, $\cos\beta = 0.8824$ **19.** $\tan\alpha = 0.3430$, $\sin\beta = 0.9459$

**Exercise Set 2.3, page A3**

**1.** $\sin\alpha = 0.3656$ $\csc\alpha = 2.7352$
$\cos\alpha = -0.9308$ $\sec\alpha = -1.0744$
$\tan\alpha = -0.3928$ $\cot\alpha = -2.5459$

**3.** $\sin\alpha = -0.9442$ $\csc\alpha = -1.0591$
$\cos\alpha = 0.3293$ $\sec\alpha = 3.0372$
$\tan\alpha = -2.8678$ $\cot\alpha = -0.3487$

**5.** $\sin\alpha = -0.3780$ $\csc\alpha = -2.6456$
$\cos\alpha = -0.9258$ $\sec\alpha = -1.0801$
$\tan\alpha = 0.4083$ $\cot\alpha = 2.4494$

**Exercise Set 2.4, page A4**

**1.** $\tan 32° = 0.6249$ $\sec 32° = 1.1792$
$\csc 32° = 1.8871$ $\cot 32° = 1.6003$

**3.** $\csc 55° = 1.2208$ $\cos 55° = 0.5738$
$\sec 55° = 1.7435$ $\cot 55° = 0.7002$

**5.** $\cos\alpha = 0.8587$ $\cot\alpha = 1.6759$
$\tan\alpha = 0.5967$ $\sec\alpha = 1.1646$
$\csc\alpha = 1.9516$

**7.** $\sin\alpha = 0.5672$ $\csc\alpha = 1.7631$
$\tan\alpha = 0.6886$ $\sec\alpha = 1.2142$
$\cot\alpha = 1.4521$

**9.** 0.5180 **11.** 0.3971 **13.** 0.6122 **15.** 1.2020 **17.** 0.8941 **19.** 1.4267 **21.** 28.00° **23.** 39.66° **25.** 43.01° **27.** 0.6691 **29.** 0.6249 **31.** −0.8090 **33.** −0.9603 **35.** 1.0457 **37.** 135.60° **39.** 275.20° **41.** 321.60° **43.** 196.30°

**Exercise Set 3.1, page A6**

**1.** IV **3.** III **5.** IV **7.** II **9.** I **11.** −0.9301 **13.** −0.8922 **15.** 0.199 **17.** 1.381 **19.** 720.00° **21.** −310.91°

**Exercise Set 3.2, page A6**

**1.** 0.8633 **3.** −0.5000 **5.** −0.8660 **7.** 1.7319 **9.** −0.4755 **11.** −1.6611 **13.** 5.5971

**Exercise Set 3.3, page A7**

**1.** $-0.8415, 0.1411, -0.8367, -0.7568, 0, -0.5806$

**3.** $-0.2130, -0.6002, -0.9789, -0.0730, -0.7071, -0.9863$

**5.** $-0.6421, -7.0153, 0.6547, -0.8637, 27224.2, 1.4024$

**Exercise Set 3.4, page A7**

**1.** $1.9191, -0.8961, 0.2001, 1.0522, -1.3088, 0.5222$ **3.** $-0.8304, 1.6436, -1.3411, 0.3972, -0.1247, 0.8625$

**Exercise Set 3.5, page A7**

**1.** 0.1874 **3.** $-1.2780$ **5.** 0.9533 **7.** 0.6405

**Exercise Set 4.2, page A7**

**1.** 0.2588 **3.** 0.2588 **5.** 0.2679 **7.** $-0.4799$ **9.** 0.3531 **11.** $-0.5470$

**15.** 0.9917 **17.** 0.4385 **19.** 2.0494

**Exercise Set 4.3, page A8**

**1.** 0.7100 **3.** 0.9917 **5.** 0.4385 **7.** 2.0494

**Exercise Set 4.5, page A8**

**1.** 19.27° **3.** 71.57° **5.** 24.09° **7.** 22.5° **9.** 43.95° or 75.52°

**Exercise Set 5.1, page A8**

**1.** $b = 36.42$, $c = 43.33$, $\angle B = 57.2°$ **3.** $a = 1422.5$, $c = 1922.3$, $\angle B = 42.27°$

**5.** $a = 9.353$, $b = 17.451$, $\angle B = 61.81°$ **7.** $c = 3.3780$, $\angle A = 43.52°$, $\angle B = 46.48°$

**9.** $c = 1.0097$, $\angle A = 12.77°$, $\angle B = 77.22°$ **11.** 48.4 ft

**Exercise Set 5.2, page A9**

**1.** $a = 135.4$, $b = 45.7$, $\angle C = 59°$ **3.** $b = 4485.6$, $c = 2535.9$, $\angle A = 71.2°$

**5.** $c = 10.55$ or $7.62$, $\angle B = 84.4°$ or $95.6°$, $\angle C = 37°$ or $25.8°$ **7.** $b = 12{,}864.9$, $\angle A = 23.5°$, $\angle B = 124.8°$

**9.** $a = 360.9$, $\angle A = 26.2°$, $\angle C = 41.5°$ **11.** 117.4 ft

**Exercise Set 5.3, page A9**

**1.** $c = 11264$, $\angle A = 29.1°$, $\angle B = 41.6°$ **3.** $a = 7.558$, $\angle B = 83.3°$, $\angle C = 45.2°$

**5.** $b = 0.9504$, $\angle A = 63{,}6°$, $\angle C = 54.8°$ **7.** $\angle A = 39.12°$, $\angle B = 24.02°$, $\angle C = 116.86°$

**9.** $\angle A = 45.01°$, $\angle B = 57.71°$, $\angle C = 77.28°$ **11.** 150.62 in **13.** 505.99 **15.** 6.329 **17.** 236.57 $\text{in}^2$

**Exercise Set 5.4, page A10**

**1.** $\langle -5.12, 5.34\rangle$, magnitude 7.40, direction $-46.2°$ **3.** $\langle 7.53, -1.25\rangle$, magnitude 7.63, direction $-9.43°$

**5.** $\langle 1.34, 12.55\rangle$ **7.** $\langle -831.82, -80.85\rangle$

**Exercise Set 5.5, page A11**

**1.** $0.52\mathbf{i}$, $9.88\mathbf{j}$ **3.** $-883.6\mathbf{i}$, $-423.4\mathbf{j}$ **5.** S 38.7 E 6.4 mph **7.** 1467 ft lb, 2741 ft lb

**Exercise Set 5.6, page A11**

**1.** at equilibrium, stationary
0.32 cm above equilibrium
0.55 cm above equilibrium
0.55 cm below equilibrium

**3.** 0.12 cm below equilibrium
0.24 cm below equilibrium
0.79 cm below equilibrium
0.23 cm below equilibrium

**Exercise Set 5.7, page A11**

**1.** $(-1.87, 1.15)$ **3.** $(3.33, 4.22)$ **5.** $(6.5, 2.2)$ **7.** $(-6.06, 0.53)$

**9.** 3, 3.84, 4, 3.78, 3, 2.24, 2, 2.08, 3 **11.** 0, 1.4, 2.1, 3.15, 4.4, 5.6, 6.6, 7.17, 8.8

**Exercise Set 7.1, page A12**

**1.** 8.574, 8.634, 8.694, 8.754, 8.815, 8.877, 8.815 **3.** 2.67 **5.** 31.71 **7.** 1.96 **9.** 0.78 **11.** 1.63

**13.** $-0.17$ **15.** 1.5

**Exercise Set 7.4, page A12**

**1.** 0.613 **3.** 3.103 **5.** 4.819 **7.** 1.627 **9.** 1.272 **11.** 0.582

**Exercise Set 7.5, page A13**

**1.** (a) $27,899.83 (b) 12 years and 12 days **3.** 231 days

# Index